Integral Equations for Real-Life Multiscale Electromagnetic Problems

The ACES Series on Computational and Numerical Modelling in Electrical Engineering

Andrew F. Peterson, PhD - Series Editor

The volumes in this series will encompass the development and application of numerical techniques to electrical and electronic systems, including the modelling of electromagnetic phenomena over all frequency ranges and closely related techniques for acoustic and optical analysis. The scope includes the use of computation for engineering design and optimization, as well as the application of commercial modelling tools to practical problems. The series will include titles for senior undergraduate and postgraduate education, research monographs for reference, and practitioner guides and handbooks.

Titles in the Series

K. Warnick, "**Numerical Methods for Engineering**," 2010.

W. Yu, X. Yang and W. Li, "**VALU, AVX and GPU Acceleration Techniques for Parallel FDTD Methods**," 2014.

A.Z. Elsherbeni, P. Nayeri and C.J. Reddy, "**Antenna Analysis and Design Using FEKO Electromagnetic Simulation Software**," 2014.

A.Z. Elsherbeni and V. Demir, "**The Finite-Difference Time-Domain Method in Electromagnetics with MATLAB® Simulations, 2nd Edition**," 2015.

M. Bakr, A.Z. Elsherbeni and V. Demir, "**Adjoint Sensitivity Analysis of High Frequency Structures with MATLAB®**," 2017.

O. Ergul, "**New Trends in Computational Electromagnetics**," 2019.

D. Werner, "**Nanoantennas and Plasmonics: Modelling, design and fabrication**," 2020.

K. Kobayashi and P. D. Smith, "**Advances in Mathematical Methods for Electromagnetics**," 2020

V. Lancellotti, "**Advanced Theoretical and Numerical Electromagnetics, Volume 1: Static, stationary and time-varying fields**," 2021.

V. Lancellotti, "**Advanced Theoretical and Numerical Electromagnetics, Volume 2: Field representations and the method of moments**," 2021.

S. Roy, "**Uncertainty Quantification of Electromagnetic Devices, Circuits, and Systems**," 2021

A. Baghai-Wadji "**Mathematical Quantum Physics for Engineers and Technologists, Volume 1: Fundamentals**," 2023.

Integral Equations for Real-Life Multiscale Electromagnetic Problems

Edited by
Francesca Vipiana and Zhen Peng

The Institution of Engineering and Technology

Published by SciTech Publishing, an imprint of The Institution of Engineering and Technology, London, United Kingdom

The Institution of Engineering and Technology is registered as a Charity in England & Wales (no. 211014) and Scotland (no. SC038698).

First published 2023

The Institution of Engineering and Technology
Futures Place
Kings Way, Stevenage
Hertfordshire, SG1 2UA, United Kingdom

www.theiet.org

British Library Cataloguing in Publication Data
A catalogue record for this product is available from the British Library

ISBN 978-1-83953-476-8 (hardback)
ISBN 978-1-83953-477-5 (PDF)

Typeset in India by MPS Limited

Cover Image: Pobytov/DigitalVision Vectorsvia Getty Images

Contents

About the editors

Francesca Vipiana is a full professor in the Department of Electronics and Telecommunications at Politecnico di Torino (POLITO), Italy.

Her main research activities concern numerical techniques based on integral equations and method of moment approaches, with a focus on multiresolution and hierarchical schemes, domain decomposition, preconditioning and fast solution methods, and advanced quadrature integration schemes. Moreover, her research interests include the modeling, design, realization, and testing of microwave imaging systems for medical and industrial applications.

Currently, Prof. Vipiana coordinates "*THERAD – Microwave Theranostics for Alzheimer's Disease*," research project funded by the "Compagnia di SanPaolo" bank foundation, and "*INSIGHT – An innovative microwave sensing system for the evaluation and monitoring of food quality and safety*," joint research project within the Executive Program of Scientific and Technological Cooperation between Italy and China, funded by the National Natural Science Foundation of China (NSFC) and the Italian Ministry of Foreign Affairs and International Cooperation. Moreover, she is the principal investigator, for the POLITO research unit, in the national project "*BEST-Food, Broadband Electromagnetic Sensing Technologies for Food quality and security assessment*," and in the Marie Skłodowska-Curie Doctoral Network "*GENIUS – Glide-symmetric mEtamaterials for iNnovative radIo-frequency commUnication and Sensing*," funded by the European Union's Horizon Europe Programme and by the UK Research and Innovation.

She has 20 years full-time equivalent research experience. Since 2007, she has been an instructor at the European School of Antennas (ESoA) courses and, since 2008, the teaching professor of the course, "Advanced Computational EM for Antenna Analysis" at the POLITO Doctoral School where she is part of the PhD advisors board. Prof. Vipiana received the Lot Shafai Mid-Career Distinguished Award from the IEEE Antennas and Propagation Society in 2017 and she is an associate editor of *IEEE Transactions on Antennas and Propagation* and of the *IEEE Antennas and Propagation Magazine*, where, in 2020, she was also a guest editor for the special issue "*Electromagnetic Imaging and Sensing for Food Quality and Safety Assessment.*"

Zhen Peng is an associate professor in the Department of Electrical and Computer Engineering at the University of Illinois at Urbana-Champaign, USA. His research interests include classical electromagnetism with scalable algorithms; statistical electromagnetics for complex environments, for example, physics-oriented statistical wave analysis integrating order and chaos, and electromagnetic information theory for wireless communication; quantum electromagnetics; and measurement and control of uncertainties in chaotic reverberation chambers. He was a guest editor of IEEE *Transactions on Components, Packaging and Manufacturing Technology* in 2023, and an associate editor of IEEE *Transactions on Microwave Theory and Techniques* from 2018 to 2020. He has won several best paper awards including Best Electromagnetics Paper Award at the 16th European Conference on Antennas and Propagation in 2022, the EPEPS Best Paper Award at the 30th Conference on Electrical Performance of Electronic Packaging and Systems, the IEEE EMC Symposium Best Paper Award at the 2019 IEEE International Symposium on Electromagnetic Compatibility, Signal & Power Integrity, the 2018 Best Transaction Paper Award-IEEE Transactions on Components, Packaging and Manufacturing Technology, the 2014 IEEE Antenna and Propagation Sergei A. Schelkunoff Transactions Prize Paper Award. He was a recipient of the National Science Foundation CAREER Award (ENG/ECCS/CCSS) in 2018.

Chapter 1

Introduction

Francesca Vipiana[1] and Zhen Peng[2]

In the context of computational electromagnetics (CEM), surface integral equation (SIE) techniques based on the method of moments (MoM) [1] offer a potent tool that has become essential for simulating and engineering a diverse range of applications. These applications encompass advanced antenna design [2,3], radar cross-section (RCS) [4], stealth technologies [5], electromagnetic compatibility and interference (EMC/EMI) [6], and nanoscience applications [7], among others. SIE methods are particularly attractive when dealing with large-scale radiation and scattering issues. Unlike volumetric approaches that require the characterization of three-dimensional (3D) structures and embedding space, SIE methods necessitate the parameterization of two-dimensional (2D) boundary surfaces only. Although they result in dense and extensive matrix systems for large-scale problems, the utilization of iterative fast solvers, such as the multilevel fast multipole algorithm (MLFMA) [8,9], enables efficient resolution of such problems.

The scope of this book "Integral Equations for Real-Life Multiscale Electromagnetic Problems" is to collect and describe the main recent available approaches for the numerical solution of SIEs to analyze real-life multiscale electromagnetic problems. In CEM, formulations based on SIEs are currently the most used for the analysis of electrically large and complex structures. Still, it is essential to have available state-of-the-art techniques to solve them in an efficient and accurate way.

The book is organized into seven scientific chapters, completed with the "Introduction" and "Conclusion and perspectives" chapters.

Chapter 2 "Surface integral equation formulations," authored by Donald R. Wilton and William A. Johnson, encompasses a concise overview of essential concepts required to comprehend, formulate, and computationally address SIEs encountered in the field of electromagnetics. Utilizing this knowledge, the prevalent integral equations employed in time-harmonic problems are established, involving linear, piecewise homogeneous, and isotropic materials. Then, numerical methods employed to solve

[1]Wavision Research Group, Department of Electronics and Telecommunications, Politecnico di Torino, Italy
[2]Electromagnetics Lab and Center for Computational Electromagnetics, Department of Electrical and Computer Engineering University of Illinois at Urbana—Champaign, USA

these integral equations are presented, including techniques for accurately evaluating the singular and near-singular integrals that emerge in the process.

The following two chapters, Chapter 3 "Kernel-based fast factorization techniques," authored by Özgür Ergül, Bahram Khalichi, and Vakur B. Ertürk, and Chapter 4 "Kernel-independent fast factorization methods for multiscale electromagnetic problems," authored by Mengmeng Li *et al.*, are both dedicated to fast factorization techniques for an efficient and accurate solution to the electromagnetic problem. Chapter 3 describes kernel-based methods, where the primary emphasis lies on the underlying kernel of the problem, adjusting its utilization to effectively handle electromagnetic interactions in more efficient manners while maintaining accurate numerical performances. In particular, it focuses on the multilevel fast multipole algorithm (MLFMA), analyzing all its properties and possible implementations to obtain accurate, efficient, and stable solutions to multi-scale problems. Instead, Chapter 4 describes kernel-independent techniques that are entirely algebraic and take advantage of the rank-deficient nature of MoM coupling matrix blocks, generated by two distinct groups of basic functions that are well separated in space. By employing low-rank factorization methods, the MoM matrix can be approximated, enabling swift evaluations of matrix–vector products in iterative solutions or rapid direct solvers.

Chapter 5, entitled "Domain decomposition method" and authored by Víctor Martín, Hong-Wei Gao, Diego M. Solís, José M. Taboada, and Zhen Peng, focuses on the application of domain decomposition (DD) methods in solving time-harmonic electromagnetic wave problems based on SIE. These methods are highly desirable due to their capacity to yield efficient and effective preconditioned iterative solution algorithms, and to their inherently parallel nature that makes them particularly attractive, according to the current trends in computer architecture. The chapter presents two classes of DD methods. One class utilizes the latest developments in the surface-based discontinuous Galerkin (DG) formulation where the continuity of currents at domain boundaries is directly enforced by employing an interior penalty DG formulation. Instead, the other class of DD methods follows the "tear-and-interconnect" approach, where transmission conditions are imposed along the tearing contours between subdomains.

The next two chapters, Chapter 6 "Multi-resolution preconditioning," authored by Francesca Vipiana, Víctor Martín and José M. Taboada, and Chapter 7 "Calderón preconditioners for electromagnetic integral equations," authored by Adrien Merlini, Simon B. Adrian, Alexandre Dély, and Francesco P. Andriulli, are both devoted to preconditioning techniques applied to the MoM matrix to improve its conditioning and so enabling a faster convergence of the used iterative solution algorithm. Chapter 6 aims to provide all the theoretical and practical knowledge for a proficient implementation of the multi-resolution (MR) preconditioner in the electromagnetic analysis of perfect electric conductor (PEC) structures with arbitrary 3D shapes via both the electric field integral equation (EFIE) and the combined field integral equation (CFIE). The objective of Chapter 7 is to offer a broad comprehension of the underlying mechanisms of Calderón preconditioning, presenting an overview of its diverse applications to commonly used electromagnetic formulations. While the chapter acknowledges the existence of intricate mathematical developments, it primarily focuses on providing references to detailed analyses rather than delving into those complexities extensively.

Finally, Chapter 8, entitled "The decoupled potential integral equation" and authored by Felipe Vico and Miguel Ferrando-Bataller, explores an experimental approach known as the decoupled potential integral equation (DPIE). The objective of this formulation is to develop a method that exhibits robustness across all frequencies, with a specific focus on low frequencies when dealing with multiple connected geometries.

References

[1] Harrington RF. *Field Computation by Moment Method*. Piscataway, NJ: IEEE Press; 1993.

[2] Wang X, Peng Z, Lim KH, *et al.* Multisolver domain decomposition method for modeling EMC effects of multiple antennas on a large air platform. *IEEE Transactions on Electromagnetic Compatibility*. 2012;54(2):375–388.

[3] Hesford AJ and Chew WC. On preconditioning and the eigensystems of electromagnetic radiation problems. *IEEE Transactions on Antennas and Propagation*. 2008;56(8):2413–2420.

[4] Blanca IGT, Rodríguez JL, Obelleiro F, *et al.* Experience on radar cross section reduction of a warship. *Microwave and Optical Technology Letters*. 2014;56(10):2270–2273.

[5] Peng Z, Lim KH, and Lee JF. Nonconformal domain decomposition methods for solving large multiscale electromagnetic scattering problems. *Proceedings of the IEEE*. 2013;101(2):298–319.

[6] Solís DM, Martín VF, Araújo MG, *et al.* Accurate EMC engineering on realistic platforms using an integral equation domain decomposition approach. *IEEE Transactions on Antennas and Propagation*. 2020;68(4):3002–3015.

[7] Obelleiro F, Taboada JM, Solís DM, *et al.* Directive antenna nanocoupler to plasmonic gap waveguides. *Optics Letters*. 2013;38(10):1630–1632.

[8] Song JM and Chew WC. Multilevel fast-multipole algorithm for solving combined field integral equations of electromagnetic scattering. *Microwave and Optical Technology Letters*. 1995;10(1):14–19.

[9] Taboada JM, Araujo MG, Bertolo JM, *et al.* MLFMA-FFT parallel algorithm for the solution of large-scale problems in electromagnetics (Invited Paper). *Progress in Electromagnetics Research*. 2010;105:15–30.

Chapter 2
Surface integral equation formulations

Donald R. Wilton[1] *and William A. Johnson*[2]

This chapter includes a brief review of fundamental material needed for understanding, formulating, and numerically solving surface integral equations appearing in electromagnetics. Using this material, we then develop the most common integral equations for time-harmonic problems involving linear, piecewise homogeneous, and isotropic materials. Methods for numerically solving the integral equations are developed and discussed, including approaches for numerically evaluating the singular and near-singular integrals that arise.

2.1 Maxwell's equations

The Maxwell equations are a set of four laws: Faraday's law, Ampere's law, and the electric and magnetic forms of Gauss's law. Each of the set of four equations may, in turn, be written in the following three different forms: integral, differential, and boundary forms. It has been argued [1] that the integral forms of Maxwell's equations are the most fundamental in the sense that all other forms derive from them. We deal here only with time-harmonic problems and assume that all source and field quantities vary as $e^{j\omega t}$, effectively replacing the operator $\partial/\partial t$ by $j\omega$ in the time-domain forms of Maxwell's equations [2].

2.1.1 Integral form of Maxwell's equations

The integral form of Maxwell's equations for $\exp(j\omega t)$ time dependence is

$$\oint_C \boldsymbol{E} \cdot d\boldsymbol{\ell} = -j\omega \int_S \mathbf{B} \cdot \hat{\boldsymbol{n}}\, dS - \int_S \boldsymbol{M} \cdot \hat{\boldsymbol{n}}\, dS, \tag{2.1}$$

$$\oint_C \boldsymbol{H} \cdot d\boldsymbol{\ell} = j\omega \int_S \boldsymbol{D} \cdot \hat{\boldsymbol{n}}\, dS + \int_S \boldsymbol{J} \cdot \hat{\boldsymbol{n}}\, dS, \tag{2.2}$$

[1]Department of Electrical and Computer Engineering, University of Houston, USA
[2]Consultant, Jemez Springs, NM, USA

where S is an *open* surface with *closed* boundary C. On the other hand, if S is a *closed* surface, enclosing a volume V, we can also write

$$\oint_S \boldsymbol{D} \cdot \hat{\boldsymbol{n}}\, dS = \int_V q\, dV, \tag{2.3}$$

$$\oint_S \boldsymbol{B} \cdot \hat{\boldsymbol{n}}\, dS = \int_V m\, dV. \tag{2.4}$$

The electric and magnetic field strength quantities, $(\boldsymbol{E}, \boldsymbol{H})$, are related to the corresponding flux density quantities $(\boldsymbol{D}, \boldsymbol{B})$ via the *constitutive equations* involving the local permittivity and permeability, (ε, μ), respectively:

$$\boldsymbol{D} = \varepsilon \boldsymbol{E}, \tag{2.5}$$

$$\boldsymbol{B} = \mu \boldsymbol{H}. \tag{2.6}$$

In the problems considered here, we assume that the material parameters (ε, μ) are linear, piecewise homogeneous, and isotropic. In the first two Maxwell's equations, as shown in Figure 2.1, S has a unit normal $\hat{\boldsymbol{n}}$ and the closed curve C has a unit tangent $\hat{\boldsymbol{\ell}}$ chosen such that the unit vector $\hat{\boldsymbol{u}} = \hat{\boldsymbol{\ell}} \times \hat{\boldsymbol{n}}$ is both normal to C and points away from S. In the last two equations, S is a *closed* surface with unit normal $\hat{\boldsymbol{n}}$, the boundary of a volume V. Volumetric electric and magnetic source currents $\boldsymbol{J}$ and $\boldsymbol{M}$, with units $[\mathrm{A/m^2}]$ and $[\mathrm{V/m^2}]$ appear in (2.1) and (2.2), respectively. While there is no experimental evidence for the existence of magnetic monopoles or currents, magnetic currents and charges not only give an elegant symmetry to Maxwell's equations but also provide flexible and mathematically convenient means for representing electromagnetic fields. Magnetic and electric currents merely comprise magnetic and electric charges in motion; the volume charge densities of those charges we define as m $[\mathrm{Wb/m^3}]$ and q $[\mathrm{C/m^3}]$, respectively. The conservation of charge principle states that in every bounded region of space, electric charge is conserved; for both convenience and mathematical symmetries, we assume that this same conservation law also holds for magnetic charges. Thus, the total charge for either charge type changes in a region only as charges cross the region's boundaries (i.e., enter or leave the region). The rate (in $[\mathrm{C/s}]$), for example, at which electric charges decrease in a region must

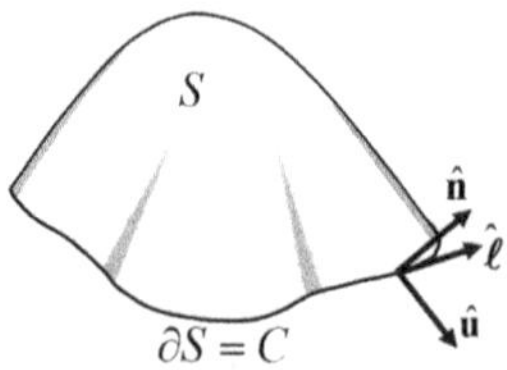

Figure 2.1 The surface S with unit normal $\hat{\boldsymbol{n}}$ is bounded by the curve $\partial S = C$. The unit vectors $\hat{\boldsymbol{\ell}}$ and $\hat{\boldsymbol{u}}$ lie in the tangent plane of S; $\hat{\boldsymbol{\ell}}$ is also tangent to C while $\hat{\boldsymbol{u}}$ is normal to C and points away from S. The three unit vectors satisfy $\hat{\boldsymbol{u}} \times \hat{\boldsymbol{\ell}} = \hat{\boldsymbol{n}}$ and form a mutually orthogonal right-handed triad along C.

equal the net electric current flux (in [C/s]) exiting the region's boundaries. Magnetic charges and currents are similarly related; both results are succinctly summarized by the *continuity equations*,

$$\oint_S \boldsymbol{J} \cdot \hat{\boldsymbol{n}}\, dS = -j\omega \int_V q\, dV, \tag{2.7}$$

$$\oint_S \boldsymbol{M} \cdot \hat{\boldsymbol{n}}\, dS = -j\omega \int_V m\, dV. \tag{2.8}$$

where, for time-harmonic quantities, $-j\omega$ plays the role of $-\partial/\partial t$, with both sides of (2.7) and (2.8) representing rates of decrease of the total charge in V. Though the continuity equations follow from physics, independent of Maxwell's equations, the latter are consistent with them. For example, if we replace the first surface integral in (2.2) with the *closed* boundary surface of (2.3), then the contour integral of (2.2) vanishes (i.e., surface S no longer has a boundary contour C), and (2.7) follows. Similarly, (2.8) follows from applying both (2.1) and (2.4) to a common closed surface S.

2.1.2 Point or differential form of Maxwell's equations

Each Maxwell equation in differential or so-called point form corresponds to an equation of the same name in integral form. Moreover, each differential form equation may be derived from a limiting procedure of its corresponding integral form. For instance, one usually applies Stokes theorem to the two curl equations, and the divergence theorem to the two scalar equations, shrinking each to a differential surface or volume, respectively, and finally obtaining

$$\nabla \times \boldsymbol{E} = -j\omega\mu \boldsymbol{H} - \boldsymbol{M}, \tag{2.9}$$

$$\nabla \times \boldsymbol{H} = j\omega\varepsilon \boldsymbol{E} + \boldsymbol{J}, \tag{2.10}$$

$$\nabla \cdot \boldsymbol{D} = q, \tag{2.11}$$

$$\nabla \cdot \boldsymbol{B} = m. \tag{2.12}$$

The integral form of the continuity equations can be similarly treated, resulting in

$$\nabla \cdot J = -j\omega\, q, \tag{2.13}$$

$$\nabla \cdot M = -j\omega\, m. \tag{2.14}$$

Note that taking the divergence of the first two Maxwell equations, using the identity $\nabla \cdot \nabla \times \boldsymbol{A} = 0$, the constitutive equations and the continuity equations, one obtains the last two Maxwell equations, provided $\omega \neq 0$. For this reason, when dealing with time-harmonic electromagnetics problems, if charges and currents are assumed to be related via the continuity equations, one needs only satisfy Faraday's and Ampere's laws, with the Gauss laws automatically following.

2.1.3 Boundary form of Maxwell's equations

The boundary forms of Maxwell's equations are also obtained from the integral forms. For the first pair, Figure 2.2, a short length, infinitesimal height ($h << a$) rectangular

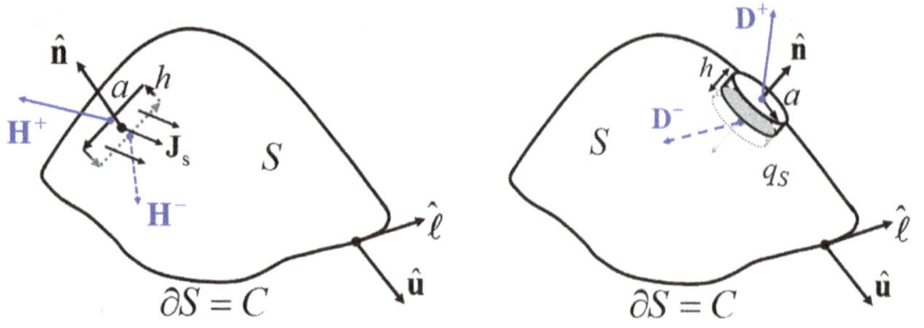

(a) Rectangular path for deriving the boundary form of Ampere's law.

(b) Pillbox surface for the boundary form of Gauss's law.

Figure 2.2 Geometries used to derive boundary forms of Maxwell's equations

path whose longest sides are parallel to and on opposite sides of a material boundary relate the field intensity components $(\boldsymbol{E}, \boldsymbol{H})$ on opposite sides to any *surface current* flux $(\boldsymbol{J}_s, \boldsymbol{M}_s)$ through the rectangle that might be present on the boundary; flux integral contributions from finite volumetric currents $(\boldsymbol{J}, \boldsymbol{M})$ or the flux fields $(\boldsymbol{B}, \boldsymbol{D})$ do not contribute to the limit as the area enclosed by the path vanishes, leaving only surface current flux contributions. This results in the equations (2.15) and (2.16) below where $\hat{\boldsymbol{n}}$ is normal to the interface and points from the $-$ to the $+$ region associated with each side of the interface. For the second pair, two such rectangular paths perpendicular to one another are the cross-sections of a cylindrical pillbox of infinitesimal height sides and radius a ($h << a$) whose circular endfaces are parallel to and on opposite sides of the interface of the two media. Integrating the two Gauss laws over the volume enclosed captures any surface charge density contributions from the interface; volume charge and cylindrical face contributions vanish in the limit as the height vanishes, whereas the area (πa^2) common to the two flux face and charge density integrals cancels, resulting in (2.17) and (2.18) below.

$$\boldsymbol{M}_s = (\boldsymbol{E}^+ - \boldsymbol{E}^-) \times \hat{\boldsymbol{n}}, \tag{2.15}$$

$$\boldsymbol{J}_s = \hat{\boldsymbol{n}} \times (\boldsymbol{H}^+ - \boldsymbol{H}^-), \tag{2.16}$$

$$q_s = \hat{\boldsymbol{n}} \cdot (\boldsymbol{D}^+ - \boldsymbol{D}^-), \tag{2.17}$$

$$m_s = \hat{\boldsymbol{n}} \cdot (\boldsymbol{B}^+ - \boldsymbol{B}^-), \tag{2.18}$$

where $\boldsymbol{M}_s$ and $\boldsymbol{J}_s$ are magnetic and electric surface current densities and q_s and m_s are electric and magnetic surface current densities, respectively. Similarly, the boundary form of the continuity equations for volumetric currents $\boldsymbol{J}$ and $\boldsymbol{M}$ at an interface is obtained as

$$\hat{\boldsymbol{n}} \cdot (\boldsymbol{J}^+ - \boldsymbol{J}^-) + \nabla_s \cdot \boldsymbol{J}_s = -j\omega\, q_s, \tag{2.19}$$

$$\hat{\boldsymbol{n}} \cdot (\boldsymbol{M}^+ - \boldsymbol{M}^-) + \nabla_s \cdot \boldsymbol{M}_s = -j\omega\, m_s, \tag{2.20}$$

where $\nabla_s \cdot$ is the surface divergence operator. We remind the reader that the two line integrals on the LHS of the integral forms of Faraday's and Ampere's laws have the units (V) and (A), respectively, as do the magnetic and electric line (filament) currents K and I passing through an enclosed path; hence, the units of *surface* current densities $\boldsymbol{M}_s$ and $\boldsymbol{J}_s$ must be [V/m] and [A/m], while *volumetric* current density units for $\boldsymbol{M}$ and $\boldsymbol{J}$ are [V/m^2] and [A/m^2]. Note also that a differential volumetric integral over a filament current produces a *dipole moment* result, $Id\ell$, where $d\ell$ is the differential length of the filament that falls inside the differential volume. Thus, in the sense that they do not radiate, point sources of current are essentially non-existent because a radiating current element must have at least some non-vanishing length, $d\ell > 0$. Since a dipole's orientation is also important, its moment is often expressed in vector form as $\mathbf{I}d\ell$ or $Id\boldsymbol{\ell}$.

2.1.4 The Helmholtz equations and potential representations

We have already noted that if $\omega \neq 0$, the two Gauss laws are implied by the differential forms of Faraday's and Ampere's laws plus the continuity equations. Together with their behavior at infinity, the two curl equations should be sufficient to determine $(\boldsymbol{E}, \boldsymbol{H})$ for a given pair of current sources $(\boldsymbol{J}, \boldsymbol{M})$. One step in this direction is to eliminate either $\boldsymbol{E}$ or $\boldsymbol{H}$ between the two. For example, the elimination of the magnetic field is easily accomplished by taking the curl of both sides of Faraday's law, (2.9), and substituting from Ampere's law, (2.10), to eliminate the magnetic field. This yields the so-called vector Helmholtz equation for the electric field:

$$\nabla \times (\nabla \times \boldsymbol{E}) - k^2\boldsymbol{E} = -j\omega\mu\boldsymbol{J} - \nabla \times \boldsymbol{M}, \tag{2.21}$$

where $k = \omega\sqrt{\mu\varepsilon} = 2\pi/\lambda$ is the *wavenumber* and where λ is the wavelength of a wave propagating in the material if both μ and ε are both real (i.e., the medium is lossless). Applying the same procedure, but reversing the roles of Faraday's and Ampere's laws, and eliminating the electric field leads to the magnetic form of the vector Helmholtz equation:

$$\nabla \times (\nabla \times \boldsymbol{H}) - k^2\boldsymbol{H} = -j\omega\varepsilon\boldsymbol{M} + \boldsymbol{J}. \tag{2.22}$$

Unfortunately, it still remains difficult to find general solutions of (2.21) and (2.22). It turns out that *potential* representations of the fields are better suited to finding solutions since they can be chosen to automatically satisfy conditions such as the vanishing of a curl or divergence. To this end, we first note that since Maxwell's equations are linear, the source pair $(\boldsymbol{J}, \boldsymbol{M})$ can be partitioned into a sum of simpler source pair sets as $(\boldsymbol{J}, \boldsymbol{0}) + (\boldsymbol{0}, \boldsymbol{M})$, each of which involves but a *single* current species (electric or magnetic, respectively). Once we obtain the fields due to each single current species acting alone, we obtain solutions for both present by superposing the results for each species. For example, for electric currents only, we set $\boldsymbol{M} = \boldsymbol{0}$ and $m = 0$ in (2.9) and (2.12), respectively, and determine the fields $(\boldsymbol{E}^J, \boldsymbol{H}^J)$, with the superscript indicating the source type. For this case, and under the mild hypothesis of derivative continuity [3], the magnetic form of Gauss's law, $\nabla \cdot \boldsymbol{B}^J = 0$, implies that $\boldsymbol{B}^J = \nabla \times \boldsymbol{A}$ (and vice versa) for a vector $\boldsymbol{A}$ we call the *magnetic vector potential* since we use it to directly determine magnetic field quantities, $\boldsymbol{H}^J = \frac{1}{\mu}\boldsymbol{B}^J = \frac{1}{\mu}\nabla \times \boldsymbol{A}$. $\boldsymbol{A}$ is not uniquely specified, however, since both its curl and divergence can be specified independently, according to the Helmholtz decomposition theorem [4]. Later we will also specify $\nabla \cdot \boldsymbol{A}$ so as to simplify the defining equation for $\boldsymbol{A}$; for the moment, however, we substitute this representation of $\boldsymbol{H}$ into (2.9), and rearrange to $\nabla \times (\boldsymbol{E}^J + j\omega \boldsymbol{A}) = \boldsymbol{0}$. The identity $\nabla \times (\nabla\Phi) = \boldsymbol{0}$ implies that if $\boldsymbol{E}^J$ has the form $\boldsymbol{E}^J = -j\omega\boldsymbol{A} - \nabla\Phi$, it would automatically satisfy this zero curl condition; but it can also be shown that under rather mild restrictions, every $\boldsymbol{E}^J$ can always be so represented [3]. The newly introduced function Φ is called the *electric scalar potential* and the negative sign is chosen to agree with the familiar electrostatic (i.e., $\omega \to 0$) representation for $\boldsymbol{E}^J$ involving scalar potential only. In summary, we have now developed the potential field representation $(\boldsymbol{E}^J, \boldsymbol{H}^J) = (-j\omega\boldsymbol{A} - \nabla\Phi, \frac{1}{\mu}\nabla \times \boldsymbol{A})$ for the fields due to electric current sources $\boldsymbol{J}$ acting alone. Similarly, for magnetic currents $\boldsymbol{M}$ acting alone, the potential field representation is $(\boldsymbol{E}^M, \boldsymbol{H}^M) = (-\frac{1}{\varepsilon}\nabla \times \boldsymbol{F}, -j\omega\boldsymbol{F} - \nabla\Psi)$, involving the *electric vector potential* $\boldsymbol{F}$ and *magnetic scalar potential* Ψ. Superposing these results, we obtain finally the so-called *mixed potential* representation of $(\boldsymbol{E}, \boldsymbol{H}) = (\boldsymbol{E}^J + \boldsymbol{E}^M, \boldsymbol{H}^J + \boldsymbol{H}^M)$:

$$\boldsymbol{E} = -j\omega\boldsymbol{A} - \nabla\Phi - \frac{1}{\varepsilon}\nabla \times \boldsymbol{F}, \tag{2.23}$$

$$\boldsymbol{H} = -j\omega\boldsymbol{F} - \nabla\Psi + \frac{1}{\mu}\nabla \times \boldsymbol{A}. \tag{2.24}$$

To obtain wave equations for the potentials, it is convenient to return to the potential-field representations generated by electric or magnetic source species, respectively, as follows:

$$\text{For } \boldsymbol{M} = \boldsymbol{0}: \quad \boldsymbol{E}^J = -j\omega\boldsymbol{A} - \nabla\Phi, \quad \boldsymbol{H}^J = \frac{1}{\mu}\nabla \times A, \tag{2.25}$$

$$\text{For } \boldsymbol{J} = \boldsymbol{0}: \quad \boldsymbol{E}^M = -\frac{1}{\varepsilon}\nabla \times F, \quad \boldsymbol{H}^M = -j\omega\boldsymbol{F} - \nabla\Psi. \tag{2.26}$$

Substituting the potential representation (2.25) for $(\boldsymbol{E}^J, \boldsymbol{H}^J)$ into Ampere's law, using the identity $\nabla \times (\nabla \times A) = \nabla(\nabla \cdot A) - \nabla^2\boldsymbol{A}$, and rearranging, we obtain

$$\nabla^2\boldsymbol{A} + k^2\boldsymbol{A} = \nabla(\nabla \cdot \boldsymbol{A} + j\omega\varepsilon\mu\Phi) - \mu\boldsymbol{J}. \tag{2.27}$$

Similarly, substituting $(\boldsymbol{E}^M, \boldsymbol{H}^M)$, (2.26), into Faraday's law, we find

$$\nabla^2\boldsymbol{F} + k^2\boldsymbol{F} = \nabla(\nabla \cdot \boldsymbol{F} + j\omega\varepsilon\mu\Psi) - \varepsilon\boldsymbol{M}. \tag{2.28}$$

Despite strong similarities between the equation pair (2.21) and (2.22) and the pair (2.27) and (2.28), the latter provide an important opportunity for simplification since the divergences of $\boldsymbol{A}$ and $\boldsymbol{F}$ remain yet to be specified. Of many possible so-called *gauge* choices here, the so-called Lorenz gauges, are obvious ones:

$$\nabla \cdot \boldsymbol{A} = -j\omega\varepsilon\mu\Phi, \tag{2.29}$$

$$\nabla \cdot \boldsymbol{F} = -j\omega\varepsilon\mu\Psi, \tag{2.30}$$

so that (2.27) and (2.28) immediately simplify to

$$\nabla^2 \boldsymbol{A} + k^2 \boldsymbol{A} = -\mu \boldsymbol{J}, \tag{2.31}$$

$$\nabla^2 \boldsymbol{F} + k^2 \boldsymbol{F} = -\varepsilon \boldsymbol{M}, \tag{2.32}$$

respectively. Equations for the scalar potentials Φ and Ψ are easily obtained by substituting (2.25) and (2.26) into the two Gauss laws, (2.11) and (2.12), respectively, yielding

$$\nabla^2 \Phi + k^2 \Phi = -q/\varepsilon \tag{2.33}$$

and

$$\nabla^2 \Psi + k^2 \Psi = -m/\mu. \tag{2.34}$$

Additional advantages of the vector-valued equations (2.31) and (2.32) are that the operator ∇^2, when applied to the rectangular components of the potential vectors, reduces to the scalar Laplacian operator so that all the rectangular components of (2.31) and (2.32) plus the two scalar potentials (2.33) and (2.34) satisfy the same generic scalar wave equation,

$$\nabla^2 \psi + k^2 \psi = -f(\boldsymbol{r}), \tag{2.35}$$

for some potential ψ with scalar volume source density $f(\boldsymbol{r})$. We note the source term $f(\boldsymbol{r})$ can be written as a superposition or convolution integral

$$f(\boldsymbol{r}) = \int_V f(\boldsymbol{r}')\delta(\boldsymbol{r} - \boldsymbol{r}')\, dV', \tag{2.36}$$

where V includes at least the *support* of $f(\boldsymbol{r})$, supp $(f(\boldsymbol{r}))$, that is, the smallest closed region outside of which the function $f(\boldsymbol{r})$ vanishes identically. Thus, the set supp $(f(\boldsymbol{r}))$ is completely contained within V. The three-dimensional delta function $\delta(\boldsymbol{r} - \boldsymbol{r}') = \delta(x - x')\delta(y - y')\delta(z - z')$ represents a *point source of unit density at* $\boldsymbol{r}'$; note its units are $[1/\text{m}^3]$. We wish to express ψ as a source-weighted sum over point source solutions, and to do so we first introduce a scalar function $G(\boldsymbol{r}, \boldsymbol{r}')$, a solution of the point-source equation

$$\nabla^2 G + k^2 G = -\delta(\boldsymbol{r} - \boldsymbol{r}'), \tag{2.37}$$

where $G(\boldsymbol{r}, \boldsymbol{r}')$ represents a generic potential field at $\boldsymbol{r}$ due to a unit density source at $\boldsymbol{r}'$. Since a (scalar) point source is both non-directional and (2.37) is coordinate system independent, we should also expect the same of G, i.e., that it depends only on the radial separation $R = |\boldsymbol{r} - \boldsymbol{r}'|$ between the observation point $\boldsymbol{r}$ and source point $\boldsymbol{r}'$. We also note that in a local spherical coordinate system with an origin at $\boldsymbol{r}'$, $\nabla^2 G = \frac{1}{R^2}\frac{d}{dR}\left(R^2 \frac{dG}{dR}\right)$, from which one easily verifies that

$$G(\boldsymbol{r}, \boldsymbol{r}') = \frac{e^{-jkR}}{4\pi R} \tag{2.38}$$

satisfies (2.37) everywhere except possibly at $R = 0$ (since the expression for ∇^2 involves division by zero there). To also verify the unit volume integral behavior of the 3D delta function at $R = 0$, we must examine the equality of volume integrals over terms on both sides of (2.37). The non-directional nature of the problem also suggests integrating over a volumetric test sphere, V_a, of radius a centered at $R = 0$, noting that $\nabla^2 = \nabla \cdot \nabla$, and evaluating the integral of the Laplacian term via the divergence theorem. Indeed, one finds $\int_{V_a} (\nabla^2 G + k^2 G)\, dV = -\int_{V_a} \delta(\boldsymbol{r} - \boldsymbol{r}'), dV = -1$, *independent* of the test sphere radius a.

Finally, the linearity of the scalar wave equation suggests that if an arbitrary source density distribution can be expressed as a weighted superposition of point sources, (2.36), then the distribution's potential can be expressed as a similarly weighted superposition of potentials due to point sources:

$$\psi = \int_V G(\boldsymbol{r}, \boldsymbol{r}') f(\boldsymbol{r}')\, dV'. \tag{2.39}$$

Indeed, that (2.39) is a solution of (2.35) is easily verified by direct substitution and use of (2.37) and (2.36). We remark that, by implication, the integral in (2.39) must also include any contributions from surface, filament, or point sources present, which one writes in the forms $\int_S G(\boldsymbol{r}, \boldsymbol{r}') f_s(\boldsymbol{r})\, dS'$, $\int_C G(\boldsymbol{r}, \boldsymbol{r}') f_\ell(\boldsymbol{r}')\, d\ell'$, or $\sum_n F_n G(\boldsymbol{r}, \boldsymbol{r}_n)$, respectively.

Using (2.23), (2.24), (2.31)–(2.35), and (2.39), we are now able to write the fields in terms of potentials as follows [2]:

$$\boldsymbol{E} = -j\omega \boldsymbol{A} - \nabla \Phi - \frac{1}{\varepsilon} \nabla \times \boldsymbol{F}, \tag{2.40}$$

$$\boldsymbol{H} = -j\omega \boldsymbol{F} - \nabla \Psi + \frac{1}{\mu} \nabla \times \boldsymbol{A}, \tag{2.41}$$

where

$$\boldsymbol{A}(\boldsymbol{r}) = \mu \int_V G(\mathbf{r}, \mathbf{r}') \boldsymbol{J}(\boldsymbol{r}')\, dV', \tag{2.42}$$

$$\boldsymbol{F}(\boldsymbol{r}) = \varepsilon \int_V G(\mathbf{r}, \mathbf{r}') \boldsymbol{M}(\boldsymbol{r}')\, dV', \tag{2.43}$$

$$\Phi(\boldsymbol{r}) = \frac{1}{\varepsilon} \int_V G(\mathbf{r}, \mathbf{r}') q(\boldsymbol{r}')\, dV', \tag{2.44}$$

$$\Psi(\boldsymbol{r}) = \frac{1}{\mu} \int_V G(\mathbf{r}, \mathbf{r}') m(\boldsymbol{r}')\, dV'. \tag{2.45}$$

Before leaving this section, we call the reader's attention to the following observations concerning (2.40)–(2.45):

- Though the integration domains and sources in the integrals above are volumetric, they implicitly include source integrals over surfaces and filaments, as well as sums over point sources.
- Note the extent to which we rely on linearity and superposition: the fields $(\boldsymbol{E}, \boldsymbol{H})$ are vector component sums, each term of which is a sum of real and imaginary parts, and constructed as a linear superposition of a vector potential, the vector-valued gradient of a scalar potential, and the curl of a vector potential, i.e., a

sum over both current and charge terms, respectively. Each of these, in turn, is a convolution integral, i.e., a sum of a point-source responses $G(\boldsymbol{r}, \boldsymbol{r}')$ weighted by source densities of the appropriate charge or current species defined on V. Later on, in discretizing a problem, we further subdivide the source domains into collections of simpler subdomains, each with simplified local source representations. In the latter case, however, the new level of sums is typically represented as matrix–vector products.

2.1.5 Far fields and far potentials

In the far field, it is convenient to express a field observation point $\mathbf{r}$ in spherical coordinates as $\boldsymbol{r} = r(\cos\phi \sin\theta \hat{\boldsymbol{x}} + \sin\phi \sin\theta \hat{\boldsymbol{y}} + \cos\theta \hat{\boldsymbol{z}})$ whose origin is at $r = 0$, assumed not far from or even within the source region. At large radial distances $r = |\boldsymbol{r}|$ from all source points $\boldsymbol{r}'$, the approximation $R = |\boldsymbol{r} - \boldsymbol{r}'| \approx r - \hat{\boldsymbol{r}} \cdot \boldsymbol{r}'$ where $\hat{\boldsymbol{r}} = \boldsymbol{r}/r$, can be used in the locally varying phase factor term in the numerator of $G(\boldsymbol{r}, \boldsymbol{r}')$ while $R \approx r$ can be used in the denominator to approximate $G(\boldsymbol{r}, \boldsymbol{r}')$ as $\frac{e^{-jkr}}{4\pi r} e^{jk\hat{\boldsymbol{r}}\cdot\boldsymbol{r}'}$ in the far vector potentials, yielding

$$\boldsymbol{A}(\boldsymbol{r}) \approx \mu \frac{e^{-jkr}}{4\pi r} \int_V \boldsymbol{J}(\boldsymbol{r}') e^{jk\hat{\boldsymbol{r}}\cdot\boldsymbol{r}'} dV' \tag{2.46}$$

$$\boldsymbol{F}(\boldsymbol{r}) \approx \varepsilon \frac{e^{-jkr}}{4\pi r} \int_V \boldsymbol{M}(\boldsymbol{r}') e^{jk\hat{\boldsymbol{r}}\cdot\boldsymbol{r}'} dV'. \tag{2.47}$$

Similarly, the far scalar potentials are

$$\Phi(\boldsymbol{r}) \approx \frac{1}{\varepsilon} \frac{e^{-jkr}}{4\pi r} \int_V q(\boldsymbol{r}') e^{jk\hat{\boldsymbol{r}}\cdot\boldsymbol{r}'} dV' \tag{2.48}$$

$$\Psi(\boldsymbol{r}) \approx \frac{1}{\mu} \frac{e^{-jkr}}{4\pi r} \int_V m(\boldsymbol{r}') e^{jk\hat{\boldsymbol{r}}\cdot\boldsymbol{r}'} dV'. \tag{2.49}$$

But since $\nabla \frac{e^{-jkr}}{4\pi r} e^{jk\hat{\boldsymbol{r}}\cdot\boldsymbol{r}'} \approx -jk \frac{e^{-jkr}}{4\pi r} e^{jk\hat{\boldsymbol{r}}\cdot\boldsymbol{r}'} \hat{\boldsymbol{r}}$, we may simply replace ∇ by $-jk\hat{\boldsymbol{r}}$ in the far field. Combined with the Lorenz gauge conditions, it then follows that the gradients of the scalar potentials are entirely radially directed and completely cancel radial components of the vector potential contributions to the far field; thus, in the far field, fields $(\boldsymbol{E}, \boldsymbol{H})$ become completely transverse to the observation vector $\boldsymbol{r}$ and can be written as

$$\begin{aligned} \boldsymbol{E} &\approx -j\omega \left(\hat{\boldsymbol{r}} \times \boldsymbol{A}\right) \times \hat{\boldsymbol{r}} + j\omega\eta(\hat{\boldsymbol{r}} \times \boldsymbol{F}) \\ &\approx -j\omega(\hat{\boldsymbol{\theta}}\hat{\boldsymbol{\theta}} + \hat{\boldsymbol{\phi}}\hat{\boldsymbol{\phi}}) \cdot \boldsymbol{A} - j\omega\eta(\hat{\boldsymbol{\theta}}\hat{\boldsymbol{\phi}} - \hat{\boldsymbol{\phi}}\hat{\boldsymbol{\theta}}) \cdot \boldsymbol{F} \end{aligned} \tag{2.50}$$

and

$$\begin{aligned} \boldsymbol{H} &\approx \frac{1}{\eta} \hat{\boldsymbol{r}} \times \boldsymbol{E} \\ &\approx -j\omega(\hat{\boldsymbol{\theta}}\hat{\boldsymbol{\theta}} + \hat{\boldsymbol{\phi}}\hat{\boldsymbol{\phi}}) \cdot \boldsymbol{F} + j\frac{\omega}{\eta}(\hat{\boldsymbol{\theta}}\hat{\boldsymbol{\phi}} - \hat{\boldsymbol{\phi}}\hat{\boldsymbol{\theta}}) \cdot \boldsymbol{A}, \end{aligned} \tag{2.51}$$

where $\eta = \sqrt{\mu/\varepsilon}$ is the local intrinsic impedance of the host medium. For compactness above, we have used a dyadic notation for terms such as $\hat{\boldsymbol{\phi}}\hat{\boldsymbol{\theta}} \cdot \boldsymbol{A}$, for example, which is assumed to be evaluated as $\hat{\boldsymbol{\phi}}(\hat{\boldsymbol{\theta}} \cdot \boldsymbol{A}) = \hat{\boldsymbol{\phi}}A_\theta$, and the unit vectors transverse to $\hat{\boldsymbol{r}}$ in spherical coordinates are

$$\begin{aligned} \hat{\boldsymbol{\theta}} &= \hat{\boldsymbol{x}}\cos\phi\cos\theta + \hat{\boldsymbol{y}}\sin\phi\cos\theta - \hat{\boldsymbol{z}}\sin\theta \\ \hat{\boldsymbol{\phi}} &= -\hat{\boldsymbol{x}}\sin\phi + \hat{\boldsymbol{y}}\cos\phi. \end{aligned} \tag{2.52}$$

We note here that there exists a second solution of the free space Green's function equation (2.37) that is of the form $\frac{e^{+jkR}}{4\pi R}$, but the sign change in the exponential phase factor implies it represents a (non-physical) incoming, rather than an outgoing wave. In unbounded regions with bounded sources, imposition of the so-called scalar or *Sommerfeld radiation condition* [5–7]

$$\lim_{r\to\infty} r\left(\frac{\partial G}{\partial r} + jkG\right) = 0, \tag{2.53}$$

guarantees a unique solution and one that involves only outgoing wave solutions of (2.37). It is clear that each vector component of (2.40) and (2.41) satisfies the Sommerfeld condition. The transverse nature and plane-wave-like relation between far-field components $(\boldsymbol{E}, \boldsymbol{H})$ as well as their outgoing behavior are captured together in the so-called *Silver–Müller radiation conditions* [8,9],

$$\lim_{r\to\infty} r\left(\hat{\boldsymbol{r}} \times \boldsymbol{H} + \boldsymbol{E}/\eta\right) = \boldsymbol{0} \quad \text{and} \quad \lim_{r\to\infty} r\left(\hat{\boldsymbol{r}} \times \boldsymbol{E} - \eta\boldsymbol{H}\right) = \boldsymbol{0}. \tag{2.54}$$

2.1.6 The duality principle

There is a symmetry between pairs of Maxwell's equations that clearly appears when magnetic current and charges are included in the equations. This is made evident in Table 2.1 which, when the quantities on the first row are replaced by the corresponding quantities on the second row, leave Maxwell's equations (and all wave and potential equations, etc. that follow from them) invariant in infinite, homogeneous media. The duality principle thus may be used either to derive or serve as a quick cross-check on derived expressions. Its most frequent use is to reduce the mathematical labor of deriving new results. For example, in Section 2.1.4, we assumed two situations, the first with electric and the second with magnetic currents present only. Once an expression for the magnetic vector potential $\boldsymbol{A}$ due to electric currents was available,

Table 2.1 If the variables in the first row of the table are replaced by those of the second row, expressions for Maxwell's equations, potentials, and wave equations for fields in infinite homogeneous media remain invariant.

$\boldsymbol{E}$	$\boldsymbol{H}$	$\boldsymbol{J}$	$\boldsymbol{M}$	q	m	ε	μ	$\boldsymbol{A}$	$\boldsymbol{F}$	Φ	Ψ
$\downarrow$	$\downarrow$	$\downarrow$	$\downarrow$	$\downarrow$	$\downarrow$	$\downarrow$	$\downarrow$	$\downarrow$	$\downarrow$	$\downarrow$	$\downarrow$
$\boldsymbol{H}$	$-\boldsymbol{E}$	$\boldsymbol{M}$	$-\boldsymbol{J}$	m	$-q$	μ	ε	$\boldsymbol{F}$	$-\boldsymbol{A}$	Ψ	$-\Phi$

we could then have used duality to directly obtain the electric vector potential $\boldsymbol{F}$ due to magnetic currents. As was done, superposition can then be used to obtain the result when both current species are present. Such uses of duality will often be implied in subsequent derivations in this chapter. When using duality, however, some care must be exercised if boundary conditions are present. For example, the dual of the perfect electric conductor (PEC) boundary condition $\hat{\boldsymbol{n}} \times \boldsymbol{E} = \mathbf{0}$ is the perfect magnetic conductor (PMC) boundary condition $\hat{\boldsymbol{n}} \times \boldsymbol{H} = \mathbf{0}$; here it is not simply the variables that are exchanged, but the physical problem parameters must also be dual to the original. Further discussion of and uses for duality may be found in [2,10,11].

2.1.7 Uniqueness theorem

Consider a region V bounded by a closed surface S with unit outward normal $\hat{\boldsymbol{n}}$, as shown in Figure 2.3(a). Our concern is the following question: "Given a set of sources $\boldsymbol{J}$ and $\boldsymbol{M}$, under what conditions are solutions to Maxwell's equations unique in V?" Uniqueness questions are important not only in establishing whether or not a given problem has a unique solution in V but also in discovering what information is needed to obtain a unique solution. Uniqueness theorems also directly connect fields in a region to their sources and vice versa. They also suggest what changes can be made to a problem outside V while maintaining the same fields inside V and on S. One of their more important uses, however, is in providing shortcut proofs or tests for proposed, hypothetical solutions of Maxwell's equations. We find them especially useful in proving image and source equivalence principles from "guesses," based perhaps on experience or intuition.

Uniqueness theorems are generally proven by contradiction. That is, assume that the fields are non-unique so that two field solutions $(\boldsymbol{E}^a, \boldsymbol{H}^a)$ and $(\boldsymbol{E}^b, \boldsymbol{H}^b)$ exist for

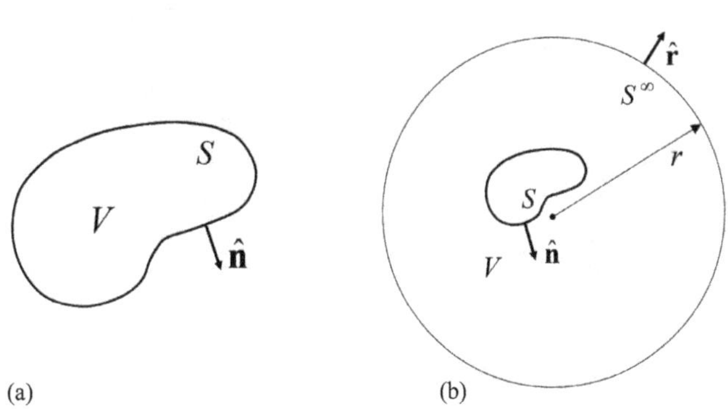

Figure 2.3 In (a), an interior region V is bounded by a surface S of bounded extent and with outward unit normal $\hat{\mathbf{n}}$. In (b), region V is exterior to S and bounded by a surface S^∞ of radius r with outward unit normal $\hat{\mathbf{n}} = \hat{\mathbf{r}}$, where $r \to \infty$. The location of the coordinate origin for r is arbitrary, but must be a finite distance from S.

the same boundary conditions on S and the same set of sources $(\boldsymbol{J}, \boldsymbol{M})$ in V. We assume $(\boldsymbol{E}^a, \boldsymbol{H}^a) \neq (\boldsymbol{E}^b, \boldsymbol{H}^b)$, and easily find that the difference fields $(\delta\boldsymbol{E}, \delta\boldsymbol{H}) = (\boldsymbol{E}^a - \boldsymbol{E}^b, \boldsymbol{H}^a - \boldsymbol{H}^b)$ must satisfy the source-free Faraday and Ampere's laws,

$$\begin{aligned} \nabla \times \delta\boldsymbol{E} &= -j\omega\mu\delta\boldsymbol{H} \\ \nabla \times \delta\boldsymbol{H} &= j\omega\varepsilon\delta\boldsymbol{E}. \end{aligned} \tag{2.55}$$

Next, we consider that the net outward complex power flow $\delta\mathbb{S}$ across S, which from the Poynting vector $\delta\boldsymbol{S} = \frac{1}{2}\delta\boldsymbol{E} \times \delta\boldsymbol{H}^*$ and Poynting's theorem, is given by

$$\begin{aligned} \delta\mathbb{S} &= \frac{1}{2}\oint_S \delta\boldsymbol{S} \cdot \hat{\boldsymbol{n}}\, dS = \frac{1}{2}\oint_S (\delta\boldsymbol{E} \times \delta\boldsymbol{H}^*) \cdot \hat{\boldsymbol{n}}\, dS \\ &= -j\omega \int_V \frac{1}{2}\left(\mu'|\delta\boldsymbol{H}|^2 - \varepsilon'|\delta\boldsymbol{E}|^2\right) dV \\ &\quad -\omega \int_V \frac{1}{2}\left(\mu''|\delta\boldsymbol{H}|^2 + \varepsilon''|\delta\boldsymbol{E}|^2\right) dV, \end{aligned} \tag{2.56}$$

where we have assumed $\mu = \mu' - j\mu''$ and $\varepsilon = \varepsilon' - j\varepsilon''$ to account for either lossless or lossy media. In obtaining (2.56), we use the divergence theorem, the identity $\nabla \cdot (\delta\boldsymbol{E} \times \delta\boldsymbol{H}^*) = \nabla \times \delta\boldsymbol{E} \cdot \delta\boldsymbol{H}^* - \delta\boldsymbol{E} \cdot \nabla \times \delta\boldsymbol{H}^*$, and (2.55). Below, we treat separately the cases where the homogeneous medium associated with V is lossy, unbounded and lossless, or bounded and lossless. We also assume one of the following three boundary conditions on S:

- The tangential electric field $\hat{\boldsymbol{n}} \times \boldsymbol{E}$ is specified (i.e., $\hat{\boldsymbol{n}} \times \delta\boldsymbol{E} = \boldsymbol{0}$) on S.
- The tangential magnetic field $\hat{\boldsymbol{n}} \times \boldsymbol{H}$ is specified (i.e., $\hat{\boldsymbol{n}} \times \delta\boldsymbol{H} = \boldsymbol{0}$) on S.
- The tangential field $\hat{\boldsymbol{n}} \times \boldsymbol{E}$ is specified (i.e., $\hat{\boldsymbol{n}} \times \delta\boldsymbol{E} = \boldsymbol{0}$) on a portion of S while $\hat{\boldsymbol{n}} \times \boldsymbol{H}$ is specified (i.e., $\hat{\boldsymbol{n}} \times \delta\boldsymbol{H} = \boldsymbol{0}$) on the remainder of S.

Bounded dissipative media: For lossy bounded media, at least one of the terms μ'' or ε'' in the last integral on the right-hand side of (2.56) is positive. If we assume fields interior to V are non-unique, then $\delta\boldsymbol{E}$ or $\delta\boldsymbol{H}$ cannot vanish everywhere in V. But for any of the above boundary conditions on the bounded surface S of Figure 2.3(a), at least one of $\hat{\boldsymbol{n}} \times \delta\boldsymbol{E}$ or $\hat{\boldsymbol{n}} \times \delta\boldsymbol{H}$ vanishes at every point of S, and, hence, so does the complex power flow $\delta\mathbb{S}$ across S in (2.56). In a lossy medium, however, this represents a contradiction since for $\omega > 0$, the last volume integral in (2.56) must be negative if either μ'' or ε'' or both are positive, whereas the integrals on the first line vanish. Hence at least one of the difference fields, $\delta\boldsymbol{E}$ or $\delta\boldsymbol{H}$, must also vanish in V. But by (2.55), if one vanishes, then both vanish. Thus fields in a lossy region V with boundary S are unique if they satisfy Maxwell's equations in V and one of the above three prescribed boundary conditions on S.

Bounded, lossless media: We consider next the bounded region of Figure 2.3(a) with a lossless interior ($\mu'' = \varepsilon'' = 0$) with boundary S enclosing V. We note first that for S bounded and closed, the above three specified boundary conditions describe a cavity with either PEC or PMC walls, or a mixture of both types of walls. Next, we note that since no sources are present, with lossless media, and with boundary fields satisfying either of the three boundary conditions specified above, the surface integral (both real and imaginary parts) of (2.56) as well as the last (real-valued) volume integral vanish identically. The next-to-last (imaginary-valued) volume integral then must also vanish, implying either that (a) $\delta \boldsymbol{E} = \delta \boldsymbol{H} = \mathbf{0}$, in turn implying uniqueness, or (b) the net time-averaged stored electric and magnetic energies in V must exactly balance in V. But this latter condition is a known property of cavity resonators with PMC or PEC conducting walls at their discrete resonance frequencies. These well-known source-free solutions of Maxwell's equations for interior problems with impenetrable boundaries can exist at discrete frequencies that depend strongly on the cavity geometry and (lossless) material properties. At cavity resonance frequencies, no source is required to sustain the field oscillations, no real average power is dissipated or radiated, and associated modal electric and magnetic fields are in phase quadrature. Finally, the time-average stored electric and magnetic energies are equal and their amplitudes are proportional to the squared magnitudes of the modal field amplitudes. The cavity problem is of interest in surface integral equation modeling in computational electromagnetics for two principal reasons:

- One wants to use a surface integral formulation to determine the resonant frequencies of a closed cavity of shape S with impenetrable boundaries.
- One's interest instead is in solving an exterior scattering or radiation problem involving a closed PEC, PMC, or mixed PMC, PEC boundary S, but due to inevitable modeling errors that occur, solutions to the external problem are severely contaminated at frequencies near the internal resonances by the weak excitation of and coupling to the highly resonant interior cavity problem. This is the so-called *internal resonance problem* often associated with surface integral equation modeling of electromagnetic scattering problems.

Unbounded media: We consider last the situation of Figure 2.3(b) with an interior boundary S and with V an exterior boundary enclosed within a spherical boundary surface S^{∞} of radius r and outward normal $\hat{\boldsymbol{n}} = \hat{\boldsymbol{r}}$. We now evaluate the integrals (2.56) in the limit as r approaches infinity. Note that we must first replace the original Poynting surface integral on $\oint_S$ in (2.56) with the two boundary integrals, $-\oint_S + \oint_{S^{\infty}}$, with the negative sign appearing in the first integral since, for V now *exterior* to S, the divergence theorem requires an outward normal to V that now points *into* S. Again, the integral on S vanishes for any of the three boundary conditions specified above. Hence, (2.56) reduces to

$$\delta \mathbb{S} = \frac{1}{2} \oint_{S^{\infty}} (\delta \boldsymbol{E} \times \delta \boldsymbol{H}^*) \cdot \hat{\boldsymbol{n}} \, dS. \tag{2.57}$$

A lemma by Rellich [7,12–14] for scalar acoustic fields has been extended to the electromagnetic case which shows that if the integral on the right above vanishes, all fields within V must vanish. That is, the vanishing of the integral (2.57) becomes the condition for uniqueness for the unbounded, lossless, or lossy case.

2.2 Equivalence principles

2.2.1 The volumetric equivalence principle

The volumetric equivalence principle allows volumetric electric current sources to be replaced by magnetic current sources or vice versa. Under the replacement, the new electric and magnetic fields $(\boldsymbol{E}', \boldsymbol{H}')$ are completely equivalent to the original fields $(\boldsymbol{E}, \boldsymbol{H})$ outside the source region, but only partially equivalent within the source region [4,15].

2.2.1.1 Replacing magnetic current ($\boldsymbol{M}$) sources with electric current ($\boldsymbol{J}'$) sources and vice versa

From (2.21), it is clear that wherever $\boldsymbol{M}$ alone is present, it may be replaced by an equivalent electric current $\boldsymbol{J}'$, where

$$\boldsymbol{J}' = \frac{1}{j\omega\mu}\nabla \times \boldsymbol{M}. \tag{2.58}$$

But in Faraday's law, since $\boldsymbol{E}$ remains unchanged, then the electric field remains unchanged $(\boldsymbol{E}' = \boldsymbol{E})$ by the transformation from magnetic to electric currents, whereas the magnetic field $\boldsymbol{H}'$ after transformation must be corrected in the source region to obtain the original field $\boldsymbol{H}$ as follows:

$$\boldsymbol{H} = \boldsymbol{H}' - \frac{\boldsymbol{M}}{j\omega\mu}. \tag{2.59}$$

Similarly, in (2.22), $\boldsymbol{J}$ acting alone may be replaced by an equivalent magnetic current $\boldsymbol{M}'$, where

$$\boldsymbol{M}' = \frac{-1}{j\omega\varepsilon}\nabla \times \boldsymbol{J}. \tag{2.60}$$

But from Ampere's law, the magnetic field remains unchanged $(\boldsymbol{H}' = \boldsymbol{H})$ by the transformation from electric to magnetic currents, while the electric field $\boldsymbol{E}'$ after transformation must be corrected in the source region as follows:

$$\boldsymbol{E} = \boldsymbol{E}' - \frac{\boldsymbol{J}}{j\omega\varepsilon}. \tag{2.61}$$

The volume equivalence principle can be used, for example, to show that an electric (magnetic) current loop is equivalent to a magnetic (electric) dipole and vice versa [2]. Several related equivalences are discussed in [16].

2.2.2 The surface equivalence principle

The surface equivalence principle is usually used in formulating surface integral equations involving *equivalent surface current* sources [2,17,18]. Figure 2.4 shows an unbounded region V^+ with material parameters (ε^+, μ^+) which contains a bounded, closed inclusion V^- with material parameters (ε^-, μ^-). The closed boundary S separates the regions, and the electric and magnetic fields in V^+ are designated $(\boldsymbol{E}^+, \boldsymbol{H}^+)$ while those in V^- are $(\boldsymbol{E}^-, \boldsymbol{H}^-)$. Either or both regions may contain impressed electric

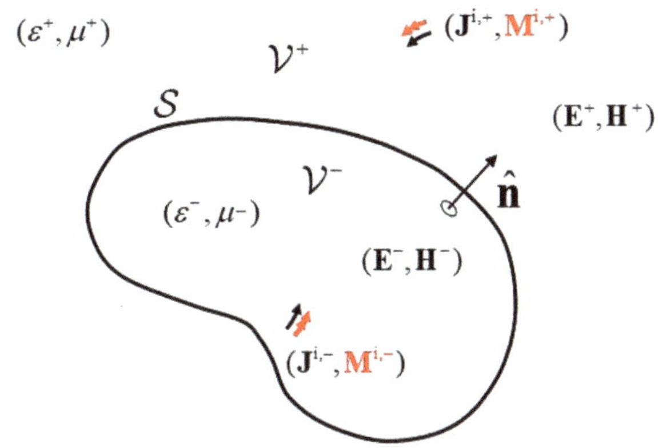

Figure 2.4 Original problem showing external and internal regions V^+ and V^- with material parameters (μ^+, ε^+) and (μ^-, ε^-), respectively. The regions are separated by the boundary surface S, and any impressed electric and magnetic sources $(\boldsymbol{J}^{i,\pm}, \boldsymbol{M}^{i,\pm})$ that may exist in either or both regions are shown; the corresponding fields in each region are designated $(\boldsymbol{E}^+, \boldsymbol{H}^+)$ and $(\boldsymbol{E}^-, \boldsymbol{H}^-)$, respectively. The unit normal $\hat{\boldsymbol{n}}$ on S points into V^+; tangential fields are assumed continuous at S, i.e., $\hat{\boldsymbol{n}} \times \boldsymbol{E}^+ = \hat{\boldsymbol{n}} \times \boldsymbol{E}^- \triangleq \hat{\boldsymbol{n}} \times \boldsymbol{E}$ and similarly for $\boldsymbol{H}$.

and magnetic sources $(\boldsymbol{J}^{i,+}, \boldsymbol{M}^{i,+})$ and $(\boldsymbol{J}^{i,-}, \boldsymbol{M}^{i,-})$. The unit normal $\hat{\boldsymbol{n}}$ on S points into V^+ and, if we assume that tangential electric and magnetic fields at S are continuous there, we may drop the $\pm$ designation, writing $(\hat{\boldsymbol{n}} \times \boldsymbol{H}^\pm, \boldsymbol{E}^\pm \times \hat{\boldsymbol{n}})$ as $(\hat{\boldsymbol{n}} \times \boldsymbol{H}, \boldsymbol{E} \times \hat{\boldsymbol{n}})$ instead. Figure 2.5 shows the *exterior equivalence* representation, wherein the original fields and sources are specified in the exterior region, V^+, whereas in V^- (the *null field* region) both the fields and sources are required to vanish, i.e., $(\boldsymbol{E}^-, \boldsymbol{H}^-) = (\boldsymbol{J}^{i,-}, \boldsymbol{M}^{i,-}) = (\mathbf{0}, \mathbf{0})$. According to the boundary form of Maxwell's equations, we require the introduction of *equivalent surface currents* $(\boldsymbol{J}^+, \boldsymbol{M}^+) = (\hat{\boldsymbol{n}} \times \boldsymbol{H}^+, \boldsymbol{E}^+ \times \hat{\boldsymbol{n}})$ on S to support the resulting discontinuity in tangential field components at S and to complete the exterior equivalence representation. As Figure 2.6 shows, we can in a dual manner also construct an *interior equivalence* representation by specifying that the original fields and sources of Figure 2.4 be retained in the interior, while the exterior becomes the null field region with vanishing sources and fields. It is easily verified that the equivalent surface currents required to support the resulting field discontinuities are merely the negatives of those needed for the exterior equivalence, i.e., $(\boldsymbol{J}^-, \boldsymbol{M}^-) = (-\hat{\boldsymbol{n}} \times \boldsymbol{H}, -\boldsymbol{E} \times \hat{\boldsymbol{n}}) = (-\boldsymbol{J}^+, -\boldsymbol{M}^+)$.

The interior and exterior equivalence forms of Figures 2.5 and 2.6, in which null fields appear in the region complementary to the equivalence region, are examples of Love's equivalence principle[19,20], and if the fields at S are tangentially continuous, we may drop the $\pm$ designation on equivalent currents and use $(\boldsymbol{J}, \boldsymbol{M})$ to designate equivalent surface currents for the representation in V^+ while $(-\boldsymbol{J}, -\boldsymbol{M})$ represent

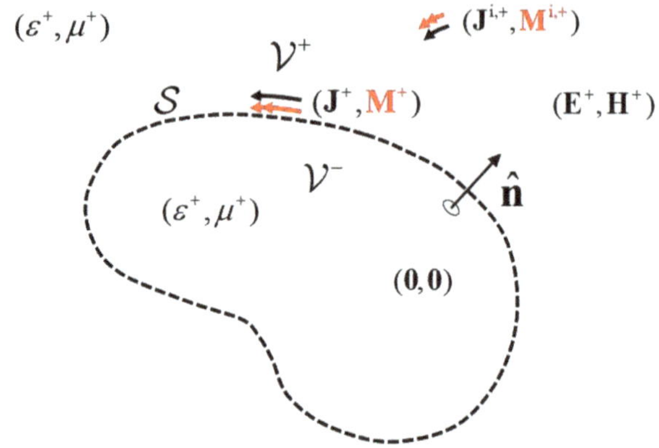

Figure 2.5 In Love's exterior equivalence [2,19], the original fields are specified and all sources retained in the exterior region V^+; all interior sources and fields, $(\boldsymbol{J}^{i,-}, \boldsymbol{M}^{i,-})$ and $(\boldsymbol{E}^-, \boldsymbol{H}^-)$, respectively, are set to $(\mathbf{0}, \mathbf{0})$. To support the resulting field tangential discontinuities at S, equivalent electric and magnetic surface currents, $(\boldsymbol{J}^+, \boldsymbol{M}^+) = (\hat{\boldsymbol{n}} \times \boldsymbol{H}^+, \boldsymbol{E}^+ \times \hat{\boldsymbol{n}})$, respectively, are introduced on S. The validity of the exterior equivalence follows from the uniqueness theorem. With no sources and with null fields in the interior, the medium parameters of the null field region may be modified if desired. For example, as shown, the exterior medium parameters are often extended to the interior to enable the use of a homogeneous medium Green's function.

those for V^-. An infinite number of equivalent forms exist for which different fields or material parameters are specified in regions complementary to that for which the equivalence is valid. It is emphasized that since the fields and sources vanish in the null field regions of both the exterior and interior regions, the material parameters in either null field region may be altered without affecting the fields in the complementary equivalence region. To illustrate, suppose V^- in Figure 2.7 is a perfectly electric conductor (PEC) with a closed boundary S and unit normal $\hat{\boldsymbol{n}}$ illuminated by an electric current "point" source, i.e. a dipole of infinitesimal length and moment $\mathbf{I}^{i,+}(\boldsymbol{r}')d\ell'$, all residing in a homogeneous medium. Using Love's equivalence, the dipole and surface equivalent currents together produce a null field in V^-. Hence, the original PEC may, for example, be removed and replaced by merely extending the exterior medium parameters into the interior such that V^+ and V^- together constitute an unbounded homogeneous medium. This permits one to use the homogeneous medium Green's functions and associated potential representations for the fields in both V^+ and V^-. Note that both the exciting dipole and the equivalent currents now each radiate in an infinite homogeneous medium, with the fields of the first constituting an *incident field* while those of the equivalent currents constitute the *scattered field*, their fields canceling in V^-. But the PEC boundary condition requires that $\boldsymbol{E}$ vanish as S is

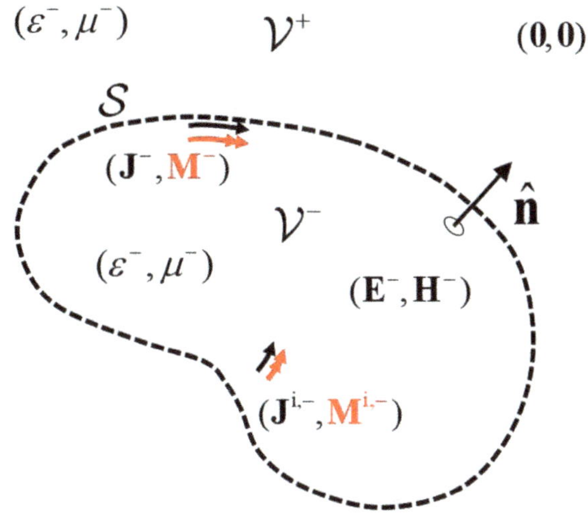

Figure 2.6 *In Love's interior equivalence [2,19], the original fields are specified and all sources retained in the interior region V^-; all exterior sources and fields, $(\boldsymbol{J}^{i,+}, \boldsymbol{M}^{i,+})$ and $(\boldsymbol{E}^+, \boldsymbol{H}^+)$, respectively, are set to $(\mathbf{0}, \mathbf{0})$. To support the resulting field tangential discontinuities at S, equivalent electric and magnetic surface currents, $(\boldsymbol{J}^-, \boldsymbol{M}^-) = (-\hat{\boldsymbol{n}} \times \boldsymbol{H}^-, -\boldsymbol{E}^- \times \hat{\boldsymbol{n}})$, respectively, are introduced on S. The validity of the exterior equivalence follows from the uniqueness theorem. With no sources and with null fields in the exterior, the medium parameters of the null field region may be modified as desired. For example, as shown, the interior medium parameters are often extended to the exterior to enable the use of a homogeneous medium Green's function.*

approached from the exterior, hence, the boundary form of Maxwell's equations yields $(\boldsymbol{J}, \boldsymbol{M}) = (\hat{\boldsymbol{n}} \times \boldsymbol{H}, \mathbf{0})$, i.e., no surface equivalent magnetic current exists for the Love PEC equivalence. It is also possible, as illustrated in Figure 2.8, to represent a PEC using a magnetic current alone, but the interior of S will no longer be a null field region. Clearly a wide variety of equivalences exist, but for numerical solutions, we generally must extend or define medium parameters in the complementary region such that a Green's function exists for sources in the equivalence region.

2.3 Boundary field representations

Surface integral equations for equivalent currents are obtained by applying appropriate boundary conditions at boundaries S separating different equivalence regions. Applying a boundary condition at S usually requires that field observation points $\boldsymbol{r}$

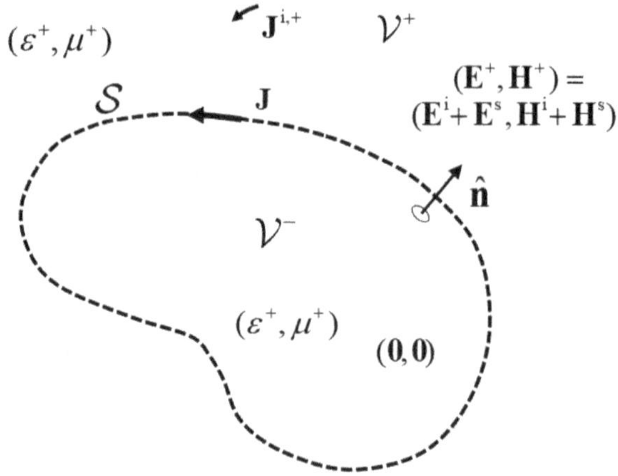

Figure 2.7 *Love's equivalence principle for scattering by a perfect electric conductor (PEC). V^- is the PEC region, with a vanishing tangential electric field, $\hat{\boldsymbol{n}} \times \boldsymbol{E}^+$ on S. The PEC is removed and V^- is specified as a null field region with the material parameters of V^+ extended to V^-. Thus, both $\hat{\boldsymbol{n}} \times \boldsymbol{E}^-$ and $\hat{\boldsymbol{n}} \times \boldsymbol{E}^+$ vanish on S, and so also must $\boldsymbol{M}$. The exterior field can also be written as $\boldsymbol{E}^+ = \boldsymbol{E}^i + \boldsymbol{E}^s$, where $\boldsymbol{E}^i$ represents the* incident field *due to $\boldsymbol{J}^{i,+}$ and $\boldsymbol{E}^s$ represents the* scattered field *due to $\boldsymbol{J}$ with both electric currents radiating in an infinite, homogeneous medium. Note here that $\boldsymbol{J}$ is not simply an equivalent current; it also represents the physical PEC surface conduction current, $\boldsymbol{J} = \hat{\boldsymbol{n}} \times \boldsymbol{H}^+$, and is unique (except possibly at resonance frequencies of the cavity with PEC boundary walls S).*

approach S from one side or the other and be evaluated there using potential representations. We write $\boldsymbol{r} \to S^+$ if $\boldsymbol{r}$ approaches S from the region pointed into by the surface normal, $\hat{\boldsymbol{n}}$, or $\boldsymbol{r} \to S^-$ if from the opposite side. Thus the electric form of Gauss's law, (2.17), at S becomes, for example, $\hat{\boldsymbol{n}} \cdot (\lim_{\boldsymbol{r}\to S^+} \boldsymbol{D} - \lim_{\boldsymbol{r}\to S^-} \boldsymbol{D}) = q_s(\boldsymbol{r})$.

Approaching a surface from a single-side contrasts with the boundary form of Maxwell's equations (2.15)–(2.18) in that the latter relates *local* sources at $\boldsymbol{r}$ on S to *differences* between *local*, opposite side field components at $\boldsymbol{r}$. The evaluation of field components on a single side of S, however, often involves both local and non-local contributions, the former directly related to a source strength at $\boldsymbol{r}$, and the latter arising from integration over any remaining equivalent sources that can contribute to fields on S. Whether or not a local source contribution exists depends on the behavior of a potential's associated Green's function, or its derivatives, as $R = |\boldsymbol{r} - \boldsymbol{r}'| \to 0$. We first examine below possible local contributions.

Assume that d measures the distance of an observation point $\boldsymbol{r}$ from a smooth point of S along its unit normal there; whether d is positive or negative depends

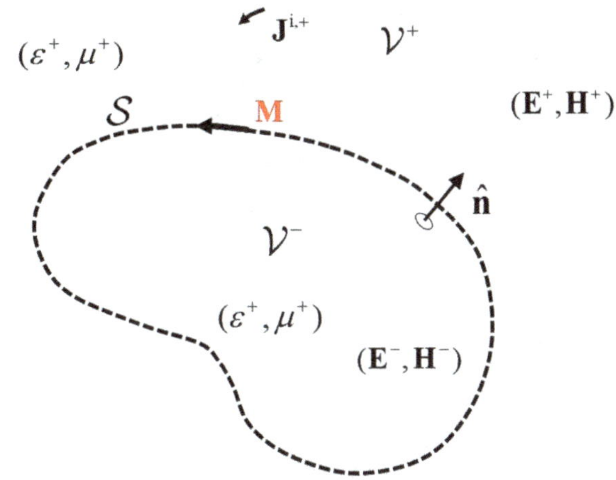

Figure 2.8 *Equivalence principle for scattering by a perfect electric conductor (PEC) using only an equivalent magnetic current, $\boldsymbol{M}$. The PEC in V^- is first removed, and with $\boldsymbol{J} = \boldsymbol{0}$ the tangential magnetic field is continuous at S, i.e., $\hat{\boldsymbol{n}} \times \boldsymbol{H}^+ = \hat{\boldsymbol{n}} \times \boldsymbol{H}^-$ while, due to the tangential field discontinuity of $\boldsymbol{M}$, $\hat{\boldsymbol{n}} \times \boldsymbol{E}^-$ no longer vanishes there. With no sources in V^-, the uniqueness theorem guarantees that the resulting fields, $(\boldsymbol{E}^-, \boldsymbol{H}^-)$ are unique in V^-, except at resonance frequencies of the cavity with perfectly magnetic conducting (PMC) boundary walls S. Here $\boldsymbol{H}^-$ can be viewed as a continuation of $\boldsymbol{H}^+$ into V^-, with $\boldsymbol{E}^-$ related to $\boldsymbol{H}^-$ there via Faraday's law.*

on whether the observation point approaches the surface from the S^+ or S^- side, respectively. We remove for separate treatment the local section δS_a of S, shown in Figure 2.9, contained in the spherical ball of radius a centered at the surface limit point of $\boldsymbol{r}$ on S. We will also assume that $|d| << a$, implying that whenever we take the limit $a \to 0$, we must first take the limit $d \to 0$. Note also that our smooth surface assumption at the $\boldsymbol{r}$ limit on S further implies that δS_a becomes a flat circular disk, while the assumed electric sources $\boldsymbol{J}$ and q_s approach constants there. Any non-vanishing potential integral contributions from the disk source thus constitute local contributions to the field there. Non-local field contributions are those due to the remaining equivalent sources on the punctured boundary surface S with δS_a removed and are merely evaluated via the potential integral representations in the limit $a \to 0$. The net fields at the surface are simply a superposition of those due to local and non-local sources.

To evaluate the local disk contributions from electric scalar and magnetic vector potentials $\delta_a \Phi$ and $\delta_a \boldsymbol{A}$ due to surface electric charge and current contributions $q(\boldsymbol{r}')$ and $\boldsymbol{J}(\boldsymbol{r}')$, respectively, we note the limits $q(\boldsymbol{r}') \to q(\boldsymbol{r})$, $\boldsymbol{J}(\boldsymbol{r}') \to \boldsymbol{J}(\boldsymbol{r})$, and $e^{-jkR} \to 1$ as $a \to 0$, remove these source limits from the integrand, and evaluate the

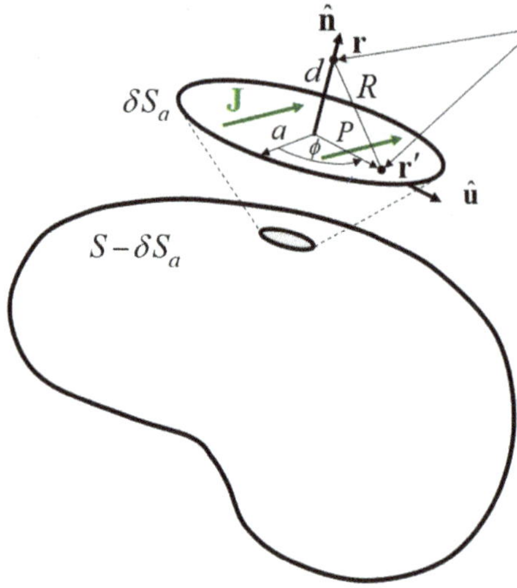

Figure 2.9 *Boundary surface S separates two equivalence regions. To separate the local and non-local contributions to fields at a smooth point* **r** *of S, assume* **r** *approaches S along its unit normal there, with d the signed distance of* **r** *from S and* $|d| << a$. *The portion* δS_a *of S within a spherical ball of radius a centered at the limit point* **r** *is removed and shown in expanded view and local polar coordinates* (P, ϕ) *may be used to parameterize source point locations* $\mathbf{r}'$ *on* δS_a. *Note that as* $a \to 0$, δS_a *approaches a circular disk of radius a.*

resulting surface integral. The integral may be parameterized in polar coordinates (P, ϕ); however, radial-angular coordinates (R, ϕ), where $R^2 = P^2 + d^2$, are often more convenient since $dS' = PdPd\phi = RdRd\phi$ and the Jacobian factor R cancels similar factors appearing in the denominator of the Green's function or its derivatives. Thus, we have as $a \to 0$,

$$\begin{bmatrix} \delta_a \Phi \\ \delta_a \boldsymbol{A} \end{bmatrix} \to \begin{bmatrix} q(\mathbf{r})/\varepsilon \\ \mu \boldsymbol{J}(\mathbf{r}) \end{bmatrix} \int_0^{2\pi} \int_{|d|}^{\sqrt{a^2+d^2}} \left(\frac{1}{4\pi R} \right) RdRd\phi \to \begin{bmatrix} 0 \\ \mathbf{0} \end{bmatrix}. \tag{2.62}$$

We may similarly evaluate the local contributions to magnetic scalar and electric vector potentials $\delta_a \Psi$ and $\delta_a \boldsymbol{F}$ due to surface magnetic charge and current contributions $m(\boldsymbol{r}')$ and $\boldsymbol{M}(\boldsymbol{r}')$, respectively, obtaining

$$\begin{bmatrix} \delta_a \Psi \\ \delta_a \boldsymbol{F} \end{bmatrix} \to \begin{bmatrix} m(\mathbf{r})/\mu \\ \varepsilon \boldsymbol{M}(\mathbf{r}) \end{bmatrix} \int_0^{2\pi} \int_{|d|}^{\sqrt{a^2+d^2}} \left(\frac{1}{4\pi R} \right) RdRd\phi \to \begin{bmatrix} 0 \\ \mathbf{0} \end{bmatrix}. \tag{2.63}$$

Field representation by potentials also generally requires the gradient of the scalar potentials and the curl of the vector potentials. Both can be expressed in terms of the gradient of the Green's function

$$\nabla G(\boldsymbol{r},\boldsymbol{r}) = -\frac{(1+jkR)e^{-jkR}(\boldsymbol{r}-\boldsymbol{r}')}{4\pi R^3} \to -\frac{(\boldsymbol{r}-\boldsymbol{r}')}{4\pi R^3} = \frac{P\hat{\boldsymbol{u}} - d\hat{\boldsymbol{n}}}{4\pi R^3}, \tag{2.64}$$

where, again, the Green's function gradient reduces to its static limit as $ka \to 0$. We have also parameterized the source point as $\boldsymbol{r}' = \boldsymbol{r} + P\hat{\boldsymbol{u}} - d\hat{\boldsymbol{n}}$ (see Figure 2.9). Note that $\hat{\boldsymbol{n}}$ is the disk (and local surface) normal whereas $\hat{\boldsymbol{u}}$ is the unit vector lying in the disk plane and normal to its edge. Hence, the required limit integrals are

$$\begin{aligned}\begin{bmatrix} \varepsilon\,\nabla(\delta_a\Phi) \\ \frac{1}{\mu}\nabla\times(\delta_a\boldsymbol{A}) \end{bmatrix} &= \int_{\delta_a S}\begin{bmatrix} \nabla(\,G(\boldsymbol{r},\boldsymbol{r}')\,q(\boldsymbol{r}')\,) \\ \nabla\times(\,G(\boldsymbol{r},\boldsymbol{r}')\boldsymbol{J}(\boldsymbol{r}')\,) \end{bmatrix} dS' = \int_{\delta_a S}\nabla G(\boldsymbol{r},\boldsymbol{r}')\begin{bmatrix} q(\boldsymbol{r}') \\ \times\boldsymbol{J}(\boldsymbol{r}') \end{bmatrix} dS' \\ &\to \begin{bmatrix} q(\boldsymbol{r}) \\ -\boldsymbol{J}(\boldsymbol{r})\times \end{bmatrix}\int_{|d|}^{\sqrt{a^2+d^2}}\left(\int_0^{2\pi}\frac{P\,\hat{\boldsymbol{u}} - d\,\hat{\boldsymbol{n}}}{4\pi R^3}d\phi\right)RdR \\ &\to \begin{bmatrix} -\frac{1}{2}\,\mathrm{sgn}\,(d)\hat{\boldsymbol{n}}\,q(\boldsymbol{r}) \\ -\frac{1}{2}\,\mathrm{sgn}\,(d)\,\hat{\boldsymbol{n}}\times\boldsymbol{J}(\boldsymbol{r}) \end{bmatrix},\end{aligned} \tag{2.65}$$

where $\mathrm{sgn}\,x$ is $+1$ if $x > 0$ and -1 if $x < 0$. In the first equality of (2.65), the gradient and curl operators are taken inside the first integral where, as in the next equality, they only appear as gradients of the Green's function since it is the only term in the integrand depending on the observation point. Furthermore, as in (2.62), $q(\boldsymbol{r}')$ and $\boldsymbol{J}(\boldsymbol{r}')$ approach their values at $\boldsymbol{r}$ as $a \to 0$ and thus are removed from the integral. Finally, since the angular integral over $\hat{\boldsymbol{u}}$ vanishes while d and $\hat{\boldsymbol{n}}$ are constants, the remaining integrals are easily evaluated. Hence, for surface magnetic currents and charges, we obtain, dual to (2.65),

$$\begin{aligned}\begin{bmatrix} \mu\;\nabla\delta_a\Psi \\ \frac{1}{\varepsilon}\nabla\times\,\delta_a\boldsymbol{F} \end{bmatrix} &= \int_{\delta_a S}\begin{bmatrix} \nabla(\,G(\boldsymbol{r},\boldsymbol{r}')\,m(\boldsymbol{r}')\,) \\ \nabla\times(\,G(\boldsymbol{r},\boldsymbol{r}')\,\boldsymbol{M}(\boldsymbol{r}')\,) \end{bmatrix} dS' \\ &\to \begin{bmatrix} -\frac{1}{2}\,\mathrm{sgn}\,(d)\,\hat{\boldsymbol{n}}\,m(\boldsymbol{r}) \\ -\frac{1}{2}\,\mathrm{sgn}\,(d)\,\hat{\boldsymbol{n}}\times\boldsymbol{M}(\boldsymbol{r}) \end{bmatrix}.\end{aligned} \tag{2.66}$$

We note that by performing the angular integral *before* the radial one, the integral on $\hat{\boldsymbol{u}}$ vanishes in (2.65) and (2.66). This is fortuitous since the radial integral with integrand P/R^2 diverges as $a \to 0$. Indeed, the integral should *not* be expected to exist for $\boldsymbol{r}$ at a non-smooth (e.g., edge or vertex) point of S where not only does the angular integral not span the full range $(0, 2\pi)$, but neither do $q(\boldsymbol{r}')$ and $\boldsymbol{J}(\boldsymbol{r}')$ approach constants as $\boldsymbol{r}' \to \boldsymbol{r}$. Even on smooth surfaces, however, the requirement that the angular integral vanish in (2.65) and (2.66) serves to warn us that some care is required in the numerical evaluation of potential integrals and their derivatives near observation points. In general, an integral over δS_a may vanish, converge, or diverge as $a \to 0$, depending on whether the kernel of the integral equation has a weak singularity, a strong singularity, or is hypersingular there, respectively. For example, (2.62) and (2.63) involve integrands with weak singularities, whereas (2.65) and (2.66) involve strong singularities. An advantage of representing fields via potentials, however, is that for smooth surfaces, one rarely must deal with hypersingular kernels. Indeed, we note that none of the above cases exhibit this behavior. On the other hand, if the integral over δS_a converges and is non-vanishing as $a \to 0$, as e.g., for (2.65) and (2.66), the integrand is said to be strongly singular. In this case, we call the non-vanishing disk contribution to the surface integral at the observation point the "residue" contribution because of its resemblance in form, interpretation, and calculation to the residues of path integrals in complex variable theory. The remaining part of the surface integral is the limit as $a \to 0$ of the integral over the punctured surface $S - \delta S_a$ and is called the *principal value integral*, often written as p.v.$\int_S ...dS'$ or $⨍_S ...dS'$; we use either notation as convenient in the following. The residue contribution typically changes sign as an observation point crosses S. But since sources have been removed at observation points, the principal value integral is continuous as S is approached from either side. Finally, if the residue integral vanishes, the integral reduces to an ordinary surface integral that is continuous across S. Using the above results, the electric and magnetic fields at a surface with surface currents and charges there may be written as

$$\begin{aligned}\boldsymbol{E} &= \pm\hat{\boldsymbol{n}} \times \frac{\boldsymbol{M}}{2} \pm \hat{\boldsymbol{n}}\,\frac{q(\boldsymbol{r})}{2\varepsilon} - j\omega\boldsymbol{A} - \text{p.v.}(\nabla\Phi) - \frac{1}{\varepsilon}\text{p.v.}(\nabla \times \boldsymbol{F}) \\ &= \pm\hat{\boldsymbol{n}} \times \frac{\boldsymbol{M}}{2} \pm \hat{\boldsymbol{n}}\,\frac{q(\boldsymbol{r})}{2\varepsilon} - j\omega\mu\int_S G(\boldsymbol{r},\boldsymbol{r}')\boldsymbol{J}(\boldsymbol{r}')\,dS' + \frac{1}{j\omega\varepsilon}⨍_S \nabla G(\boldsymbol{r},\boldsymbol{r}')\nabla'\cdot\boldsymbol{J}(\boldsymbol{r}')\,dS' \\ &\quad - ⨍_S \nabla G(\boldsymbol{r},\boldsymbol{r}') \times \boldsymbol{M}(\boldsymbol{r}')\,dS', \quad \boldsymbol{r} \in S^{\pm}, \end{aligned} \tag{2.67}$$

and

$$\begin{aligned}\boldsymbol{H} &= \mp\hat{\boldsymbol{n}} \times \frac{\boldsymbol{J}}{2} \pm \hat{\boldsymbol{n}}\,\frac{m(\boldsymbol{r})}{2\mu} - j\omega\boldsymbol{F} - \text{p.v.}(\nabla\Psi) + \frac{1}{\mu}\text{p.v.}(\nabla \times \boldsymbol{A}) \\ &= \mp\hat{\boldsymbol{n}} \times \frac{\boldsymbol{J}}{2} \pm \hat{\boldsymbol{n}}\,\frac{m(\boldsymbol{r})}{2\mu} - j\omega\varepsilon\int_S G(\boldsymbol{r},\boldsymbol{r}')\boldsymbol{M}(\boldsymbol{r}')\,dS' + \frac{1}{j\omega\mu}⨍_S \nabla G(\boldsymbol{r},\boldsymbol{r}')\nabla'\cdot\boldsymbol{M}(\boldsymbol{r}')\,dS' \\ &\quad + ⨍_S \nabla G(\boldsymbol{r},\boldsymbol{r}') \times \boldsymbol{J}(\boldsymbol{r}')\,dS', \quad \boldsymbol{r} \in S^{\pm}. \end{aligned} \tag{2.68}$$

In forming integral equations, we generally need the tangential components of $(\boldsymbol{E}, \boldsymbol{H})$ on boundaries; hence, we introduce *rotated* tangential components $(\hat{\boldsymbol{n}} \times \boldsymbol{E}, \hat{\boldsymbol{n}} \times \boldsymbol{H})$ and find that the above equations become

$$\begin{aligned}\hat{\boldsymbol{n}} \times \boldsymbol{E} &= \hat{\boldsymbol{n}} \times \lim_{\boldsymbol{r}\to S^{\pm}} \left(-j\omega\boldsymbol{A} - \nabla\Phi - \frac{1}{\varepsilon}\nabla \times F\right) \\ &= \eta\mathscr{L}\boldsymbol{J} - \left(\pm\frac{1}{2}\mathscr{I} - \mathscr{K}\right)\boldsymbol{M}, \quad \boldsymbol{r} \in S^{\pm} \end{aligned} \tag{2.69}$$

and

$$\begin{aligned}\hat{\boldsymbol{n}} \times \boldsymbol{H} &= \hat{\boldsymbol{n}} \times \lim_{\boldsymbol{r}\to S^{\pm}} \left(-j\omega\boldsymbol{F} - \nabla\Psi + \frac{1}{\mu}\nabla \times \boldsymbol{A}\right) \\ &= \eta^{-1}\mathscr{L}\boldsymbol{M} + \left(\pm\frac{1}{2}\mathscr{I} - \mathscr{K}\right)\boldsymbol{J}, \quad \boldsymbol{r} \in S^{\pm}, \end{aligned} \tag{2.70}$$

where, following [21], we define the dimensionless linear operators $\mathscr{L}$ and $\mathscr{K}$ as follows:

$$\begin{aligned}\mathscr{L}X &= -jk\hat{n} \times \left(\int_S G(r,r')X(r')\,dS' + \frac{1}{k^2}\nabla \int_S G(r,r')\nabla' \cdot X(r')\,dS' \right) \\ &= -jk\hat{n} \times \int_S \left[\left(\mathscr{I} + \frac{\nabla\nabla}{k^2} \right) G(r,r') \right] \cdot X(r')\,dS' \\ &= -jk\hat{n} \times \int_S \mathcal{G}(r,r') \cdot X(r')\,dS', \quad r \to S, \end{aligned} \tag{2.71}$$

and

$$\begin{aligned}\mathscr{K}X &= -\hat{n} \times \fint_S \nabla G(r,r') \times X(r')\,dS' \\ &= -\hat{n} \times \nabla \times \fint_S \mathcal{G}(r,r') \cdot X(r')\,dS', \quad r \to S, \end{aligned} \tag{2.72}$$

respectively, where

$$\mathcal{G} = \left(\mathscr{I} + \frac{\nabla\nabla}{k^2} \right) G \tag{2.73}$$

is a *dyadic* Green's function operator, i.e., a linear operator that operates on a vector, generating a new vector from it. One notes that the integral associated with the operator $\mathscr{K}$ remains a principal value integral whose "residue" appears explicitly as the $\pm\frac{1}{2}\mathscr{I}$ term, while that associated with $\mathscr{L}$ is no longer written as a principal value integral since the normal component of its residue disappears under the $\hat{n}\times$ operation.

Here we make some observations concerning the operators $\mathscr{L}$ and $\mathscr{K}$ defined in (2.71) and (2.72). In (2.71), for example, the first- and second-line equalities can be shown to be equivalent using the identity $G\nabla' \cdot J = \nabla' \cdot (GJ) - \nabla' G \cdot J$ in the first equality, applying the surface divergence theorem to the term involving $\nabla' \cdot (GJ)$, and noting that $\nabla' G = -\nabla G$ in the second. (Note there is no boundary contribution from the divergence theorem since S is assumed closed.) Though analysis shows that the two forms are analytically equivalent, they are very different from a numerical perspective. For example, the first equality requires that $\nabla' \cdot J$ exist, while the second does not. However, the second requires that we handle a much more singular Green's function derivative, of order $1/R^3$, which is easily shown to be non-integrable. It is possible to define such integrals, however—by reversing the analysis, returning to essentially the same form given in the first line of (2.71)! The reasonably equitable distribution of differentiability requirements appearing in various integral equation terms is one of the primary reasons our focus here is heavily weighted towards potential representations of electric and magnetic fields.

Regarding the $\mathscr{K}$ operator of (2.72), we note that as r' approaches r on S, both ∇G and J become tangential to S, their cross-product approaches $\hat{n}$, and this in turn vanishes in the cross-product with $\hat{n}$ outside the integral. This analysis, of course, fails when $r' = r$, but that case is specifically eliminated by the principal value designation; it resulted in the residue contribution $\pm\frac{1}{2}\mathscr{I}$ that usually appears together with $\mathscr{K}$. Finally, we note that we are able to employ the *same* dyadic Green's function $\mathcal{G}$ in defining $\mathscr{K}$ and $\mathscr{L}$ only because $\nabla \times (\mathcal{G} \cdot \boldsymbol{J}) = \nabla \times [(\mathscr{I} + \frac{\nabla\nabla}{k^2})G \cdot \boldsymbol{J}] = \nabla \times (G\boldsymbol{J})$, i.e. since $\nabla \times \nabla = \mathbf{0}$. In summary, while the "equivalent" representations appearing in (2.71) and (2.72) are listed in increasing order of notional compactness, they are essentially in decreasing order of practicality for computation!

2.3.1 The Calderón identities

We next derive a useful set of identities relating to the $\mathscr{L}$ and $\mathscr{K}$ operators. We first rewrite (2.69) and (2.70) relating surface currents $\boldsymbol{J}$ and $\boldsymbol{M}$ to tangential fields in a slightly rearranged matrix format as

$$\begin{bmatrix} \hat{\boldsymbol{n}} \times \boldsymbol{E} \\ \hat{\boldsymbol{n}} \times \eta \boldsymbol{H} \end{bmatrix} = \begin{bmatrix} (\mp \frac{1}{2}\mathscr{I} + \mathscr{K}) & \mathscr{L} \\ \mathscr{L} & \pm\frac{1}{2}\mathscr{I} - \mathscr{K} \end{bmatrix} \begin{bmatrix} \boldsymbol{M} \\ \eta \boldsymbol{J} \end{bmatrix}, \quad \boldsymbol{r} \in S^{\pm}. \tag{2.74}$$

Equation (2.74) represents the fields $(\boldsymbol{E}, \boldsymbol{H})$ observed at $S^{\pm}$ due to an arbitrarily defined set of surface current sources $(\boldsymbol{J}, \boldsymbol{M})$ on S radiating in a homogeneous medium with no other excitation present. According to Love's equivalence principle (see Figures 2.5 and 2.6), specified surface currents $(\boldsymbol{J}, \boldsymbol{M}) = (\pm \hat{\boldsymbol{n}} \times \boldsymbol{H}, \pm \boldsymbol{E} \times \hat{\boldsymbol{n}})$ would produce the *same* tangential fields on both $S^{\pm}$, respectively. Substituting these currents into the right-hand side of (2.74) and rearranging, we thus obtain

$$\begin{aligned} \begin{bmatrix} \hat{\boldsymbol{n}} \times \boldsymbol{E} \\ \hat{\boldsymbol{n}} \times \eta \boldsymbol{H} \end{bmatrix} &= \begin{bmatrix} \frac{1}{2}\mathscr{I} \mp \mathscr{K} & \pm \mathscr{L} \\ \mp \mathscr{L} & \frac{1}{2}\mathscr{I} \mp \mathscr{K} \end{bmatrix} \begin{bmatrix} \hat{\boldsymbol{n}} \times \boldsymbol{E} \\ \hat{\boldsymbol{n}} \times \eta \boldsymbol{H} \end{bmatrix} \\ &= \mathcal{P}^{\pm} \begin{bmatrix} \hat{\boldsymbol{n}} \times \boldsymbol{E} \\ \hat{\boldsymbol{n}} \times \eta \boldsymbol{H}, \end{bmatrix}, \quad \boldsymbol{r} \in S^{\pm}, \end{aligned} \tag{2.75}$$

where the operator matrices

$$\mathcal{P}^{\pm} = \begin{bmatrix} \frac{1}{2}\mathscr{I} \mp \mathscr{K} & \pm \mathscr{L} \\ \mp \mathscr{L} & \frac{1}{2}\mathscr{I} \mp \mathscr{K} \end{bmatrix}, \quad \boldsymbol{r} \in S^{\pm} \tag{2.76}$$

are called *projectors* since, by (2.75) and (2.76), they have the projection property that they transform or map onto themselves, i.e., $(\mathcal{P}^{\pm})^2 = \mathcal{P}^{\pm}$, as is seen by operating with $\mathcal{P}^{\pm}$ on both sides of (2.75). On rearranging, it then follows that

$$\mathcal{P}^{\pm}\left(\mathcal{P}^{\pm} - \mathcal{I}\right) \begin{bmatrix} \hat{\boldsymbol{n}} \times \boldsymbol{E} \\ \hat{\boldsymbol{n}} \times \eta \boldsymbol{H} \end{bmatrix} = \begin{bmatrix} \mathbf{0} \\ \mathbf{0} \end{bmatrix}, \quad \boldsymbol{r} \in S^{\pm}, \tag{2.77}$$

where $\mathcal{I}$ is the identity operator matrix corresponding to (2.76) and hence

$$\mathcal{P}^{\pm}\left(\mathcal{P}^{\pm} - \mathcal{I}\right) = \begin{bmatrix} -\frac{1}{4}\mathscr{I} + \mathscr{K}^2 - \mathscr{L}^2 & -\mathscr{K}\mathscr{L} - \mathscr{L}\mathscr{K} \\ \mathscr{L}\mathscr{K} + \mathscr{K}\mathscr{L} & -\frac{1}{4}\mathscr{I} + \mathscr{K}^2 - \mathscr{L}^2 \end{bmatrix}, \quad \boldsymbol{r} \in S^{\pm}. \tag{2.78}$$

Since (2.77) holds for arbitrary $\hat{\boldsymbol{n}} \times \boldsymbol{E}$ and $\hat{\boldsymbol{n}} \times \boldsymbol{H}$, all the operator terms of $\mathcal{P}^{\pm}\left(\mathcal{P}^{\pm} - \mathcal{I}\right)$ in (2.78) must vanish, yielding the *Calderón identities* [21],

$$\mathscr{L}^2 = -\frac{1}{4}\mathscr{I} + \mathscr{K}^2 \text{ and } \mathscr{L}\mathscr{K} = -\mathscr{K}\mathscr{L}. \tag{2.79}$$

Generally speaking, the identity operator $\mathscr{I}$, which is its own inverse, is often considered the ideal invertible operator. Recall also that the operator $\mathscr{K}$ may in a sense be considered to be a well-behaved perturbation of the identity operator when it appears in combination with it, for example, as in $\frac{1}{2}\mathscr{I} \mp \mathscr{K}$. On the other hand, the characteristics of the operator $\mathscr{L}$, particularly, are less than ideal, often enhancing errors and particularly resulting in slow convergence of iterative solution schemes. Hence, it seems remarkable that the product of operator $\mathscr{L}$ with itself yields $-\frac{1}{4}\mathscr{I} + \mathscr{K}^2$, which has a similar "identity plus perturbation" form as $\frac{1}{2}\mathscr{I} \mp \mathscr{K}$. In recent years, this observation has proved invaluable in suggesting better-performing numerical formulations for problems involving the operator $\mathscr{L}$ [22,23].

2.4 The Lorentz reciprocity theorem

The Lorentz reciprocity theorem relates two problems in which the medium, geometry, and frequency remain the same, but the sources and, hence, the resulting fields are different. Let $(\boldsymbol{J}^a, \boldsymbol{M}^a)$ be a set of sources of finite extent producing fields $(\boldsymbol{E}^a, \boldsymbol{H}^a)$ in an unbounded linear medium; similarly, $(\boldsymbol{J}^b, \boldsymbol{M}^b)$ is a set of sources of finite extent producing fields $(\boldsymbol{E}^b, \boldsymbol{H}^b)$ in the same environment and at the same frequency. Using the identity $\nabla \cdot (\boldsymbol{a} \times \boldsymbol{b}) = \boldsymbol{b} \cdot \nabla \times \boldsymbol{a} - \boldsymbol{a} \cdot \nabla \times \boldsymbol{b}$ plus Faraday's and Ampere's laws, (2.9) and (2.10), the expression $\nabla \cdot (\boldsymbol{E}^a \times \boldsymbol{H}^b - \boldsymbol{E}^b \times \boldsymbol{H}^a)$ easily reduces to

$$\nabla \cdot (\boldsymbol{E}^a \times \boldsymbol{H}^b - \boldsymbol{E}^b \times \boldsymbol{H}^a) = (\boldsymbol{E}^b \cdot \boldsymbol{J}^a - \boldsymbol{H}^b \cdot \boldsymbol{M}^a) - (\boldsymbol{E}^a \cdot \boldsymbol{J}^b - \boldsymbol{H}^a \cdot \boldsymbol{M}^b). \quad (2.80)$$

Integrating over all space and applying the divergence theorem to the left-hand side yields a surface integral over the far-field sphere at infinity. Using the $1/R$ behavior along with the local, plane wave-like nature of fields at infinity shows that while the surface area of the far-field sphere grows as R^2, the crossflux term $(\boldsymbol{E}^a \times \boldsymbol{H}^b - \boldsymbol{E}^b \times \boldsymbol{H}^a)$ vanishes faster than $1/R^2$, and, hence, the crossflux integral vanishes as $R \to \infty$. On rearranging the resulting integral, we obtain the Lorentz reciprocity theorem for unbounded regions,

$$\int_V (\boldsymbol{E}^a \cdot \boldsymbol{J}^b - \boldsymbol{H}^a \cdot \boldsymbol{M}^b)\, dV = \int_V (\boldsymbol{E}^b \cdot \boldsymbol{J}^a - \boldsymbol{H}^b \cdot \boldsymbol{M}^a)\, dV. \quad (2.81)$$

If, for example, $(\boldsymbol{J}^b, \boldsymbol{M}^b)$ are actually surface or lineal currents, $(\boldsymbol{J}_s^b, \boldsymbol{M}_s^b)$ or $(\mathbf{I}_e^b, \mathbf{I}_m^b)$, respectively, the volumetric integral over those sources collapses to the corresponding surface or line integrals, i.e., $\int_V \boldsymbol{E}^a \cdot \boldsymbol{J}^b\, dV \to \int_{S^b} \boldsymbol{E}^a \cdot \boldsymbol{J}_s^b\, dS$ or $\int_{C^b} \boldsymbol{E}^a \cdot \mathbf{I}_e^b\, d\ell$, and similarly for the $\boldsymbol{H}^a \cdot \boldsymbol{M}^b$ term.

For bounded regions, additional cross-flux surface integrals arise from the left-hand side of (2.80). In view of the identities $\hat{\boldsymbol{n}} \cdot (\boldsymbol{E}^a \times \boldsymbol{H}^b) = (\hat{\boldsymbol{n}} \times \boldsymbol{E}^a) \cdot \boldsymbol{H}^b = \boldsymbol{E}^a \cdot (\boldsymbol{H}^b \times \hat{\boldsymbol{n}})$, however, either the "$a$" or the "$b$" fields can be replaced by boundary equivalent surface currents, and then included in (2.81) merely as a superposition of volume source integrals but specialized to surface (equivalent) sources. The integral on the left-hand side of the equality (2.81) is known as the *reaction integral* of fields a on sources b, often written $< a, b >$, while the right-hand side integral is the reaction $< b, a >$ of fields b on sources a. The reciprocity theorem shows that if, say, $\boldsymbol{J}^a$ is a point source of unit moment $\hat{\boldsymbol{a}}$ at $\boldsymbol{r}^a$, i.e., $\boldsymbol{J}^a = \hat{\boldsymbol{a}}\delta(\boldsymbol{r} - \boldsymbol{r}^a)$, and $\boldsymbol{J}^b$ is a similar point source of unit moment $\hat{\boldsymbol{b}}$ at $\boldsymbol{r}^b$, then it easily follows that $\boldsymbol{E}^a(\boldsymbol{r}^b) \cdot \hat{\boldsymbol{b}} = \boldsymbol{E}^b(\boldsymbol{r}^a) \cdot \hat{\boldsymbol{a}}$, and, hence, that the *transmission and receiving problems are reciprocal*. More generally, however, the reciprocity theorem relates the field response at one source set due to a second source set, to the response at the second source set due to the first set.

One may also view one source set as a primary excitation while the other serves specifically as a *test* set to probe or measure (in a weighted average sense) the fields elsewhere, generalizing our setup for proving reciprocity for point source pairs. The reaction viewpoint of field probing (testing) using secondary sources is not only closely akin to the manner in which fields are actually measured using calibrated probes in the laboratory, but the testing-fields-due-to-sources viewpoint is also employed extensively in the computational solution of surface integral equations. Indeed, the most fundamental and common calculation in solving surface integral equations involves the evaluation of reaction integrals between assumed sources on surfaces and multiple test source sets spanning many observation points (or regions) on equivalent source surfaces.

Here we make another important observation concerning the reaction integrals (2.81). We note each dot product involves a *field intensity* quantity with a (surface or volumetric) *flux density* quantity. These quantities are, in reality, different vector types, as evidenced by the fact that one typically applies the divergence operator only to flux quantities, such as $(\boldsymbol{J}, \boldsymbol{M}, \boldsymbol{B}, \boldsymbol{D})$, but the curl operator only to the intensity quantities $(\boldsymbol{E}, \boldsymbol{H})$. Ideally, one should always attempt to ensure that, for vector quantities, such reaction integrals involve only dot products between intensity and flux quantities. However, we also note that in such expressions such as $\boldsymbol{D} = \varepsilon \boldsymbol{E}$, $\boldsymbol{B} = \mu \boldsymbol{H}$ and $\nabla \times \boldsymbol{E} = -j\omega\mu \boldsymbol{H} - \boldsymbol{M}$ multiplication of intensity quantities $\boldsymbol{E}, \boldsymbol{H}$ by material parameters ε, μ, apparently not only rescales them but also *transforms them into flux-type quantities* $(\boldsymbol{D}, \boldsymbol{B})$, respectively. The curl operator, $\nabla\times$ or $\hat{\boldsymbol{n}}\times$ at a boundary, applied to an intensity vector transforms it into a flux quantity, as with the terms in Ampere's and Faraday's laws. The divergence operator, $\nabla\cdot$ or $\hat{\boldsymbol{n}}\cdot$ on a boundary, applied to flux densities $(\boldsymbol{D}, \boldsymbol{B})$, however, converts them to scalar charge densities as in the Gauss laws. Similarly, the scalar potentials are scalar quantities, which, when operated on by ∇, become field intensity quantities. The scalar field analog of the reaction theorem involves integrals over products of scalar potentials Φ and source density quantities q. For electric fields, for example, it reads

$$\int_V \Phi^a q^b \, dV = \int_V \Phi^b q^a \, dV. \tag{2.82}$$

Similarly, for magnetic sources and potentials, we find

$$\int_V \Psi^a m^b \, dV = \int_V \Psi^b m^a \, dV. \tag{2.83}$$

In summary, the natural (symmetric) inner products for vector quantities are of the form $< \boldsymbol{A}; \boldsymbol{B} >$ where $\boldsymbol{A}$ is a flux and $\boldsymbol{B}$ is an intensity vector type or vice versa, while natural inner products of scalar quantities are of the form $< A, B >$ where A is a scalar potential and B is a scalar source density quantity or vice versa. We will see later that a typical indicator of a less-than-ideal numerical formulation is one in which the "natural" reaction integral forms ("projections," or "symmetric inner products") are not used.

2.5 Surface integral equation formulations and solutions by moment methods

2.5.1 Surface representation by triangulation

In this section, we discuss the use of planar triangles to model a closed surface S. This is one of the simplest and most common modeling schemes and, in principle, there are very few surfaces, including those with edges and corners, that cannot be accurately modeled using a sufficiently dense triangular mesh. Furthermore, once a surface integral problem is converted to a system of linear equations, the solution scheme is often essentially independent of the meshing scheme. In Figure 2.10, the surface S on the left is approximated at right by a triangulation $\tilde{S}$ using planar, triangular elements, T^e, $e = 1, 2, \cdots, E$. Note that here and in what follows, we assume simple structures S that can only be approximated by beginning with a single planar triangle and adding triangles *only* along boundary edges such that no edge is shared by more than two triangles. The vertices are located by globally indexed vertex position vectors $\boldsymbol{r}_v$, $v = 1, 2, \cdots, V$, where V is the total number of vertices, numbered arbitrarily. In addition to referencing a vertex by its global index, i.e., as the vth vertex or position vector $\boldsymbol{r}_v$ of $\tilde{S}$, it is often convenient to use a local reference, e.g., to refer to the ith vertex vector of triangle T^e, $\boldsymbol{r}_i^e$, where $i = 1, 2,$ or 3 is a local indexing scheme particular to the eth triangle, as shown in Figure 2.11. When $\boldsymbol{r}_v$ and $\boldsymbol{r}_i^e$ refer to the same vertex, either designation may be used, as convenient. Since a vertex of T^e is usually common to several adjacent triangles, the local designation is not unique. We adopt the same notation convention for any quantity for which it is convenient to have both a local and global designation, i.e., the single subscript v refers to the globally indexed quantity whereas quantities with *both* a subscript and a superscript, e.g., i and e, respectively, refer to the ith local vertex or (opposite) edge of element e. Thus $\boldsymbol{r}_v$ corresponds to both $\boldsymbol{r}_i^e$ and $\boldsymbol{r}_j^f$ if the ith node (edge) of T^e is shared with the jth node of T^f. Similarly, the nth interior edge of S, ℓ_n may also be called ℓ_i^e representing the ith edge of T^e and is always assumed to be opposite the ith vertex. Typically one stores tables for each element $e = 1, 2, \ldots, E$ that list for T^e a local-to-global vertex mapping $i \mapsto v$, $i = 1, 2, 3$ and a local-to-global edge mapping for *interior*

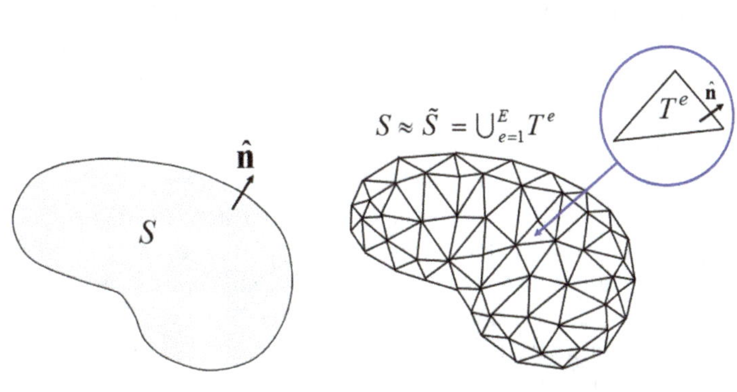

Figure 2.10 Modeling a closed surface S as a collection of planar triangular elements.

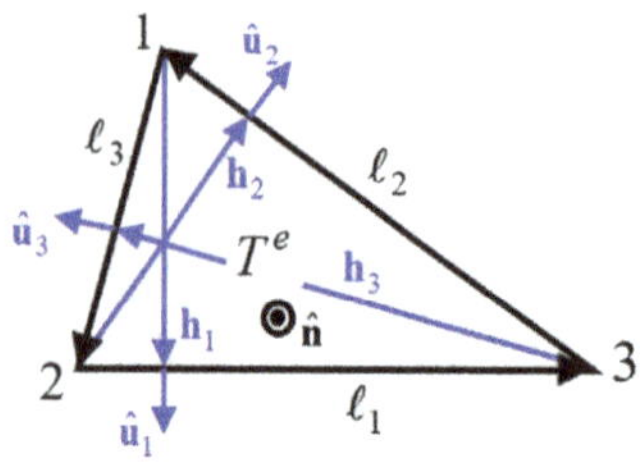

Edge lengths, edge vectors, unit edge vectors:

$$\boldsymbol{\ell}_i = \mathbf{r}^e_{i-1} - \mathbf{r}^e_{i+1}, \quad \ell_i = |\boldsymbol{\ell}_i|, \quad \hat{\boldsymbol{\ell}}_i = \frac{\boldsymbol{\ell}_i}{\ell_i}, \qquad i = 1,2,3$$

Triangle unit normal, area:

$$\hat{\mathbf{n}} = \frac{2(\boldsymbol{\ell}_i \times \boldsymbol{\ell}_{i+1})}{A^e}, \quad 2A^e = |\boldsymbol{\ell}_i \times \boldsymbol{\ell}_{i+1}| = \ell_i h_i, \quad i = 1,2, \text{ or } 3$$

Heights, height vectors, unit height vectors:

$$h_i = \frac{2A^e}{\ell_i}, \quad \hat{\mathbf{u}}_i = \hat{\boldsymbol{\ell}}_i \times \hat{\mathbf{n}} = \frac{\boldsymbol{\ell}_i \times \hat{\mathbf{n}}}{\ell_i}, \quad \mathbf{h}_i = \hat{\mathbf{u}}_i h_i, \quad i = 1,2,3$$

Note: $\nabla \xi_i = -\dfrac{\hat{\mathbf{u}}_i}{h_i} = -\dfrac{\mathbf{h}_i}{h_i^2}, \quad i = 1,2,3$

Figure 2.11 Triangle T^e showing its local vertex indexing, $i = 1, 2, 3$. Vertex i is located at $\mathbf{r}^e_i$ in the local *triangle indexing scheme. A mapping from local index i of T^e to the corresponding global vertex index, v_i, i.e., $i \mapsto v$, should be tabulated for each triangular element. Vertices should be numbered with the local index increasing as one traverses the boundary in a counterclockwise fashion with respect to its local normal $\hat{\mathbf{n}}$, inherited from S. Vector $\hat{\mathbf{u}}$ is the outward normal at the boundary of T^e lying in the tangent plane of T^e. Typical geometry parameters needed in computations are given in the table at the right of the triangle.*

edges $i \mapsto n, i \in (1, 2, 3)$. A typical data structure for the mesh geometry is shown in Table 2.2 and discussed further below [24]. Also needed is a local set of coordinates for each triangle T^e. Consider an arbitrary point $\boldsymbol{r} = (x, y, z)$ in T^e, as shown in Figure 2.12. Triangle T^e is subdivided into three subtriangles (dashed lines in the figure), with the point $\boldsymbol{r}$ serving as their common vertex, and with one edge of each shared with an edge of T^e. The subtriangle opposite each vertex i has area $A_i, i = 1, 2, 3$, and the total triangle area of T^e is $A^e = A_1 + A_2 + A_3$, which when divided by A^e simply implies that the fractional areas $\xi_i = A_i/A^e$ sum to unity, i.e., $\xi_1 + \xi_2 + \xi_3 = 1$. Thus the locally defined coordinates (ξ_1, ξ_2, ξ_3) on T^e are its so-called *area* (or *homogeneous*, *barycentric*, or *simplex*) coordinates. Note that the coordinate ξ_i vanishes at edge i (opposite vertex i), is unity at vertex i, and, as illustrated in the figure, its constant coordinate lines are parallel to edge i of T^e. This should be obvious since for $\boldsymbol{r}$ on a line parallel to edge i (opposite vertex i), A_i is constant since its base length is the fixed edge length, while its height is always a constant distance from edge i. Also note that the unity sum constraint implies that *only two of the three area coordinates are independent*. This should be expected since we can parameterize a 2D a surface with two independent variables; hence, we can choose any pair of the dependent variables (ξ_1, ξ_2, ξ_3) as independent variables, e.g., for interpolation or integration on T^e. For example, we often choose different independent variable pairs to parameterize each subtriangle of T^e as we cycle through vertex indices. For example, we may choose to parameterize the qth subtriangle using ξ_{q+1} and ξ_{q-1} as coordinates for $q = 1, 2, 3$, with the arithmetic for vertex indices performed modulo 3 (i.e., $i + 1 = 1$ if $i = 3$, and $i - 1 = 3$ if $i = 1$). Dashed constant coordinate lines associated with the point

Table 2.2 *The mesh connectivity table lists, for each element and local vertex or edge index, the corresponding global vertices and* degrees of freedom (DoFs). *The data shown for element e* = 17 *corresponds to the vertices and DoFs shown in the inset of Figure 2.13. If edge i is a boundary edge of $\tilde{S}$, the global DoF entry is set to zero.*

Element e	Local vertex or edge i	Global vertex	Global DoF
⋮	⋮	⋮	⋮
17	1	36	−13
	2	16	47
	3	23	−33
⋮	⋮	⋮	⋮

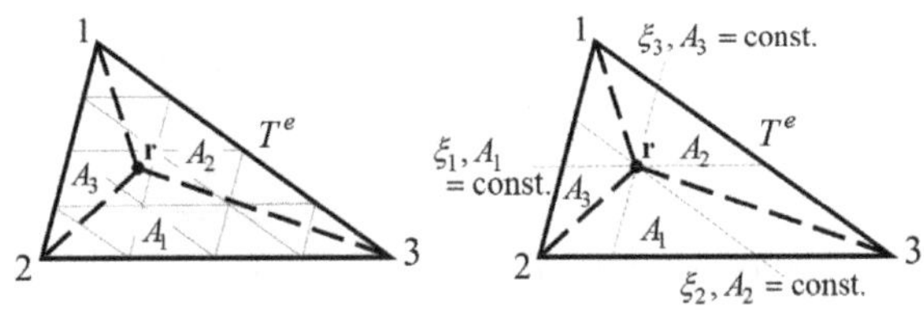

Figure 2.12 *The figure at left shows an arbitrary point* **r** *located in* T^e *(or its planar extension). The local area coordinates of the point are obtained by first subdividing* T^e *into the three subtriangles shown (dashed lines).* A_1 *is the area of the subtriangle sharing edge* 1 *(opposite vertex 1) with* T^e, *and similarly for subtriangles* A_2 *and* A_3. *Note in the right figure that when* **r** *varies along a line parallel to an edge, the height (and, hence, the area and area coordinate) of the subtriangle formed by* **r** *and the edge remain constant. For* **r** *inside* T^e, *all three area coordinates are positive and sum to unity. Area coordinates apply to the entire extended plane of* T^e *if one merely assumes that a subtriangle's height is always measured from its associated (extended) edge and becomes negative (along with its area and area coordinate) whenever* **r** *crosses the extended edge.*

r are shown in the figure at the right of Figure 2.12. Note that area coordinate ξ_i vanishes along edge i and is unity at vertex i, and, hence, each coordinate lies in the interval (0,1) if the point ***r*** is interior to T^e, but if ***r*** is outside the triangle, but still in the (extended) plane of T^e, then at least one of the area coordinates becomes negative.

Clearly area coordinates on a triangle are neither orthogonal nor independent. As will become clear, however, they provide a simple and convenient means for expressing vertex- and edge-indexed quantities in a homogeneous form essentially independent of triangle shape and edge or vertex indexing. As Figure 2.12 shows, the

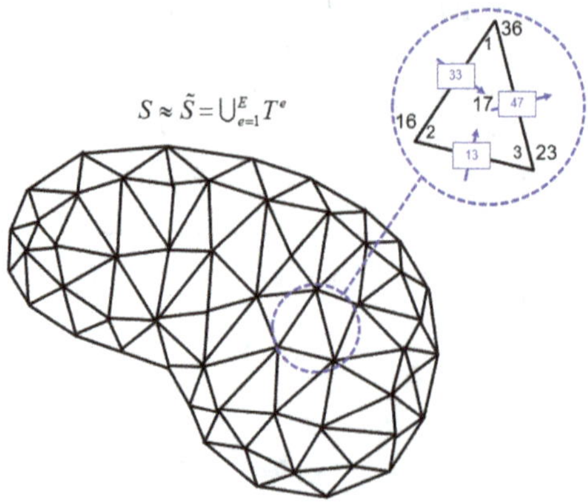

Figure 2.13 Connectivity information for a mesh is typically stored in a mesh connectivity table (see Table 2.2). The surface triangulation $\tilde{S}$ shown approximates S and consists of indexed triangle elements $T^e, e = 1, \cdots, E$. Triangle T^{17} is shown in the inset figure; also labeled are its local and global vertex indices, $i = 1, 2, 3$ and 36, 16, and 23, respectively. Equivalent surface currents are usually modeled by interpolating sampled values of their normal or tangential components defined at triangle edge mid-points. The sampled values, also called degrees of freedom *(DoFs), are often indexed $n = 1, 2, \cdots, N$, where N is the number of DoFs (unknowns). The figure shows DoF indices 13, 47, and 33 associated with local edges $i = 1, 2$ and 3 (opposite vertices 1,2, and 3), respectively, of T^{17}. Arrows represent assigned current reference directions for current components, e.g., normal to the edges of T^{17}, as shown. The current reference direction is assumed positive if the arrow points* out *of the triangle; if it is* into *the triangle, this may be indicated by appending a negative sign to the DoF number as in Table 2.2. To indicate an edge with no associated DoF, a DOF index of zero may be assigned to it in Table 2.2.*

ξ_i coordinates vary linearly over T^e, are unity at vertex i, and vanish at the remaining vertices of T^e; hence, they provide a linear interpolation of any scalar function $f(\boldsymbol{r})$ defined on T^e in terms of its values $f_i = f(\boldsymbol{r}_i^e)$ at vertices of T^e, i.e., $f(\boldsymbol{r}) \approx \sum_{i=1}^{3} f_i \xi_i$ on T^e. This observation applies even to the (scalar) rectangular components x, y, z of the position vector $\boldsymbol{r} = (x, y, z)$ itself, i.e., $x = \sum_{i=1}^{3} x_i^e \xi_i$, and similarly for the y and z components. The three resulting scalar equations are more simply expressed in vector form as

$$\boldsymbol{r} = \sum_{i=1}^{3} \boldsymbol{r}_i^e \xi_i, \tag{2.84}$$

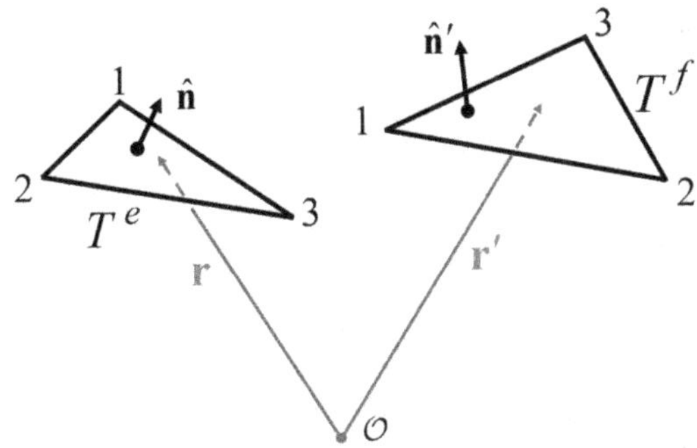

Figure 2.14 A pair of interacting triangles. A surface current source might exist on T^f and its field is observed on T^e, for example.

thus, providing a local parameterization of observation triangle T^e in terms of its vertices and associated area coordinates. The local source triangle T^f is similarly parameterized as

$$r' = \sum\nolimits_{j=1}^{3} r_j^f \xi_j', \tag{2.85}$$

with the primes on both r and ξ signaling the use of different local coordinates and indices for source *vs.* observation points.

Area coordinates are extensively used to parameterize reaction integrals involving source and test triangle pairs, T^f and T^e, respectively. The integral over a triangle T^e of a function $f(r)$ may be expressed in area coordinates as [24]

$$\int_{T^e} f(\boldsymbol{r})\, dS = 2|T^e| \int_0^1 \int_0^{1-\xi_j} f(\boldsymbol{r})\, d\xi_i\, d\xi_j, \quad i \neq j \in 1, 2, 3, \tag{2.86}$$

where $|T^e|$ is the area of the triangle with $\boldsymbol{r}$ expressed as in (2.84), and where we recall that $\xi_k = 1 - \xi_i - \xi_j,\ k \neq i \neq j$. For $f(\boldsymbol{r})$ a polynomial in area coordinates $\boldsymbol{\xi} = (\xi_1, \xi_2, \xi_3)$, the integrals may be performed analytically. On the other hand, just as each Gauss–Legendre scheme can exactly integrate any polynomial of a certain order in one-dimension, "Gauss-triangle" (GT) quadrature schemes exist that can exactly integrate any polynomial of certain orders in the two independent variables ξ_1 and ξ_2 over triangles with polynomial-like integrands $f(\boldsymbol{r})$. A K-point GT numerical quadrature approximation to (2.86) is thus

$$\int_{T^e} f(\boldsymbol{r})\, dS \approx 2|T^e| \sum_{k=0}^{K} w_k f(\boldsymbol{r}^{(k)}), \quad \boldsymbol{r}^{(k)} = \xi_1^{(k)} \boldsymbol{r}_1 + \xi_2^{(k)} \boldsymbol{r}_2 + \xi_3^{(k)} \boldsymbol{r}_3, \tag{2.87}$$

where $\xi_3^{(k)} = 1 - \xi_1^{(k)} - \xi_2^{(k)}$. Specific sample points and weights, $(\boldsymbol{\xi}^{(k)}, w_k), k = 1, 2, \cdots, K$ for $K = 1, 3, 7$ and associated error orders for GT quadrature are reported in Table 2.3. Note that while triangular-faceted surface representations can exactly represent the geometries of planar, polygonal surfaces, if the mesh vertices are placed on a curved surface S or curved boundary of S, then the associated triangulation always

Table 2.3 *Area coordinate sample points and weights for evaluating surface integrals on triangle T for surface integrals of the form* $\int_T f(\boldsymbol{r})\,dS$ *where* $f(\boldsymbol{r})$ *is assumed smooth. The error order of the 1, 3, and 7 point rules is* $O(|\boldsymbol{\xi}|^2)$, $O(|\boldsymbol{\xi}|^3)$, *and* $O(|\boldsymbol{\xi}|^6)$, *respectively. Higher order rules are also available [25].*

No. points, K	$\xi_1^{(k)}$	$\xi_2^{(k)}$	w_k
$K = 1$	0.33333333333333	0.33333333333333	0.50000000000000
$K = 3$	0.66666666666667	0.16666666666667	0.16666666666667
	0.16666666666667	0.66666666666667	0.16666666666667
	0.16666666666667	0.16666666666667	0.16666666666667
$K = 7$	0.33333333333333	0.33333333333333	0.11250000000000
	0.79742698535309	0.10128650732346	0.06296959027241
	0.10128650732346	0.79742698535309	0.06296959027241
	0.10128650732346	0.10128650732346	0.06296959027241
	0.47014206410512	0.47014206410512	0.06619707639425
	0.47014206410512	0.05971587178977	0.06619707639425
	0.05971587178977	0.47014206410512	0.06619707639425

under-represents true areas and boundary arc lengths of S. Such surface representation errors inevitably propagate into the evaluation of integrals over planar-faceted approximations of S. We assume in the following that *every* surface S is eventually (though non-uniquely) approximated using a triangular mesh, and use S to denote *either* a discretized or an un-discretized surface, as context dictates.

2.5.2 Defining electromagnetic quantities on a mesh

Figure 2.15 shows a *primary* rectangular surface mesh and a *secondary* or *dual* mesh shifted a half cell in either direction from the primary mesh so that the cell centroids and vertices of the two meshes coincide. We assume a rectangular mesh for simplicity; the concept can, however, be generalized to triangular (or other) meshes (compare Figures 2.15 and 2.21). Notice that electric source density quantities (surface charge density q_s and surface current density $\boldsymbol{J}$) are in black on the primary mesh, while the electric field and potential quantities are in gray on the secondary mesh. The offset meshes are needed to adequately approximate Maxwell's equations. For example, assume that the endpoints of a mesh edge of length ℓ along which $\boldsymbol{E}$ and $\boldsymbol{A}$ are defined are at $\boldsymbol{r}_1$ and $\boldsymbol{r}_2$. Then the average value $\bar{E}$ of the component of $\boldsymbol{E}$ parallel to and along the edge is given by

$$\bar{E} = \frac{1}{\ell}\int_{r_1}^{r_2} \boldsymbol{E}\cdot d\boldsymbol{\ell} = -j\omega\bar{A} - \frac{\Phi_2 - \Phi_1}{\ell}$$

where $\bar{A}$ is the corresponding average value of the parallel component of $\boldsymbol{A}$ along the edge integration path,

$$\bar{A} = \frac{1}{\ell}\int_{r_1}^{r_2} \boldsymbol{A}\cdot d\boldsymbol{\ell}.$$

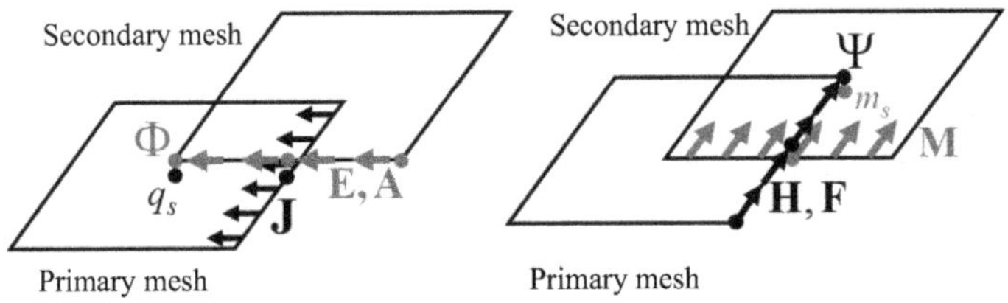

*Figure 2.15 Adjacent cells extracted from a rectangular primary and secondary mesh. Note cells are shifted with respect to one another such that vertices of one are located at centroids of the other. At left, electric surface charge density q_s is defined at the centroid of the primary mesh while electric current density **J** is defined as normal to its edges. Line integral (intensity) quantities **E** and **A** are defined parallel to edges of the secondary (dual) mesh, while the electric scalar potential point function Φ is defined at its vertices. In the figure at right, (or by duality), magnetic surface charge density m_s is defined at the centroid of the secondary mesh, and magnetic current density **M** is defined normal to its edges. Line integral quantities **H** and **F** are defined parallel to edges of the primary mesh, while the magnetic scalar potential point function Ψ is defined at its vertices. Designating which mesh is primary or secondary is arbitrary and often reversed; generalization to a 3D mesh results in the Yee lattice [26].*

(If $\boldsymbol{M}$ is non-vanishing at mesh edges, an additional term in the path integral expression will be present with contributions from $\boldsymbol{F}$.) Note that if we choose a *closed* path of integration on the grid and sum the products $\bar{E}\ell$ around cell edges of the secondary grid, we find

$$\oint \boldsymbol{E} \cdot d\boldsymbol{\ell} = -j\omega \oint \boldsymbol{A} \cdot d\boldsymbol{\ell}$$

since intermediate scalar potential differences cancel and the scalar potentials are the same at path endpoints. This result holds, of course, not only for path edge integrals around edges of the single secondary mesh cell shown but also for the edge integral sums for *any* closed path on the mesh of a connected surface. Hence, if $\omega = 0$, the conservative property of static electric fields is exactly preserved for every such closed path on the mesh. For $\omega \neq 0$, we may use $\mu\boldsymbol{H} = \nabla \times \boldsymbol{A}$ and Stokes's theorem to verify that the vector potential term on the right above is simply the magnetic flux linking the path. Preserving such global physics properties down to the mesh scale is important for ensuring problem accuracy and suggests that some attention should be given to mesh details as well as to our numerical representations of fields and sources.

We note that by duality, analogous properties hold for the magnetic fields and potentials, but they should be defined on edges of the *primary* mesh, while magnetic sources are associated with the *secondary* mesh. Finally, we note that the offset mesh scheme of Figure 2.15 additionally guarantees that nodes of the primary and secondary edge midpoints are coincident. This often significantly improves the accuracy while simplifying the computation of inner products with test sources of the forms appearing in reaction integrals (2.81)–(2.83) (specialized to surface sources) and involving products of $\boldsymbol{J}$ with $\boldsymbol{E}$, $\boldsymbol{M}$ with $\boldsymbol{H}$, q with Φ, and m with Ψ.

2.5.3 The electric field integral equation (EFIE)

For a PEC, the total (incident plus scattered) tangential electric field must vanish on S, i.e., $\boldsymbol{E}^{\mathrm{s}}_{\tan} = -\boldsymbol{E}^{\mathrm{i}}_{\tan}$, where $\boldsymbol{E}_{\tan} = \hat{\boldsymbol{n}} \times (\boldsymbol{E} \times \hat{\boldsymbol{n}})$. Using the mixed potential representation for $\boldsymbol{E}^{\mathrm{s}}$, this may be written as

$$j\omega \boldsymbol{A}_{\tan} + \nabla_{\tan}\Phi = \boldsymbol{E}^{\mathrm{i}}_{\tan}, \quad \boldsymbol{r} \in S, \tag{2.88}$$

or in more concise operator notation,

$$\eta \mathscr{L}\boldsymbol{J} = \boldsymbol{E}^{\mathrm{i}} \times \hat{\boldsymbol{n}}, \quad \boldsymbol{r} \in S. \tag{2.89}$$

The so-called mixed potential representation,

$$\left[j\omega\mu \int_S G(\boldsymbol{r},\boldsymbol{r}')\,\boldsymbol{J}(\boldsymbol{r}')\,dS' - \frac{1}{j\omega\varepsilon}\nabla \int_S G(\boldsymbol{r},\boldsymbol{r}')\nabla' \cdot \boldsymbol{J}(\boldsymbol{r}')\,dS' \right]_{\tan} = \boldsymbol{E}^{\mathrm{i}}_{\tan}, \quad \boldsymbol{r} \in S, \tag{2.90}$$

however, is usually the most convenient form for implementing numerical methods. In (2.90), $\boldsymbol{E}^{\mathrm{i}}_{\tan}$ is assumed given and $\boldsymbol{J}$ is an unknown surface current, and the equation is presumed to hold at each and every point $\boldsymbol{r}$ of S, i.e., in the *strong* sense of the equality. But that would require enforcing an infinite number of constraint points $\boldsymbol{r}$ as well as determining an infinite number of *degrees of freedom* (DoFs), i.e., $\boldsymbol{J}$ values at points $\boldsymbol{r}'$. A numerical solution, however, can only deal with a finite number of DoFs and a finite number of constraints, resulting in a so-called *weak* form of the equation that constitutes a finite linear system of equations for determining some finite representation of the surface current. Along the way, we discretize (a) the surface geometry S, (b) the surface current $\boldsymbol{J}$, and (c) both sides of (2.90) to obtain a system of linear equations approximating the original integral equation. Each such discretization introduces approximations; furthermore, (d) additional error will usually result from any numerical integrations that must be performed, as well as from (e) the numerical process of solving the linear system. At least some attention must usually be given to evaluating and/or controlling the effects of each of these five error sources.

To reduce the dimensionality of the space in which (2.90) holds we instead require that it does not hold in a point-wise sense, but rather in some weighted-average sense for a finite number of so-called weight or *testing functions* defined on S. That is, assuming a finite set of testing functions $\boldsymbol{\Lambda}_m(\boldsymbol{r}), m = 1, \cdots, N$, form a weighted average of both sides of (2.90) with $\boldsymbol{\Lambda}_m(\boldsymbol{r})$ for $m = 1, \cdots, N$, yielding a corresponding so-called *weak* form of (2.90),

$$j\omega < \boldsymbol{\Lambda}_m; \boldsymbol{A} > + < \boldsymbol{\Lambda}_m; \nabla\Phi > \; = \; < \boldsymbol{\Lambda}_m; \boldsymbol{E}^{\mathrm{i}} >, \;\; m = 1, \cdots, N, \tag{2.91}$$

where we introduce the *symmetric inner product* notation $< \boldsymbol{A}; \boldsymbol{B} >= \int_S \boldsymbol{A}(\boldsymbol{r}) \cdot \boldsymbol{B}(\boldsymbol{r})\, dS$. We note that symmetric inner products are often further generalized, for example, by writing the potential integrals in the forms

$$\boldsymbol{A}(\boldsymbol{r}) = \mu \int_S G(\boldsymbol{r}, \boldsymbol{r}')\boldsymbol{J}(\boldsymbol{r}')\, dS' \overset{\Delta}{=} \mu < G, \boldsymbol{J} >, \tag{2.92}$$

involving the product of a vector and scalar, and

$$\Phi(\boldsymbol{r}) = \frac{1}{-j\omega\varepsilon} \int_S G(\boldsymbol{r}, \boldsymbol{r}')\nabla' \cdot \boldsymbol{J}(\boldsymbol{r}')\, dS' \overset{\Delta}{=} \frac{1}{-j\omega\varepsilon} < G, \nabla \cdot \boldsymbol{J} >, \tag{2.93}$$

involving the product of two scalars, so that (2.91) becomes

$$j\omega\mu < \boldsymbol{\Lambda}_m; < G, \boldsymbol{J} >> - \frac{1}{j\omega\varepsilon} < \boldsymbol{\Lambda}_m; < \nabla G, \nabla \cdot \boldsymbol{J} >> = < \boldsymbol{\Lambda}_m; \boldsymbol{E}^{\mathrm{i}} > . \tag{2.94}$$

However, we prefer a slightly different notation wherein the Green's function operator is sandwiched between the testing function and the unknown and where any dot or cross-product operations are shown explicitly above the comma. The comma also signifies integration between the quantities on either side of it. Hence, instead of (2.94), we write the tested form of the EFIE as

$$j\omega\mu < \boldsymbol{\Lambda}_m; G, \mathbf{J} > - \frac{1}{j\omega\varepsilon} < \boldsymbol{\Lambda}_m; \nabla G, \nabla \cdot \mathbf{J} > = < \boldsymbol{\Lambda}_m; \boldsymbol{E}^{\mathrm{i}} > . \tag{2.95}$$

Note that in the above notation, a comma between two quantities indicates an integration between a product of the two quantities. If a "dot" or "cross" appears above the comma, then both quantities must be vectors with the indicated product. All integrations are assumed to be over S, but may actually reduce to just the supports of the quantities integrated. Generally only a Green's function form (i.e., the *kernel* of the integral equation) appears between two commas, and it is assumed to be a function of both $\boldsymbol{r}$ and $\boldsymbol{r}'$ with the outer integral assumed to be integrated over unprimed (observation) coordinates and the inner over primed (source) coordinates.*

In (2.95), if we further assume that the test functions are *divergence-conforming*, i.e., they satisfy

1. $\boldsymbol{\Lambda}_m \cdot \hat{\boldsymbol{u}} = 0$ on any boundary ∂S of S, and
2. $\nabla \cdot \boldsymbol{\Lambda}_m$ exists everywhere interior to S,

then the gradient operator on G may be conveniently transferred onto $\boldsymbol{\Lambda}_m$ using the surface divergence theorem and integration by parts to arrive at the identity

$$< \boldsymbol{\Lambda}_m; \nabla\Phi > = - < \nabla \cdot \boldsymbol{\Lambda}_m, \Phi > + \int_{\partial S} \Phi\, \boldsymbol{\Lambda}_m \cdot \hat{\boldsymbol{u}}\, d\ell'. \tag{2.96}$$

*This three-term symmetric inner product notation is an adaptation of the common "bra-ket" notation of quantum physics.

where the line integral in (2.96) vanishes because of the first condition. With this result, we may now rewrite (2.91) as

$$j\omega < \mathbf{\Lambda}_m; \boldsymbol{A} > - < \nabla \cdot \mathbf{\Lambda}_m, \Phi > = < \mathbf{\Lambda}_m; \boldsymbol{E}^{\mathrm{i}} >, \tag{2.97}$$

or, paralleling the more explicit form (2.95), as

$$j\omega\mu < \mathbf{\Lambda}_m; G, \boldsymbol{J} > + \frac{1}{j\omega\varepsilon} < \nabla \cdot \mathbf{\Lambda}_m, G, \nabla \cdot \boldsymbol{J} > = < \mathbf{\Lambda}_m; \boldsymbol{E}^{\mathrm{i}} > . \tag{2.98}$$

Note that (2.98) requires the existence of divergences for both $\boldsymbol{J}$ and $\mathbf{\Lambda}_m$. Furthermore, if S is an open surface, $\boldsymbol{J} \cdot \hat{\boldsymbol{u}}$ must vanish at the boundaries of S so that a line charge does not accumulate on the boundaries of S. Hence, both $\boldsymbol{J}$ and the test functions must be divergence conforming. If, for convenience, we add the condition that $\mathbf{\Lambda}_m$ also be *interpolatory*, i.e., the component of $\mathbf{\Lambda}_m$ normal to (internal) edge m be unity and vanish at all other edges, we may use $\mathbf{\Lambda}_m$ both as an expansion or basis function to represent the current and as a testing function. This dual use of a basis function as a testing function is called *Galerkin's method.* The so-called RWG bases are a useful set of such bases with all these properties [27]. Each one is defined on the two triangles adjacent to every interior edge of S and vanishes outside these two triangles. Hence, if S has a total of E edges, with B of them along boundaries, then the total number of bases or DoFs is $N = E - B$. The nth such basis is normalized so as to produce a unit flux density normal to and along the nth interior edge of S. It is non-vanishing only on the adjacent triangles T_n^+ and T_n^- having edge n in common and has an assumed positive flux reference direction from T_n^+ to T_n^-. Often it is convenient to assign the DoF number "0" to identify boundary edges, where normal current components vanish and hence have no assigned DoF. The RWG bases are defined as [24,27]

$$\mathbf{\Lambda}_n(\boldsymbol{r}) = \begin{cases} \frac{\boldsymbol{\rho}_n^\pm}{h_n^\pm} = \pm\frac{\boldsymbol{r}-\boldsymbol{r}_n^\pm}{h_n^\pm}, & \boldsymbol{r} \in T_n^\pm, \\ \mathbf{0}, & \text{otherwise} \end{cases}, \quad n = 1, 2, \cdots, N, \tag{2.99}$$

and their divergence is

$$\nabla \cdot \mathbf{\Lambda}_n(\boldsymbol{r}) = \begin{cases} \pm\frac{2}{h_n^\pm}, & \boldsymbol{r} \in T_n^\pm, \\ 0, & \text{otherwise} \end{cases}, \quad n = 1, 2, \cdots, N. \tag{2.100}$$

In the above, $\boldsymbol{r}_n^\pm$ is the vertex in $T_n^\pm$ opposite edge n, as shown in Figure 2.16. Note that the support of a single basis is given by supp $\mathbf{\Lambda}_n(\boldsymbol{r}) = T_n^+ \cup T_n^-$. It is easily shown that $\mathbf{\Lambda}_n$ has the interpolatory property of having a unit flux density along the edge common to the triangle pair T_n^+ and T_n^-, while the flux density vanishes at all other edges of S.

Rather than always integrating over test element pairs, it is more convenient and efficient to deal with a single test element at a time and to simultaneously integrate over those test bases whose supports overlap that element. Only three test bases at most can overlap onto element T^e, one each associated with each edge $i = 1, 2, 3$ of element e whose associated DoF is non-vanishing. For every element $e = 1, 2, \cdots, N$, it is then convenient to define local test bases $\mathbf{\Lambda}_i^e(\boldsymbol{r})$ assigned to local edges $i = 1, 2, 3$ with non-vanishing DoFs and having an assumed reference direction out of T^e. On element e, the local basis $\mathbf{\Lambda}_i^e(\boldsymbol{r})$ is identical to the portion of the global test basis $\mathbf{\Lambda}_i^e(\boldsymbol{r})$

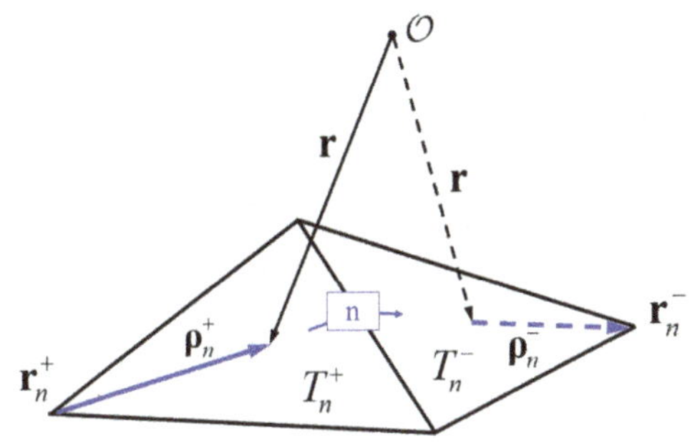

Figure 2.16 *Edge n is the common edge between two adjacent triangles, T_n^+ and T_n^-. For $\boldsymbol{r}$ in $T_n^{\pm}$, $\boldsymbol{\Lambda}_n = \boldsymbol{\rho}_n^{\pm}/h_n^{\pm}$ where $\boldsymbol{\rho}_n^{\pm} = \pm(\boldsymbol{r} - \boldsymbol{r}_n^{\pm})$ and $h_n^{\pm}$ is the height of vertex $\boldsymbol{r}_n^{\pm}$ above common edge n. The rectangular box labels edge n and indicates the implied current reference direction.*

overlapping T^e there, except possibly for their reference directions. If the location and height of the ith vertex of triangle T^e are $\boldsymbol{r}_i^e$ and h_i^e, respectively, the local basis function may defined in local coordinates $\boldsymbol{\xi}$ as

$$\boldsymbol{\Lambda}_i^e(\boldsymbol{r}) = \frac{\boldsymbol{r} - \boldsymbol{r}_i^e}{h_i^e} = \frac{\xi_{i+1}\boldsymbol{\ell}_{i-1} - \xi_{i-1}\boldsymbol{\ell}_{i+1}}{h_i^e}, \quad i = 1, 2, 3, \quad e = 1, 2, \cdots, N, \tag{2.101}$$

which follows from (2.84) and the edge vector definitions of Figure 2.14 [27]. Hence, if say, T_n^+ corresponds to the ith edge of element e while T_n^- corresponds to the jth edge of adjacent element T_j^f, then the global and local bases are related as follows:

$$\boldsymbol{\Lambda}_n(\boldsymbol{r}) = \begin{cases} +\boldsymbol{\Lambda}_i^e(\boldsymbol{r}), & \boldsymbol{r} \in T^e \ (=T_n^+), \\ -\boldsymbol{\Lambda}_j^f(\boldsymbol{r}), & \boldsymbol{r} \in T^f \ (=T_n^-), \\ \boldsymbol{0}, & \text{otherwise.} \end{cases} \tag{2.102}$$

Also note the similar properties required of both the testing function $\boldsymbol{\Lambda}_m$ and the assumed current $\boldsymbol{J}$ in (2.98). In particular, the divergence of both must exist and both $\boldsymbol{\Lambda}_m$ and $\boldsymbol{J}$ and their divergences are integrated against the same Green's function. Their symmetric roles are a consequence of reciprocity, and further suggest we might use the testing RWG bases to represent the unknown surface current as well. That is, we expand the current as

$$\boldsymbol{J}(\boldsymbol{r}') = \sum_{n=1}^{N} I_n \, \boldsymbol{\Lambda}_n(\boldsymbol{r}') \tag{2.103}$$

which may then be substituted into (2.98) to obtain the $N \times N$ system matrix equation

$$[Z_{mn}]\,[I_n] = [V_m], \tag{2.104}$$

where

$$[Z_{mn}] = j\omega\,[L_{mn}] + \frac{1}{j\omega}\,[S_{mn}] = \eta < \mathbf{\Lambda}_m \times \hat{\boldsymbol{n}}, \mathscr{L}\mathbf{\Lambda}_n > \tag{2.105}$$

is known as the *impedance matrix*. The matrix elements

$$L_{mn} = \mu \int_S \int_S G(\boldsymbol{r}, \boldsymbol{r}')\,\mathbf{\Lambda}_m(\boldsymbol{r}) \cdot \mathbf{\Lambda}_n(\boldsymbol{r}')\,dS'dS \;\triangleq\; \mu < \mathbf{\Lambda}_m; G, \mathbf{\Lambda}_n > \tag{2.106}$$

and

$$S_{mn} = \frac{1}{\varepsilon} \int_S \int_S \nabla \cdot \mathbf{\Lambda}_m(\boldsymbol{r})\,G(\boldsymbol{r}, \boldsymbol{r}')\nabla' \cdot \mathbf{\Lambda}_n(\boldsymbol{r}')\,dS'dS \;\triangleq\; \frac{1}{\varepsilon} < \nabla \cdot \mathbf{\Lambda}_m, G, \nabla \cdot \mathbf{\Lambda}_n > \tag{2.107}$$

form the system *inductance* and system *elastance* matrices [24], respectively, while

$$V_m = \int_S \mathbf{\Lambda}_m(\boldsymbol{r}) \cdot \boldsymbol{E}^{\mathrm{i}}(\boldsymbol{r})\,dS \;\triangleq < \mathbf{\Lambda}_m; \boldsymbol{E}^{\mathrm{i}} > = < \mathbf{\Lambda}_m \times \hat{\boldsymbol{n}}; \boldsymbol{E}^{\mathrm{i}} \times \hat{\boldsymbol{n}} > \tag{2.108}$$

are the column entries of the N-component system *excitation* or *voltage* column vector $[V_m]$ and $G(\boldsymbol{r}, \boldsymbol{r}') = \frac{e^{-jkR}}{4\pi R}$, where $R = |\boldsymbol{r} - \boldsymbol{r}'|$. We note that the dS integrals over S reduce to integrals over the support of $\mathbf{\Lambda}_m$ while the dS' integrals reduce to integrals over the support of $\mathbf{\Lambda}_n$. Note also in (2.106) and (2.107) that since test bases and their derivatives are functions of $\boldsymbol{r}$ alone, they may be removed from under the source (inner) integral and performed first for fixed $\boldsymbol{r}$ values, with the test (outer) integration following.

2.5.4 Fill and assembly of element and system matrices and column excitation vectors

In an implementation of the discretized equations, the integration domains of each of the double integrals over S implicit in the inner products above reduce simply to integration over the supports of $\mathbf{\Lambda}_m$ and $\mathbf{\Lambda}_n$, i.e., to $T_m^+ \cup T_m^-$ and $T_n^+ \cup T_n^-$, respectively. However, we gain even greater integration efficiency if we further reduce the double-integral domains to integrals over *single* pairs of interacting test and source elements, T^e and T^f, respectively, adding all their contributions to the system matrix once they are computed for the pair. Since each source triangle T^f involves $j = 1, 2, 3$ independent bases while each test triangle T^e involves $i = 1, 2, 3$ independent test functions, we can store the results as 3×3 matrices, which in view of (2.100) and (2.101), can be easily synthesized from scalar integrands involving just the nine combinations of constant or linear area coordinates $1, \xi_1$, and ξ_2 defined for each triangle of the interacting pair T^e and T^f. These integrals can then be used to synthesize the integrals over the vector bases and scalar potentials, which in turn can be assembled into a 3×3 *element matrix* as eventual partial contributions to the system inductance and elastance matrices in (2.110) and (2.111) below. The system matrix may then be assembled by adding these partial contributions (modified by ± 1 sign factors, σ_i^e, σ_j^f, to account for reference directions) directly to appropriate elements of the system matrix [27]. The element matrices are explicitly defined as follows:

$$\left[Z_{ij}^{ef}\right] = j\omega\left[L_{ij}^{ef}\right] + \frac{1}{j\omega}\left[S_{ij}^{ef}\right] = \eta < \boldsymbol{\Lambda}_i^e \times \hat{\boldsymbol{n}}, \mathscr{L}\boldsymbol{\Lambda}_j^f >,$$

$$i,j = 1,2,3, \quad e,f \in (1,2,\cdots,E), \tag{2.109}$$

where $[Z_{ij}^{ef}]$ is the *element impedance matrix* and E is the number of elemental triangles. The element matrix entries are

$$L_{ij}^{ef} = \mu \int_{T^e}\int_{T^f} G(\boldsymbol{r},\boldsymbol{r}')\,\boldsymbol{\Lambda}_i^e(\boldsymbol{r})\cdot\boldsymbol{\Lambda}_j^f(\boldsymbol{r}')\,dS'dS \triangleq \mu < \boldsymbol{\Lambda}_i^e; G, \boldsymbol{\Lambda}_j^f > \tag{2.110}$$

and

$$S_{ij}^{ef} = \frac{1}{\varepsilon}\int_{T^e}\int_{T^f} \nabla\cdot\boldsymbol{\Lambda}_i^e(\boldsymbol{r})G(\boldsymbol{r},\boldsymbol{r}')\nabla'\cdot\boldsymbol{\Lambda}_j^f(\boldsymbol{r}')\,dS'dS \triangleq \frac{1}{\varepsilon} < \nabla\cdot\boldsymbol{\Lambda}_i^e, G, \nabla\cdot\boldsymbol{\Lambda}_j^f > \tag{2.111}$$

forming the element *inductance* and element *elastance* matrices, respectively, while

$$V_i^e = \int_{T^e} \boldsymbol{\Lambda}_i^e(\boldsymbol{r})\cdot\boldsymbol{E}^{\mathrm{i}}(\boldsymbol{r})\,dS \triangleq < \boldsymbol{\Lambda}_i^e; \boldsymbol{E}^{\mathrm{i}} > (= < \boldsymbol{\Lambda}_i^e \times \hat{\boldsymbol{n}}; \boldsymbol{E}^{\mathrm{i}} \times \hat{\boldsymbol{n}} >) \tag{2.112}$$

are column entries in the element *excitation* or *voltage* vector $\left[V_i^e\right]$. For well-separated elements T^e and T^f, repeated use of the GT rules of (2.87) and Table 2.3 may be used to evaluate (2.110), (2.111), and (2.112).

The filling of an element matrix for each element pair, and the distributing of its contents to (partially) fill, i.e., to *assemble*, the system matrix (2.105) are illustrated in Figure 2.17. That is, we loop over all the element pairs $e = 1,...E$ and $f = 1,...E$; for each, we add to the partially assembled system matrix $[Z_{mn}]$ the entries $\sigma_i^e\sigma_j^f Z_{ij}^{ef}$ from the element matrix for $i,j = 1,2,3$ as described below. Similarly, for the right-hand side element column vector, for each e, we add to the system excitation column vector $[V_m]$ the entries $\sigma_i^e V_i^e$ for $i = 1,2,3$. The intermediate element matrices and column vectors not only eliminate the possibility of repeating expensive numerical integrations over the same element pairs, but, until the system matrix and column vectors are complete, they are protected from any operations other than adding partial contributions resulting from the element pair integrations. When evaluation of each element matrix entry of Z_{ij}^{ef} has been completed, the entry $\sigma_i^e\sigma_j^f Z_{ij}^{ef}$ of the element matrix is added to row m and column n of the system matrix where m is the global DoF index corresponding to the ith local DOF of element e, and n is the global DoF corresponding to the jth local DOF of element f. Let m_i^e be the global DoF number of the ith edge of element e. Similarly, let n_j^f be the DoF number of the jth edge of element f. If edge i of element e is a boundary edge of S and, therefore, has no DoF, set $m_i^e = 0$ and similarly for n_j^f. Then accumulate contributions

$$\sigma_i^e\sigma_j^f Z_{ij}^{ef}, \quad i,j = 1,2,3; \quad e,f = 1,2,\cdots,E$$

in global element $Z_{m_i^e,n_j^f}$ unless either $m_i^e = 0$ or $n_j^f = 0$.

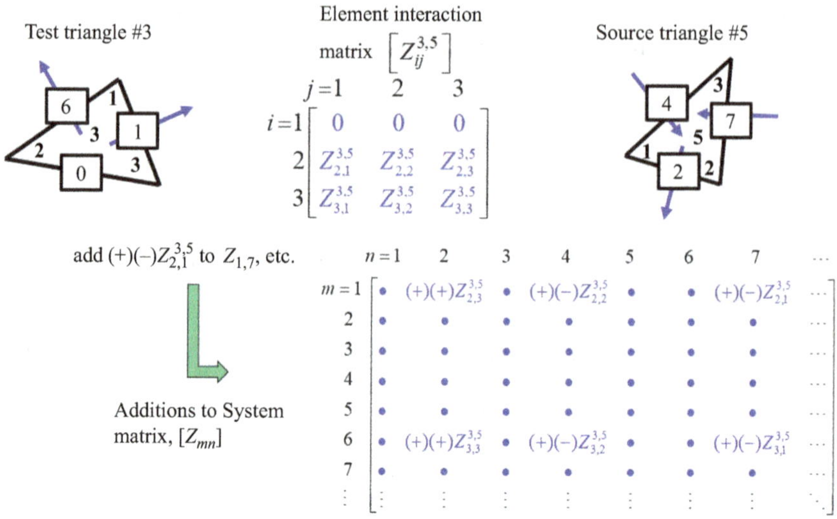

Figure 2.17 Consider the interaction between source triangle n = 5 and test triangle m = 3: numbered boxes on edges show the edge's assigned global unknown index (degree of freedom (DoF)), with "0" assigned to structure edges of S without a DoF, as for edge 1 (opposite vertex 1) of (test) triangle 3. The interaction matrix element $\left[Z_{i,j}^{e,f}\right]$ *is filled with interactions between the ith basis function of test triangle e and the jth basis function of source triangle f, where* $i,j = 1, 2, 3$. *Once the element interaction matrix is filled, the partial contributions of its elements are appropriately distributed by adding them to appropriate elements of the system matrix* $\left[Z_{m,n}\right]$. *Thus, element* $\sigma_i^e \sigma_j^f Z_{i,j}^{e,f}$, $i,j = 1, 2, 3$, *is added to* $Z_{m,n}$, *where m, if non-zero, is the global DoF corresponding to the ith edge of element e, and n, if non-zero, is the global DoF corresponding to the jth edge of element f; both m and n can be obtained, for example, from Table 2.2. The sign factor* σ_i^e *is* $+1$ *if the reference direction for the ith basis of element e is out of the element and is* -1 *otherwise; the sign factor* σ_j^f *is similarly defined.*

The procedure for assembling the right-hand system vector $[V_m]$ is similar to the system matrix assembly scheme: for m_i^e defined as earlier, then accumulate contributions $\sigma_i^e V_i^e$, $i = 1, 2, 3$; $e = 1, 2, \cdots, E$ in global element $V_{m_i^e}$ unless $m_i^e = 0$.

2.5.4.1 Numerical evaluation of EFIE element matrices

The element matrices can be evaluated using the GT rules of Table 2.3 as long the elements are sufficiently well-separated, usually at least 2–3 times the largest edge

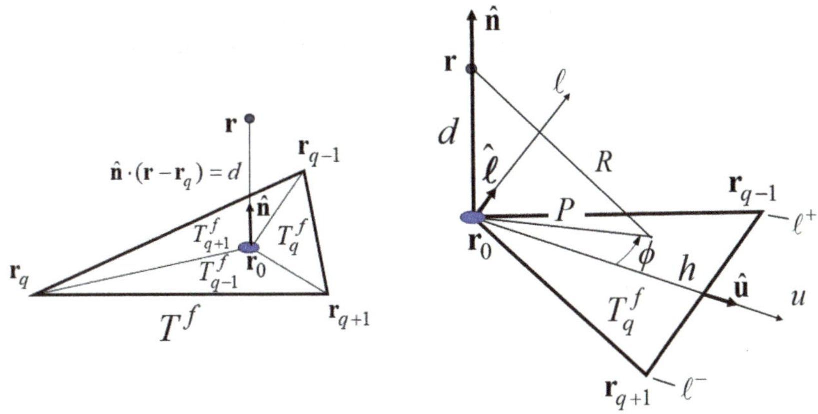

(a) Source triangle T^f. (b) Source subtriangle T_q^f.

Figure 2.18 In (a), a nearby observation point $\mathbf{r}$ *is projected normally onto the extended plane of* T^f. *The projection point* $\mathbf{r}_0$ *may fall either interior or exterior to* T^f. *By connecting vertices of* T^f *to* $\mathbf{r}_0$*, the source triangle is subdivided into three subtriangles* T_q^f, $q = 1, 2, 3$, *with a common vertex at* $\mathbf{r}_0$. *In (b), local rectangular, polar, and radial-angular coordinates,* (u, ℓ, d), (P, ϕ, d), *and* (R, ϕ, d), *respectively, are defined on subtriangle* T_q^f.

length of the two interacting triangles. For example, using the same K-point GT rule with weights w_k^{GT} and area coordinate sample point triples $(\xi_1^{GT,(k)}, \xi_2^{GT,(k)}, \xi_3^{GT,(k)})$, $k = 1, 2, \ldots, K$, for both test and source integrations, the inductance element matrix (2.110) may be approximated as

$$L_{ij}^{ef} \approx 4|T^e||T^f|\mu \sum_{k=1}^{K} \sum_{l=1}^{K} w_k^{GT} w_l^{GT} G(\boldsymbol{r}^{(k)}, \boldsymbol{r}^{(l)})\, \boldsymbol{\Lambda}_i^e(\boldsymbol{r}^{(k)}) \cdot \boldsymbol{\Lambda}_j^f(\boldsymbol{r}^{(l)}), \tag{2.113}$$

where $|T^e|$ is the area of test element T^e, $\boldsymbol{r}^{(k)} = \boldsymbol{r}_1^e \xi_1^{GT,(k)} + \boldsymbol{r}_2^e \xi_2^{GT,(k)} + \boldsymbol{r}_3^e \xi_3^{GT,(k)}$; similar definitions apply to the sum on the index l for sample points on the source triangle T^f.

For nearby and coincident element pairs, a more delicate treatment of at least the source integral is required. Consider the source triangle T^f of Figure 2.18. The nearby observation point $\boldsymbol{r}$ is projected onto the (extended) plane of T^f at $\boldsymbol{r}_0$. Triangle T^f is then subdivided into three subtriangles, T_q^f, $q = 1, 2, 3$, with a common vertex at $\boldsymbol{r}_0$, as shown in the figure. Note if $\boldsymbol{r}_0$ falls outside T^f, then at most one or two subtriangles will lie entirely outside T^f and have negative area coordinates. Since the minimum value of $R = |\boldsymbol{r} - \boldsymbol{r}'|$ is d, $\boldsymbol{r}_0$ also locates the point on the plane where the magnitude of the Green's function has its largest value; thus, the splitting of T^f ensures that (near-)singularities can occur *only* at the common vertex of each subtriangle. The relevant vector and scalar potential source integrals on T_q^f are, therefore,

$$\begin{aligned}
&\int_{T_q^f} \frac{e^{-jkR}}{4\pi R} \begin{bmatrix} \mathbf{\Lambda}_j^f(\mathbf{r}') \\ \nabla \cdot \mathbf{\Lambda}_j^f(\mathbf{r}') \end{bmatrix} dS' \\
&= \int_0^h \int_{u \tan \phi^-}^{u \tan \phi^+} \frac{e^{-jkR}}{4\pi R} \begin{bmatrix} \mathbf{\Lambda}_j^f(\mathbf{r}') \\ \nabla \cdot \mathbf{\Lambda}_j^f(\mathbf{r}') \end{bmatrix} d\ell du, \quad R^2 = u^2 + \ell^2 + d^2, \; \ell^{\pm} = h \tan \phi^{\pm} \\
&= \int_{\phi^-}^{\phi^+} \int_0^{h/\cos\phi} \frac{e^{-jkR}}{4\pi R} \begin{bmatrix} \mathbf{\Lambda}_j^f(\mathbf{r}') \\ \nabla \cdot \mathbf{\Lambda}_j^f(\mathbf{r}') \end{bmatrix} PdP\, d\phi, \; P^2 = R^2 - d^2 = u^2 + \ell^2 \\
&= \int_{\phi^-}^{\phi^+} \int_{|d|}^{R(\phi)} \frac{e^{-jkR}}{4\pi R} \begin{bmatrix} \mathbf{\Lambda}_j^f(\mathbf{r}') \\ \nabla \cdot \mathbf{\Lambda}_j^f(\mathbf{r}') \end{bmatrix} RdR\, d\phi, \quad R(\phi) = \sqrt{d^2 + h^2/\cos^2\phi},
\end{aligned} \tag{2.114}$$

where the geometrical parameters are defined in (2.114) and shown on the right-hand side figure of Figure 2.18. Note the integral (2.114) is expressed successively in rectangular, polar, and so-called angular-radial coordinates, respectively, where in the latter, the Jacobian R conveniently cancels the singular or near-singular $1/R$ factor in the Green's function; this use of a transformation's Jacobian to cancel rapidly varying terms and smooth the integrand is the essential idea behind the so-called *singularity cancellation* methods to accelerate numerically evaluated integrals. The cancellation results in the last integral of (2.114) having a bounded, smooth integrand amenable to the use of Gauss–Legendre (GL) quadrature. However, it is first convenient to transform the R-integration interval from $R \in (|d|, R(\phi))$ to the unit interval $\xi \in (0,1)$ via the transformation $R = R(\phi)\xi + |d|(1 - \xi)$, where $R(\phi)$ is defined in (2.114). This last transform results in the integral form

$$\begin{aligned}
&\int_{T_q^f} \frac{e^{-jkR}}{4\pi R} \begin{bmatrix} \mathbf{\Lambda}_j^f(\mathbf{r}') \\ \nabla \cdot \mathbf{\Lambda}_j^f(\mathbf{r}') \end{bmatrix} dS' \\
&\qquad = \frac{1}{4\pi} \int_{\phi^-}^{\phi^+} \left[\left(\sqrt{d^2 + \frac{h^2}{\cos^2\phi}} - |d| \right) \int_0^1 e^{-jkR} \begin{bmatrix} \mathbf{\Lambda}_j^f(\mathbf{r}') \\ \nabla \cdot \mathbf{\Lambda}_j^f(\mathbf{r}') \end{bmatrix} d\xi \right] d\phi.
\end{aligned} \tag{2.115}$$

Though the ξ-integral has been smoothed, the difference-of-limits term, $(R(\phi) - |d|)$, behaves as $h/\cos\phi$ for $d << h$, and may also vary rapidly with ϕ, especially for poorly shaped subtriangles for which $\phi^{\pm}$ approaches $\pm\pi/2$, respectively. However, if the outer integral is transformed to a new variable, $v = v(\phi)$, with $d\phi = dv/v'(\phi)$, then choosing $v'(\phi) = 1/\cos\phi$ should cancel this variation. Integration of $v'(\phi)$

yields $v(\phi) = \log\left|\tan\left(\frac{\pi}{4} + \frac{\phi}{2}\right)\right| = \sinh^{-1}(\tan\phi)$ [28,29] and, noting this implies both $\tan\phi = \sinh v$ and $1/\cos\phi = \cosh v$, we obtain the following result [30]:

$$\int_{T_q^f} \frac{e^{-jkR}}{4\pi R}\begin{bmatrix}\mathbf{\Lambda}_j^f(\mathbf{r}') \\ \nabla\cdot\mathbf{\Lambda}_j^f(\mathbf{r}')\end{bmatrix} dS' = \frac{1}{4\pi}\int_{v^-}^{v^+}\left(\frac{\sqrt{d^2+h^2\cosh^2 v}-|d|}{\cosh v}\right)\int_0^1 e^{-jkR}\begin{bmatrix}(\mathbf{r}'-\mathbf{r}_j)/h_j \\ 2/h_j\end{bmatrix} d\xi\, dv, \quad (2.116)$$

where $v^{\pm} = \sinh^{-1}(\tan\phi^{\pm})$. Assuming, for simplicity, K-point GL rules with weights and sample points $(w_k, \xi^{(k)})$ for both the ξ- and v-integrals in (2.116), the result is

$$\int_{T_q^f} \frac{e^{-jkR}}{4\pi R}\begin{bmatrix}(\mathbf{r}'-\mathbf{r}_j)/h_j \\ 2/h_j\end{bmatrix} dS' \approx \frac{(v^+-v^-)}{4\pi}\sum_{r=1}^{K} w_r\left(\frac{\sqrt{d^2+h^2\cosh^2 v^{(r)}}-|d|}{\cosh v^{(r)}}\right)\sum_{s=1}^{K} w_s e^{-jkR^{(r,s)}}\begin{bmatrix}(\mathbf{r}^{(r,s)}-\mathbf{r}_j)/h_j \\ 2/h_j\end{bmatrix}, \quad (2.117)$$

where

$$\begin{aligned}
\phi^{(r)} &= \phi^-(1-\xi^{GL,(r)}) + \phi^+\xi^{GL,(r)}, \\
v^{(r)} &= \sinh^{-1}\tan\phi^{(r)} \\
R^{(r,s)} &= |d|(1-\xi^{GL,(s)}) + \sqrt{d^2+h^2\cosh^2 v^{(r)}}\,\xi^{GL,(s)}, \\
P^{(r,s)} &= \sqrt{(R^{(r,s)})^2-d^2}, \\
\mathbf{r}^{(r,s)} &= \mathbf{r}_0 + P^{(r,s)}(\cos\phi^{(r)}\hat{\mathbf{u}} + \sin\phi^{(r)}\hat{\boldsymbol{\ell}}).
\end{aligned} \quad (2.118)$$

Recall that $\mathbf{r}_j$ is the jth vertex of the original source triangle T^f, and h_j is its height; h is the height of the qth subtriangle, T_q^f. Note that to obtain the vector and scalar potential-related integrals for the entire source triangle T^f, we must sum the contributions to (2.117) from each subtriangle, $q = 1, 2, 3$.

The approach to (near-)singular integral evaluation described here and in [30] has been found to be quite robust [29,31], and appropriate combinations of radial and angular GL orders to guarantee four significant digit accuracies for the potential integrals are provided in [30]. Related schemes such as those described in [32] maintain similar simplicity and generality while potentially offering additional accuracy and efficiency.

We note finally that in the course of developing this approach for solving the EFIE, we have encountered all the main sources of error common to most surface integral equation formulations mentioned in introducing the EFIE, Section 2.5.3. We now briefly expand on them as follows:

- Geometry error due to surface discretization. Such error occurs, for example, when curved bodies such as spheres are modeled using planar triangles. Various higher order geometry modeling schemes have been developed to address this source of error.

- Discretization of the unknown equivalent current(s). Effects of this source of error may also be improved by using higher order basis functions.
- Discretization of the integral equation itself due to the testing scheme. Again, higher order testing, usually of the same order as the bases, and involving higher order quadrature, is sometimes employed to reduce the effects of this error source. A general scheme for reducing geometry modeling, basis, and testing scheme errors using higher order methods may be found in [33].
- Errors in moment matrix elements resulting from their numerical evaluation using numerical quadrature. In particular, it is important to properly recognize, treat, and evaluate singular integrals numerically without compromising computation time. This topic continues to receive extensive treatment in the literature.
- Errors in numerically solving the system matrix equation. Solution errors also depend on the solution scheme employed. Electrically small problems may often be solved using straightforward Gaussian elimination or LU decomposition. But for electrically large problems, iterative methods are often employed, in which case the most important question is the convergence rate of the approach. The convergence of iterative schemes is mostly determined by the conditioning of the matrix and the algorithm used. Such solution methods can often be improved by reformulating the method or by preconditioning the system equations. Other error sources that affect a solution are the use of fast schemes that, in effect, compress, sparsify, or locally diagonalize the system matrix.

In short, code testing and error control are important subtopics in both the use and development of electromagnetic modeling codes.

2.5.5 The magnetic field integral equation (MFIE)

There are two approaches to deriving the MFIE for scattering by a closed PEC with bounding surface S embedded in a homogeneous background medium. For both, we first use the equivalence principle to replace the conductor surface with an equivalent tangential current $\boldsymbol{J}$ and we specify a null magnetic field within S. The background medium can then be extended to the interior of S. The boundary form of Ampere's law applied at S implies that

$$\boldsymbol{J} = \hat{\boldsymbol{n}} \times \boldsymbol{H}^{\mathrm{i}} + \lim_{r \to S^+} \hat{\boldsymbol{n}} \times \boldsymbol{H}^{s}, \tag{2.119}$$

where an integral equation may then be obtained by using the surface integral representation for $\hat{\boldsymbol{n}} \times \boldsymbol{H}^{s}$ in terms of the surface source $\boldsymbol{J}$ on S^+. On the other hand, for this last step, we can instead simply enforce the condition that the total magnetic field vanish on S^-:

$$\hat{\boldsymbol{n}} \times \boldsymbol{H}^{\mathrm{i}} + \lim_{r \to S^-} \hat{\boldsymbol{n}} \times \boldsymbol{H}^{\mathrm{s}} = \boldsymbol{0}, \tag{2.120}$$

and again use the surface integral representation for $\hat{\boldsymbol{n}} \times \boldsymbol{H}^s$, but this time on S^-. That the two are equivalent follows easily from the source jump condition

$$\boldsymbol{J} = \lim_{r\to S^+} \hat{\boldsymbol{n}} \times \boldsymbol{H}^s - \lim_{r\to S^-} \hat{\boldsymbol{n}} \times \boldsymbol{H}^s, \tag{2.121}$$

but the null field condition is usually simpler and more convenient to apply. To complete the derivation, we substitute the surface potential representation of $\lim_{r\to S^-} \hat{\boldsymbol{n}} \times \boldsymbol{H}^s$, (2.70), into (2.121):

$$\frac{1}{2}\boldsymbol{J}(\boldsymbol{r}) - \hat{\boldsymbol{n}} \times \fint_S \nabla G(\boldsymbol{r},\boldsymbol{r}') \times \boldsymbol{J}(\boldsymbol{r}')\, dS' = \hat{\boldsymbol{n}} \times \boldsymbol{H}^{\mathrm{i}}, \quad \boldsymbol{r} \in S, \tag{2.122}$$

or in the operator form,

$$\left(\tfrac{1}{2}\mathscr{I} + \mathscr{K}\right)\boldsymbol{J} = \hat{\boldsymbol{n}} \times \boldsymbol{H}^{\mathrm{i}}, \quad \boldsymbol{r} \in S. \tag{2.123}$$

Expanding the unknown surface current in the RWG basis expansion, (2.103), and testing the resulting equation with $\boldsymbol{\Lambda}_m(\boldsymbol{r})$, $m = 1, 2, \cdots, N$ (as for the EFIE) leads to a discretized form of the MFIE:

$$[\beta_{mn}]\,[I_n] = \left[I^{\mathrm{i}}_m\right], \tag{2.124}$$

where the MFIE system matrix is

$$\begin{aligned}\beta_{mn} &= \tfrac{1}{2} < \boldsymbol{\Lambda}_m; \boldsymbol{\Lambda}_n > - < \boldsymbol{\Lambda}_m\,;\, \hat{\boldsymbol{n}} \times \nabla G \overset{\times}{;} \boldsymbol{\Lambda}_n > \\ &= < \boldsymbol{\Lambda}_m; \left(\tfrac{1}{2}\mathscr{I} + \mathscr{K}\right)\boldsymbol{\Lambda}_n >, \quad m, n \in 1, 2, \cdots, N\end{aligned} \tag{2.125}$$

and the tested excitation column vector is

$$I^{\mathrm{i}}_m =< \boldsymbol{\Lambda}_m; \hat{\boldsymbol{n}} \times \boldsymbol{H}^{\mathrm{i}} >, \quad m \in 1, 2, \cdots, N. \tag{2.126}$$

Note that here, since S is assumed closed, the total number of edges and the number of DoFs are the same, $E = N$. As with the EFIE, the domains of all the double integrals over S implicit in the inner product can be contracted to the supports of $\boldsymbol{\Lambda}_m$ and $\boldsymbol{\Lambda}_n$, i.e., to $T^+_m \cup T^-_m$ and $T^+_n \cup T^-_n$, respectively. Also, as with the EFIE, even greater integration efficiency is achieved if we further contract the integration domains to single test (T^e_i) and source (T^f_j) element pairs at a time, assembling the system matrix by adding these partial pair contributions (including sign factors σ^e_i, σ^f_j to account for reference directions) directly to appropriate elements of the system matrix as they are computed. For each triangular element pair e,f, we generate an element matrix entry,

$$\beta^{ef}_{ij} = \begin{cases} \tfrac{1}{2} < \boldsymbol{\Lambda}^e_i; \boldsymbol{\Lambda}^f_j >, & e = f, \\ < \boldsymbol{\Lambda}^e_i\,; \mathscr{K}\boldsymbol{\Lambda}^f_j >, & e \neq f, \end{cases} \tag{2.127}$$

where $i,j \in 1, 2, 3$ and $e,f \in 1, 2, \cdots, E$. The corresponding element excitation vector is

$$I^{\mathrm{i},e}_i = < \boldsymbol{\Lambda}^e_i\,; \hat{\boldsymbol{n}} \times \boldsymbol{H}^{\mathrm{i}} > . \tag{2.128}$$

Note in (2.127), when $e = f$ (a "self-interaction"), the inner product involving the integral disappears since both the gradient of G and basis function are tangential to

S_i^e. Hence, their cross-product is normal for S_i^e and disappears under the subsequent cross-product with $\hat{\boldsymbol{n}}$; on the other hand, when $e \neq f$ the inner product between the bases disappears since neither of their supports overlap, leaving only the integral contribution.

2.5.5.1 Numerical evaluation of MFIE element matrices

The numerical evaluation of the element matrix (2.127) parallels that of the EFIE. For well-separated element pairs T^e and T^f, the GT approach of (2.113) leads to

$$\beta_{ij}^{ef} \approx -4|T^e||T^f|\sum_{k=1}^{K}\sum_{l=1}^{K} w_k^{GT} w_l^{GT} \boldsymbol{\Lambda}_i^e(\boldsymbol{r}^{(k)})\left[\cdot \hat{\boldsymbol{n}} \times \left(\nabla G(\boldsymbol{r}^{(k)}, \boldsymbol{r}^{(l)}) \times \boldsymbol{\Lambda}_j^f(\boldsymbol{r}^{(l)})\right)\right], \tag{2.129}$$

where

$$\nabla G(\boldsymbol{r}, \boldsymbol{r}') = -\frac{(jkR+1)e^{-jkR}}{4\pi R^3}\mathbf{R} \tag{2.130}$$

with $\mathbf{R} = \boldsymbol{r} - \boldsymbol{r}'$ and $R = |\mathbf{R}|$.

For nearby and self-elements ($e = f$), as with the EFIE, extra care is required. Namely, we may keep the test integral portion of (2.129) (i.e., terms involving the index k and test triangle T^e) and replace the inner sum with the more sophisticated scheme described in [34], which is essentially an extension of the EFIE approach to the gradient kernel. We thus consider the inner integral related to the curl of vector potential, $-\nabla \times \boldsymbol{A}$:

$$\begin{aligned}
\int_{T_q^f} \boldsymbol{\Lambda}_j^f(\boldsymbol{r}') \times \nabla G(\boldsymbol{r}, \boldsymbol{r}')\, dS' &= -\int_{T_q^f} \boldsymbol{\Lambda}_j^f(\boldsymbol{r}') \times \mathbf{R}\frac{(1+jkR)e^{-jkR}}{4\pi R^3}\, dS' \\
&= \mathbf{R}_j \times \int_{T_q^f} \boldsymbol{\Lambda}_j^f(\boldsymbol{r}')\frac{(1+jkR)e^{-jkR}}{4\pi R^3}\, dS', \\
&= \mathbf{R}_j \times \int_{T_q^f} \boldsymbol{\Lambda}_j^f(\boldsymbol{r}')\frac{(1+jkR)e^{-jkR}-1}{4\pi R^3}\, dS' \\
&\quad + \mathbf{R}_j \times \int_{T_q^f} \boldsymbol{\Lambda}_j^f(\boldsymbol{r}')\frac{1}{4\pi R^3}\, dS',
\end{aligned} \tag{2.131}$$

where we have defined $\mathbf{R}_j = \boldsymbol{r} - \boldsymbol{r}_j$, and noted that with $\mathbf{R} = \mathbf{R}_j - \boldsymbol{\rho}_j$ and $\boldsymbol{\Lambda}_j^f(\boldsymbol{r}') = \boldsymbol{\rho}_j/h_j$, the cross-product becomes $\boldsymbol{\Lambda}_j^f(\boldsymbol{r}') \times \mathbf{R} = \boldsymbol{\Lambda}_j^f(\boldsymbol{r}') \times (\mathbf{R}_j - \boldsymbol{\rho}_j) = -\mathbf{R}_j \times \boldsymbol{\Lambda}_j^f(\boldsymbol{r}')$. Substituting these results into (2.131) yields the second equality. The third results from both subtracting and adding the last integral, forming a difference integrand in the first that cancels the leading order singularity. It is easily verified that the difference $(1+jkR)e^{-jkR} - 1 \xrightarrow{R\to 0} O(R^2)$, reducing the integrand singularity order to $O(1/R)$. As in Section 2.5.4.1, the radial angular transform may thus be applied to the difference integral. The last integral in (2.131) is a static integral whose evaluation is described in [35]; its integrand is similar to those appearing in (2.65) and (2.66), but now integrated in closed form over a triangular domain rather than over a circular disc.

Hence, it still includes the residue contribution of the original integral that is usually explicitly extracted from it as $\boldsymbol{r} \to T^f$.

The source integration approach described is a hybrid approach, with *singularity subtraction* or *extraction* used first to reduce the order of the singularity of the difference integral and singularity cancellation using the radial-angular scheme to evaluate the reduced order difference integral. Both steps rely on the existence of a closed-form integral on T^f of an asymptotic approximation to the integrand. We have merely assumed the testing schemes for EFIE and MFIE are adequate, but this is often not the case; for higher accuracy in computing EFIE and MFIE matrix element entries representing interactions between source and test triangles that are not well-separated, one may follow the testing scheme of [36].

2.5.5.2 Open bodies and the MFIE

Unlike the EFIE, the MFIE cannot be applied to open conducting bodies; the reasons for this are discussed in the following two sections. We note, however, the MFIE is usually better-conditioned than the EFIE since it is an integral equation of the second kind. Also, there are a fewer restrictions on the basis and testing functions that may be used and a number of alternatives to using RWG bases alone have been reported; some are discussed in this book. Such choices may strongly influence both sensitivity to computational error (matrix conditioning) and solution accuracy of the MFIE.

2.5.5.3 Specialization of MFIE to infinite, flat PEC surface

It is illuminating to attempt to specialize the MFIE (2.122) to an infinite, planar PEC surface. In that case, the normal $\hat{\boldsymbol{n}}$ approaches a constant and it is easily seen that the integral term in (2.122) vanishes. This occurs because both $\boldsymbol{J}$ and ∇G (noting that $\boldsymbol{r} \neq \boldsymbol{r}'$ in the principal value integral) lie in the plane of S. Hence their cross-product is either parallel or anti-parallel to $\hat{\boldsymbol{n}}$, and disappears under the cross-product with $\hat{\boldsymbol{n}}$. Thus on an infinite planar PEC surface, a solution to (2.122) is readily seen to be

$$\boldsymbol{J} = 2\hat{\boldsymbol{n}} \times \boldsymbol{H}^{\mathrm{i}}. \tag{2.132}$$

This well-known property of scattering by an infinite PEC plane results because the integral term essentially constitutes a perturbation term accounting for surface curvature and that disappears as S approaches planarity. Integral equations like the MFIE in which the unknown appears both inside and outside an integral are known as *integral equations of the second kind*, in contrast to *integral equations of the first kind*, such as the EFIE, in which the unknown appears only inside an integral with a high-order singularity. In general, the former is more numerically robust than the latter because of the presence of the unknown outside the integral, with the principal value integral acting as a less-dominant "correction" term. The robustness of a numerical algorithm is typically expressed by its *condition number*, which is a measure of the discretized integral operator's tendency to magnify modeling errors of the types discussed in Section 2.5.4.1. The condition number or error sensitivity of the second kind equations (e.g., the MFIE) is generally lower than that of the first kind (e.g., EFIE) equations.

The view of second-kind equations as consisting of a zeroth-order approximation plus a perturbation term that tends to disappear with decreasing curvature is also partial justification for using the term (2.132) in the so-called *physical optics approximation* for surface current on the illuminated side of a large, smooth PEC scattering surface S satisfying the inequality $\kappa \ll \lambda$ where λ is the wavelength and κ is the surface curvature at points on the illuminated side of S [37].

2.5.6 Conducting sheets and the EFIE and MFIE

The MFIE generally applies to closed, but not to open conductors. Figure 2.19 shows two nearly parallel conducting surfaces whose maximum separation is τ. We assume that as $\tau \to 0$, the conducting surface S^+ and its normal $\hat{\boldsymbol{n}}^+$ remain fixed while S^- and its normal $\hat{\boldsymbol{n}}^-$ approach S^+ and $-\hat{\boldsymbol{n}}^+$, respectively. Thus the original closed surface collapses onto itself, forming a thin conducting (PEC) sheet. We then compare the EFIE and MFIE in this limiting process.

First consider the EFIE for the thin conductor case. By the equivalence principle, the two conductor surfaces S^+ and S^- are replaced by two equivalent surface currents $\boldsymbol{J}^+$ and $\boldsymbol{J}^-$, respectively, that gradually become coincident as $\tau \to 0$. Thus, in the limit, the EFIE, (2.90), takes the form

$$\left[j\omega\mu \int_{S^+} G(\boldsymbol{r}, \boldsymbol{r}')\boldsymbol{J}(\boldsymbol{r}')\, dS' - \frac{1}{j\omega\varepsilon}\nabla \int_{S^+} G(\boldsymbol{r}, \boldsymbol{r}')\nabla' \cdot \boldsymbol{J}(\boldsymbol{r}')\, dS' \right]_{\text{tan}} = \boldsymbol{E}^{\text{i}}_{\text{tan}}(\boldsymbol{r}), \quad \boldsymbol{r} \in S, \tag{2.133}$$

where $\boldsymbol{J} = \boldsymbol{J}^+ + \boldsymbol{J}^-$ is the sum of the two coincident currents, treated as a single equivalent current. We note also we must have $\hat{\boldsymbol{u}} \cdot \boldsymbol{J} = \hat{\boldsymbol{u}} \cdot (\boldsymbol{J}^+ + \boldsymbol{J}^-) = 0$ along the boundary of S^+ as $\tau \to 0$ so that a flux imbalance on opposite sides of the boundary of S does not deposit a line source charge there. Since (2.133) is essentially equivalent to

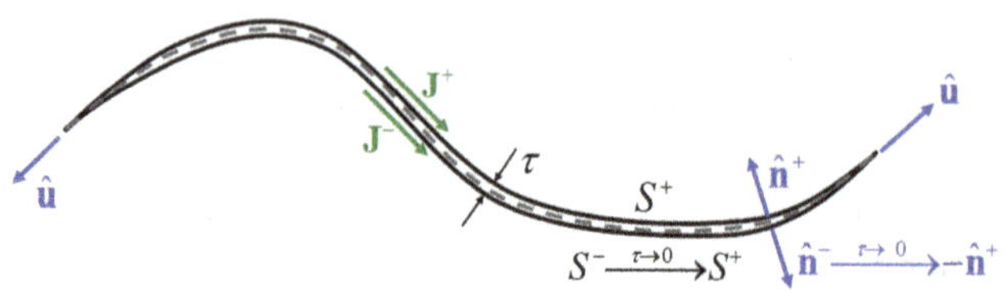

Figure 2.19 Cross-section of a closed, thin conducting surface S of maximum conductor thickness τ. S is partitioned into "top" and "bottom" surfaces, S^+ and S^-, that support surface currents $\boldsymbol{J}^+$ and $\boldsymbol{J}^-$, respectively. We assume that as $\tau \to 0$, $S^- \to S^+$ while $\hat{\boldsymbol{n}}^- \to -\hat{\boldsymbol{n}}^+$. The vector $\hat{\boldsymbol{u}}$ is a unit outward normal vector along the boundaries of S^+ and $\lim_{\tau\to 0} S^-$ and is normal to both $\hat{\boldsymbol{n}}^+$ and the unit vector $\hat{\boldsymbol{\ell}}$ tangent to the boundary of S^+. Unit vectors $\hat{\boldsymbol{n}}^+$, $\hat{\boldsymbol{u}}$, and $\hat{\boldsymbol{\ell}}$ form an orthogonal triad with $\hat{\boldsymbol{n}}^+ = \hat{\boldsymbol{u}} \times \hat{\boldsymbol{\ell}}$.

(2.90), we can now understand how the EFIE is used and interpreted on open surfaces: because the electric field is continuous across electric currents we are unable to distinguish between $\boldsymbol{E}_{\tan}$ measured on S^+ or S^- as $\tau \to 0$, yet for determining fields, it is unnecessary to do so, since the two independent but coincident currents on S^+ and S^- become superimposed, with their sum acting as a single equivalent unknown current $\boldsymbol{J}$. Note that a subdivision of S^+ into planar triangles, for example, induces an identical, coincident subdivision of S^- where, if the nth DoF is a boundary edge, the associated basis functions must satisfy $\hat{\boldsymbol{u}} \cdot \boldsymbol{\Lambda}_n^+ = -\hat{\boldsymbol{u}} \cdot \boldsymbol{\Lambda}_n^-$ in view of the boundary constraint. If we have assigned "0" as the DoF index for all boundary edges, it is then an easy matter when filling the impedance matrix to simply skip over the associated row (test) or column (source) elements of the element interaction matrix.

We can attempt to use the same approach with the MFIE to treat the geometry of Figure 2.19, i.e., a closed conductor $S^+ \cup S^-$ collapsing to a thin sheet S^+. The modified MFIE becomes

$$\frac{1}{2}\left[\boldsymbol{J}^+ - \boldsymbol{J}^-\right] - \hat{\boldsymbol{n}}^+ \times \fint_{S^+} \nabla G(\boldsymbol{r}, \boldsymbol{r}') \times \left[\boldsymbol{J}^+ + \boldsymbol{J}^-\right] dS' = \hat{\boldsymbol{n}}^+ \times \boldsymbol{H}^{\mathrm{i}}, \quad \boldsymbol{r} \in S^+ \tag{2.134}$$

where the two current contributions outside the integral have opposite signs since the rotated magnetic fields arising from currents just above and below observation points on the dashed line between conductors in Figure 2.19, have opposite signs. The integrals are over the same domain S^+ in the limit as $\tau \to 0$, but the source currents now add under the principal value integral. Hence, though we reduce the integral to a single integral and domain, two unknowns remain in the equation since both currents, or equivalently, their sums and differences, remain in the equation. Thus, (2.134) has been reduced to an *identity* relating the two opposite side currents, and incident field, but without sufficient information to allow both currents to be determined. The identity may be used in conjunction with (2.133), however, to separate the total current into its two opposite side components. That is, we might first use the EFIE to determine the sum current, $\boldsymbol{J}^\Sigma = \boldsymbol{J}^+ + \boldsymbol{J}^-$, then use (2.134) to determine the difference current, $\boldsymbol{J}^\Delta = \boldsymbol{J}^+ - \boldsymbol{J}^-$, with the result

$$\boldsymbol{J}^\Delta(\boldsymbol{r}) = 2\hat{\boldsymbol{n}}^+ \times \boldsymbol{H}^{\mathrm{i}}(\boldsymbol{r}) + 2\hat{\boldsymbol{n}}^+ \times \fint_{S^+} \nabla G(\boldsymbol{r}, \boldsymbol{r}') \times \boldsymbol{J}^\Sigma(\boldsymbol{r}')\, dS', \quad \boldsymbol{r} \in S^+, \tag{2.135}$$

and where $\boldsymbol{J}^+$ and $\boldsymbol{J}^-$ are finally determined as $\boldsymbol{J}^+ = \frac{1}{2}\left(\boldsymbol{J}^\Sigma + \boldsymbol{J}^\Delta\right)$ and $\boldsymbol{J}^- = \frac{1}{2}\left(\boldsymbol{J}^\Sigma - \boldsymbol{J}^\Delta\right)$. We summarize the situation for an infinitesimally thick conductor with electric currents $\boldsymbol{J}^+$ and $\boldsymbol{J}^-$ on the two opposite sides as follows:

- On an open surface, we can write only a single independent equation for the EFIE since it is continuous across the two coincident tangential electric currents on opposite sides. Since the sum of the two currents acts as one, however, their sum can be found using the EFIE. The sum currents, in turn, may be used to determine either near or far fields. Indeed, they can be used to find $\hat{\boldsymbol{n}} \times \boldsymbol{H}$ on both S^+ and S^-, which data, together with the incident magnetic field, would then yield the two opposite side surface currents.

- For a closed, thin conductor of maximum thickness $\tau \neq 0$, the MFIE samples the excitation and scattered magnetic fields just inside S^+ and S^-. The sampled data for each side is independent, in principle (for $\tau \neq 0$), permitting a solution for the two independent surface currents. As $\tau \to 0$, however, the two field observation surfaces S^+ and S^- collapse to a single surface, the sampled field data on them is no longer independent, and hence insufficient for solving the integral equation for both $\boldsymbol{J}^+$ and $\boldsymbol{J}^-$.

2.5.7 Internal resonances and the CFIE

For a closed PEC body with boundary S, the EFIE is unable to distinguish whether the sources of the excitation arise from the exterior or the interior of S. The inability of the integral equation to distinguish between (exterior) scattering problems or (interior) cavity excitation problems means, for example, that at cavity resonance frequencies, even small errors in the exterior problem may act to excite the interior cavity resonator, whose resonant wall currents can, in principle, be arbitrarily large. Since the equivalent current found is actually a superposition of any interior and exterior induced currents, very large errors in scattering currents often result near cavity resonance frequencies.

Surprisingly, a solution to this problem is to simply form a linear combination of the EFIE and MFIE. First, we write the EFIE and MFIE in their compact forms

$$\hat{\boldsymbol{n}} \times \mathscr{L}\boldsymbol{J} = \frac{\boldsymbol{E}^{\mathrm{i}}_{\tan}}{\eta}, \quad \boldsymbol{r} \in S,$$

$$\left(\tfrac{1}{2}\mathscr{I} + \mathscr{K}\right)\boldsymbol{J} = \hat{\boldsymbol{n}} \times \boldsymbol{H}^{\mathrm{i}}, \quad \boldsymbol{r} \in S, \tag{2.136}$$

respectively. We then form a combined field integral equation (CFIE) by multiplying the MFIE by a weighting factor α and adding the resulting equations as follows:

$$\left(\hat{\boldsymbol{n}} \times \mathscr{L} + \alpha\left(\tfrac{1}{2}\mathscr{I} + \mathscr{K}\right)\right)\boldsymbol{J} = \tfrac{1}{\eta}\boldsymbol{E}^{\mathrm{i}}_{\tan} + \alpha\hat{\boldsymbol{n}} \times \boldsymbol{H}^{\mathrm{i}}. \tag{2.137}$$

The corresponding discretized version of (2.137) is obtained by merely taking the corresponding linear combinations of (2.104) and (2.124), yielding

$$\left[\tfrac{1}{\eta}Z_{mn} + \alpha\beta_{mn}\right][I_n] = \left[\tfrac{1}{\eta}V_m + \alpha I^{\mathrm{i}}_m\right]. \tag{2.138}$$

The uniqueness of solutions $\boldsymbol{J}$ of (2.137) follows by first assuming non-uniqueness, then showing this assumption leads to a contradiction. That is, if we assume there exists at least two non-vanishing solutions, $\boldsymbol{J}^a \neq \boldsymbol{J}^b$ of (2.137), then the difference in current $\delta\boldsymbol{J} = \boldsymbol{J}^a - \boldsymbol{J}^b$ must satisfy the homogeneous equation obtained by separately substituting $\boldsymbol{J}^a$ and $\boldsymbol{J}^b$ into (2.137), subtracting the two resulting equations, and using the linearity properties of the operators to obtain

$$-\tfrac{1}{\eta}\delta\boldsymbol{E}_{\tan} - \alpha\hat{\boldsymbol{n}} \times \delta\boldsymbol{H} = \left(\hat{\boldsymbol{n}} \times \mathscr{L} + \alpha\left(\tfrac{1}{2}\mathscr{I} + \mathscr{K}\right)\right)\delta\boldsymbol{J} = \boldsymbol{0}, \tag{2.139}$$

where, for convenience, we have defined and introduced the tangential fields $-\frac{\delta\boldsymbol{E}_{\tan}}{\eta} = \hat{\boldsymbol{n}} \times \mathscr{L}\delta\boldsymbol{J}$ and $-\alpha\hat{\boldsymbol{n}} \times \delta\boldsymbol{H} = \alpha\left(\tfrac{1}{2}\mathscr{I} + \mathscr{K}\right)\delta\boldsymbol{J}$ radiated by the homogeneous solution $\delta\boldsymbol{J}$ with $\boldsymbol{r}$ on S^-. Next we dot multiply the terms to the left of the first equality in

(2.139) by their complex conjugate and integrate the resulting expression over S, obtaining

$$\int_S \left(\left| \frac{1}{\eta} \delta \boldsymbol{E}_{\tan} \right|^2 + |\alpha|^2 \, |\delta \boldsymbol{H}_{\tan}|^2 \right) dS + \frac{2\alpha}{\eta} \mathrm{Re} \int_S (\delta \boldsymbol{E} \times \delta \boldsymbol{H}^*) \cdot (-\hat{\boldsymbol{n}})\, dS = 0, \boldsymbol{r} \in S^-, \tag{2.140}$$

where α/η is chosen positive and real with $|\alpha|$ near unity so that electric to magnetic field ratios are scaled comparable to the background medium intrinsic impedance.

Since $-\hat{\boldsymbol{n}}$ is directed into S, the last integral of (2.140) represents real power flowing into S, hence, must be non-negative, as must the square magnitudes of the integrands of the first integral. But the right-hand side is zero, hence both integrand terms of the first integral must vanish. In the first integral, $\delta \boldsymbol{E}_{\tan}$ thus vanishes on both S^- and S^+ since $\delta \boldsymbol{E}_{\tan}$ is continuous across $\delta \boldsymbol{J}$. By the uniqueness theorem, however, this further implies that $\hat{\boldsymbol{n}} \times \delta \boldsymbol{H}|_{S^+}$ also vanishes. Then, since $\hat{\boldsymbol{n}} \times \delta \boldsymbol{H}|_{S^-}$ also vanishes, $\delta \boldsymbol{J} = \hat{\boldsymbol{n}} \times (\delta \boldsymbol{H}|_{S^+} - \delta \boldsymbol{H}|_{S^-}) = \boldsymbol{0}$, and uniqueness is established [38,39].

We note that if the interior of S vanishes, with one of its sides collapsing to the other, as discussed in 2.5.6, the MFIE approaches the form (2.134). And if a CFIE is again formed by linearly combining the EFIE and MFIE as above for both limit surfaces, a sign change between them occurs due to reversal of one of the surface's normal. The resulting sum-and-difference equations are independent and sufficient to enable solving for the independent currents on the two opposing limit surfaces [40]. Thus, the CFIE is unique among PEC formulations discussed that not only eliminates internal resonances but applies to thin structures as well [40] (though with twice the number of unknown currents as the EFIE).

In the general case, the interior resonance frequencies must be the same for both the EFIE and MFIE; it might seem they would also be resonant frequencies of the homogeneous CFIE, (2.140), since it is merely a linear combination of the other two. This is not the case, however, since the two homogeneous equations do not have the *same* homogeneous solutions [38,39]. Instead, they satisfy separate homogenous EFIE and MFIE equations, $\hat{\boldsymbol{n}} \times \mathscr{L}\, \delta \boldsymbol{J}^{\mathrm{efie}} = \boldsymbol{0}$, and $\left(\frac{1}{2}\mathscr{I} + \mathscr{K}\right) \delta \boldsymbol{J}^{\mathrm{mfie}} = \boldsymbol{0}$, respectively, which equations cannot be linearly combined to form (2.139) since $\delta \boldsymbol{J}^{\mathrm{efie}} \neq \delta \boldsymbol{J}^{\mathrm{mfie}}$. It is not difficult to see why this is the case: the EFIE requires $\boldsymbol{J}^{\mathrm{efie}}$ to produce $\boldsymbol{E}_{\tan} = \boldsymbol{0}$ on S^-, but since $\boldsymbol{E}_{\tan}$ is continuous across electric current sources, it vanishes on S^+ as well. By the uniqueness theorem, then, all the external fields vanish. On the other hand, the MFIE requires $\boldsymbol{J}^{\mathrm{mfie}}$ to produce $\boldsymbol{H}_{\tan} = \boldsymbol{0}$ on S^-, but causes a jump in $\boldsymbol{H}_{\tan}$, implying that the exterior magnetic (and electric) fields do not vanish. Thus, the EFIE and MFIE formulations produce *different* magnetic fields exterior to S at resonances; hence, $\boldsymbol{J}^{\mathrm{efie}} \neq \boldsymbol{J}^{\mathrm{mfie}}$ [39].

2.5.8 Integral equation formulations for dielectrics

In this section, we derive two integral equations for homogeneous dielectric objects embedded in a homogeneous host (background) medium of infinite extent. The dielectric body is assumed bounded by a closed surface S separating the interior and exterior regions with media parameters μ^-, ε^- and μ^+, ε^+, respectively. For generality, we assume that either or both regions may contain sources that produce incident fields $(\boldsymbol{E}^{\mathrm{i},-}, \boldsymbol{H}^{\mathrm{i},-})$ and $(\boldsymbol{E}^{\mathrm{i},+}, \boldsymbol{H}^{\mathrm{i},+})$. If the incident fields are due to current sources $(\boldsymbol{J}^{\mathrm{i},\pm}, \boldsymbol{M}^{\mathrm{i},\pm})$ exterior or interior to S, respectively, they may be generated from their sources using potential representations as if they radiated in an infinite homogeneous medium with the parameters of the media in which they reside. The scattered fields exterior to S are represented in terms of equivalent currents $(\boldsymbol{J}, \boldsymbol{M})$ on S, as shown in Figure 2.5. The medium parameters of the outside medium are extended to the interior of S, and a null field is assumed inside and tangential to S^-. For the interior equivalence, the representation of Figure 2.6 is used, with the interior medium extended to the exterior where a null field is assumed. Because tangential fields are continuous at a dielectric interface, jump conditions require that the interior equivalent currents are $(-\boldsymbol{J}, -\boldsymbol{M})$, i.e., they are simply the negatives of the exterior equivalent currents.

Scattered fields on either side of S may be represented in terms of the generic equations (2.69) and (2.70) for the fields radiated by surface currents $(\boldsymbol{J}, \boldsymbol{M})$. We arrange these in the operator matrix form as

$$\begin{bmatrix} \hat{\boldsymbol{n}} \times \boldsymbol{E} \\ \hat{\boldsymbol{n}} \times \boldsymbol{H} \end{bmatrix} = \begin{bmatrix} \eta \mathscr{L} & \mp\frac{1}{2}\mathscr{I} + \mathscr{K} \\ \pm\frac{1}{2}\mathscr{I} - \mathscr{K} & \eta^{-1}\mathscr{L} \end{bmatrix} \begin{bmatrix} \mathbf{J} \\ \mathbf{M} \end{bmatrix}, \quad \boldsymbol{r} \in S^{\pm}. \tag{2.141}$$

Note that (2.141) indicates that only the signs of the identity operators change for the fields radiated by $(\boldsymbol{J}, \boldsymbol{M})$ for some fixed medium parameters when we merely change the surface on which fields are evaluated from S^+ to S^-. We want, however, to use and appropriately modify (2.141) to enforce null field conditions on both tangential electric and magnetic fields at S^- for the exterior equivalence and similarly at S^+ for the interior equivalence. This requires the following modifications to the field expressions taken from (2.141) and applied in the appropriate equivalence region:

- In both equivalence regions, we add together the appropriate incident and scattered fields, with the latter in the operator form of (2.141), and apply the null field condition. This is done for both the electric and magnetic fields resulting in four null field equations.
- In each null field expression, one should add $+$ or $-$ superscripts for all medium parameters as well as the operators $\mathscr{L}$ and $\mathscr{K}$ to indicate on which side of S the associated equivalence applies. Since we are applying a null field condition, we note that the observation point $\boldsymbol{r}$ is on the opposite side of S.
- The currents $(\boldsymbol{J}, \boldsymbol{M})$ are used in all the exterior equivalences, but $(-\boldsymbol{J}, -\boldsymbol{M})$ replace the sources for the interior equivalence cases.

This procedure results in the following four equations in the two unknowns, $(\boldsymbol{J}, \boldsymbol{M})$:

$$\begin{aligned}
\hat{\boldsymbol{n}} \times \boldsymbol{E}^+\Big|_{\boldsymbol{r}\in S^-} + \hat{\boldsymbol{n}} \times \boldsymbol{E}^{\mathrm{i},+} &= \eta^+\mathscr{L}^+\boldsymbol{J} + (\tfrac{1}{2}\mathscr{I} + \mathscr{K}^+)\boldsymbol{M} + \hat{\boldsymbol{n}} \times \boldsymbol{E}^{\mathrm{i},+} = \mathbf{0}, \ \boldsymbol{r} \in S^- \\
\hat{\boldsymbol{n}} \times \boldsymbol{E}^-\Big|_{\boldsymbol{r}\in S^+} + \hat{\boldsymbol{n}} \times \boldsymbol{E}^{\mathrm{i},-} &= -\eta^-\mathscr{L}^-\boldsymbol{J} - (-\tfrac{1}{2}\mathscr{I} + \mathscr{K}^-)\boldsymbol{M} + \hat{\boldsymbol{n}} \times \boldsymbol{E}^{\mathrm{i},-} = \mathbf{0}, \ \boldsymbol{r} \in S^+ \\
\hat{\boldsymbol{n}} \times \boldsymbol{H}^+\Big|_{\boldsymbol{r}\in S^-} + \hat{\boldsymbol{n}} \times \boldsymbol{H}^{\mathrm{i},+} &= (-\tfrac{1}{2}\mathscr{I} - \mathscr{K}^+)\boldsymbol{J} + \tfrac{1}{\eta^+}\mathscr{L}^+\boldsymbol{M} + \hat{\boldsymbol{n}} \times \boldsymbol{H}^{\mathrm{i},+} = \mathbf{0}, \ \boldsymbol{r} \in S^- \\
\hat{\boldsymbol{n}} \times \boldsymbol{H}^-\Big|_{\boldsymbol{r}\in S^+} + \hat{\boldsymbol{n}} \times \boldsymbol{H}^{\mathrm{i},-} &= -(\tfrac{1}{2}\mathscr{I} - \mathscr{K}^-)\boldsymbol{J} - \tfrac{1}{\eta^-}\mathscr{L}^-\boldsymbol{M} + \hat{\boldsymbol{n}} \times \boldsymbol{H}^{\mathrm{i},-} = \mathbf{0}, \ \boldsymbol{r} \in S^+.
\end{aligned} \tag{2.142}$$

Clearly, we only need two independent equations to determine the two unknowns $\boldsymbol{J}$ and $\boldsymbol{M}$. Indeed, essentially all known dielectric formulations are merely some linear combination of these equations that arrives at two independent equations. Of these, perhaps the most natural is the PMCHWT formulation, which is equivalent to simply enforcing continuity of tangential $\boldsymbol{E}$ and $\boldsymbol{H}$ across S. The acronym PMCHWT derives from the initials of the authors of the original publications that employed the formulation [41–43]. It has been shown that the formulation is free of internal resonance difficulties, and, hence, always yields a unique solution. We obtain the PMCHWT formulation from (2.142) by equating the first and second as well as the third and fourth equations of the system, resulting in the following matrix operator system:

$$\begin{bmatrix} \eta^+\mathscr{L}^+ + \eta^-\mathscr{L}^- & \mathscr{K}^+ + \mathscr{K}^- \\ -\mathscr{K}^+ - \mathscr{K}^- & \frac{1}{\eta^+}\mathscr{L}^+ + \frac{1}{\eta^-}\mathscr{L}^- \end{bmatrix} \begin{bmatrix} \boldsymbol{J} \\ \boldsymbol{M} \end{bmatrix} = \begin{bmatrix} -\hat{\boldsymbol{n}} \times \boldsymbol{E}^{\mathrm{i},+} + \hat{\boldsymbol{n}} \times \boldsymbol{E}^{\mathrm{i},-} \\ -\hat{\boldsymbol{n}} \times \boldsymbol{H}^{\mathrm{i},+} + \hat{\boldsymbol{n}} \times \boldsymbol{H}^{\mathrm{i},-} \end{bmatrix}, \quad \mathbf{r} \in S, \tag{2.143}$$

where, expressed in progressively more compact notation,

$$\begin{aligned} \mathscr{L}^{\pm}\boldsymbol{X} &= -jk^{\pm}\hat{\boldsymbol{n}} \times \left(\int_S G^{\pm}(\boldsymbol{r},\boldsymbol{r}')\boldsymbol{X}(\boldsymbol{r}')\,dS' + \frac{\nabla}{(k^{\pm})^2} \int_S G^{\pm}(\boldsymbol{r},\boldsymbol{r}')\nabla' \cdot \boldsymbol{X}(\boldsymbol{r}')\,dS' \right) \\ &= -jk^{\pm}\hat{\boldsymbol{n}} \times \int_S \left(\mathscr{I} + \frac{\nabla\nabla}{(k^{\pm})^2} \right) G^{\pm}(\boldsymbol{r},\boldsymbol{r}') \cdot \boldsymbol{X}(\boldsymbol{r}')\,dS' \\ &= -jk^{\pm}\hat{\boldsymbol{n}} \times \int_S \mathscr{G}^{\pm}(\boldsymbol{r},\boldsymbol{r}') \cdot \boldsymbol{X}(\boldsymbol{r}')\,dS' \\ &= -jk^{\pm}\hat{\boldsymbol{n}} \times \int_S \mathscr{G}^{\pm} \cdot \boldsymbol{X}\,dS', \end{aligned} \tag{2.144}$$

and

$$\begin{aligned} \mathscr{K}^{\pm}\boldsymbol{X} &= -\hat{\boldsymbol{n}} \times \fint_S \nabla G^{\pm}(\boldsymbol{r},\boldsymbol{r}') \times \boldsymbol{X}(\boldsymbol{r}')\,dS' \\ &= -\hat{\boldsymbol{n}} \times \nabla \times \fint_S \mathscr{G}^{\pm}(\boldsymbol{r},\boldsymbol{r}') \cdot \boldsymbol{X}(\boldsymbol{r}')\,dS' \\ &= -\hat{\boldsymbol{n}} \times \nabla \times \fint_S \mathscr{G}^{\pm} \cdot \boldsymbol{X}\,dS'. \end{aligned} \tag{2.145}$$

Another well-known alternative formulation for dielectric scattering problems is the Müller formulation [44]. In principle, the Müller formulation is better conditioned than the PMCHWT formulation since it is a second-kind integral equation (and involves no first kind operators). For problems involving materials with relatively small dielectric contrasts, the Müller formulation has often been found to be more accurate than the PMCHWT formulation. For our model two-region problem, if the relative permittivities of the interior and exterior media are ε_r^- and ε_r^+, respectively, then the Müller formulation is obtained by multiplying the first of the null field equations (2.142) by ε_r^+ and the second by ε_r^- and adding the two equations. Combining the two scalar potential terms under the same integral results in a difference of the two Green's functions such that their dominant singularities cancel, i.e.,

$\lim_{R\to 0} 4\pi(G^+ - G^-) = -j(k^+ - k^-)$. Similarly, we multiply the third equation of (2.142) by μ_r^+ and the fourth by μ_r^- and add the two equations to cancel the dominant singularity in the magnetic scalar potential. The resulting equations may be arranged in the operator matrix form as

$$\begin{bmatrix} \frac{\varepsilon_r^+ + \varepsilon_r^-}{2}\mathcal{I} + \varepsilon_r^+\mathcal{K}^+ - \varepsilon_r^-\mathcal{K}^- & \varepsilon_r^+\eta^+\mathcal{L}^+ - \varepsilon_r^-\eta^-\mathcal{L}^- \\ -\frac{\mu_r^+}{\eta^+}\mathcal{L}^+ + \frac{\mu_r^-}{\eta^-}\mathcal{L}^- & \frac{\mu_r^+ + \mu_r^-}{2}\mathcal{I} + \mu_r^+\mathcal{K}^+ - \mu_r^-\mathcal{K}^- \end{bmatrix} \begin{bmatrix} \mathbf{M} \\ \mathbf{J} \end{bmatrix}$$

$$= \begin{bmatrix} -\hat{\boldsymbol{n}} \times \varepsilon_r^+ \boldsymbol{E}^{\mathrm{i},+} - \hat{\boldsymbol{n}} \times \varepsilon_r^- \boldsymbol{E}^{\mathrm{i},-} \\ \hat{\boldsymbol{n}} \times \mu_r^+ \boldsymbol{H}^{\mathrm{i},+} + \hat{\boldsymbol{n}} \times \mu_r^- \boldsymbol{H}^{\mathrm{i},-} \end{bmatrix}. \quad (2.146)$$

Note that in the Müller formulation, (2.146), the diagonal operator blocks are dominated by material-weighted combinations of identity operators and principal value integrals, whereas the off-diagonal blocks involve strictly non-singular integrals. This is in sharp contrast to the PMCHWT formulation, Equation (2.143), in which material-weighted forms of the more singular $\mathcal{L}$ operators appear on the diagonal, with only weak coupling arising from the off-diagonal operators.

2.6 Surface integral equation challenges

The solution integral solution methods examined earlier are historically among the first and simplest approaches. However, in using them to solve surface integral equations, they all present various challenges. Often the difficulties are embedded within characteristics of the integral operators themselves and often depend on the geometrical or frequency domains over which they are used. In this section, we address some of the more recent analytical and numerical methods used to mitigate many of these problems.

2.6.1 Vector norms, matrix norms, and condition number

Condition numbers are very useful detectors and indicators of problems with ill-conditioning, internal or spurious resonances, and low-frequency breakdown. These are among the most common and significant problems encountered in solving surface integral equations. Condition numbers are expressed in terms of vector and matrix norms, which are defined and described first.

2.6.1.1 Vector norms

Consider the column vector,

$$x = \begin{bmatrix} x_1 \\ x_2 \\ \vdots \\ x_N \end{bmatrix}, \quad (2.147)$$

and define the p-norm of x as

$$||x||_p = \left(\sum_{i=1}^{N} |x_i|^p \right)^{\frac{1}{p}}, \tag{2.148}$$

with the most common choices of p being the following:

1. 1-norm: $||x||_1 = \sum_{i=1}^{N} |x_i|$
2. 2-norm: $||x||_2 = \sqrt{\sum_{i=1}^{N} |x_i|^2}$
3. ∞-norm: $||x||_\infty = \max_i |x_i|$.

The 1-norm is often called the "taxicab" or "Manhattan" norm since, like taxi fares between two points on a map of a city laid out in a north-south rectangular grid, it is computed by summing distance magnitudes over the net horizontal (east–west) and vertical (north–south) distances traveled between two points on the map. This contrasts with the 2-norm or "Euclidean" norm representing the shortest (i.e., "as the crow flies") distance between them. The last norm, the infinity norm, selects the magnitude of the largest component of the vector $x = [x_i]$ as representative of its size. Vector norms have the following properties:

- $||x|| \geq 0$ with equality if and only if $x = 0$
- $||\lambda x|| = |\lambda| ||x||$ (λ real or complex)
- $||x + y|| \leq ||x|| + ||y||$. (triangle inequality)

For the above properties, we assume the same vector norm is used on both sides of the (in)equality. The following inequalities relate to different norms of the same vector x:

- $||x||_\infty \leq ||x||_1 \leq N||x||_\infty$
- $||x||_\infty \leq ||x||_2 \leq \sqrt{N}||x||_\infty$
- $||x||_2 \leq ||x||_1 \leq \sqrt{N}||x||_2$.

2.6.1.2 Matrix norms

The norm of a square matrix

$$A = [A_{ij}],\ i = 1, 2, \ldots, N,\ j = 1, 2, \ldots, N$$

is defined such that for a matrix system $Ax = b$ relating a vector x to a vector b, we can also relate their norms. We require that a matrix norm of A satisfy the following properties:

- $||A|| \geq 0$ with equality if and only if $a_{ij} = 0$ for all i and j
- $||\lambda A|| = |\lambda| ||A||$ (λ real or complex)
- $||A + B|| \leq ||A|| + ||B||$ (triangle inequality)
- $||AB|| \leq ||A|| \, ||B||$.

The following matrix p-norms satisfy these properties,

1. 1-norm: $\|A\|_1 = \max\limits_{1 \leq j \leq N} \sum_{i=1}^{N} |A_{ij}|$ (column-sum norm)
2. 2-norm: $\|A\|_2 = \sigma_{\max}$

3. ∞-norm: $\|A\|_\infty = \max_{1 \leq i \leq N} \sum_{j=1}^{N} |A_{ij}|,$ (row-sum norm)

where $\sigma_{\max}$ is the largest *singular value* of A, i.e., the square root of the largest eigenvalue σ_k^2 of $A^H A x_k = \sigma_k^2 x_k$, $k = 1, \ldots, N$. Some inequalities relating to different norms for the same matrix $A = [a_{ij}]$ are as follows:

- $\frac{1}{\sqrt{N}}||A||_\infty \leq ||A||_2 \leq \sqrt{N}||A||_\infty$
- $\frac{1}{\sqrt{N}}||A||_1 \leq ||A||_2 \leq \sqrt{N}||A||_1$.

2.6.1.3 Matrix condition number

The principal use of the condition number is to roughly indicate relative error in solutions of the matrix equation $Ax = b$ due to relative errors in elements of the matrix and vectors A and b, respectively. That is, if the entries of A and b are actually $A + \delta A$ and $b + \delta b$, where matrix δA and column vector δb represent the respective matrix and column vector element errors, then the perturbed solution, $x + \delta x$, satisfies

$$(A + \delta A)(x + \delta x) = Ax + A\delta x + \delta Ax + \delta A\delta x = b + \delta b.$$

We may use $Ax = b$ to cancel the corresponding terms on both sides of the above equation and, assuming that both error terms δA and δx are small, neglect their product $\delta A\delta x$. Rearranging the result and dividing by $||x||$, we obtain

$$\frac{\delta x}{||x||} = \frac{A^{-1}}{||x||}(\delta b - \delta Ax).$$

Next, noting that $Ax = b$ implies $1/||x|| \leq ||A||/||b||$, and using vector and matrix norm properties, we find the *normed relative error* in the solution, $||\delta x||/||x||$, is bounded as follows:

$$\frac{||\delta x||}{||x||} \leq \frac{||A^{-1}||}{||x||}(||\delta b|| + ||\delta A||||x||) \leq \kappa(A)\left(\frac{||\delta b||}{||b||} + \frac{||\delta A||}{||A||}\right), \qquad (2.149)$$

where $\kappa(A) = \text{cond } A = ||A^{-1}||\,||A||$ is defined as the *condition number* of matrix A. Note the relation (2.149) involves dimensionless relative error terms only, and the normed relative error in the solution is bounded by the product of the condition number and the sum of relative errors of the elements of A and b. We also observe that $\kappa(A) \geq 1$ since $1 = ||I|| \leq ||A^{-1}||\,||A||$, where I is the identity matrix. Hence, in solving $Ax = b$, solution error-bound estimates are always greater than the error in the initial data, i.e., the sum of relative errors in A and b. Another interpretation of the condition number is that, given an equation with data accurate to a certain number of significant digits, $\log_{10} \kappa(A)$ is the upper bound on the number of additional significant digits that may be lost in solving the system equation. Thus, our interest is actually in the error *order*, not its absolute magnitude. This is an important point to keep in mind since relative error bound estimates are often much higher than the actual error; unfortunately, however, sharper bounds cannot be found since there always exists an excitation and solution, that exactly satisfies the bound equality. Finally, we should note that for solution efficiency, condition numbers are in practice often *estimated* during the matrix-solving process rather than computed directly from norm definitions. For these reasons, as a solution error metric, a condition number sometimes must be treated more as a fire alarm than a thermometer.

2.6.2 The EFIE and $\mathscr{L}$ operator

In Section 2.5.3, we formulated the EFIE using the classical approach, and noted that it applies to both open or closed structures, generally exhibiting high accuracy using a straightforward application of Galerkin's method with divergence-conforming RWG basis and testing functions. Its principal difficulties are with internal resonances, low-frequency breakdown, and poor matrix conditioning. We examine each of these in turn.

2.6.2.1 Internal resonances

As discussed in Section 2.1.7, internal resonances are frequencies $f = f_p$ (or wavenumbers $k = k_p$), $p = 1, 2, \cdots, \infty$, such that a closed PEC surface S becomes a resonant cavity whose cavity mode wall currents $\boldsymbol{J}$ on S do not radiate, yet for which no excitation is needed. These currents also support nontrivial modal fields in the cavity interior. Hence, at a resonant frequency with wavenumber k_p, the excitation voltage column vector may be set to zero, with the discretized form of the EFIE taking the form $[Z_{mn}][I_n] = [0]$. The following equivalent conditions must hold for nontrivial solutions to exist for this homogeneous matrix equation:

- A non-vanishing solution of the homogeneous matrix equation $[Z_{mn}][I_n] = [0]$ exists for $k = k_p$.
- The matrix inverse of $[Z_{mn}]$ does not exist.
- The system matrix determinant vanishes, i.e., $\det[Z_{mn}]|_{k=k_p} = 0$.
- The condition number is infinite: $\text{cond}[Z_{mn}]|_{k=k_p} = \infty$ or $1/\text{cond}[Z_{mn}]|_{k=k_p} = 0$.

In practice, none of these conditions can be met exactly for *real* wavenumbers or frequencies since the problem discretization essentially assures that a typical structure is never sufficiently well-discretized nor surface current sufficiently well-modeled or well-sampled that no boundary radiation occurs; rather the zeros of $\det[Z_{mn}]$ typically contain a small imaginary component representing the radiation loss that inevitably occurs due to such errors. Hence, each of the four conditions is merely *approached* as the actual resonant wavenumber is approached, i.e., $k \to k_p$. Monitoring the condition number (or its reciprocal) over a range of wavenumbers is usually the most cost-effective and practical means of identifying internal resonances. Usually, very sharp, steep changes (peaks or minima) of several orders of magnitude identify these resonant frequencies. For the EFIE, essentially the only approach for dealing with internal resonances is to change the problem formulation, i.e., to use the combined field integral equation (CFIE), (2.137), which does not allow homogeneous solutions for real frequencies.

Above we assumed that one's interest is in solving exterior problems; instead one may want to use the EFIE to determine a cavity's internal resonance frequencies. For this, one could, in principle, use a root-finder to locate minima of the system determinant or reciprocal condition number. Using either of these particular indicators can be problematic, however. System determinants, for example, often vary over enormous numerical ranges, frequently under- or over-flowing with relatively small changes in wavenumbers; reciprocal condition numbers, on the other hand (especially when *estimated*), are often not sufficiently smoothly varying with wavenumber to work well with typical numerical root-finders. Hence, the usual difficulty in using integral equations to determine resonant frequencies is deciding on a smooth, well-scaled resonance indicator. In any case, however, some improvement in modeling resonators is often gained by replacing the free-space Green's function with the non-radiating Green's function $G(\boldsymbol{r}, \boldsymbol{r}') = \cos(kR)/(4\pi R)$ which guarantees that (a) all roots will be real and (b) the homogeneous integral equation can be trivially modified to eliminate the need for complex arithmetic. For the interior problem, this simple replacement of the Green's function is possible since the replacement form still satisfies (2.37), but is no longer required to be an outgoing solution satisfying (2.53).

2.6.2.2 Low-frequency breakdown

For either open or closed PEC surfaces S, and for sufficiently low frequencies, the condition number begins to increase monotonically with decreasing frequency, suggesting that the EFIE possibly has a homogeneous solution at $k = 0$. In fact, it generally has *many* homogeneous solutions, all unrelated to any internal resonances that may exist. Furthermore, these homogeneous solutions increasingly contaminate and cause errors in the expected surface current solutions as frequency decreases. This sort of current error can also begin to occur with decreasing wavenumber within any *subregion* of S such that $k\delta << 1$, where δ is an appropriate bound on local cell edge lengths in the subregion. Cells in the subregion essentially experience a local quasi-electrostatic environment that allows them to become increasingly contaminated with localized, erroneous current loops as frequency decreases. To better understand this, it is useful to explicitly derive and exhibit homogeneous electric current solutions of the EFIE for $k = 0$; this task is undertaken next. We first note that as $\omega \to 0$, the vector potential term of (2.90) decreases while the scalar potential term increases. This implies that below a certain frequency—typically when the vector potential term and the absolute error in the computed scalar potential term are of similar order—then the EFIE reduces approximately to

$$\left[-\frac{1}{j\omega\varepsilon}\nabla\int_S G(\boldsymbol{r}, \boldsymbol{r}')\nabla'\cdot\boldsymbol{J}(\boldsymbol{r}')\,dS'\right]_{\tan} \approx \boldsymbol{E}^{\mathrm{i}}_{\tan}, \;\boldsymbol{r}\in S. \tag{2.150}$$

Clearly, any *divergenceless* surface current density $\boldsymbol{J}$ on S constitutes a homogeneous solution of (2.150). The discretized version of (2.150), on specializing (2.103)–(2.108) by simply dropping the vector potential contributions (i.e., inductance matrix elements (2.106)), takes the form

$$\frac{1}{j\omega\varepsilon}\sum_{n=1}^{N} I_n \int_S\int_S \nabla\cdot\boldsymbol{\Lambda}_m(\boldsymbol{r})G(\boldsymbol{r}, \boldsymbol{r}')\nabla'\cdot\boldsymbol{\Lambda}_n(\boldsymbol{r}')\,dS'dS \approx \int_S \boldsymbol{\Lambda}_m(\boldsymbol{r})\cdot\boldsymbol{E}^{\mathrm{i}}(\boldsymbol{r})\,dS,$$
$$m = 1, 2, \cdots N. \tag{2.151}$$

Since flux lines of a divergenceless current close on themselves, we attempt to construct the smallest such (*elemental*) loops on S using $\boldsymbol{\Lambda}_n$, where for simplicity, we assume S is an open, simply-connected surface. Such elemental loops (representing a basis for modeling divergenceless fields on S) obviously may be expressed as a linear combination of RWG bases representing a constant net flux circulating about an interior vertex v of S, as illustrated in Figure 2.20. Thus, we define an elemental loop basis, $\boldsymbol{\mathcal{O}}_v(\boldsymbol{r})$, circulating about vertex v, $v = 1, 2, \cdots, V$, as

$$\boldsymbol{\mathcal{O}}_v(\boldsymbol{r}) = \sum_{p=1}^{P_v} \sigma_{n(p,v)} \boldsymbol{\Lambda}_{n(p,v)}(\boldsymbol{r}) / \ell_{n(p,v)}, \quad v = 1, 2, \cdots V, \tag{2.152}$$

where V is the number of vertices interior to S. As shown in Figure 2.20, the index p, $p = 1, 2, \cdots, P_v$, numbers the (interior) edges attached to vertex v in counterclockwise fashion. In (2.152), $n(p, v)$ is the global DoF (interior edge) index for the pth edge attached to vertex v, $\ell_{n(p,v)}$ is its edge length, and $\sigma_{n(p,v)} = -1$ if the reference direction for $\boldsymbol{\Lambda}_{n(p,v)}$ is directed clockwise about vertex v'; it is $+1$ otherwise. Note that each RWG basis appearing in (2.152) is divided by its associated edge length so that it carries exactly one unit of total flux across each edge; hence, no flux accumulates within the triangles whose associated edges share vertex v, i.e., $\nabla \cdot \boldsymbol{\mathcal{O}}_v = 0$. Each of the loops $\boldsymbol{\mathcal{O}}_v$ is an independent homogeneous solution of the static ($k = 0$) form of the EFIE; alternatively, on tracing the development of (2.151) each is also found to be a test function that annihilates the gradient of the scalar potential operator—regardless of the frequency. The latter result also follows from performing the test integral by adding contributions from each of the closed flux line paths of $\boldsymbol{\mathcal{O}}_v$ (Figure 2.20) and invoking the conservative nature of $\nabla\Phi$ on each closed path.

The ability of linear combinations of the RWG bases to exactly model homogeneous solutions of the EFIE when $\omega = 0$ is of such importance that it is often stated as an essential requirement for div-conforming bases. A basic requirement of the physics of electromagnetism is that in the static limit, $\omega \to 0$, electric and magnet

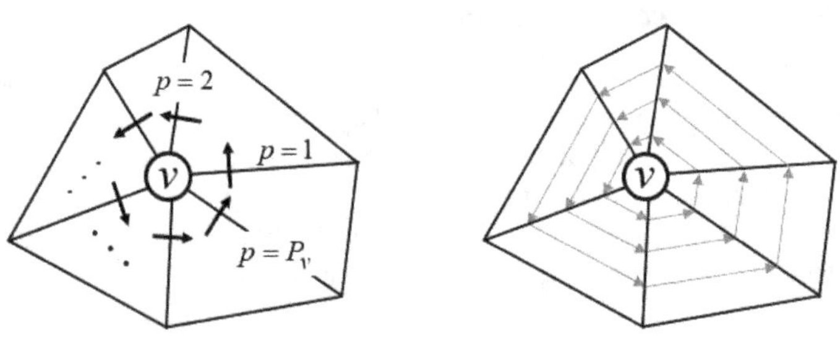

Figure 2.20 *Vertex v is an interior vertex of S. Index p numbers counterclockwise the edges attached to vertex v. Basis function* $\sigma_{n(p,v)}\boldsymbol{\Lambda}_{n(p,v)}/\ell_{n(p,v)}$ *is an RWG basis associated with global edge n corresponding to the pth edge of vertex v; sign factor* $\sigma_{n(p,v)} = \pm 1$ *corrects the sign of the basis if it is not already directed counterclockwise, while* $\ell_{n(p,v)}$ *is the corresponding edge length and normalizes the RWG to carry unit flux across the edge. Arrows in the figure at left show the edge reference directions to form a divergenceless loop of RWG bases circulating about (interior) vertex v. Flux lines of the resulting divergenceless basis function,* $\boldsymbol{\mathcal{O}}_v(\boldsymbol{r})$*, are shown at right.*

fields completely decouple, with the surface electric current $\boldsymbol{J}$ exciting only static magnetic fields; $\boldsymbol{J}$ also decouples from the electric surface charge, q, which becomes the sole source of the static electric field. All the V-independent divergenceless current loops together represent V zeros of the system determinant, det $[Z_{mn}]$, at $\omega = 0$. Without this zero-divergence or divergence *null space* modeling property, these zeros would necessarily be shifted away from $\omega = 0$, and instead appear at real or complex ω-plane locations as spurious determinant zeros—allowing the non-physical spurious homogeneous solutions to pollute solutions for frequencies near these zeros in the complex frequency plane. Such unanticipated appearances of these nonphysical spurious resonances at non-zero frequencies are a much more serious problem than dealing with them when they are properly confined to $\omega = 0$ by use of div-conforming bases.

2.6.2.3 Poor matrix conditioning

At frequencies well away from resonances and for mesh edges of moderate electrical lengths, the surface currents found from the EFIE are generally quite accurate even when condition numbers are relatively high. However, high condition numbers are not conducive to solving the matrix system via iterative methods, since the number of iterations needed for convergence also increases with condition number. Either another formulation is needed, or some form of matrix preconditioner must often be used. In recent years, the Calderon identities have been used to successfully precondition the EFIE, though with more effort expended on the details of discretization [22,45]. The scheme of [22] also takes advantage of the favorable properties of Buffa–Christiansen (BC) bases, which have also found use in improving several of the classic formulations. In the next section, which discusses the properties of the MFIE, a brief introduction to the BC bases will be given.

2.6.3 The MFIE and $\mathcal{K}$ operator

The classical MFIE discussed in Section 2.5.5 is a second kind integral equation that applies to closed surfaces S with PEC boundary conditions. The second kind integral equations have characteristically low condition numbers, except near resonance frequencies of the interior cavity formed by S. Thus they may usually be solved by iterative methods. However, the MFIE has typically been found to be less accurate than the classical EFIE, especially at low frequencies [46,47]. Recently it has been found, however, that the MFIE can be improved in both conditioning and accuracy by employing appropriate testing schemes [48–52], with perhaps the most significant improvements resulting from the appropriate use of BC bases in their curl-conforming form [50,51]. Since their introduction, and despite their increased complexity, BC bases have been found useful in various ways for improving most of the classic formulations, hence, it seems worthwhile to briefly introduce and discuss their properties. Since they make explicit use of a dual mesh, we first define the mesh dual to the usual triangular mesh.

2.6.3.1 Mesh dual to a triangular mesh

In Section 2.15, we introduced the concept of a mesh dual to a *rectangular* mesh; here we extend the idea to a triangular mesh. Figure 2.21 represents several adjacent

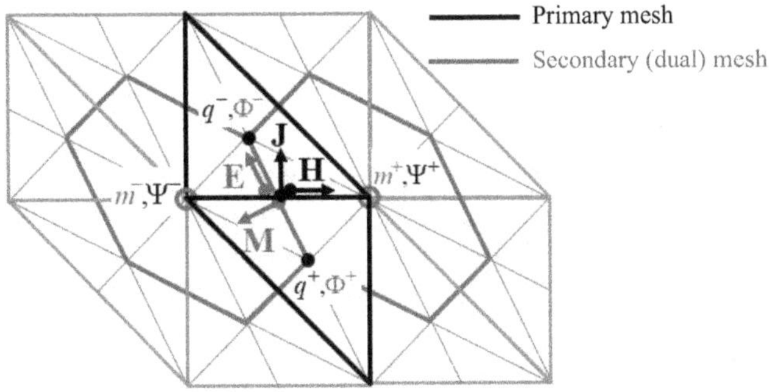

Figure 2.21 Planar representation of a local triangle group from S mapped onto a regular triangular primary mesh on surface S. The dual (secondary) mesh is formed by connecting each triangle's centroid to its vertices and to the mid-points of its edges. A reference triangle pair (in black) belonging to the primary mesh supports an RWG basis $\mathbf{\Lambda}_n$ associated with the dual reference edge, also indexed as n. The heavy gray lines highlight the support of the dual basis function $\mathbf{\Omega}_n^{BC}$; it consists of the pair of adjacent polygonal domains with a common edge as shown that may also labeled as (dual) reference edge n. Electric and magnetic currents, $\boldsymbol{J}$ and $\boldsymbol{M}$, are defined such that their fluxes cross intersecting edges of the primary and secondary meshes, respectively, depositing electric and magnetic surface charges of density q and m which we associate with centroids of the corresponding mesh cells. The associated potentials ($\boldsymbol{A}$,$\boldsymbol{F}$) and the fields ($\boldsymbol{E}$,$\boldsymbol{H}$) arising from them reside on the mesh dual to that of their sources ($\boldsymbol{J}$,$\boldsymbol{M}$), respectively.

triangles extracted from a surface triangulation of S and mapped onto the plane in a regular grid to simplify visualization. A *reference pair* of adjacent triangles on the primary (original triangular) grid is shown in black. The pair provides the support for an RWG basis $\mathbf{\Lambda}_n$ associated with the common or *reference* edge for the nth DOF. The reference pair is merely repeated in medium-width gray to represent nearby triangles of the primary grid. The thin gray lines in the figure represent the *barycentric subdivision* of the primary grid created by connecting the centroid of each triangle to the mid-points of its edges (black or medium gray lines) and to its vertices. The barycentric subdivision divides each triangle of the original mesh into six subtriangles. The support of the reference BC basis consists of all subtriangles that share a common vertex with one of the vertices of the common edge of the reference triangle. Figure 2.21 outlines in heavy gray lines the boundary of this support region, which consists of two polygons connected along a common edge that crosses the common edge of the reference triangle—or simply the reference edge. Note that each vertex of the primary (triangular) mesh is associated with a centroid of the dual (polygonal) mesh and vice versa. Furthermore, each edge of the secondary mesh crosses an edge of the primary mesh and vice versa; it is precisely in these two senses that the two meshes are considered dual.

2.6.3.2 BC bases

To provide an explicit example of a Buffa–Christiansen (BC) basis function, we construct a BC function on an irregular mesh assumed to be on a closed surface S. (The case for open surfaces, where a vertex or edge of the primary mesh may be on a surface boundary, is treated in [22].) Hence, the construction can be generalized to treat MFIE problems.

Figure 2.22 shows a section of a mesh representing a closed surface S. The primary mesh is made up of triangles, with an RWG basis presumed defined at every triangle edge of the mesh. One edge, labeled "reference" edge for discussion, is presumed to have an associated RWG basis, $\mathbf{\Lambda}_{\text{ref}}^{\text{RWG}}$ defined on it, and we construct the corresponding BC basis $\mathbf{\Omega}_{\text{ref}}^{\text{BC}}$ associated with the reference edge using the labels and coefficients listed in the figure. Starting just above the rightmost cut edge and proceeding in the reference direction, the next edge is edge "a" with weight coefficient $1/10$; thus, its contribution to the BC basis is $\frac{1}{10}\mathbf{\Lambda}_a^{\text{RWG}}/\ell_a$, where ℓ_a is the edge length. We proceed similarly, but display for discussion below only the next five local RWG basis contributions to $\mathbf{\Omega}_{\text{ref}}^{\text{BC}}$, i.e.,

$$\mathbf{\Omega}_{\text{ref}}^{\text{BC}} = \tfrac{1}{10}\mathbf{\Lambda}_a^{\text{RWG}}/\ell_a + \tfrac{2}{10}\mathbf{\Lambda}_b^{\text{RWG}}/\ell_b + \tfrac{3}{10}\mathbf{\Lambda}_c^{\text{RWG}}/\ell_c + \tfrac{4}{10}\mathbf{\Lambda}_d^{\text{RWG}}/\ell_d + \tfrac{1}{2}\mathbf{\Lambda}_e^{\text{RWG}}/\ell_e + \cdots,$$

where ℓ_α is the local barycentric edge length for $\alpha = a, b, \cdots, t$. Note the coefficients of the local RWG bases increase uniformly from zero at the cut to $\frac{1}{2}$ along the 5 edges a, b, c, d, e between the cut and the barycentric mesh edge e, with the endpoint at the reference edge midpoint. Continuing in the positive direction around the barycentric mesh edges, we encounter edges e, f, g, h, i, j whose coefficients *decrease* uniformly from $\frac{1}{2}$ to 0 over six edges. Crossing the cut, the coefficients become negative and continue decreasing uniformly to $-\frac{1}{2}$ at edge p. Recall that negative fluxes simply indicate the associated fluxes are opposite to the positive orientation indicated by the dashed gray arrows (i.e., counter-clockwise with respect to the local surface normal). Continuing, we find the flux coefficients begin increasing from $-\frac{1}{2}$ at edge p to $-\frac{1}{10}$ just below the rightmost branch edge. Collecting these results, the single BC basis is written as a superposition of local RWG bases defined on the barycentric mesh as

$$\begin{aligned}\mathbf{\Omega}_{\text{ref}}^{\text{BC}} = {} & \tfrac{1}{10}\mathbf{\Lambda}_a^{\text{RWG}}/\ell_a + \tfrac{2}{10}\mathbf{\Lambda}_b^{\text{RWG}}/\ell_b + \tfrac{3}{10}\mathbf{\Lambda}_c^{\text{RWG}}/\ell_c + \tfrac{4}{10}\mathbf{\Lambda}_d^{\text{RWG}}/\ell_d + \tfrac{1}{2}\mathbf{\Lambda}_e^{\text{RWG}}/\ell_e \\ & + \tfrac{5}{12}\mathbf{\Lambda}_f^{\text{RWG}}/\ell_f + \tfrac{4}{12}\mathbf{\Lambda}_g^{\text{RWG}}/\ell_g + \tfrac{3}{12}\mathbf{\Lambda}_h^{\text{RWG}}/\ell_h + \tfrac{2}{12}\mathbf{\Lambda}_i^{\text{RWG}}/\ell_i + \tfrac{1}{12}\mathbf{\Lambda}_j^{\text{RWG}}/\ell_j \\ & - \tfrac{1}{12}\mathbf{\Lambda}_k^{\text{RWG}}/\ell_k - \tfrac{2}{12}\mathbf{\Lambda}_l^{\text{RWG}}/\ell_l - \tfrac{3}{12}\mathbf{\Lambda}_m^{\text{RWG}}/\ell_m - \tfrac{4}{12}\mathbf{\Lambda}_n^{\text{RWG}}/\ell_n - \tfrac{5}{12}\mathbf{\Lambda}_o^{\text{RWG}}/\ell_o \\ & - \tfrac{1}{2}\mathbf{\Lambda}_p^{\text{RWG}}/\ell_p - \tfrac{4}{10}\mathbf{\Lambda}_q^{\text{RWG}}/\ell_q - \tfrac{3}{10}\mathbf{\Lambda}_r^{\text{RWG}}/\ell_r - \tfrac{2}{10}\mathbf{\Lambda}_s^{\text{RWG}}/\ell_s - \tfrac{1}{10}\mathbf{\Lambda}_t^{\text{RWG}}/\ell_t.\end{aligned} \tag{2.153}$$

A similar construction applies for each edge of the primary mesh on S. We summarize by noting that $\mathbf{\Omega}_n$ carries a unit flux across the dual mesh edge that crosses reference edge n of the primary mesh; the two adjacent polygons sharing that edge each carries

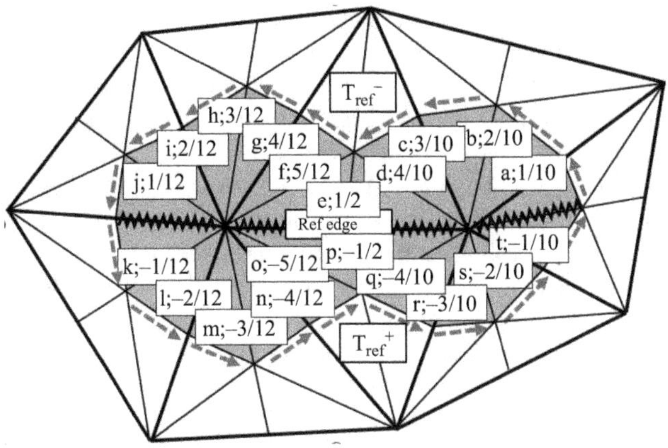

Figure 2.22 A Buffa–Christiansen (BC) basis is constructed as a linear combination of RWG bases on the barycentric *refinement shown in the figure for a closed surface S. The heavy lines identify a primary triangular mesh section. The reference edge identified in the figure is the common edge for an RWG basis whose support is the two reference triangles labeled* T_{ref}^{+} *and* T_{ref}^{-}. *A barycentric subdivision of the primary mesh is formed by connecting the midpoint of each triangle edge to its opposite vertex (light lines), and subdivides each primary mesh triangle into six subtriangles. The set of subtriangles with a vertex in common with one of the vertices of the reference edge together form the support (in gray) of the BC basis associated with the reference edge. (We assume that neither vertex of the reference edge is on a boundary of S.) The zigzag lines highlight the reference edge and extend past its end vertices. They represent surface "cuts" dividing the support into an equal number of subtriangles above and below the cut; no flux is allowed to cross a cut line. Also shown for each edge is (1) its label,* $\alpha = a, b, c, \cdots, t$, *and (2) a coefficient representing the local RWG flux crossing the common edge of an adjacent subtriangle pair. A positive flux indicates the flow direction is in the positive reference direction indicated by the dashed gray arrows, i.e., counter-clockwise around the support boundary of the BC basis with respect to* $\hat{\boldsymbol{n}}$, *the local normal to S. A negative coefficient indicates the local flux is directed opposite the reference direction.*

an equal but opposite-signed total charge, just as do adjacent triangles for RWG bases. Unlike RWG bases, however, it is the total charge—not the charge density—that is constant within each subtriangle. Thus, BC bases have similar interpolatory properties for tangential components at the edges of the dual mesh as RWG bases have for normal components on the primary mesh. Furthermore, by arranging three of them about the edges of a primary mesh triangle, they form an elemental loop such that the total charge in each polygon vanishes. Hence, like the RWG bases, they are able to model divergenceless fields.

We have now defined at each primary mesh edge two div-conforming bases, one (RWG) with an edge-normal DoF, and the other (BC) with an edge-tangential DoF. But by taking their cross-products with $\hat{\boldsymbol{n}}$, two more useful bases can also be associated with the edge. Though $\hat{\boldsymbol{n}} \times \boldsymbol{\Lambda}_n$ and $\boldsymbol{\Omega}_n \times \hat{\boldsymbol{n}}$ are no longer div-conformable, they are, however, *curl-conformable*, i.e., they have continuous tangential components at edge n. Such bases are both suitable and extensively used, e.g., in solving the vector Helmholtz equation where electric or magnetic fields typically should have continuous *tangential* fields at triangle or volumetric element boundaries. The four bases are depicted and described in the caption of Figure 2.23.

2.6.3.3 Improved MFIE testing

The logic for the classical MFIE formulation of Section 2.5.5 was that since the electric current $\boldsymbol{J}$ appeared outside the integral operator and, as a flux quantity, was expanded in terms of div-conforming RWG bases, $\boldsymbol{\Lambda}$, testing the MFIE with bases from the same set would be appropriate since all interactions vanish for element pairs with no vertices or edges in common, and for self-terms become integrals over the squared magnitudes of the vector basis, thus seemingly well-approximating the identity matrix of the MFIE operator. But this logic leads to a "variational crime" [53] that ignores our observations of Section 2.4 that reaction integrals between vector quantities should involve dot products between intensity (curl-conforming) and flux (div-conforming) quantities. To right this wrong, we first note that since the equivalent current $\boldsymbol{J}$ in the MFIE is a flux quantity, the choice of div-conforming bases for the surface current is entirely appropriate. The scattered magnetic field it produces, however, is an intensity, i.e., a curl-conforming quantity. But the MFIE actually involves the *rotated* magnetic field, $\hat{\boldsymbol{n}} \times \boldsymbol{H}$, which, therefore, should have continuous *normal* components and therefore be div-conforming—not curl-conforming. The idea that the operator $\hat{\boldsymbol{n}}\times$ converts an intensity quantity to a flux quantity may be reinforced by noting that as the surface S is approached, the term $\frac{1}{2}\boldsymbol{J}$ appears alone outside the integral—and is also a flux (div-conforming) quantity. Hence, it might seem appropriate to test the MFIE with a curl-conforming quantity, such as $\hat{\boldsymbol{n}} \times \boldsymbol{\Lambda}$. This choice, however, does not result in a well-tested integral equation since the self-term vanishes identically, which poorly-models the identity operator. But the *rotated* BC basis, $\boldsymbol{\Lambda}_{\mathrm{BC}} = \boldsymbol{\Omega}_{\mathrm{BC}} \times \hat{\boldsymbol{n}}$ of Figure 2.23, is both curl-conforming and (being quasi-div-conforming) well-aligned vectorially with $\boldsymbol{\Lambda}$ for testing. Furthermore the gradient operator appearing in the principal value integral may be transferred onto the curl-conforming test function, considerably improving accuracy in evaluating the integral [48]. These seemingly minor changes in the testing approach (though perhaps requiring significantly more effort to implement) are found to greatly improve both the accuracy and low frequency behavior of the MFIE [50,51]. Other variational misdemeanors have also been committed with the goal of at least partially improving the original deficiencies of the MFIE but with less complexity than the described approach [49,52].

It is instructive at this point to reflect on why the original EFIE scheme, though typically more ill-conditioned than the MFIE, at least maintains good accuracy for

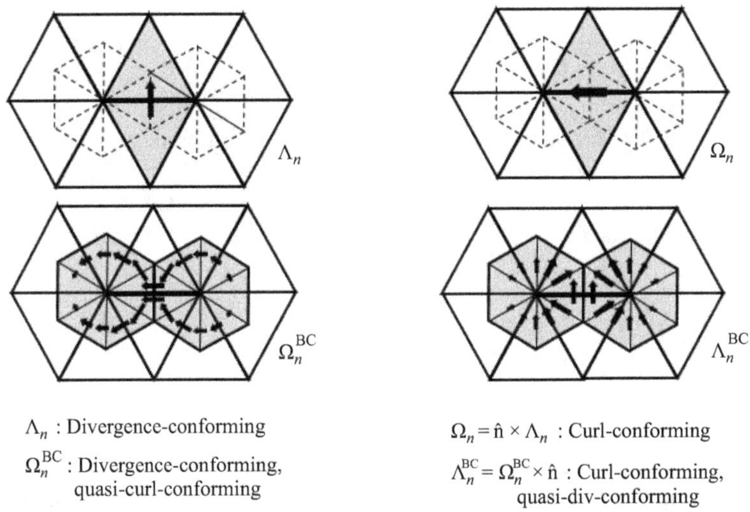

Figure 2.23 The figure shows a primary triangular mesh (black lines) and the dual mesh *(dashed lines) associated with it. The top row of figures depicts both an RWG basis and the basis* $\hat{\boldsymbol{n}} \times$ RWG *derived from it; they have adjacent triangle pairs as their support (in gray).* $\boldsymbol{\Lambda}_n$ *is div-conforming (i.e., with continuous normal components) with DoF defined normal to the common edge of the pair, while* $\boldsymbol{\Omega}_n = \hat{\boldsymbol{n}} \times \boldsymbol{\Lambda}_n$, *is curl-conforming (i.e., with continuous tangential components) with DoF tangential to the common edge. The bottom row depicts the two bases that can be formed from the BC bases. Both have as their support the two adjacent polygons (in gray) with a common edge. The BC basis may be entirely synthesized as a linear superposition of RWG bases defined on the* subtriangles *of the dual mesh, hence, the BC basis* $\boldsymbol{\Omega}_n^{BC}$ *is necessarily div-conforming, but its DoF is a single unit of flux (*not *flux density) tangential to the reference edge about which it is defined; hence it may also be called quasi-curl-conforming. Alternatively, we may think of the flux as flow through (but not necessarily normal to) the dual mesh edge(s) common to the polygonal pair. (Note that for less regular meshes, the common edge of the polygons may be split into* two *edges, with neither necessarily perpendicular to the reference edge of the triangle pair. This is easily visualized by sliding the top or bottom three vertices to the left or right.) Finally, the rotated BC basis* $\boldsymbol{\Lambda}_n^{BC} = \boldsymbol{\Omega}_n^{BC} \times \hat{\boldsymbol{n}}$ *is curl-conforming since it is a superposition of the curl-conforming bases* $\boldsymbol{\Omega}_n = \boldsymbol{\Lambda}_n \times \hat{\boldsymbol{n}}$ *but, since its DoF is defined to be normal to its reference edge, it is also said to be quasi-div-conforming.*

PECs of moderate electrical size. Here the EFIE transforms a div-conforming current (needed because of the divergence operator in the scalar potential term), represented in terms of RWG bases, into an electric field. The electric field is associated with the dual mesh, however; it is defined by components parallel to dual mesh edges. Hence, the electric field is curl-conforming on the dual mesh. Well-formed reaction integrals involving $\boldsymbol{E}$ should also involve an electric test current $\boldsymbol{J}^t$ on the primary mesh and be div-conforming; fortuitously, then, for the EFIE, the same RWG bases can, therefore, serve as test functions.

2.6.4 Mixed operator integral equations

In mixed operator integral equations, a linear combination of the $\mathscr{L}$ and $\mathscr{K}$ operator types appears within the same equation. This form includes the CFIE for conducting scatterers and the various dielectric formulations such as the PMCHWT or Müller integral equations. The principal difficulty with these formulations is that the two operator terms, for example, map the same current flux vector to *different* spaces in the same equation. This dissimilarity in the operators often leads to complexities in testing each operator-valued term appearing in the integral equation. Various approaches have been explored, including those of [52,54]. Formulations using Calderon preconditioning for both the CFIE and the PMCHWT have also been developed [23,55].

References

[1] Schelkunoff SA. On teaching the undergraduate electromagnetic theory. *IEEE Transactions on Education*. 1972;15(1):15–25.

[2] Harrington RF. *Time-Harmonic Electromagnetic Fields*. 2nd ed. Piscataway, NJ: Wiley-IEEE Press; 2001.

[3] Marsden JE and Tromba AJ. *Vector Calculus*. New York, NY: W. H. Freeman; 1981.

[4] Van Bladel JG. *Electromagnetic Fields*. 2nd ed. Piscataway, NJ: Wiley-IEEE Press; 2007.

[5] Sommerfeld A. Die Greensche Funktion der Schwingunsgleichung. *Jahresbericht der Deutschen Mathematiker-Vereinigung*. 1912;21:309–353.

[6] Sommerfeld A. *Partial Differential Equations in Physics*. New York, NY: Academic Press; 1949.

[7] Schot SH. Eighty years of Sommerfeld's radiation condition. *History of Math*. 1992;19:385–401.

[8] Silver S. *Microwave Antenna Theory and Design*. New York, NY: McGraw-Hill; 1949.

[9] Müller C. Randwertprobleme der Theorie Elektromagnetischer Schwingungen. *Math Z*. 1952;56(3):261–270.

[10] Jin JM. *Theory and Computation of Electromagnetic Fields*. Piscataway, NJ: Wiley-IEEE Press; 2010.

[11] Balanis CA. *Advanced Engineering Electromagnetics*. 2nd ed. Piscataway, NJ: Wiley; 2012.

[12] Rellich F. Über das asymptotische Verhalten der Lösungen von $\Delta u + \Lambda u = 0$ in unendlichen Gebieten. *Jahresbericht der Deutschen Mathematiker-Vereinigung*. 1943;53(6):57–65.

[13] Wilcox CH. Wave operators and asymptotic solutions of wave propagation problems of classical physics. *Archive for Rational Mechanics and Analysis*. 1956;22(1):37–76.

[14] Jones DS. *Acoustic and Electromagnetic Waves*. New York, NY: Oxford; 1986.

[15] Mayes PE. The equivalence of electric and magnetic sources. *IRE Transactions on Antennas and Propagation*. 1958;6:295–296.

[16] Van Bladel J. *Singular Electromagnetic Fields and Sources*. New York, NY: Oxford-IEEE Press; 1995.

[17] Schelkunoff SA. Some equivalence theorems of electromagnetics and their application to radiation problems. *Bell Labs Technical Journal*. 1936;15: 92–112.

[18] Schelkunoff SA. Kirchhoff's formula, its vector analogue, and other field equivalence theorems. *Communications on Pure and Applied Mathematics*. 1951;4:43–59.

[19] Love AEH. The integration of equations of propagation of electric waves. *Philosophical Transactions of the Royal Society A: London*. 1901;Ser. A(197): 1–45.

[20] Rengarajan SR and Rahmat-Samii Y. The field equivalence principle: illustration of the establishment of the non-intuitive null fields. *IEEE Antennas and Propagation Magazine*. 2000;42(4):122–128.

[21] Hsiao GC and Kleinman RE. Mathematical foundations for error estimation in numerical solutions of integral equations in electromagnetics. *IEEE Transactions on Antennas and Propagation*. 1997;45(3):316–328.

[22] Andriulli FP, Cools K, Bağci H, *et al.* A multiplicative Calderón preconditioner for the electric field integral equation. *IEEE Transactions on Antennas and Propagation*. 2008;56:2398–2412.

[23] Bağci H, Andriulli FP, Cools K, *et al.* A Calderón multiplicative preconditioner for the combined field integral equation. *IEEE Transactions on Antennas and Propagation*. 2009;57:3387–3392.

[24] Wilton DR. Computational methods. In Pike ER and Sabatier PC (eds), *Scattering: Scattering and Inverse Scattering in Pure and Applied Science*, Chapter 1.5.5, pp. 316–365. Amsterdam: Academic Press; 2002.

[25] Zhang L, Cui T, and Liu H. A set of symmetric quadrature rules on triangles and tetrahedra. *Journal of Computational Mathematics*. 2009;27(1):89–96.

[26] Yee KS. Numerical solution of initial boundary value problems involving Maxwell's equations in isotropic media. *IEEE Transactions on Antennas and Propagation*. 1966;14(3):302–307.

[27] Rao S, Wilton D, and Glisson A. Electromagnetic scattering by surfaces of arbitrary shape. *IEEE Transactions on Antennas and Propagation*. 1982;30(3):409–418.

[28] Dwight HB. *Tables of Integrals and Other Mathematical Data*. New York, NY: MacMillan; 1957.

[29] Botha M. A family of augmented Duffy transformations for near-singularity cancellation quadrature. *IEEE Transactions on Antennas and Propagation*. 2013;(61):3123–3134.

[30] Khayat M, Wilton D, and Fink P. An improved transformation and optimized sampling scheme for the numerical evaluation of singular and near-singular potentials. *IEEE Antennas and Wireless Propagation Letters*. 2008;7:377–380.

[31] Botha M and Rylander T. Analysis and estimation of quadrature errors in weakly singular source integrals of the method of moments. *International Journal of Numerical Modelling*. 2018;(31):e2269.

[32] Chang RR, Wang Z, and Xie Q. Fast convergent quadrature method for evaluating the RWG- and SWG-related convolutional integrals. *IEEE Transactions on Antennas and Propagation*. 2021;69:8583–8592.

[33] Graglia RD, Wilton DR, and Peterson AF. Higher order interpolatory vector bases for computational electromagnetics. *IEEE Transactions on Antennas and Propagation*. 1997;45:329–342.

[34] Rivero J, Vipiana F, Wilton D, *et al*. Hybrid integration scheme for the evaluation of strongly singular and near-singular integrals in surface integral equations. *IEEE Transactions on Antennas and Propagation*. 2019; 67:6532–6540.

[35] Graglia R. On the numerical integration of the linear shape functions times the 3-D Green's function or its gradient on a plane triangle. *IEEE Transactions on Antennas and Propagation*. 1993;41:1448–1455.

[36] Rivero J, Vipiana F, Wilton DR, *et al*. Acceleration of the surface test integral using vertex functions. In: *IEEE Symposium on Antennas and Propagation and USNC-URSI Radio Science Meeting*. Singapore: IEEE; 2021.

[37] Ufimtsev PY. New insight into the classical Macdonald physical optics approximation. *IEEE Antennas and Propagation Magazine*. 2008;50(3): 11–20.

[38] Mautz JR and Harrington RF. *Electromagnetic Scattering from a Homogeneous Body of Revolution*. Syracuse, NY: Syracuse University; 1977. TR-77-10.

[39] Mautz JR and Harrington RF. H-field, E-field, and combined-field solutions for conducting bodies of revolution. *Archiv für Elektronik und Übertragungstechnik*. 1978;32(4):157–164.

[40] Khayat MA. *Numerical Modeling of Thin Materials in Electromagnetic Scattering Problems*, Ph.D. Dissertation University of Houston, Houston, TX; 2003.

[41] Poggio AJ and Miller EK. *Integral Equation Solutions of Three-Dimensional Scattering Problems*. Oxford: Pergamon; 1973. p. 159–264.

[42] Chang Y and Harrington RF. A surface formulation for characteristic modes of material bodies. *IEEE Transactions on Antennas and Propagation*. 1977;25:789–795.

[43] Wu TK and Tsai LL. Scattering from arbitrarily-shaped lossy dielectric bodies of revolution. *Radio Science*. 1977;12(5):709–718.

[44] Müller C. *Foundations of the Mathematical Theory of Electromagnetic Waves*. Berlin: Springer-Verlag; 1969.

[45] Adams R. Physical and analytical properties of a stabilized electric field integral equation. *IEEE Transactions on Antennas and Propagation*. 2004;52: 362–372.

[46] Zhang Y, Cui TJ, Chew WC, *et al.* Magnetic field integral equation at very low frequencies. *IEEE Transactions on Antennas and Propagation*. 2003;51(8):1864–1871.

[47] Ergül Ö and Gürel L. Improving the accuracy of the magnetic field integral equation with the linear-linear basis functions. *Radio Science*. 2006;41(4): 1–15.

[48] Ylä-Oijala P and Taskinen M. Calculation of CFIE impedance matrix elements with RWG and nxRWG functions. *IEEE Transactions on Antennas and Propagation*. 2003;51(8):1837–1846.

[49] Ergül O and Gürel L. The use of curl-conforming basis functions for the magnetic-field integral equation. *IEEE Transactions on Antennas and Propagation*. 2006;54(7):1917–1926.

[50] Cools K, Andriulli FP, De Zutter D, *et al.* Accurate and conforming mixed discretization of the MFIE. *IEEE Antennas Wireless Propagat Letters*. 2011;10:528–531.

[51] Yan S, Jin JM, and Nie Z. Accuracy improvement of the second-kind integral equations for generally shaped objects. *IEEE Transactions on Antennas and Propagation*. 2013;61:788–788.

[52] Kornprobst J and Eibert TF. An accurate low-order discretization scheme for the identity operator in the magnetic field and combined field integral equations. *IEEE Transactions on Antennas and Propagation*. 2018;66(11):6146–6157.

[53] Strang G. *Variational Crimes in the Finite Element Method*. New York, NY: Academic Press; 1972.

[54] Ylä-Oijala P, Taskinen M, and Järvenpää S. Surface integral equation formulations for solving electromagnetic scattering problems with iterative methods. *Radio Science*. 2005;40(6):RS6002.

[55] Cools K, Andriulli FP, and Michielssen E. A Calderon multiplicative preconditioner for the PMCHWT integral equation. *IEEE Transactions on Antennas and Propagation*. 2011;59:4579–4587.

Chapter 3

Kernel-based fast factorization techniques

Özgür Ergül[1], Bahram Khalichi[2] and Vakur B. Ertürk[2]

3.1 Introduction

In computational electromagnetics, fast factorization, compression, expansion, and diagonalization techniques are developed to analyze challenging problems, including those with multi-scale characteristics, discretized with large numbers of unknowns [1]. Such techniques are developed for differential-equation and integral-equation solvers that operate in both frequency and time domains, to analyze problems with controllable accuracy, as opposed to low-frequency or high-frequency asymptotic techniques. Considering the state-of-the-art techniques and implementations in the literature, we may categorize such fast solvers into two groups, namely, kernel-based and kernel-independent techniques, depending on how they formulate and solve discretized electromagnetic problems (matrix equations). In a kernel-based technique, the underlying kernel of the problem is the main focus to be considered, and the strategy is to modify its usage to handle electromagnetic interactions in some more efficient forms, without sacrificing the accuracy of numerical results. The kernel, which is usually Green's function that depends on the type of the problem, typically requires one-to-one interactions between discretization elements such that the aim of the developed and used kernel-based technique becomes re-representation of such interactions in alternative ways. From this point of view, kernel-based methods provide optimal ways to numerically compute kernel operations, at the cost of reduced freedom in terms of applicability and generalizability in comparison to kernel-independent techniques.

Among a variety of kernel-based fast factorization techniques, accelerated Cartesian expansion (ACE) method [2], adaptive integral method (AIM) [3], and similar methods based on the fast Fourier transform (FFT) [4–6], as well as the fast multipole method (FMM) [7] and the multilevel fast multipole algorithm (MLFMA) [8–10] are well-known in the literature. As one of the prominent methods, MLFMA has been implemented by numerous researchers as noncommercial and commercial software, has been employed to analyze complex structures with diverse characteristics (including large-scale platforms with huge numbers of unknowns), and is still a topic

[1]Department of Electrical and Electronics Engineering, Middle East Technical University, Turkey
[2]Department of Electrical and Electronics Engineering, Bilkent University, Turkey

of active research on electromagnetic simulations. This chapter focuses on this important algorithm, from its conventional forms to broadband versions that are developed for accurate, efficient, and stable solutions to multi-scale problems. We consider its applications to large-scale problems, to structures with alternative materials properties, and finally, to complex geometries with dense and nonuniform discretizations. As it is impossible to cover all aspects of MLFMA and its alternative versions in a single chapter, we limit our scope to frequency-domain solutions (even though MLFMA is also used in the time domain) and certain types of problems/properties/approaches (see next section for the full description of the context) that can still be representative to demonstrate the superior functionality of this algorithm for a wide range of electromagnetic problems, including those with multi-scale characteristics.

3.2 Multilevel fast multipole algorithm

Before discussing MLFMA, we may describe the context and boundaries of this chapter. Although MLFMA is applicable to layered media [11], our focus will be a set of problems involving three-dimensional structures located in infinitely large homogeneous spaces, i.e., when the kernel to be considered is the homogeneous-space Green's function. We also limit our discussions to homogeneous or piecewise homogeneous objects that can be formulated via surface-integral-equation formulations [12], such as the electric-field integral equation (EFIE), magnetic-field integral equation (MFIE), combined-field integral equation (CFIE) [13], potential integral equations (PIEs) [14–17], or their hybridizations [18–20] for perfect electric conductors (PECs), in addition to various *suitable* formulations for penetrable (dielectric, magnetic, plasmonic, single-negative, double-negative, near-zero-index) materials [21–29]. Surfaces are assumed to be discretized via triangles, on which the discretization (basis and testing) functions are defined. In all results, vector and scalar unknowns (e.g., equivalent currents and charges) are expanded in terms of the Rao–Wilton–Glisson (RWG) [30] and pulse functions, respectively*. The core methodology is the method of moments (MoM) [35], where electromagnetic interactions are defined between discretization elements based on radiation integrals (employing the homogeneous-space Green's function) and numerical problems are constructed via boundary conditions (tested by testing functions).

Using triangles for discretization, multi-scale problems typically involve dense triangulations, either uniformly (on an entire structure) or non-uniformly (on some portion of the given structure). The conventional MLFMA described below cannot handle such cases efficiently and/or accurately; hence, low-frequency or broadband

*Therefore, the focus of this chapter is MLFMA itself, while electromagnetic problems are assumed to be formulated via proper surface-integral-equation formulations for accurate, efficient, and stable solutions. In addition, we restrict our discussion with the conventional discretizations (RWG and pulse functions), without considering improved discretizations for zero-frequency or dense-discretization problems, e.g., multiresolution [31], Buffa–Christiansen [32],[33], or linear–linear [34] functions for EFIE, MFIE, CFIE, etc., which are extensively discussed in the literature.

MLFMA implementations are needed. These implementations can be described better in reference to the conventional MLFMA; therefore, we first consider the conventional MLFMA in detail to construct a base and to describe the general *concept of MLFMA* that is common in all versions. We emphasize that structures with regular discretizations (e.g., involving the rule-of-thumb $\lambda/10$ triangles, where λ is the wavelength) are assumed to be analyzed via the conventional MLFMA, i.e., when this version operates properly without needing low-frequency treatments and other approaches to be introduced later.

3.2.1 Conventional MLFMA based on plane waves

Before considering MLFMA versions that can be more suitable for dense and nonuniform discretizations involved in multi-scale simulations, we first consider the conventional MLFMA, particularly at a formulation level to establish a basic understanding of this important algorithm. By *conventional,* we imply typical tree structures constructed by regularly dividing discretized objects into subdomains, as well as computations of electromagnetic interactions between these subdomains via plane-wave diagonalization. Although diagonalization by plane waves brings many constraints in terms of electrical dimensions of subdomains and other interaction mechanisms, they are commonly used in most of the developed implementations of MLFMA. We also assume that the conventional MLFMA is applied to structures with regular discretizations (not dense or nonuniform), e.g., those with $\lambda/10$ triangles, as other versions of MLFMA are needed to tackle more challenging dense (uniform or nonuniform) discretizations.

Consider a three-dimensional structure involving single or multiple disconnected surfaces discretized by using triangles. Application of MoM leads to a matrix equation in the form of $[Z] \cdot a = b$, even though this equation is never constructed explicitly in MLFMA. In this equation, $[Z]$ represents electromagnetic interactions between discretization elements (impedance matrix), b represents tested incident fields (excitation vector), and a is the vector of unknown coefficients to be found. In MLFMA, the aim is to efficiently perform matrix-vector multiplications $y = [Z] \cdot x$ within an iterative solver, such as the biconjugate gradient stabilized (BiCGStab) [36] method or the generalized minimal residual (GMRES) [37] method, such that solution a is reached after a number of iterations (after a number of matrix–vector multiplications). The idea is that multiplication $[Z] \cdot x$ can be performed without constructing $[Z] \cdot a = b$, i.e., without fully computing and storing the impedance matrix $[Z]$.

MLFMA is a kernel-based method since it involves factorization and diagonalization of the formulation kernel, which is the homogeneous-space Green's function in the context of this chapter. A typical factorization is Gegenbauer's addition theorem [1,7,10] (monopole-to-monopole translation), i.e.,

$$g(r, r') = \frac{\exp(ik|r - r'|)}{4\pi|r - r'|} \tag{3.1}$$

$$= \frac{ik}{4\pi} \sum_{t=0}^{\infty} (-1)^t (2t + 1) j_t(kd) h_t(kD) P_t(\hat{d} \cdot \hat{D}), \tag{3.2}$$

where $k = \omega\sqrt{\mu}\sqrt{\varepsilon}$ is the wavenumber of the medium, $D = \hat{D}D$ and $d = \hat{d}d$ are arbitrary vectors satisfying $D > d$, and $D + d = r - r'$ is the vector from the source point (r') to the observation point (r). In this factorization, Green's function is written in terms of the spherical Bessel function of the first kind (j_t), the spherical Hankel function of the first kind (h_t), and Legendre polynomial (P_t). The diagonalization, as the next step, is based on the expansion of spherical waves in terms of plane waves [10], i.e.,

$$j_t(kd) = \frac{(-i)^t}{4\pi}\int d^2\hat{k}\exp(ik\cdot d)P_t(\hat{k}\cdot\hat{d}) \tag{3.3}$$

$$j_t(kd)P_t(\hat{d}\cdot\hat{D}) = \frac{(-i)^t}{4\pi}\int d^2\hat{k}\exp(ik\cdot d)P_t(\hat{k}\cdot\hat{D}) \tag{3.4}$$

where the angular integration ($d^2\hat{k} = \sin\theta d\theta d\phi$) indicates that infinitely many plane waves are required to exactly expand a spherical wave. Inserting (3.4) into (3.2), we arrive at

$$g(r,r') = \frac{ik}{(4\pi)^2}\sum_{t=0}^{\infty} i^t(2t+1)h_t(kD)\int d^2\hat{k}\exp(ik\cdot d)P_t(\hat{k}\cdot\hat{D}). \tag{3.5}$$

Then, changing the order of the summation and integration, we obtain the diagonalized form of Green's function as

$$g(r,r') = \frac{ik}{(4\pi)^2}\int d^2\hat{k}\beta(k,d)\alpha(k,D), \tag{3.6}$$

where $\beta(k,r) = \exp(ik\cdot r)$ is the diagonal (plane-wave-to-plane-wave) shift function and

$$\alpha(k,r) = \sum_{t=0}^{\infty} i^t(2t+1)h_t(kr)P_t(\hat{k}\cdot\hat{r}) \tag{3.7}$$

is the diagonal (plane-wave-to-plane-wave) translation function.

Obviously, the diagonal form in (3.6) is more expensive (in fact, involving infinitely many operations if no truncation is used) than the direct calculation of $g(r,r')$. However, the arbitrariness of D and d makes it suitable to perform interactions (computations of Green's function) in a group-by-group manner. On the other hand, an important issue regarding (3.6) is that this expression is not low-frequency-stable, i.e., it encounters numerical instabilities (in *finite* precision) when $|r - r'| = |D - d| \ll \lambda = 2\pi/k$. This, so-called low-frequency breakdown, can be explained from many perspectives, and it is impossible to cover all in this short text. From a mathematical point of view [7], changing the order of the summation and integration, i.e., the step from (3.5) to (3.6), is the source of the breakdown. This change leads to the isolation of the exponential part [shift function $\beta(k,d)$] from the rest, i.e., translation function $\alpha(k,D)$, which is problematic since computations of $\alpha(k,D)$ involve addition and subtraction of large numbers when $|D| \ll \lambda$. Hence, the summation in (3.7) does not converge into the correct value, in fact it diverges after some point as more terms are added. In the same context, $\beta(k,d)$ is almost unity for $|d| \ll \lambda$, indicating that plane waves are not distinguished among themselves, addressing a *resolution* issue. All these are obviously consistent with a physical explanation that plane waves are indeed not suitable for expanding short-distance interactions (Green's function with small arguments).

In practice, independent of the low-frequency breakdown, the summation in (3.7) must be truncated as

$$\alpha(k,r) \approx \sum_{t=0}^{\tau} i^t(2t+1)h_t(kr)P_t(\hat{k}\cdot\hat{r}), \tag{3.8}$$

where τ is called the truncation number. Similarly, the angular integration in (3.6) must be performed by using a finite number of samples on the unit sphere. As a common (and systematically consistent) choice, $S^\theta = (\tau + 1)$ and $S^\phi = 2(\tau + 1)$ samples are selected along the θ and ϕ axes [1]. While ϕ samples are usually regular, θ samples are selected as Gauss–Legendre quadrature points to improve the integration accuracy. In such a setup, the accuracy of the whole diagonalization is controlled by a single parameter, i.e., the truncation number τ, while the relationship between τ and the accuracy is not straightforward [38]. Increasing τ means adding more terms in the translation operator and using more samples for the angular integration, while the accuracy is improved in a limited region, until the low-frequency breakdown. Specifically, as a typical behavior, the accuracy improves up to a certain value of τ, and then it deteriorates with a divergent behavior. There is an extensive literature on the selection of truncation numbers (e.g., see [39]), leading to alternative strategies, such as the excess bandwidth formula [40] that may provide estimations on the relationship between the truncation number, interaction distance, and maximum encountered error. It is remarkable that inaccuracies caused by the low-frequency breakdown depend on not only $|r - r'| = |D - d|$ but also D and d individually, which make such analyses be based on worst-case scenarios for given implementation strategies. For example, using a one-box-buffer scheme (see below), box (group) sizes cannot be smaller than $\lambda/4$ if the Green's function needs to be computed with maximum of 1–2% error[†]. This means that such an implementation of the conventional MLFMA cannot be used for objects smaller than λ.

At this stage, we can turn our attention to the implementation of the conventional MLFMA using (3.6). The key property of (3.6) that makes it suitable for group-by-group interactions in a *multilevel* scheme is the divisible characteristics of the shift function, i.e., $\beta(k,r) = \beta(k,r_1)\beta(k,r_2)$ for any $r = r_1 + r_2$. Hence, a tree structure is constructed to compute electromagnetic interactions between groups of discretization elements, where a discretization element interacts with other discretization elements via various levels depending on the distances between them. To organize all these operations, the structure to be analyzed is placed in a computational box, which is divided into sub-boxes, and their sub-boxes, etc., down to a certain level (clustering), which is usually determined by the constraints due to the low-frequency breakdown. In the conventional MLFMA, cubic boxes are preferred, whereas box centers are

[†]In fact, worst-case analyses show that $\lambda/2$ boxes are required to reach such error levels [39].

considered as critical locations where radiated and incoming fields are collected. In addition, divisions of boxes into sub-boxes are performed regularly, i.e., if a box is divided into eight sub-boxes, other boxes of the same size (boxes at the same level) are also divided into sub-boxes. This way, levels are formed from the largest box (level L) that encloses the object to the smallest boxes (level 1). An example demonstrating a simple *clustering* for the *Flamme* geometry is depicted in Figure 3.1. We note that empty boxes that do not include any part of the structure are simply discarded. Although it is an implementation issue, it should be emphasized that a clustering algorithm (simply dividing the structure into boxes at different levels) with linear time and memory complexity can be a challenging task [10].

Given a clustered object, considering the relationship between boxes and their sub-boxes, we obtain a tree structure with L levels labeled as $l = 1, 2, \ldots, L$ from the bottom to the top. For a structure of almost equal size in all three dimensions, the number of nonempty boxes at level l can be written as $N_l = N_{l-1}/4$ for $l = 2, 3, \ldots, L$, i.e., it is reduced by fourfold from a level to the next higher level[‡]. This means, $N_l \approx 4^{1-l} N_1$ with $N_1 = \mathcal{O}(N)$, where N is the number of unknowns (e.g., the number discretization elements). Using cubic boxes, for which the box size is multiplied by two from a level to the next higher level, we also have the number of samples required for radiated and incoming waves as $S_l = S_l^\theta S_l^\phi = 2(\tau_l + 1)^2 \approx 4S_{l-1}$ for $l = 2, 3, \ldots, L$ since τ_l is proportional to the box size (*in terms of the wavelength*) for a fixed level of accuracy. Then, we further have $S_l = 4^{l-1} S_1$ with $S_1 = \mathcal{O}(1)$, i.e., the lowest level boxes are selected simply in the order of wavelength, independent of the structure size.

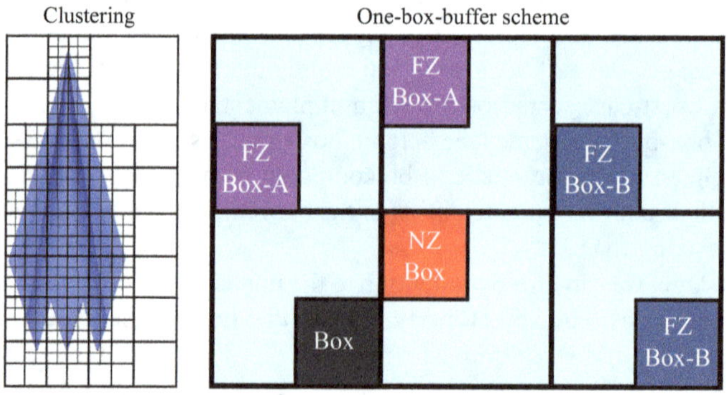

Figure 3.1 Clustering of the Flamme geometry, where the object is divided into sub-domains (sub-boxes) of different sizes (at different levels). The one-box-buffer scheme is illustrated on the right-hand side, where nonempty boxes are categorized with respect to a given box (labeled as "Box").

[‡]Obviously, for a structure elongated in one direction, we may have $N_l = N_{l-1}/2$, or some different multipliers for other shapes, while this does not change the discussion here.

This means that the computational load per level is proportional to $N_l S_l \approx 4^{1-l} N_1 4^{l-1} S_1 = N_1 S_1 = \mathcal{O}(N)$, leading to $\mathcal{O}(N \log N)$ overall complexity, since $L = \mathcal{O}(\log N)$ for a regular discretization. Consequently, each level of MLFMA is equally important in terms of computational load, which must be considered carefully for its efficient parallelization.

To organize electromagnetic interactions, near-zone and far-zone boxes are defined based on a selected scheme. As a common (and probably the most efficient) strategy, the one-box-buffer scheme categorizes boxes depending on whether they touch each other or not[§] [40]. Specifically, given a box at level l, other boxes *at the same level* are defined as either near-zone or far-zone, indicating how these boxes interact within MLFMA. Near-zone boxes are those that directly touch the box, by sharing a face, an edge, or a corner. Hence, for a given box, there are maximum 27 near-zone boxes, including the box itself. The rest of the boxes are defined to be in the far zone, while MLFMA interactions are performed not with all of them. In general, as depicted in Figure 3.1, far-zone boxes can be categorized as Type-A and Type-B, and MLFMA interactions (specifically translations) are performed only with Type-A boxes, as electromagnetic interactions regarding Type-B boxes are carried out at a higher level. Specifically, MLFMA interactions are performed between boxes that are in the far zone of each other, if and only if their *parents* are in the near-zone of each other. Hence, for a given box C at level $l = 1, 2, \ldots, L-2$, if near-zone and far-zone boxes are defined as $\mathrm{NZ}(C)$ and $\mathrm{FZ}(C)$, respectively (by categorizing boxes at the same level), *far-zone interaction list* can be defined as $\mathrm{FZI}(C) = \{C' : C' \in \mathrm{FZ}(C) \text{ and } P(C') \in \mathrm{NZ}(P(C))\}$, where $P(C)$ represents the parent of C.

Based on the one-box-buffer scheme (or any other scheme used to organize interactions), a matrix–vector multiplication $y = [Z] \cdot x$ performed in an MLFMA implementation can be written as

$$y = [Z]^{\mathrm{NZ}} \cdot x + [Z]^{\mathrm{FZ}} \cdot x, \tag{3.9}$$

where $[Z]^{\mathrm{NZ}}$ and $[Z]^{\mathrm{FZ}}$ represent near-zone and far-zone interactions, respectively, in terms of discretization elements. Interactions between discretization elements that are located in two near-zone boxes *at the lowest level* are included in $[Z]^{\mathrm{NZ}}$; they must be calculated directly and stored in memory to be used multiple times during iterations. Computations of all other interactions (corresponding to $[Z]^{\mathrm{FZ}} \cdot x$) are performed on-the-fly via MLFMA stages, namely, aggregation (from bottom to top), translation (within levels), and disaggregation (from top to bottom) [10]. We now consider these stages with their concise formulations.

[§]We note that the one-box-buffer scheme automatically satisfies the condition for Gegenbauer's addition theorem, i.e., $D > d$ in (3.2). However, it creates challenging observation/source points in terms of the low-frequency breakdown of MLFMA.

- In the aggregation stage, radiated fields are computed at box centers from bottom ($l = 1$) to top ($l = L - 2$). At the lowest level, these fields are obtained by adding the radiation patterns of the basis functions (weighted by coefficients in x) as

$$F_C^{\mathrm{rad}}(k, r_C) = \sum_{n \in C} x[n] F_n^{\mathrm{rad}}(k, r_C) \qquad (l = 1), \tag{3.10}$$

where r_C represents the center of box C. In this equation, F_n^{rad} represents the radiation pattern of the nth basis function ($n = 1, 2, \ldots, N$), $x[n]$ is the corresponding current coefficient (provided by the iterative solver), and F_C^{rad} represents the radiated field created by all sources (radiating basis functions) in C. At the higher levels, we have

$$F_C^{\mathrm{rad}}(k, r_C) = \sum_{C' \in C} \beta(k, r_C - r_{C'}) F_{C'}^{\mathrm{rad}}(k, r_{C'}) \\ (l = 2, 3, \ldots, L - 2), \tag{3.11}$$

i.e., the radiated field at the center of a box C is the summation of the radiated fields of its sub-boxes (after shifting due to the relocation of radiation centers). As radiated fields are sampled (on the unit sphere), aggregation (as well as translation and disaggregation) operations are performed in discrete forms. Since the sampling rate increases from a level to the next higher level, the sampled radiated field of a box is up-sampled (interpolated) before being used in (3.11). Such an interpolation can be performed in diverse ways, e.g., using the Lagrange interpolation method [41,42].
- Once the aggregation stage is completed, translations are performed to convert radiated fields into incoming fields between far-zone boxes. For a box C at any level l, this can be written as

$$F_C^{\mathrm{inc}}(k, r_C) = \sum_{C' \in \mathrm{FZI}(C)} \alpha(k, r_C - r_{C'}) F_{C'}^{\mathrm{rad}}(k, r_{C'}) \\ (l = 1, 2, \ldots, L - 2), \tag{3.12}$$

where FZI(C) represents the far-zone interaction list (of C), as defined earlier. Using the one-box-buffer scheme and cubic (identical) boxes, there can be maximum $7^3 - 3^3 = 316$ different translation vectors ($r_C - r_{C'}$) per level [43]. Therefore, the translation operator in (3.12) can be computed efficiently during the setup of MLFMA and stored in a very compact form [44].
- Following the translation stage, the tree structure is traced from top to bottom to compute the total incoming fields for all boxes. This can be written as

$$F_C^{\mathrm{inc},+}(k, r_C) = \beta(k, r_C - r_{P(C)}) F_{P(C)}^{\mathrm{inc},+}(k, r_{P(C)}) \\ + F_C^{\mathrm{inc}}(k, r_C) \qquad (l = L - 3, L - 2, \ldots, 1), \tag{3.13}$$

which indicates that the total incoming field for a box C is the sum of incoming fields due to translations (F_C^{inc}) and a shifted version of the total incoming field for its parent $P(C)$. As the incoming field for a parent box is sampled with a higher rate than that of its child box, down-sampling is needed before the summation in (3.13). This can be done via transpose interpolation (anterpolation) [45] that maintains the accuracy of high sampling (as opposed to the accuracy that can be obtained after a down-sampling interpolation). Finally, at the lowest level, incoming fields are received by the testing functions to complete the related part of the matrix–vector multiplication as

$$\{[Z]^{\mathrm{FZ}} \cdot x\}[m] \propto \int d^2\hat{k} F_m^{\mathrm{rec}}(k, r_C) \cdot F_C^{\mathrm{inc},+}(k, r_C) \qquad (m \in C) \tag{3.14}$$

for $m = 1, 2, \ldots, N$, where F_m^{rec} represents the receiving pattern of the mth testing function.

An aggregation–translation–disaggregation cycle is performed for a given (fixed) medium, which can be a host medium or an internal medium if a penetrable structure is considered. Computations regarding different media of a problem, even for the same set of coefficients, cannot be combined into a single cycle, as all parameters (truncation numbers and sampling rates) depend on the wavelength. Depending on the problem, different formulations based on alternative combinations of integro-differential operators may be used. For such different operators, MLFMA stages detailed above are performed exactly the same way, while the type of the operator and discretization elements used for them change only radiation and receiving patterns defined in (3.10) and (3.14). Consequently, in a combined-type formulation, where two or more operators are combined and applied on the same source, it may be possible to perform a single aggregation–translation–disaggregation cycle for a combination of operators (if they are related to the same medium) by properly defining radiation/receiving patterns.

3.2.2 Low-frequency and broadband MLFMA implementations

In multi-scale problems, dense triangulations are needed either locally (non-uniformly) or generally (uniformly) for accurate solutions. Such discretizations may be required to properly model fine geometric details, to accurately expand equivalent currents at critical locations, or for both. In any case, these problems lead to numerical issues when they are attempted to be solved via the conventional MLFMA, as the smallest box size that can be used is limited. Keeping the box size within the required limits (e.g., larger than $\lambda/4$ for 1–2% maximum error) leads to highly populated boxes (due to small triangles) and inefficient solutions. In fact, for such solutions, the complexity of MLFMA (a matrix–vector multiplication) is typically higher than the linearithmic ($\mathcal{O}(N \log N)$) level. Unsurprisingly, the history of computational electromagnetics has seen enormous efforts in developing low-frequency and broadband MLFMA implementations that can provide accurate, efficient, and stable solutions for densely discretized structures [46–80].

While it is impossible to list all low-frequency or broadband MLFMA applications, we may describe some common properties of different approaches to mitigate the low-frequency breakdown of the conventional diagonalization.

3.2.2.1 Multipole-based methods

Although its name suggests, the conventional MLFMA does not use multipoles explicitly due to the diagonalization via plane waves. Hence, if the plane-wave expansion is avoided and the *original* multipole-to-multipole interactions are employed to factorize Green's function, we reach multipole-based methods that are stable at arbitrarily low frequencies [10],[46–51]. Since multipoles cannot be used efficiently for large boxes, they can be transformed into plane waves at the higher levels of tree structures, leading to broadband implementations [52]. Such low-frequency and broadband solvers can have linearithmic complexity, while *the constant in the front* is typically very large due to inefficiencies in multipole-to-multipole shift and translation operations, whose accelerations may not be trivial [53]. Consequently, multipole-based methods are generally known to provide excellent accuracy and stability at the cost of efficiency.

3.2.2.2 Methods based on inhomogeneous plane waves

Another set of low-frequency-stable MLFMA versions are obtained by replacing (ordinary) plane waves with other types of waves, particularly inhomogeneous plane waves [46,54–57]. In these implementations, such as the low-frequency fast inhomogeneous plane-wave algorithm (LF-FIPWA), stability and accuracy are maintained by including evanescent waves‖ in computations of short-distance interactions. This is possible, particularly by considering spectral representations of Green's function to derive various expansion forms that are stable at arbitrarily low frequencies. These schemes can also be combined with a multipole-based MLFMA or the conventional MLFMA (based on conventional plane waves) to obtain broadband solvers [58–63]. However, low-frequency implementations based on inhomogeneous plane waves typically involve direction-dependent (expressions of) translations, which significantly increase the computational load, on top of implementation difficulties and challenges.

3.2.2.3 Non-directional methods using complex-domain shifts

Inhomogeneous-plane-wave-based methods, in fact, involve diagonalized expansions, where the angular integration is *deformed* in a way that plane waves in the conventional diagonalization turn into inhomogeneous versions. Implementations involving similar shifts of the integration path into a complex plane, but without introducing any directional dependency, can also be found in the literature [59,64]. A remarkable method, called the non-directive stable plane wave multilevel fast multipole algorithm (NSPWMLFMA), that is based on QR compression of translation matrices provides accurate, efficient, and stable computations for arbitrary box sizes, enabling efficient broadband simulations [65–68].

3.2.2.4 Approximate methods

A type of broadband MLFMA, called the uniform MLFMA (UMLFMA), is based on complex-domain shifts with a non-directional property [59,64], while its accuracy is known the be limited due to the numerical nature of the resulting translation operators.

‖FFT is often involved in these implementations to handle evanescent waves.

These types of sacrifices in accuracy to obtain efficient and useful (and often relatively straightforward) broadband MLFMA implementations are also common in the literature [69,70]. One example is MLFMA with approximate diagonalization (AD-MLFMA) [71]–[73], where Gegenbauer's addition theorem is used in a different way to scale spherical functions, and then, to approximate the resulting shift functions as scaled plane waves. AD-MLFMA provides broadband simulations by enabling stable computations of interactions between arbitrarily small boxes and by automatically becoming the conventional MLFMA for large boxes, while its accuracy is limited (but satisfactory for most practical applications).

3.2.2.5 Precision-based methods

From one point of view, the low-frequency breakdown of MLFMA is due to the use of finite precision arithmetic; hence, it can be mitigated without changing the original expansion in terms of plane waves, if the precision is sufficiently increased. These types of *brute-force* approaches are successfully developed and can be found in the literature [74]; but the feasibility of such low-frequency-stable and broadband implementations depends on the availability of a mixed-precision environment on the computer platform, as well as the efficiency of how mixed-precision operations are carried out. In most cases, such capabilities are provided by a third-party software, which may create bottlenecks in mixed-precision MLFMA implementations. Nevertheless, the use of mixed precision adds another dimension on the accuracy–efficiency trade-off of MLFMA [75], which may need further investigation, depending on future developments in hardware technologies.

3.2.2.6 Combinations with other methods

MLFMA can also be combined or integrated with other methods that may result in rigorous broadband implementations. One popular track is based on using favorable properties of ACE to stabilize MLFMA at low frequencies, leading to broadband implementations when the resulting algorithm is combined with the conventional MLFMA [76]–[78].

3.2.2.7 Modified tree structures for nonuniform discretizations

The low-frequency-stable expansion methods described earlier are not sufficient to efficiently analyze structures involving nonuniform discretizations. The conventional MLFMA, as well as most low-frequency and broadband MLFMA implementations involve regular tree structures based on simultaneous divisions of boxes into sub-boxes at all levels. Obviously, this is not practical for a nonuniform discretization that involves a large variety of elements with different sizes. Regular divisions based on the smallest elements lead to many complications, depending on the implementation, as larger discretization elements may not fit into lower-level boxes. Keeping the number of levels small (to properly locate larger discretization elements) leads to over-populated boxes and many near-zone interactions that can even increase the computational complexity. A commonsense approach to tackle nonuniform discretizations is *adaptively* dividing boxes into sub-boxes, where needed, to keep box populations under control. On the other hand, for such an irregular tree structure, definitions of interactions may not be straightforward, and robust algorithms are needed for efficient simulations [79]. Recently, incomplete-leaf (IL) tree structures were proposed [68,73,80], leading to broadband IL-MLFMA implementations, to efficiently and accurately solve multi-scale problems involving highly nonuniform discretizations.

3.3 Large-scale simulations and parallel computing

The first two decades of the twenty-first century saw a remarkable competition between different research groups all over the world on large-scale electromagnetic simulations [43,81–98] (see Figure 3.2 for sample matrix dimensions). The competition was mainly constructed on efficient parallelization of (the conventional) MLFMA by employing this superior algorithm on multiple processors and cores (processes), which can be significantly challenging. An efficient parallelization technique, so-called the hybrid strategy, enabled solutions to canonical problems involving more

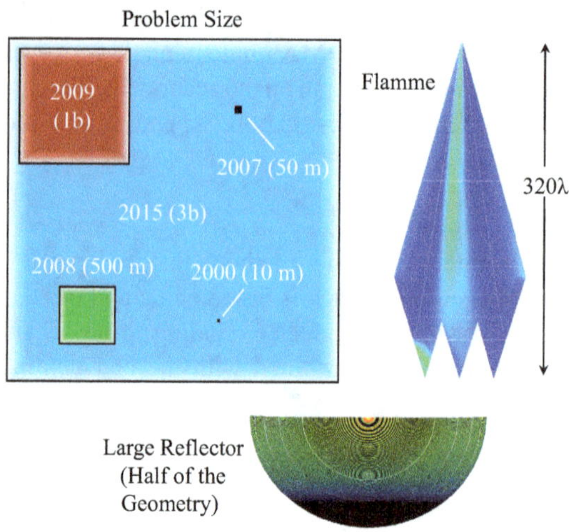

Processing Time (Sphere)

Unknowns	Total time (min)
1,462,854	4
5,851,416	16
23,405,664	61
33,791,232	107
53,112,384	183
93,622,656	333
135,164,928	471
204,823,296	647
307,531,008	1080
374,490,624	1430
540,659,712	1816

Figure 3.2 *A comparison of some matrix dimensions solved by using parallel implementations of MLFMA. The boxes are correctly scaled to demonstrate the extraordinary growth of the solvable matrix sizes in 15 years. Solutions to large-scale problems have challenging natures at all steps, including post-processing stages; even the visualization of surface currents on the Flamme geometry or on an antenna reflector at a high frequency can be a complicated task. On the right-hand side, processing times are listed when perfectly conducting spheres with increasing electrical sizes are analyzed on 128 cores of Intel Nehalem-L7555 (1.87 GHz) processors (on a computer of 16 nodes, each with 8 cores).*

than 10 million unknowns [43]. This strategy is based on partitioning a tree structure constructed in MLFMA in two different and coexisting forms, i.e., distributing groups and field samples at lower and higher levels, respectively, among processes. This way, the parallelization efficiency can significantly be increased in comparison to those obtained by simple partitioning strategies that distribute only groups or field samples. In 2007, the hierarchical strategy was introduced to further improve the parallelization of MLFMA, increasing the solvable problem size to 50 million and beyond [83]. Considering the complexity of MLFMA tree structures, the hierarchical strategy employs an adaptive partitioning of groups and field samples, attempting a balance considering their highly varying numbers at different levels. Although the parallelization itself becomes more complicated, e.g., due to different types of communications between processes during aggregation, translation, and disaggregation stages, the hierarchical strategy was able to provide higher than 70% parallelization efficiency on more than 100 processes for the first time in the literature. Consequently, the problem size was increased rapidly to more than 500 million in a few years [88] (see Figure 3.2). All these advances in algorithms occurred hand-in-hand with quickly developing computer technologies, naturally leading to diversity in techniques to adapt available resources [84,90]. In 2010, a parallel FMM-FFT was used on a massively parallel computer to solve problems with more than 1 billion unknowns [91]. Following this milestone, although different research groups presented a variety of large-scale simulations [92–96], involving huge objects in terms of wavelength, complex geometries, and structures with unique properties, a remarkable peak in terms of the number of unknowns was reached in 2015. Problems discretized with *billions* of unknowns were solved using a three-dimensional hierarchical strategy [97], which is considered to be an optimal way of partitioning MLFMA trees on the state-of-the-art computers involving multi-core CPUs¶.

As a few examples of large-scale simulations, Figure 3.3 presents solutions to scattering problems involving three different geometries, i.e., sphere, *NASA Almond*, and *Flamme*, all of which are modeled as perfect electric conductors. Each object is located in a vacuum and illuminated by plane waves: the sphere (0.3 m radius) at 340 GHz, the *NASA Almond* (approximately 0.2524 m length) at 1.8 THz, and the *Flamme* (0.6 m length) at 820 GHz. Hence, the maximum dimensions of the geometries correspond to $680\lambda_0$, $1514\lambda_0$, and $1640\lambda_0$, respectively, where λ_0 is the wavelength in vacuum. All three problems are formulated via CFIE, and each is discretized with more than 550 million unknowns. Solutions are performed by using MLFMA parallelized via the hierarchical partitioning strategy [88,93,96]. The implementation is employed on 64 cores of Intel Nehalem-L7555 processors (1.87 GHz), where the cores are located in 16 nodes connected via Infiniband. Using 11-level MLFMA ($L = 11$), the sphere problem is solved in 44 h (34 BiCGStab iterations to reach 0.001 residual error) using 1.9 TB peak memory. Figure 3.3 presents the normalized bistatic radar cross-section (RCS) on the E-plane, where 0° and 180° correspond to forward-scattering and back-scattering directions, respectively. Numerical

¶MLFMA implementations are generally less suitable for GPUs and there are few studies that demonstrate successful parallelization of MLFMA on such platforms [99].

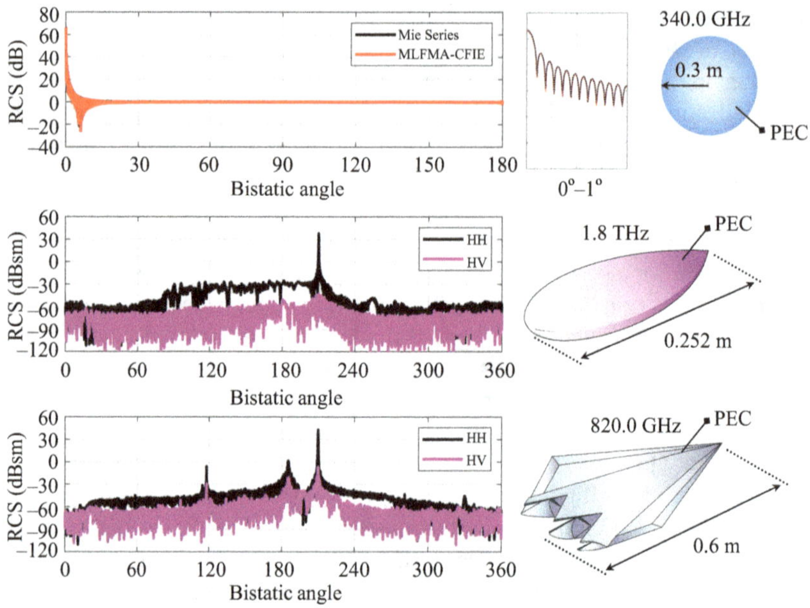

Figure 3.3 Three examples of solutions to large-scale problems involving perfectly conducting objects

results obtained with MLFMA are perfectly consistent with the analytical Mie-series solution, which is observed more clearly in the focused plot at around the forward-scattering direction, i.e., from 0° to 1°. On the other side, with 70 BiCGStab iterations to reach 0.001 residual error, the *NASA Almond* problem is solved in 57 h using 13-level MLFMA. In the bistatic RCS plot (E-plane) depicted in Figure 3.3, 210° corresponds to the forward-scattering direction, since the plane wave has an oblique incidence with 30° angle from the nose of the target. The *Flamme*, which is excited similarly, is solved in 60 h, again with 13-level MLFMA. As shown in Figure 3.3, significant RCS values are obtained at certain directions, e.g., at around 120° and 185° due to reflections from the wings of the target.

Analyses of some large-scale canonical problems using parallel MLFMA can be useful by providing reference solutions for asymptotic high-frequency techniques. Such techniques are frequently used in many industrial and scientific projects, where a large number of simulations of large-scale platforms are required. Lack of controllable accuracy is the major issue in high-frequency techniques as they rely on certain asymptotic approximations to reach fast results. Consequently, verification of a newly developed technique or a new implementation becomes a critical task, while this is possible by using full-wave solvers for such large dimensions (domain of the high-frequency technique). One interesting example is depicted in Figure 3.4, where a three-dimensional object with sharp edges and corners located in a vacuum is considered.

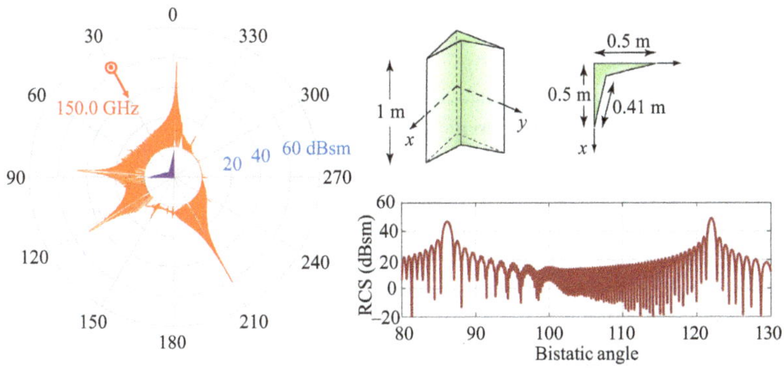

Figure 3.4 Analysis of a scattering problem involving a large-scale perfectly conducting structure with a canonical geometry. The structure is illuminated by a z-polarized plane wave, leading to various types of reflections from the target.

The perfectly conducting structure with the dimensions given in the figure is excited by a plane wave at 150 GHz, i.e., when its maximum dimension (along the z directions) corresponds to $500\lambda_0$. The problem is again formulated by CFIE (discretized with more than 170 million unknowns) and solved via MLFMA on 64 cores of Intel Harpertown-X5472 processors. A polar plot of RCS is included in Figure 3.4, in addition to a linear plot from 80° to 130°. Considering the direction of the plane wave, 210° corresponds to the forward-scattering direction, whereas the peaks at 122° and 358° are due to direct reflections from the flat surfaces. On the other hand, the peak at 86° is caused by a secondary reflection mechanism (involving both flat plates). There are also many spikes, which seem to be caused by alternative reflection mechanisms, on the top of a smooth oscillatory pattern. It is remarkable that the RCS results in Figure 3.4 are estimated to contain a maximum of 1% relative error.

3.4 Material modeling

The competition for solutions to large-scale problems summarized above mainly involved perfectly conducting objects with regular triangulations that do not possess further challenges in terms of material modeling and nonuniform discretizations. Nevertheless, the extensive capabilities of MLFMA were also used for large-scale simulations of dielectric, magnetic, and other penetrable bodies** [56,100–119]. In such a simulation, tree structures associated with different media (of the same problem) must

**The related literature also includes many studies on combining MLFMA with other methods to solve problems involving penetrable objects [120–126], as well as on MLFMA implementations involving dielectric half spaces [127,128] and solvers employing (partially or fully) volume integral equations [129–135].

Table 3.1 Different material characteristics encountered in MLFMA simulations of complex problems

Material	Permittivity and permeability	Wavenumber
Ordinary	$\lvert\varepsilon\rvert \sim \varepsilon_0$ and $\lvert\mu\rvert \sim \mu_0$	$\lvert k\rvert \sim k_0$
Single negative (Low loss)	$\mathrm{Re}\{\varepsilon\} \sim -\varepsilon_0$ or $\mathrm{Re}\{\mu\} \sim -\mu_0$	$k \sim ik_0$
Double negative (Low loss)	$\mathrm{Re}\{\varepsilon\} \sim -\varepsilon_0$ and $\mathrm{Re}\{\mu\} \sim -\mu_0$	$k \sim -k_0$
Highly conductive	$\mathrm{Im}\{\varepsilon\} \gg \varepsilon_0$ and $\lvert\mu\rvert \sim \mu_0$	$\mathrm{Re}\{k\} \approx \mathrm{Im}\{k\} \gg k_0$
Near-zero	$\lvert\varepsilon\rvert \ll \varepsilon_0$ and/or $\lvert\mu\rvert \ll \mu_0$	$\lvert k\rvert \ll k_0$

We use $a \sim b$ to indicate that a/b is a numerically *reasonable* value, depending on implementation details, programming, and computing environment (e.g., precision). As reference values, ε_0 and μ_0 represent vacuum permittivity and permeability, respectively, whereas $k_0 = \omega\sqrt{\mu_0}\sqrt{\varepsilon_0}$ represents the wavenumber in vacuum.

be separately parallelized for optimal efficiency, particularly since these tree structures are fundamentally different from each other. For example, if the overall grouping strategy (e.g., the number of levels in a recursive clustering) is fixed, the number of field samples increases for all boxes when computing electromagnetic interactions in a medium with a higher contrast (permittivity/permeability). Consequently, MLFMA operations during aggregation, translation, and disaggregation stages (e.g., interpolations and anterpolations), as well as their parallelization (from minor parametric issues to major partitioning mechanisms) can change within a single problem, depending on different electromagnetic parameters of its parts. In some cases, extra levels may be introduced (again within the same problem) due to larger electrical sizes with increasing wavenumber values. More issues arise if the object has near-zero, negative, or complex permittivity and/or permeability values [107,108,116,136].

Basically, the use of MLFMA and its types for objects with different material properties depend on the possible values of the wavenumber. In Table 3.1, we provide a short list of different cases, which are illustrative particularly when considering a single object (with those parameters) located in a vacuum so that the standard MLFMA can still be used for the outer medium while computations regarding the inner medium may possess challenges. We emphasize that partitioning in the parallelization of MLFMA (conventional or other versions) depends on the medium, while such differently partitioned tree structures in a single problem never interact with each other, as the whole solution is reduced into matrix–vector multiplications to maintain iterations.

3.4.1 Material simulations with the conventional MLFMA

The conventional MLFMA based on the plane-wave diagonalization can be used to analyze structures involving ordinary materials (see Table 3.1), as well as certain objects, such as homogenized bodies with double-negative properties. Specifically, for electromagnetic problems involving multiple regions with moderate wavenumber values (with positive or negative real parts), tree structures associated with the material media are controlled at a parametric level, without a fundamental change in interaction mechanisms. Hence, the efficiency and the accuracy of such simulations are mostly controlled at the formulation/discretization level, i.e., selections of proper formulations and their discretizations [23–29], to reach accurate solutions via efficient usage of computational resources.

Figure 3.5 presents a set of simulations involving penetrable spheres of radius 0.3 m in vacuum [105,107,108]. Each sphere is illuminated by a plane wave propagating in the z direction and polarized in the x direction. Numerical solutions are obtained by using MLFMA, when the problems are formulated via electric–magnetic current combined-field integral equation (JMCFIE) [23,26] discretized with the RWG functions (leading to 40–70 million unknowns per problem). In all cases, computational RCS values on the z–x plane (in dBsm) are compared with reference values provided

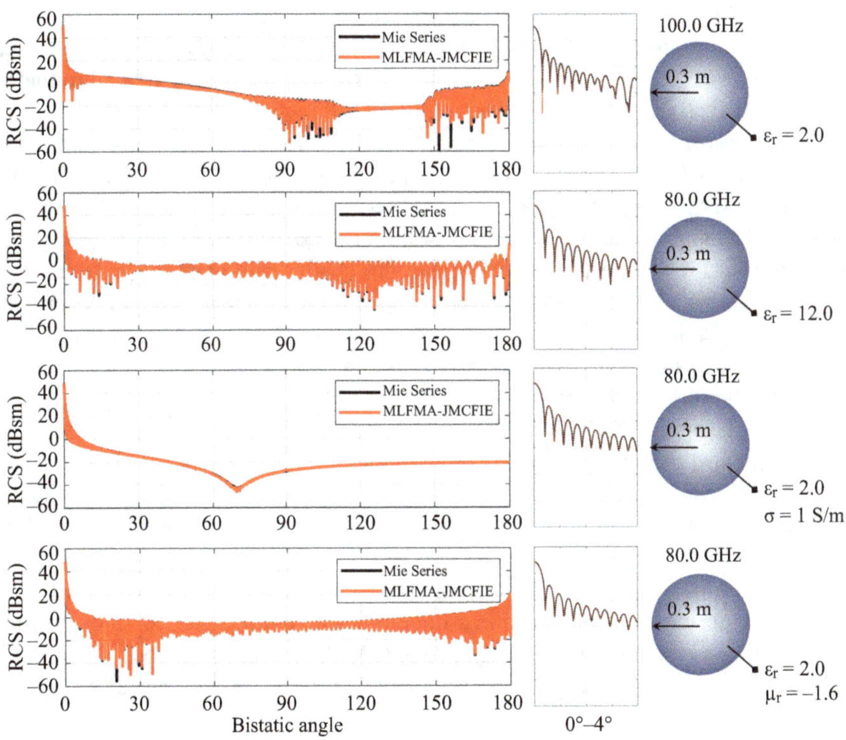

Figure 3.5 Solutions to scattering problems involving electrically large spheres illuminated by plane waves in a vacuum. Bistatic RCS values are plotted in the E-plane, where 0° and 180° correspond to forward-scattering and back-scattering directions, respectively.

by Mie-series solutions, where 0° corresponds to the forward-scattering direction. In the first example, a lossless dielectric sphere with 2.0 relative permittivity is excited at 100 GHz, i.e., when the radius corresponds to approximately $100\lambda_0$. In the second, third, and fourth examples, the excitations are at 80 GHz, while the material properties are changed to illustrate the accuracy of the implementation for various cases. Specifically, the second example involves a lossless dielectric sphere with 12.0 relative permittivity, the third one involves a lossy dielectric sphere with 2.0 relative permittivity and 1.0 S/m conductivity, and, finally, the fourth one involves a double-negative sphere with −2.0 relative permittivity and −1.6 relative permeability. We note that, in each solution, two different tree structures (for the outer and inner media) and their partitioning in the context of parallelization are needed. Using JMCFIE and the conventional MLFMA, we observe excellent accuracy (maximum 1% RMS[††] error) as the computational values overlap with the reference analytical curves. Using BiCGStab as the iterative solver, the fastest solution is achieved for the lossy sphere (involving 37 iterations to reach 0.005 residual error, completed in 14 h on 64 cores of Intel Nehalem-X5560 (2.8 GHz) processors), while the others need several days if the computational resources are not increased further.

Figure 3.6 presents numerical examples involving spherical objects with composite material properties [112,114]. Specifically, each geometry involves a spherical core of diameter 0.5 m located inside a spherical layer of diameter 1.0 m, illuminated by an x-polarized plane wave propagating in the z direction. Once again, Mie-series solutions are available to assess the accuracy of numerical simulations, which are performed via MLFMA-JMCFIE that is suitable for composite structures. In the first two examples, the frequency is fixed to 19.2 GHz, i.e., when the sizes of the objects correspond to approximately $64\lambda_0$. A lower frequency of 9.6 GHz is considered in the third and fourth examples, which involve more challenging material properties. In the first example, a perfectly conducting core is located inside a dielectric layer with 2.0 relative permittivity. Using the same type of layer, the second example involves a dielectric core with 3.0 relative permittivity. These problems are discretized with more than 50 million unknowns and solved in 15/24 h on 64 cores of Intel Nehalem-X5560 processors (performing 48/66 BiCGStab iterations to reach 0.005 residual error). The third example involves a relatively high-permittivity core (10.0 relative permittivity) inside another high-contrast lossy layer ($20.0 + 0.2i$ complex relative permittivity). Although discretized with less than 15 million unknowns, the problem is solved in 16 h (performing 58 iterations to reach 0.005 residual error) using the same computational resources. Finally, the fourth example contains a dielectric core (2.0 relative permittivity) enclosed in a double-negative layer with complex permittivity ($-10.0 + i$ relative permittivity and -1.0 relative permeability). Performing 48 BiCGStab iterations, this problem is solved in approximately 7 h. As depicted in

[††]The root-mean-square (RMS) error is calculated by considering arrays f that contain field samples on the bistatic plane and computing

$$\text{RMS error} = \frac{\|f^{\text{Computational}} - f^{\text{Mie}}\|_2}{\|f^{\text{Mie}}\|_2}, \tag{3.15}$$

where $\|\cdot\|_2$ represents the two-norm of an array.

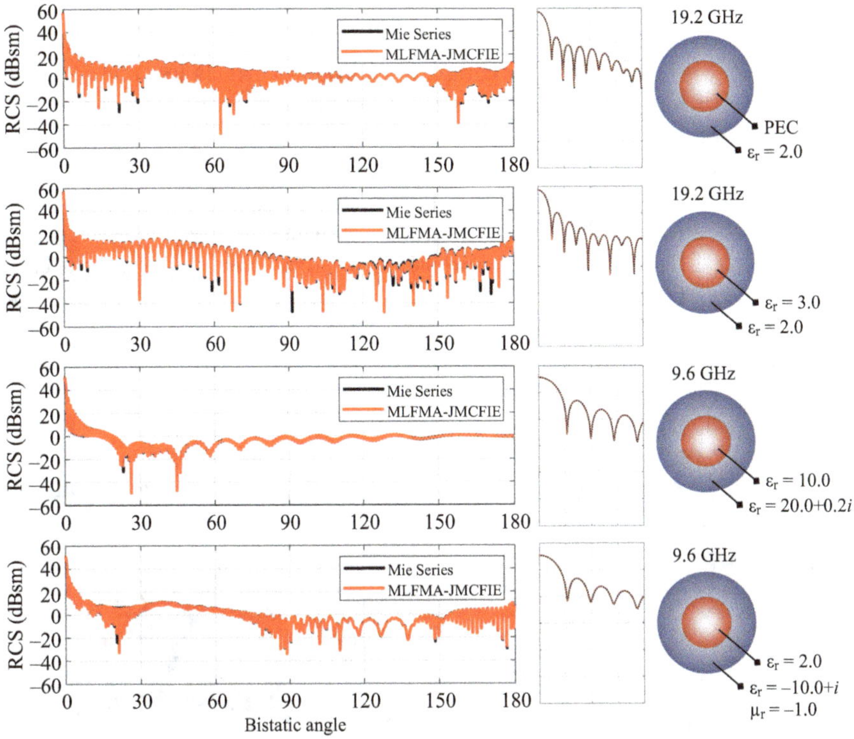

Figure 3.6 *Solutions to scattering problems involving electrically large spherical structures (spherical cores and layers) illuminated by plane waves in vacuum. Bistatic RCS values are plotted in the E-plane, where 0° and 180° correspond to forward-scattering and back-scattering directions, respectively.*

Figure 3.6, all numerical simulations provide accurate results (bistatic RCS values on the *z*–*x* plane, as in Figure 3.5) that are highly consistent with Mie-series values. For all solutions, the RMS error is successfully below 1%.

Obviously, spherical problems are useful to examine and assess the accuracy of numerical solutions, although the implementations are generally developed to analyze realistic scenarios involving complex structures with arbitrary geometries. Unfortunately, even for relatively simple bodies, it can be difficult to verify the accuracy of results, especially when the investigated objects have large electrical dimensions that make them challenging to analyze. One strategy for penetrable bodies can be considering near-zone field distributions. For a given problem, once expansion coefficients for equivalent currents are obtained via MLFMA, they can be used to compute electric and magnetic field intensities at arbitrary locations. On the other hand, medium parameters to be used in radiation integrals can be any of those involved in the analyzed structure. As an example, for a single-material object, outer and inner regions are defined by permittivity/permeability values of the host medium and the object, respectively, which can be considered separately during post-processing. Then, according to the equivalence principle, when the computed equivalent currents are allowed to radiate in a homogeneous space with the parameters of the host medium (outer problem), we expect null fields inside the object, unless the solution is contaminated by internal resonances. Similarly, homogeneous-space radiations with the object parameters should lead to null fields outside the object. Then, these theoretically expected characteristics form a ground to test (at least indirectly) computational results by examining internal/external fields in outer/inner problems, where nonzero values are attributed to numerical errors (mostly involved in expansion coefficients). It is remarkable that inner/outer problems are already considered automatically when near-zone characteristics of a penetrable structure need to be investigated (radiation integrals must be used separately for different media) so that the described error analyses do not require extra computational load.

Figure 3.7 presents the results of a set of computational problems involving a hemispherical lens structure with 4.8 relative permittivity located in a vacuum. The

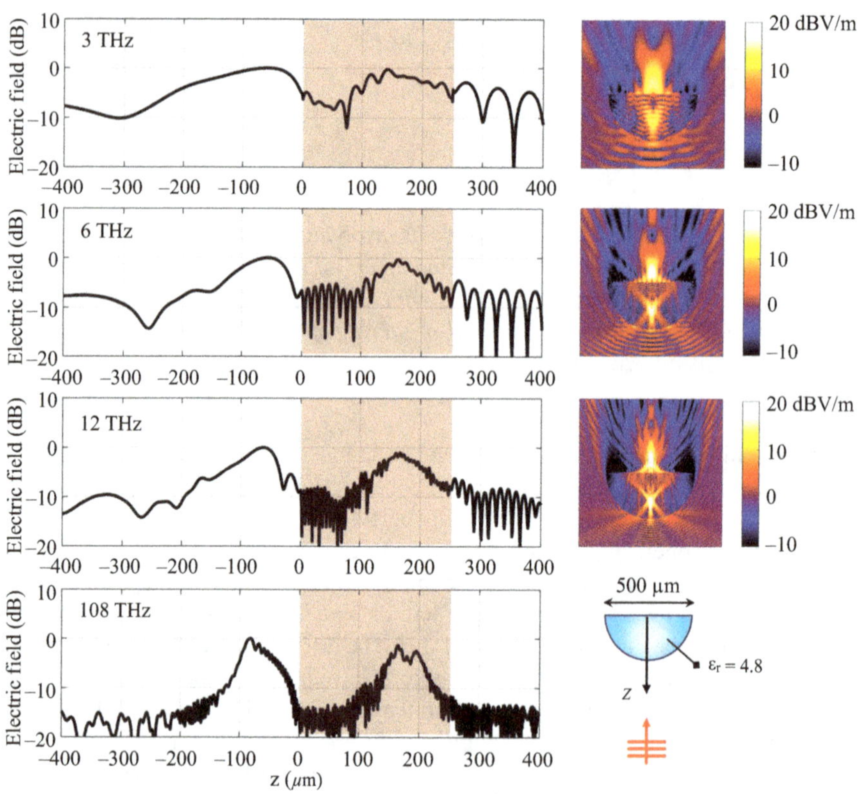

Figure 3.7 Solutions to transmission problems involving a hemispherical dielectric lens at various frequencies

diameter of the object is 500 μm and it is excited by plane waves with normal incidence onto the spherical surface. In addition to relatively low frequencies, i.e., 3.0 THz, 6.0 THz, and 12 THz (when diameter corresponds to approximately $5\lambda_0$, $10\lambda_0$, and $20\lambda_0$, respectively), the structure is analyzed at 108 THz, i.e., when its diameter corresponds to $180\lambda_0$. For solutions via MLFMA-JMCFIE using the RWG functions, the largest problem (108 THz) is discretized with nearly 50 million unknowns and solved in 19 h on 64 cores. Figure 3.7 depicts the near-zone electric field intensity values along the lens axis (z-axis) from -400 μm to $+400$ μm, where the spherical center coincides with $z = 0$ and negative values indicate locations in the transmission region. In addition, two-dimensional near-zone plots are presented for the lower frequencies to demonstrate how the lens structure behaves. As the frequency increases, we clearly observe the focusing characteristics, with a peak at around 90 μm away (at -90 μm) from the planar surface of the lens. Intensity is also increased in an interior region due to the internal reflections of some of the waves from the planar surface. Obviously, even at 108 THz, the distribution of the electric field intensity is quite different from those predicted via simple ray tracing or some asymptotic techniques, as there are many effects (e.g., diffractions) from sharp edges, as well as infinitely many complex reflections interior of the lens[‡‡].

Solutions of composite objects that involve multiple dielectric/magnetic and/or metallic regions may possess numerical challenges in terms of both efficiency and accuracy. Junctions, where three or more regions intersect, have always been implementation issues to be handled carefully to accurately satisfy boundary conditions [25]. On the other hand, there is no extra complexity caused by junctions in terms of *MLFMA implementation* or parallelization. Figure 3.8 presents simulations of a fishnet structure involving nine layers, i.e., four dielectric layers of thickness 50 nm sandwiched between five perfectly conducting layers of thickness 30 nm (as shown in the side view). The overall size of the structure, which contains a total of 10×10 holes (each 295×595 nm), is $9.165 \times 8.865 \times 0.35$ μm. The structure is located in a vacuum so that all edges where dielectric medium, metal, and vacuum meet correspond to junctions. The excitation is a beam with an oblique incidence (30°), created by a complex–source–point (CSP) approach, at 170 THz. The problem is solved via MLFMA-JMCFIE discretized with the RWG functions. Figure 3.8 depicts the electric field intensity distribution on the E-plane, where the complex electromagnetic response of the structure is observed.

The fishnet structure in Figure 3.8 is an example of metamaterials, which often contain multiple parts with different material properties. Exotic properties, which make metamaterials useful in a plethora of applications, also lead to computational challenges in their numerical analyses. Unit-cell simulations based on infinity (infinitely large periodic structure) assumptions can be useful when designing metamaterials, e.g., when optimizing unit-cell dimensions and geometries, assuming that they are used in large numbers when constructing the associated metamaterials. In real

[‡‡]Snell's law predicts no focus point but estimates that approximately 30% of the *rays* intersect at -90 ± 5 μm, while totally reflected rays (from the planar surface) intersect on a line from 180 μm to 250 μm with a variety of intensity levels.

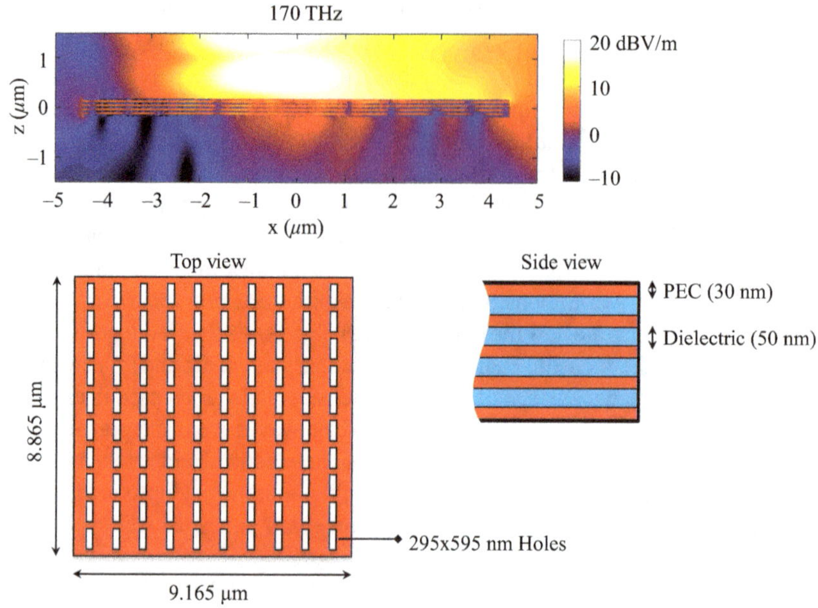

Figure 3.8 Simulation of a fishnet structure involving nine layers of dielectric and metallic slabs, and 10× 10 holes

life, however, metamaterials have finite dimensions such that their full-wave analyses can be crucial to understand how these structures operate with their edges and corners, on the top of strong electromagnetic interactions between unit cells. In fact, such interactions are responsible for difficulties in iterative convergences. Figure 3.9 presents the simulation of a metamaterial designed to operate at optical frequencies [114]. The structure involves 101×101 metallic rods arranged periodically in a dielectric slab. The metals are modeled as penetrable objects with $-8.0 + 1.0i$ relative permittivity (see Section 3.4.2 for further details on plasmonic simulations), while the relative permittivity of the slab is $1.2 + 0.1i$. The dimensions of the overall structure (dielectric slab) is $12.18 \times 12.18 \times 1.92$ μm, and it is investigated at 417 THz. Figure 3.9 depicts the electric field intensity distribution on the H-plane when the structure is excited by a CSP beam with an oblique incidence. We observe that the center of the beam is shifted to the left due to the effective negative refractive index generated by the metamaterial. The solution to the problem is performed by using MLFMA-JMCFIE discretized with the RWG functions, leading to a matrix equation involving more than 8 million unknowns. It takes 11 h to complete the solution (less than 100 iterations) on 64 cores of Intel Nehalem-X5560 (2.8 GHz) processors.

Figure 3.10 presents further examples for simulations of complex structures with various material properties. All problems are solved via MLFMA-JMCFIE discretized with the RWG functions. In the solar-cell simulation [137], the structure (made of c-Si) involves 11×11 inverted pyramids to capture incoming waves at 375 THz. We

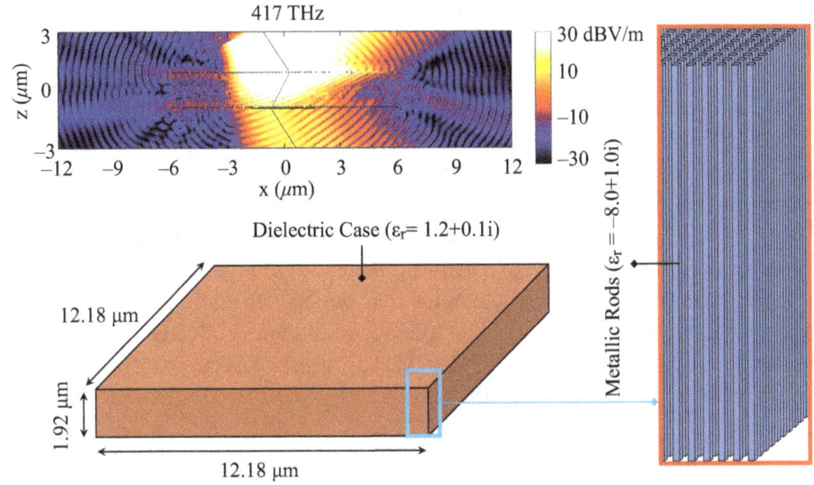

Figure 3.9 Simulation of an optical metamaterial involving 101 × 101 metallic rods inserted into a dielectric slab

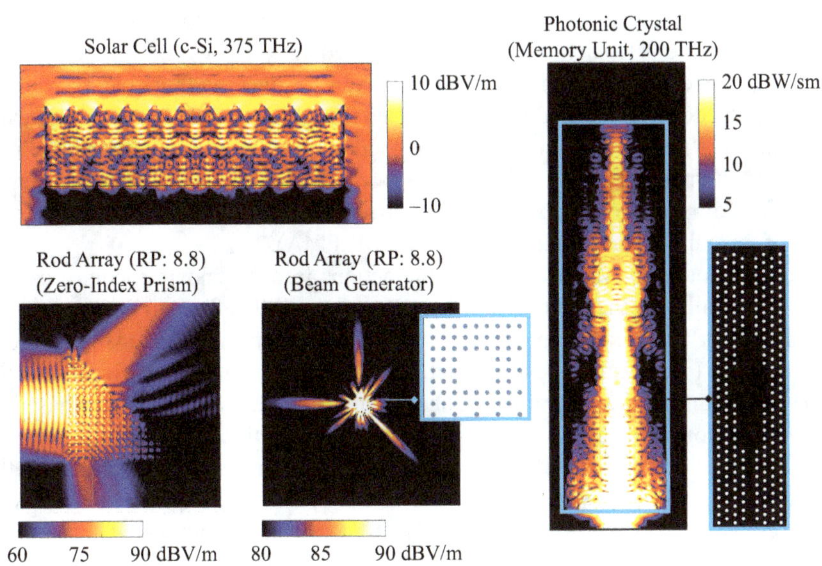

Figure 3.10 Simulations of various complex structures with different material properties

observe that the electric field intensity is successfully trapped by the cavities and forwarded inside the structure to be absorbed. The zero-index prism involves alumina rods (8.8 relative permittivity) arranged in a 15 × 15 grid. By carefully determining the radius of the rods and the distances between them, near-zero-index (NZI) behavior is successfully obtained at 10.3 GHz. As shown in the plot of the electric field intensity distribution, the captured incoming fields leave the prism perpendicularly (e.g., normal to the hypothetical oblique surface), due to the NZI characteristics of the prism. In the beam generator simulation, similar alumina rods are used as a shell structure to create directive beams from an isotropic source [138]. When a Hertzian dipole radiating at 10.3 GHz is located at the center of the designed arrangement, we obtain a tridirectional radiation, again due to the NZI characteristics of the structure together with specially designed irregularities at two sides. Finally, a photonic crystal structure that operates as a memory unit (in a nano-optical system) is presented in Figure 3.10. A well-designed arrangement of air holes that are opened through a dielectric slab with 10.0 relative permittivity enables the accumulation of electromagnetic power at the center of the structure. With 3.3×12.15 μm cross-sectional dimensions, the structure operates successfully at 200 THz.

3.4.2 Simulations of plasmonic structures

Some metals at optical frequencies possess plasmonic properties [139] that make them excellent materials to build nano-optical systems for a variety of applications. Penetrable models of these metals involve complex permittivity values with negative real parts, which are particularly large in the lower ranges of the optical spectrum. As listed in Table 3.1, such a single-negative medium (also with a low loss) leads to a wavenumber with a dominant imaginary value, indicating that electromagnetic waves rapidly decay in the medium. As opposed to highly conductive media, however, such a decay is not accompanied by strong oscillations (associated with the real part of the wavenumber). Hence, from one perspective, plasmonic structures are easier to analyze compared to highly conductive structures, while challenges still exist in terms of formulation and solution algorithm (MLFMA).

In the context of MLFMA, some interactions in a plasmonic medium can be extremely small such that their computation may have little effect on the accuracy of the final result. This means that, in the most simple way, some interactions can directly be eliminated (never calculated). In fact, considering the given error limits for the final solution and medium parameters, analytical expressions can be derived based on Green's function to identify which interactions can be omitted without deteriorating the accuracy of results [116]. For a given geometry, this means sparsification of the associated tree structure, which becomes increasingly *thin* as the negative permittivity increases. For the interactions that still must be calculated, number of harmonics (number of field samples to represent radiated/incoming fields) can also become very small. Even though these seem to provide computational advantages, parallelization becomes a tricky issue, as both number of boxes and number of field samples may have unusual values. In many cases, it becomes better to assign a smaller number of cores to compute matrix–vector multiplications associated with the inner (plasmonic) medium, while using the original (available) number of cores for the outer medium (e.g., vacuum) to maintain solution efficiency. Then, scheduling (reservation of cores, distribution of processes, combinations of numbers, etc.) can become a major issue in parallel MLFMA simulations of plasmonic structures.

In the context of formulation, none of the conventional surface integral equations is really designed for negative real permittivity, which explains their relatively poor performances for such cases. Although JMCFIE can still lead to accurate, efficient, and stable iterative solutions for mild values (e.g., see Figure 3.9), increasing negative permittivity values cannot be handled properly via this formulation. The weakened interactions in a plasmonic medium aforementioned above are meaningful in the sense that electromagnetic interactions tend to be localized as the negative real permittivity increases, until the limiting case of only self-point interactions. Such a limiting scenario merely corresponds to the perfectly conducting case, for which equations for the inner medium become self-consistent (and paradoxically useless), while the magnetic current density is enforced to zero. From this point of view, a stable formulation should be able to capture this plasmonic-to-PEC transition naturally, without numerical overflows and other issues, such as an imbalance of operators. An investigation on this issue led to a novel formulation called the modified combined tangential formulation (MCTF) [29], which appears to be stable for wide ranges of negative permittivity values. MCTF converges into EFIE in the limiting case of $\mathrm{Re}\{\varepsilon\} \rightarrow -\infty$ without any external regularization. Using MCTF instead of JMCFIE or other formulations does not change the general structure of MLFMA, while MCTF solutions often require strong preconditioning, particularly those based on multilayer strategies employing approximate forms of MLFMA [140].

Figure 3.11 presents various simulations involving plasmonic objects, particularly nano-optical structures for various applications. All solutions are carried out by using MLFMA-MCTF discretized with the RWG functions. Nano-optical lenses involve thin Ag layers, on which hexagonal holes are opened. The electric current density distributions on these structures are depicted when they are excited via plane waves with normal incidence. By changing hole dimensions or center-to-center distances, it is possible to control active regions of the structures such that various focusing characteristics can be obtained. As another interesting structure, the isolated

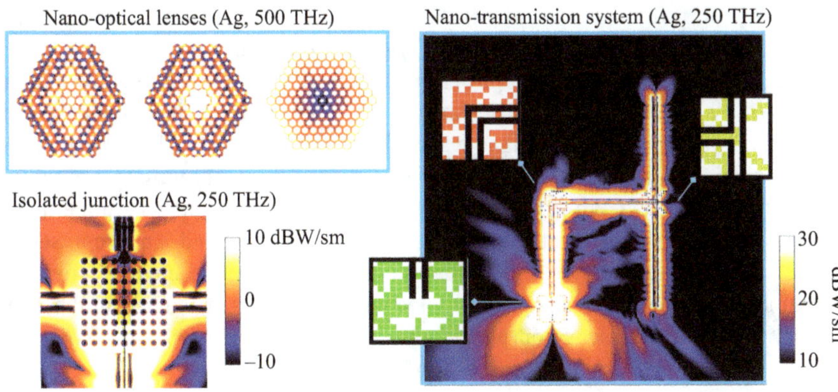

Figure 3.11 Simulations of various complex structures made of a plasmonic material (Ag)

junction involves an array of spherical Ag nanoparticles located at a four-way junction of Ag nanowire pairs. The nanowires do not have any physical contact with each other and with the nanoparticles (electrical isolation), while optical transmission is still maintained thanks to the array designed for this purpose. The figure demonstrates the power density distribution at the junction region when one of the nanowire pairs is excited (from the other side—not visible in the plot) at 250 THz. Finally, Figure 3.11 includes results for a nano-optical transmission system [141] involving pairs of Ag nanowires and well-designed couplers made of Ag nano-cubes, all located in a vacuum. The system, which operates as a power splitter, includes a corner and a three-way junction (where the power is divided), as well as an input region where the nanowires are excited via dipoles. The couplers are designed via optimization by genetic algorithms (GAs) [142] to maximize the power density at the two outputs, while minimizing the difference between them, at 250 THz. For efficient optimization attempts, GAs and MLFMA-MCTF are integrated into a single mechanism, where various strategies, such as dynamic accuracy control [143], are employed to accelerate optimization and improve the quality of the results[§§]. The overall structure covers a 5×10 μm area and the electromagnetic problem is not large in terms of the number of unknowns (less than 50,000), while an optimization requires nearly 8,000 simulations (*embarrassingly* parallelized on multiple cores). We emphasize that, in spite of the small number of unknowns, even a single (accurate and efficient) solution to the nano-optical structure is a challenge, considering that Ag has a relative permittivity of approximately $-61 + 4.3i$ at 250 THz.

Simulations of nanoantennas are particularly important to understand various behaviors and characteristics of these interesting structures [144–146]. These nanoscale antennas are commonly made of metals with plasmonic properties and they operate at various frequencies in the optical spectrum. As material properties are frequency-dependent, analyzing the overall response of a nanoantenna geometry requires a parametric study, where metric dimensions and frequency values should be scanned. One example is depicted in Figure 3.12 for a simple bowtie nanoantenna geometry made of Ag located in a vacuum. The original dimensions of the nanoantenna are $214 \times 100 \times 20$ nm, and it is desired to find the frequencies, at which this particular design operates efficiently with original dimensions or when it is scaled. The plot in Figure 3.12 shows the power enhancement factor (observed power density divided by the incident power density) at the center of the nanoantenna when it is excited by normally incident plane waves from 250 THz to 600 THz. In addition to the original dimensions (corresponding to 1.0), geometric scaling is applied by using different factors from 0.5 to 4.0. In the same plot, positions where the maximum dimension of the nanoantenna corresponds to $\lambda/4$, $\lambda/2$, and λ are represented by three different curves. We observe that relatively good enhancement factors can be obtained for various cases, e.g., the nanoantenna operates efficiently in the 450–500 THz range for 0.5 scale factor (when the nanoantenna is 107 nm), while it operates

[§§] Employing dynamic accuracy control, approximate versions of MLFMA are used for the required solutions during an optimization process, while the solution accuracy becomes a parameter of the optimization itself.

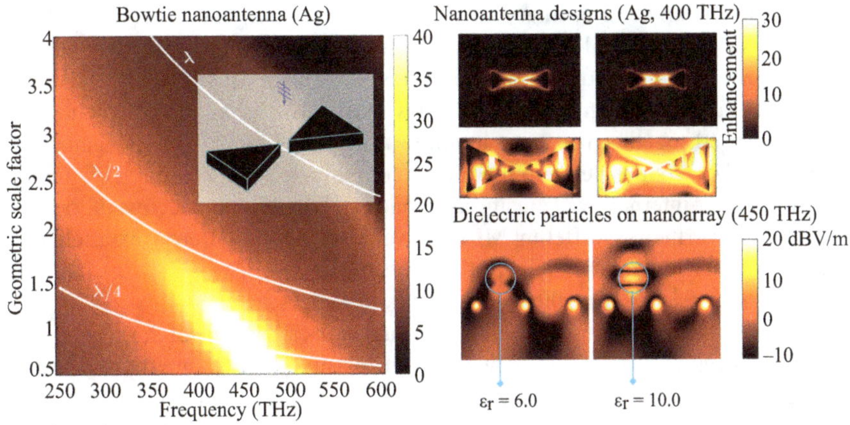

Figure 3.12 Solutions to various problems involving nanoantennas with plasmonic properties

efficiently above 400 THz when the scale factor is 1.0–1.5. We also observe that there are frequencies, at which this design does not operate efficiently simply by scaling such that either geometrical or material properties must be changed.

In addition to the parametric study, Figure 3.12 presents further simulations of nanoantennas with various properties. First, as shown in the enhancement plots, the geometric design of a nanoantenna is crucial for its performance when collecting electromagnetic power. Even with similar bowtie shapes, *hot spots* can be engineered via minor modifications to obtain high enhancement factors at gap locations. More dramatic changes, like log-periodic splits, may lead to extraordinary capabilities in terms of power enhancement, as well as multiband/wideband operations. Figure 3.12 also depicts the electric field intensity distributions when dielectric particles are placed above nanoantenna arrays. These scenarios represent particle-sensing applications [147], where nanoantennas enable the detection of small nanoparticles (which cannot be detected otherwise), thanks to their plasmonic properties and consequently elevated sensitivities to nearby objects.

3.4.3 Simulations of near-zero-index (NZI) structures

In real life, NZI structures are realized by periodically arranging ordinary objects, e.g., by using alumina rods as depicted in Figure 3.10. On the other hand, simulations of their homogenized models (where near-zero permittivity and/or permeability are directly used as material properties) can be extremely useful to understand general behaviors of different geometries and to analyze their electromagnetic characteristics in alternative scenarios before considering actual structures. Similar to those for plasmonic objects, the challenges that arise in NZI simulations should be categorized into formulation issues and algorithmic (MLFMA) difficulties. In the context of MLFMA, a small wavenumber means an electrically small inner problem, despite that the object itself can be of ordinary size in the host medium (outer problem). Hence, instead of the conventional MLFMA based on plane waves, low-frequency versions should be employed to compute inner interactions. This means that MLFMAs with different expansion/diagonalization schemes (basically, different implementations) need to be used together for the solution to a single problem. This is again feasible as MLFMA versions perform matrix–vector multiplications required by iterative solvers and their internal mechanisms are irrelevant from the perspective of the iterative solver. Nevertheless, different MLFMA versions require different parallelization strategies or different applications of the same strategy, which may bring challenges again in terms of scheduling the workload among processes.

In terms of formulation, also similar to plasmonic cases, none of the conventional formulations (as well as MCTF developed for plasmonic structures) provide good performance for NZI problems. Formulations either fail completely (non-convergence and/or inaccurate results) or provide slow iterative convergence, in comparison to their performances for ordinary materials. JMCFIE can handle a wide variety of NZI structures, while it fails as the relative permittivity and/or permeability approach zero. We note that the performance of a formulation also depends on the NZI characteristics. Specifically, only when the relative permittivity has a near-zero value, we consider an epsilon-near-zero (ENZ) case, where $|\varepsilon| \ll \varepsilon_0$, $|\mu| \sim \mu_0$, $|k| \ll k_0$, and $\eta \gg \eta_0$. On the other hand, a mu-near-zero (MNZ) case occurs when $|\mu| \ll \mu_0$, $|\varepsilon| \sim \varepsilon_0$, $|k| \ll k_0$, and $\eta \ll \eta_0$. Finally, when both permittivity and permeability have near-zero values (an EMNZ case), the parameters can be written as $|\varepsilon| \ll \varepsilon_0$, $|\mu| \ll \mu_0$, $|k| \ll k_0$, and $\eta \sim \eta_0$. Consequently, in a simulation of an NZI object, accuracy, efficiency, and/or stability issues may arise due to numerical overflows or imbalances when the permittivity, permeability, wavenumber, and/or intrinsic impedance, or their certain combinations go to zero/infinity. Based on extensive analyses, novel formulations can be developed to accurately and efficiently solve NZI problems, when the objects have different material properties (ENZ, MNZ, and EMNZ).

Figure 3.13 presents several examples for simulations of NZI structures using a mixed formulation [148], which is stable for all NZI cases, discretized with the RWG functions. Solutions are accelerated via AD-MLFMA, which automatically turns into the conventional MLFMA for ordinary media, e.g., vacuum that is the host medium in all examples. The EMNZ lens is a structure obtained by a spherical extraction from a $16\lambda \times 16\lambda \times 16\lambda$ cube at 1 GHz. It has EMNZ characteristics with 0.01 relative permittivity and 0.01 relative permeability. The problem is discretized with less than one million unknowns and solved via GMRES (without preconditioning and restart) that requires 127 iterations to reach 0.001 residual error. The distribution of the electric field intensity on three perpendicular (main) planes is depicted in Figure 3.13, where we observe a strong focusing thanks to the NZI characteristics of the lens. The EMNZ beam generator involves a cylindrical cavity opened through a $5\lambda \times 5\lambda \times 5\lambda$ NZI cube with 0.005 relative permittivity and 0.005 relative permeability. A Hertzian dipole, which is located at the center of the cavity, is aligned in the cylinder direction and radiates at 10.3 GHz. Absorber-like corrugations are created at two surfaces of the cube such that the structure generates two beams perpendicular to each other. Finally, Figure 3.13 depicts a waveguide (WR90) with two sharp turns, whose transmission

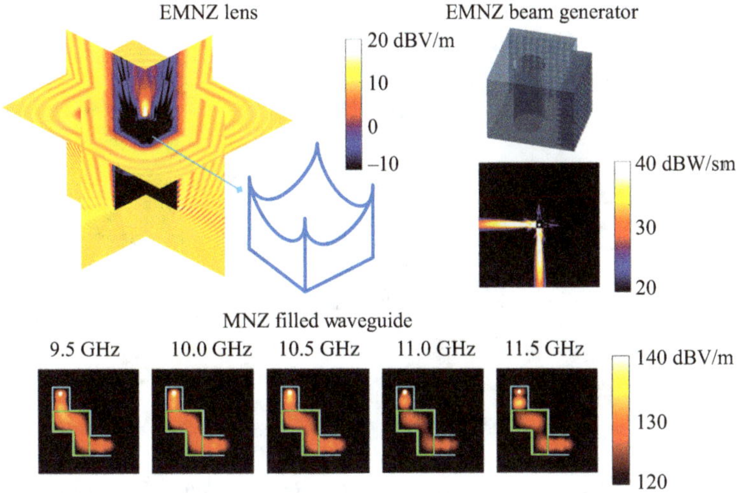

Figure 3.13 Solutions to various problems involving structures with NZI properties

is enhanced by using an MNZ material inside. A dipole source is located at one side of the waveguide, which is closed by a perfectly conducting plate. The cross-section of the entire geometry covers an area of 79 × 79 mm, while its center (only the region that includes the two turns) is filled by an MNZ material with 0.1 relative permeability and unity relative permittivity. The electric field intensity distributions shown in Figure 3.13 illustrate excellent transmission properties of the structure from 9.5 to 11.5 GHz.

3.5 Problems with dense discretizations

Using surface-integral-equation formulations, the discretization size is not *directly* affected by material properties (e.g., low/high contrasts), as opposed to volume formulations. Therefore, complex material properties of some of the problems considered in the previous section do not necessarily bring difficulties in terms of discretization or computational challenges (e.g., a large number of unknowns, nonuniform discretizations, etc.) regarding discretization size ($\lambda/10$ is used as a rule-of-thumb). On the other hand, dense discretizations[|||] are often necessary for real-life problems, due to at least two reasons: (1) geometric concerns, i.e., to accurately model the given geometry in the discretized form; (2) accuracy concerns, i.e., to accurately model

[|||]Here, *dense* is used to describe meshes consisting of discretization elements that are small not only in terms of wavelength but also in terms of the overall size of the given geometry. Consequently, for a densely discretized object, a relatively large number of unknowns is required for the overall modeling of the associated electromagnetic problem.

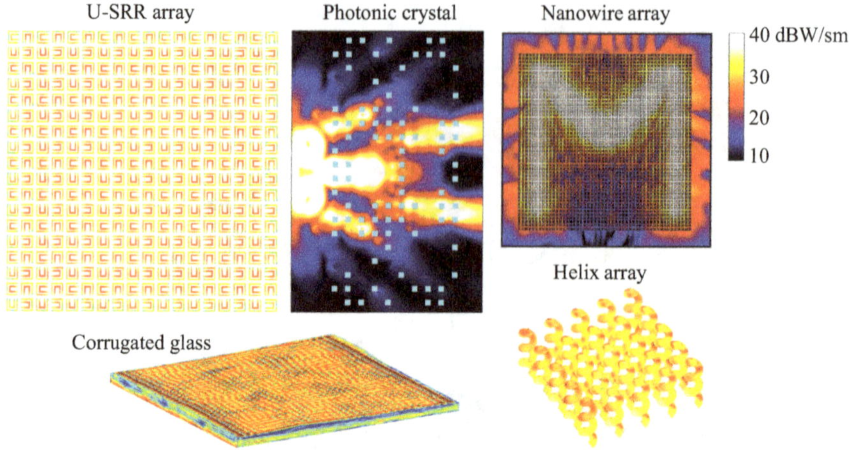

Figure 3.14 Solutions to various problems that involve dense discretizations (small triangles) with respect to wavelength

electromagnetic quantities (currents/charges). These two concerns may coexist in a single problem, while they may also lead to nonuniform discretizations, i.e., when certain parts of an object require such dense discretizations while others do not.

It is remarkable that uniform or nonuniform dense discretizations may be avoided via certain approaches, e.g., by employing higher-order basis and testing functions, curved elements or similar conforming discretizations, etc. But in the context of triangulations and low-order basis functions (RWG functions) considered throughout this chapter, such discretizations are automatically involved in multi-scale simulations that aim for accurate analyses of structures with varying electromagnetic *scales*. Subsequently, dense and nonuniform discretizations have multi-scale characteristics *themselves*, as they need solution techniques (so-called broadband solvers) that employ different methodologies at different scales within a single problem. At the bottom line, this complex and rather ambiguous definition of *multi-scaled-ness* is irrelevant to the challenging characteristics of such problems that need rigorous applications of alternative solution algorithms (MLFMA versions in our case) and suitable formulations to reach accurate, efficient, and stable solutions.

Figure 3.14 presents various examples, in which dense discretizations are required for different reasons. As the structures and associated problems differ significantly, we summarize some main aspects (all structures are located in a vacuum).

- The U-SRR array [149] with 267 × 267 × 3.2 mm dimensions involves two layers of 18 × 18 elements, each of which is a pair of U-type split-ring resonators (SRRs). The resonators are periodically oriented in different directions to create a polarization-independent frequency-selective surface. They are modeled as perfect electric conductors with zero thicknesses, such that EFIE is used to formulate the problem, while MLFMA or AD-MLFMA simulations are accelerated via multi-layer preconditioning [140] for fast and accurate solutions. The plot in Figure 3.14 presents the electric current density on the surfaces of the U-SRRs located on one of the two layers, when the structure is excited via a plane wave with a left-hand circular polarization at 7.65 GHz.

- The photonic crystal involves an optimized arrangement of dielectric rods on a 10×20 grid [150]. The presence/absence of the rods are found via an on–off optimization that aims to obtain a double beam at the output when the structure is excited by a single beam. Each rod has a 0.15×0.15 μm cross-section and 7.5 μm length, corresponding to 5λ at the operating frequency of 200 THz, while their relative permittivity is 4.0. The problem is formulated by using JMCFIE and solved via MLFMA or AD-MLFMA, depending on triangulations used to model the rods. In the plot included in Figure 3.14, we observe the normalized power density in a linear scale ([0,1]), as well as the rods that form the structure.
- The nanowire array contains a total of 50×50 nanowires of length 4.8 μm. The cross-sectional area of the entire structure is 9.9×9.9 μm, and it is excited by a total of 344 Hertzian dipoles that provide an M-shaped radiation pattern just above the array. Each nanowire has a 0.1×0.1 μm square cross-section, while they are modeled as perfectly conducting objects that can be formulated via EFIE. A discretization with small triangles leads to a matrix equation involving more than one million unknowns. The plot in Figure 3.14 depicts the power density distribution at 250 THz in the vicinity of the array on the transmission side, where the M shape is clearly recognized thanks to the excellent transmission properties of the nanowires.
- The helix array contains 5×5 Ag helices that are automatically generated by a computer program via optimization. Dense triangulations are used to represent smooth surfaces of the helices, as required for their proper analysis. When excited by a linearly polarized wave from top or bottom, the structure makes the transmitted wave almost circular thanks to its carefully shaped geometry. The problem is formulated by using MCTF and solved via MLFMA or AD-MLFMA. The plot in Figure 3.14 illustrates the electric current density on the surfaces of the helices when the array is excited by a plane wave at 600 THz.
- Finally, the corrugated glass is a $16.0 \times 16.0 \times 1.0$ μm slab, whose top surface is deformed via shape optimization to minimize reflections [151]. At 567 THz, the relative permittivity of the glass is set to 2.25. The problem is formulated by using JMCFIE and solved via MLFMA or AD-MLFMA. Figure 3.14 depicts the electric current density on the corrugated side of the slab when it is excited by a plane wave.

For the U-SRR problem, $\lambda/30$ triangles are required to represent U-SRR geometries, while a denser mesh is used to accurately represent currents. For the photonic crystal and nanowire array problems, although $\lambda/10$ triangles are more than sufficient to represent the rods and the nanowires (square prisms), denser triangulations are necessary to model the electric (and magnetic) currents on them. On the other hand, for the helix array, the main concern is the accurate representation of the geometry (*hence*, currents on it) that is fine-tuned via optimization. Similarly, for the corrugated glass,

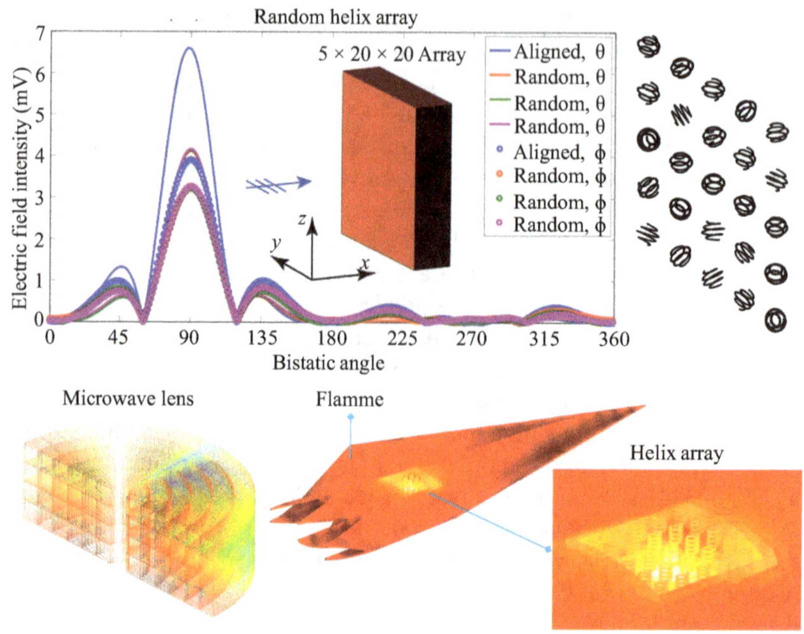

Figure 3.15 Solutions to various problems that involve dense discretizations (small triangles) with respect to wavelength

random but optimized deformations need to be modeled accurately to reach accurate results at the end.

Figure 3.15 presents further examples involving densely discretized structures, whose simulations possess numerical challenges. The random helix array involves helical elements, similar to the 5×5 array in Figure 3.14. This time, a total of $5 \times 20 \times 20 = 2{,}000$ helical elements are arranged in a regular grid enclosed by a $\lambda/2 \times 2\lambda \times 2\lambda$ volume at 3 GHz. The elements are modeled as PEC, while they are excited by a linearly polarized (y-polarized) plane wave in a vacuum. The problem is formulated by using MFIE and solved via AD-MLFMA. To accurately model the helices, $\lambda/500$ triangles are used, leading to a matrix equation involving nearly 6 million unknowns. When the helices are aligned vertically (hence, identically) in the z direction, the transmitted wave at the back of the array is mainly polarized in the θ direction (corresponding to z on the array axis), while there is also a significant component in the ϕ direction (corresponding to y on the array axis). When the helices are randomly oriented, however, the difference between polarizations is reduced, as shown via three different examples. The microwave lens in Figure 3.15 is another challenging problem involving a very dense discretization. Although the structure seems like being made of wires, it actually involves very thin PEC surfaces discretized via triangles, on which the RWG functions are defined (once again) to expand the electric current density. The plot in Figure 3.15 depicts the electric current density induced on the structure when it is excited by a Hertzian dipole located at its center at 1.0 GHz. Finally, Figure 3.15 illustrates an example, where a 5 × 5 helix array (involving PEC helices) is placed into a cavity opened on the *Flamme* geometry (also PEC). The structure is investigated at 5 GHz, at which the *Flamme* is 10λ (not very challenging), while the presence of helices leads to challenges as their accurate geometric models need dense discretizations. The problem could be solved by discretizing the whole geometry with triangles of equal size (i.e., suitable for the helices), while the other (more efficient) option is preferred by making the mesh size λ/10 on the *Flamme*, leading to a nonuniform discretization (see the next section). Hence, the number of unknowns is kept at nearly 100,000, despite the fact that an 11-level AD-MLFMA is needed to efficiently solve the problem. We also note that the problem is formulated via a hybrid integral equation [20] that is suitable for this kind of nonuniform discretizations of PEC objects. The plot in Figure 3.15 illustrates the electric current density when a Hertzian dipole is located just beneath the helix array.

As a final set of examples for simulations involving densely discretized objects, Figure 3.16 presents solutions to scattering problems involving a geometry that marks the years when this chapter is being written. A generic coronavirus geometry (PEC) with 100 × 100 × 102 nm dimensions is considered to be excited by a plane wave from 2,000 THz to 6,000 THz. Using 2 nm triangles (corresponding to λ/75–λ/25 sizes), the structure is discretized with nearly 50,000 triangles. The problem is formulated by using a combined potential-field formulation [152] and solved via AD-MLFMA. Figure 3.16 presents the power density distributions in the vicinity of the structure on the cross-sectional plane. We note that, due to the PEC modeling, internal fields are vanishingly small, demonstrating the excellent accuracy of the simulations.

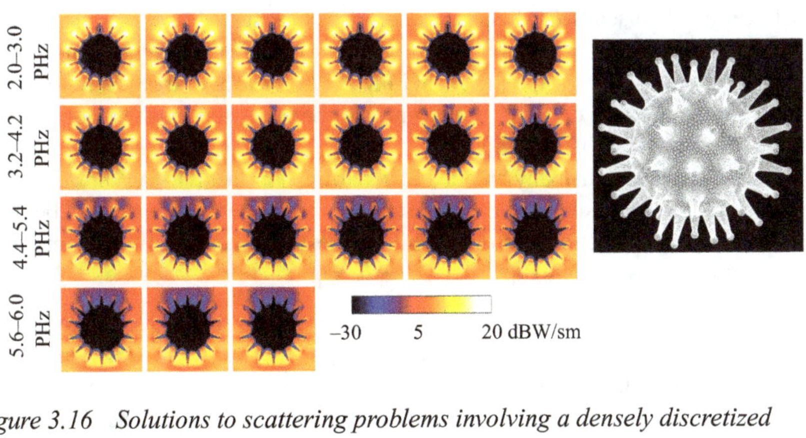

Figure 3.16 Solutions to scattering problems involving a densely discretized generic coronavirus geometry

3.6 Problems with non-uniform discretizations

Finally, we consider fast and accurate simulations of complex structures involving non-uniform discretizations. Such structures are electrically large, while, at the same time, they need dense discretizations at certain parts for accurate modeling of geometries, current/charge distributions, or both, as discussed in Section 3.5. As also mentioned in Section 3.2.2.7, we need to consider different aspects when developing truly broadband solvers to analyze such multi-scale problems.

- Formulation: The discretized form of the surface integral equation used to formulate the problem must be stable for dense discretizations. This means employing low-frequency-stable formulations, such as PIEs [14–17] or certain approaches that similarly incorporate charges as unknowns [153,154] for perfect electric conductors, or stabilizing conventional formulations via suitable discretization schemes [31–33]. Hybridization may be required [20] if the used low-frequency-stable formulations and/or discretizations have efficiency issues when they are applied onto coarsely discretized parts of the analyzed objects.
- Solution algorithm: A stable version of MLFMA must be used to efficiently and accurately perform electromagnetic interactions between small elements located on densely discretized parts. As discussed in Section 3.2.2, there are many alternatives, such as implementations based on multipoles [46–51], inhomogeneous plane waves [54–57], alternative shift mechanisms leading to non-directional translations [59,64–68], and certain approximations that enable stability [69–73]. As the analyzed structures are typically large in terms of wavelength, the MLFMA implementations to be employed must be broadband [52,53,58–63] by combining low-frequency and high-frequency expansions within the same solution mechanisms, if the approach does not already reduce into the conventional MLFMA or similarly efficient form for long-distance interactions [72]. Implementations of the conventional MLFMA in high-precision or mixed-precision environments [74,75], or its combination with other types of solution algorithms [76–78] are further alternatives to reach broadband solvers.
- Clustering: As discussed in Section 3.2.2.7, in the context of MLFMA, the strategy to construct tree structures must be modified to efficiently handle non-uniform discretizations. For example, in IL-MLFMA [73,80,133], a truly *recursive* clustering is applied by dividing boxes into sub-boxes based on their populations, without a regular division of all boxes at all levels (that is practiced in the conventional MLFMA and in many broadband versions). This leads to *leaf* boxes located at different levels such that near-zone and far-zone interactions need to be organized in more sophisticated forms, while the resulting computational complexity can remain at $\mathcal{O}(N \log N)$.

Obviously these aspects should be considered depending on the strategy to design and implement the solver, as well as the type of the target problems involving nonuniform discretizations. Combinations of MLFMA with some alternative solvers may provide stable computations of short-distance interactions, while sufficient levels of efficiency may be achieved without devising irregular tree structures but rather by employing the low-frequency-stable solver (that may not even require any tree structure) to analyze densely discretized parts. On the other hand, in such hybrid implementations, reaching an efficient combination of MLFMA with the other solver can become the main issue to obtain accurate, efficient, and stable simulations.

Figures 3.17–3.19 present solutions to three challenging problems involving nonuniform discretizations. Although the total number of unknowns is relatively small, these problems are difficult (if not impossible) to handle via conventional MLFMA implementations. Involving perfectly conducting objects, all problems are formulated via PIEs [14]–[17] discretized via the RWG and pulse functions. Solutions are performed by using NSPWMLFMA [65]–[68] employing IL tree structures. The first two problems involve computer components, i.e., a microchip (Figure 3.17) and a CPU fan (Figure 3.18), both of which are excited via Hertzian dipoles. Figure 3.17 illustrates the discretization of the microchip from different perspectives (in addition to dimensions); pins and edges of the main body are discretized with very small triangles to accurately represent current distributions in these locations, leading to approximately 140,000 unknowns. The Hertzian dipole is located at $z = 3$ cm (above the microchip), is oriented in the $\hat{h} = (\hat{x} + \hat{z})/\sqrt{2}$ direction, and radiates at 4.2 GHz. A seven-level IL tree structure is constructed by limiting the maximum box population at 75. The solution, which involves 55 GMRES iterations to reach 10^{-4} residual error, is completed in 46 h (on a single core). Figure 3.17 also depicts the current density induced on the microchip with an excellent level of detail for the densely discretized parts. Figure 3.18 presents the discretization of the CPU fan (in addition to dimensions), as well as the current distribution when it is excited by an x-oriented Hertzian dipole located at $(x, y, z) = (4, 4, 6)$ cm (above the fan) and radiating at 3 GHz. This problem, which is discretized with approximately 90,000 unknowns, is also solved

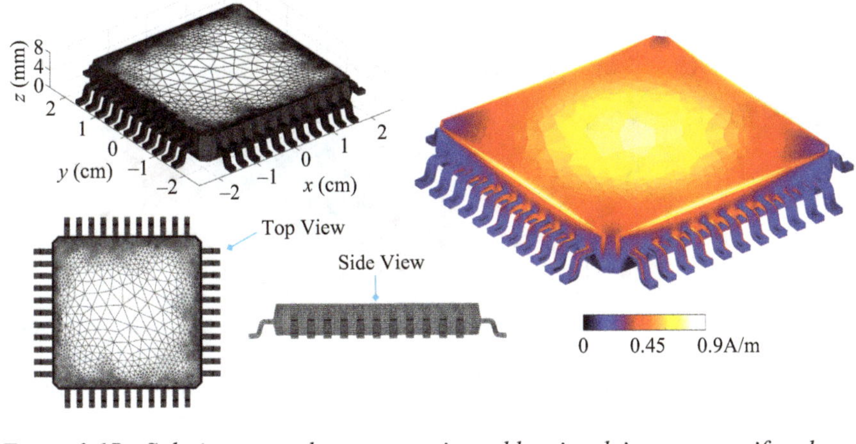

Figure 3.17 Solution to an electromagnetic problem involving a non-uniformly discretized microchip. The pins and edges of the main body are densely discretized.

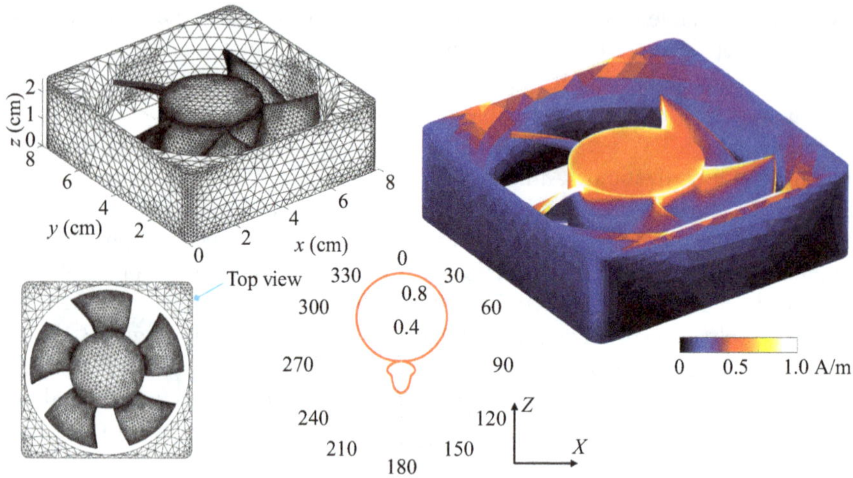

Figure 3.18 *Solution to an electromagnetic problem involving a non-uniformly discretized CPU fan. The edges of the blades and center cylinder are densely discretized.*

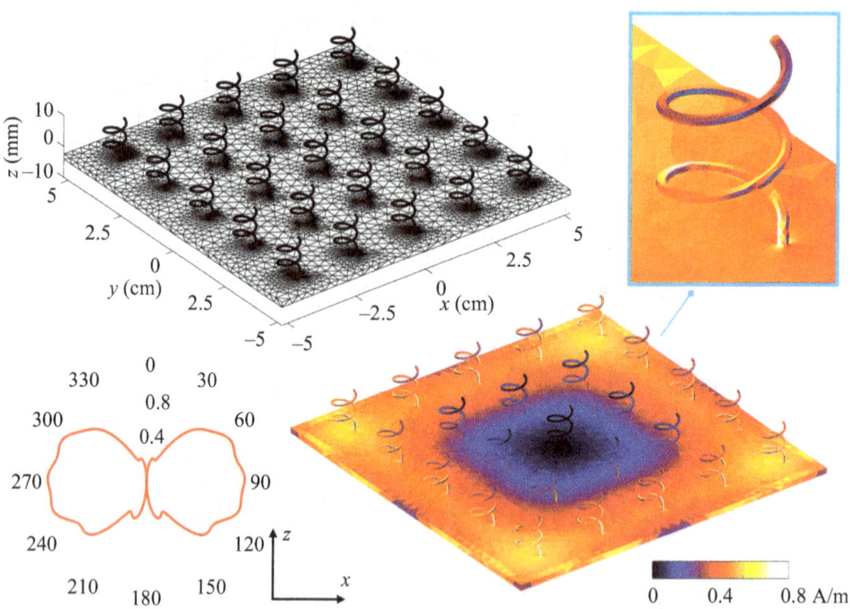

Figure 3.19 *Solution to an electromagnetic problem involving a non-uniformly discretized structure. A total of 5 × 5 helical elements are attached to a thin slab. The helices and the locations where they are connected to the slab are densely discretized.*

via a 7-level NSPWMLFMA, while the number of GMRES iterations (to reach 10^{-4} residual error) and the total solution time are 58 and 63 h (on a single core), respectively. Figure 3.18 further presents the far-zone electric field intensity radiated by the overall structure (Hertzian dipole + CPU fan) on the z–x plane.

Figure 3.19 presents the solution to the third problem involving a highly nonuniform discretization. A total of 5×5 helical elements are attached to a thin slab, and the structure is excited by a z-oriented Hertzian dipole located at $z = 12$ cm (above structure) and radiating at 5 GHz. As shown in Figure 3.19, very dense discretizations are used to model helical elements at critical locations where they are connected to the slab. The resulting number of unknowns is approximately 170,000 and the problem is solved by using an eight-level tree structure constructed by limiting the maximum population size by 100. Performing 110 GMRES iterations, the total solution time is 78 h on a single core. Figure 3.19 again illustrates the electric current density induced on the structure. The zoomed picture demonstrates how small triangles are used to accurately represent helices and expand the currents induced on them. The plot of the far-zone electric field intensity (on the z–x plane) shows the radiation of the overall structure, demonstrating the disturbance of the dipole radiation by a nearby structure.

3.7 Conclusions and new trends

This chapter has focused on MLFMA as a representative kernel-based fast factorization technique. To construct a basis for further discussion, we first considered the conventional MLFMA, which is based on the plane-wave expansion of electromagnetic waves, at a formulation level. To solve multi-scale problems involving dense (uniform or non-uniform) discretizations of electrically large objects, alternative MLFMA versions are needed since the conventional MLFMA suffers from a low-frequency breakdown. We listed a variety of ways to implement low-frequency-stable MLFMAs, such as based on multipoles, inhomogeneous plane waves, coordinate shifts, and approximation techniques. We showed how MLFMA implementations can be used to solve extremely large problems via parallelization, while they can be applied to complex structures with different material properties, including plasmonic and NZI objects. Examples were given for solutions of densely discretized objects to demonstrate how MLFMA can handle such complicated problems that possess modeling challenges. Finally, problems with non-uniform discretizations that naturally arise in multi-scale simulations were considered. A rigorous implementation for stable, accurate, and efficient solutions of these problems requires a well-designed combination of a suitable formulation/discretization, an effective solution algorithm (MLFMA version), and a carefully designed clustering mechanism.

MLFMA has been one of the major techniques in the last several decades, providing accurate analyses of complex structures. Today further research is conducted on how to extend the capabilities of this technique, particularly by combining it with other solution mechanisms that may enable not only multi-scale but also multiphysics simulations. Another area of intensive research is on closing the persistent

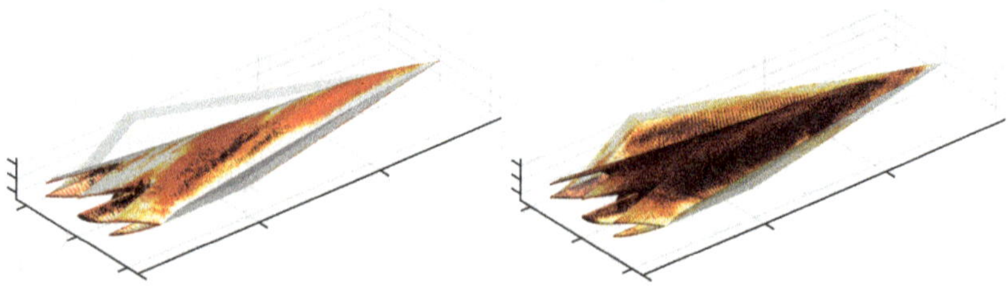

Figure 3.20 Using machine learning in MLFMA simulations of a scattering problem involving the Flamme. The trimmed (eliminated) basis [left] and testing [right] functions at the 40th iteration of the solution are shown by yellow, red, or black colors, while the remaining functions are white.

gap between full-wave solvers (like MLFMA) and high-frequency asymptotic techniques. It is not surprising that MLFMA has been combined with high-frequency techniques in numerous studies, while the resulting implementations do not necessarily meet the expectations (the accuracy of MLFMA combined with the speed of high-frequency techniques), or they may possess such favorable properties for a very restricted class of problems for which the implementations are customized. Introducing approximations in MLFMA to let it be less accurate (non-full-wave) but more efficient is a challenging task that can be realized by all MLFMA coders. Its optimal nature makes MLFMA extremely difficult to modify, at least fundamentally, to reach a more accurate or efficient solver without upsetting the balance between them.

Since MLFMA already has a linearithmic computational complexity, one attractive idea is to reduce the number of interactions to accelerate a solution without too much deteriorating the final accuracy. Specifically, one may simply drop weak (or ineffective) electromagnetic interactions and use strong (or effective) ones to reach a solution. But, even this seemingly simple task can become a major challenge in the context of MLFMA. In fact, a majority of matrix elements are never calculated explicitly in MLFMA so that the user does not have any access to interactions (one-by-one) to categorize and keep/discard them. Nevertheless, advances in computer science provide new opportunities that can be beneficial to upgrade even MLFMA. As demonstrated in a recent work [155], briefly illustrated in Figure 3.20, machine learning can be employed to eliminate electromagnetic interactions during an iterative solution. In this figure, basis and testing functions that are eliminated until a certain iteration are depicted. Although MLFMA starts with a full set of basis/testing functions, as usual, machine learning is used to identify ineffective interactions[¶¶]

[¶¶]It should be emphasized that these ineffective interactions are not necessarily between discretization elements that are far away from each other; they strictly depend on the overall problem and the structure under focus.

(that have little effect on the results of the matrix–vector multiplications) and systematically eliminate them as the iterative solution continues. This way, MLFMA is continuously accelerated during the solution to a problem, resulting in a significant overall reduction in the processing time without deteriorating the final accuracy.

Considering rapid advances in both computer software and hardware, computational electromagnetics will remain to be an active research area in the future. Multi-scale and multi-physics problems are among the topics to be considered in more detail, as novel solvers with unprecedented levels of accuracy and efficiency are continuously developed and implemented. With its unique characteristics, MLFMA is likely to be one of the focal points in the future of computational electromagnetics.

Acknowledgments

Author contributions: Özgür Ergül generated the results in Figures 3.1–3.16 and 3.20, and wrote the draft of the chapter. Bahram Khalichi and Vakur B. Ertürk generated the results in Figures 3.17–3.19. All authors contributed to the writing of the final version of the chapter.

Özgür Ergül thank the former and current members of the Computational Electromagnetics Research Group at METU (CEMMETU); particularly, Aşkın Altınoklu, Barışcan Karaosmanoğlu, Doğa Yücalan, Gökhan Karaova, Göktuğ Işıklar, Hamza Eray, Hande İbili, Mustafa Algun, Ömer Eroğlu, Özgür Eriş, Sadri Güler, Şirin Yazar, Uğur Meriç Gür, and Yeşim Koyaz, for their contributions in developing implementations and/or generating results.

Özgür Ergül work was supported by the Scientific and Technical Research Council of Turkey (TUBITAK) under the Research Grant 118E243 and by the Turkish Academy of Sciences (TUBA) in the framework of the YSA Program.

References

[1] Chew WC, Jin JM, Michielssen E, *et al. Fast and Efficient Algorithms in Computational Electromagnetics*. Norwood: Artech House; 2001.

[2] Shanker B and Huang H. Accelerated Cartesian expansions - a fast method for computing of potentials of the form $R^{-\nu}$ for all real ν. *J. Comput. Phys.* 2007;226(1):732–53.

[3] Bleszynski E, Bleszynski M, and Jaroszewicz T. AIM: adaptive integral method for solving large-scale electromagnetic scattering and radiation problems. *Radio Sci.* 1996;31(5):1225–51.

[4] Phillips JR and White JK. A precorrected-FFT method for electrostatic analysis of complicated 3-D structures. *IEEE Trans. Comput.-Aided Design Integr. Circuits Syst.* 1997;16(10):1059–72.

[5] Seo SM and Lee JF. A fast IE-FFT algorithm for solving PEC scattering problems. *IEEE Trans. Magn.* 2005;41(5):1476–79.

[6] Okhmatovski V, Yuan M, Jeffrey I, *et al.* A three-dimensional precorrected FFT algorithm for fast method of moments solutions of the mixed-potential

integral equation in layered media. *IEEE Trans. Microw. Theory Tech.* 2009;57(12):3505–17.

[7] Coifman R, Rokhlin V, and Wandzura S. The fast multipole method for the wave equation: a pedestrian prescription. *IEEE Antennas Propag. Mag.* 1993;35(3):7–12.

[8] Song J, Lu CC, and Chew WC. Multilevel fast multipole algorithm for electromagnetic scattering by large complex objects. *IEEE Trans. Antennas Propag.* 1997;45(10):1488–93.

[9] Sheng XQ, Jin JM, Song J, *et al.* Solution of combined-field integral equation using multilevel fast multipole algorithm for scattering by homogeneous bodies. *IEEE Trans. Antennas Propag.* 1998;46(11):1718–26.

[10] Ergül Ö and Gürel L. *The Multilevel Fast Multipole Algorithm (MLFMA) for Solving Large-Scale Computational Electromagnetics Problems*. Wiley-IEEE; 2014.

[11] Vande Ginste D, Michielssen E, Olyslager F, *et al.* An efficient perfectly matched layer based multilevel fast multipole algorithm for large planar microwave structures. *IEEE Trans. Antennas Propag.* 2006;54(5):1538–48.

[12] Poggio AJ and Miller EK. Integral equation solutions of three-dimensional scattering problems. In: Mittra R, editor. *Computer Techniques for Electromagnetics*. Oxford: Pergamon Press; 1973. Chapter 4.

[13] Mautz JR and Harrington RF. H-field, E-field, and combined field solutions for conducting bodies of revolution. *AEÜ*. 1978;32(4):157–64.

[14] Chew WC. Vector potential electromagnetics with generalized gauge for inhomogeneous media: formulation. *Prog. Electromag. Res*. 2014;149:69–84.

[15] Vico F, Ferrando M, Greengard L, *et al.* The decoupled potential integral equation for time-harmonic electromagnetic scattering. *Commun. Pur. Appl. Math.* 2014;69(4):771–812.

[16] Gür UM, Ergül Ö. Accuracy of sources and near-zone fields when using potential integral equations at low frequencies. *IEEE Antennas Wireless Propag. Lett.* 2017;16:2783–86.

[17] Liu QS, Sun S, and Chew WC. A potential-based integral equation method for low-frequency electromagnetic problems. *IEEE Trans. Antennas Propag.* 2018;66(3):1413–26.

[18] Rusch WVT and Pogorzelski RJ. A mixed-field solution for scattering from composite bodies. *IEEE Trans. Antennas Propag*. 1986;34(7):955–58.

[19] Ergül Ö and Gürel L. Iterative solutions of hybrid integral equations for coexisting open and closed surfaces. *IEEE Trans. Antennas Propag.* 2009;57(6):1751–58.

[20] Karaosmanoğlu B and Ergül Ö. Generalized hybrid surface integral equations for finite periodic perfectly conducting objects. *IEEE Antennas Wireless Propag. Lett.* 2017;16:1068–71.

[21] Rao SM and Wilton DR. E-field, H-field, and combined field solution for arbitrarily shaped three-dimensional dielectric bodies. *Electromagnetics*. 1990;10(4):407–21.

[22] Sheng XQ, Jin JM, Song J, *et al.* Solution of combined-field integral equation using multilevel fast multipole algorithm for scattering by homogeneous bodies. *IEEE Trans. Antennas Propag.* 1998;46(11):1718–26.

[23] Ylä-Oijala P and Taskinen M. Application of combined field integral equation for electromagnetic scattering by dielectric and composite objects. *IEEE Trans. Antennas Propag.* 2005;53(3):1168–73.

[24] Ylä-Oijala P and Taskinen M. Well-conditioned Müller formulation for electromagnetic scattering by dielectric objects. *IEEE Trans. Antennas Propag.* 2005;53(10):3316–23.

[25] Ylä-Oijala P, Taskinen M, and Sarvas J. Surface integral equation method for general composite metallic and dielectric structures with junctions. *Prog. Electromagn. Res.* 2005;52:81–108.

[26] Ergül Ö and Gürel L. Comparison of integral-equation formulations for the fast and accurate solution to scattering problems involving dielectric objects with the multilevel fast multipole algorithm. *IEEE Trans. Antennas Propag.* 2009;57(1):176–87.

[27] Solis DM, Taboada JM, and Obelleiro F. Surface integral equation method of moments with multiregion basis functions applied to plasmonics. *IEEE Trans. Antennas Propag.* 2015;63(5):2141–52.

[28] Gomez-Sousa H, Rubinos-Lopez O, and Martinez-Lorenzo JA. Comparison of iterative solvers for electromagnetic analysis of plasmonic nanostructures using multiple surface integral equation formulations. *J. Electromagn. Waves Appl.* 2016;30(4):456–72.

[29] Karaosmanoğlu B, Yılmaz A, and Ergül Ö. A comparative study of surface integral equations for accurate and efficient analysis of plasmonic structures. *IEEE Trans. Antennas Propag.* 2017;65(6):3049–57.

[30] Rao SM, Wilton DR, and Glisson AW. Electromagnetic scattering by surfaces of arbitrary shape. *IEEE Trans. Antennas Propag.* 1982;30(3):409–18.

[31] Vipiana F, Pirinoli P, and Vecchi G. Spectral properties of the EFIE-MoM matrix for dense meshes with different types of bases. *IEEE Trans. Antennas Propag.* 2007;55(11):3229–38.

[32] Andriulli FP, Cools K, BagcıH, *et al.* A multiplicative Calderon preconditioner for the electric field integral equation. *IEEE Trans. Antennas Propag.* 2008;56(8):2398–412.

[33] Cools K, Andriulli FP, De Zutter D, *et al.* Accurate and conforming mixed discretization of the MFIE. *IEEE Antennas Wireless Propag. Lett.* 2014;10:528–31.

[34] Ergül Ö and Gürel L. Linear-linear basis functions for MLFMA solutions of magnetic-field and combined-field integral equations. *IEEE Trans. Antennas Propag.* 2007;55(4):1103–10.

[35] Harrington RF. *Field Computation by Moment Methods*. Wiley-IEEE; 1993.

[36] Van der Vorst H. Bi-CGSTAB: a fast and smoothly converging variant of Bi-CG for the solution of nonsymmetric linear systems. *SIAM J. Sci. Stat. Comput.* 1992;13(2):631–644.

[37] Saad Y and Schultz MH. GMRES: a generalized minimal residual algorithm for solving nonsymmetric linear systems. *SIAM J. Sci. Stat. Comput.* 1986;7(3):856–69.

[38] Koc S, Song JM, and Chew WC. Error analysis for the numerical evaluation of the diagonal forms of the scalar spherical addition theorem. *SIAM J. Numer. Anal.* 1999;36(3):906–21.

[39] Kalfa M, Ertürk VB, and Ergül Ö. Error analysis of MLFMA with closed-form expressions. *IEEE Trans. Antennas Propag.* 2021;69(10):6618–6623.

[40] Ohnuki S and Chew WC. Truncation error analysis of multipole expansions. *SIAM J. Sci. Comput.* 2003;25(4): 1293–306.

[41] Ergül Ö and Gürel L. Enhancing the accuracy of the interpolations and anterpolations in MLFMA. *IEEE Antennas Wireless Propag. Lett.* 2006;5:467–70.

[42] Ergül Ö and Van den Bosch I, Gürel L. Two-step Lagrange interpolation method for the multilevel fast multipole algorithm. *IEEE Antennas Wireless Propag. Lett.* 2009;8:69–71.

[43] Velamparambil S, Chew WC, and Song J. 10 million unknowns: Is it that big? *IEEE Antennas Propag. Mag.* 2003;45(2):43–58.

[44] Ergül Ö and Gürel L. Optimal interpolation of translation operator in multilevel fast multipole algorithm. *IEEE Trans. Antennas Propag.* 2006;54(12):3822–26.

[45] Brandt A. Multilevel computations of integral transforms and particle interactions with oscillatory kernels. *Comput. Phys. Commun.* 1991;65:24–38.

[46] Greengard L, Huang J, Rokhlin V, *et al.* Accelerating fast multipole methods for the Helmholtz equation at low frequencies. *IEEE Comput. Sci. Eng.* 1998;5:32–38.

[47] Zhao JS and Chew WC. Three dimensional multilevel fast multipole algorithm from static to electrodynamic. *Microw. Opt. Technol. Lett.* 2000;26(1):43–48.

[48] Zhao JS and Chew WC. Applying matrix rotation to the three-dimensional low-frequency multilevel fast multipole algorithm. *Microw. Opt. Technol. Lett.* 2000;26(2):105–10.

[49] Zhao JS and Chew WC. Applying LF-MLFMA to solve complex PEC structures. *Microw. Opt. Technol. Lett.* 2001;28(3):155–60.

[50] Chu YH and Chew WC. A multilevel fast multipole algorithm for electrically small composite structures. *Microw. Opt. Technol. Lett.* 2004;43(3):202–7.

[51] Ergül Ö and Gürel L. Efficient solutions of metamaterial problems using a low-frequency multilevel fast multipole algorithm. *Prog. Electromagn. Res.* 2010;108:81–99.

[52] Jiang LJ and Chew WC. A mixed-form fast multipole algorithm. *IEEE Trans. Antennas Propag.* 2005;53(12):4145–56.

[53] Järvenpää S, and Wallén H. Faster conversion between multipole series and plane waves for broadband MLFMA. *IEEE Trans. Antennas Propag.* 2019;67(6):3987–95.

[54] Jiang LJ and Chew WC. Low-frequency fast inhomogeneous plane-wave algorithm (LF-FIPWA). *Microw. Opt. Technol. Lett.* 2004;40(2):117–22.

[55] Darve E and Have P. Efficient fast multipole method for low-frequency scattering. *J. Comput. Phys*. 2004;197(1):341–63.

[56] Wallén H, Järvenpää S, Ylä-Oijala P, *et al*. Broadband Müller-MLFMA for electromagnetic scattering by dielectric objects. *IEEE Trans. Antennas Propag*. 2007;55(5):1423–1430.

[57] Aronsson J and Okhmatovski V. Vectorial low-frequency MLFMA for the combined field integral equation. *IEEE Antennas Wireless Propag. Lett*. 2011;10:532–35.

[58] Darve E and Have P. A fast multipole method for Maxwell equations stable at all frequencies. *Phil. Trans. R. Soc. Lond. A*. 2004;362:603–28.

[59] Wallén H and Sarvas J. Translation procedures for broadband MLFMA. *Prog. Electromagn. Res*. 2005;55:47–78.

[60] Cheng H, Crutchfield WY, Gimbutas Z, *et al*. A wideband fast multipole method for the Helmholtz equation in three dimensions. *J. Comput. Phys*. 2006;216:300–25.

[61] Wulf D and Bunger R. An efficient implementation of the combined wideband MLFMA/LF-FIPWA. *IEEE Trans. Antennas Propag*. 2009;57(2):467–74.

[62] Dufva T and Sarvas J. Broadband MLFMA with plane wave expansions and optimal memory demand. *IEEE Trans. Antennas Propag*. 2009;57(3):742–53.

[63] Järvenpää S, Markkanen J, and Ylä-Oijala P. Broadband multilevel fast multipole algorithm for electric-magnetic current volume integral equation. *IEEE Trans. Antennas Propag*. 2013;61(8):4393–97.

[64] Xuan L, Zhu A, Adams RJ, *et al*. A broadband multilevel fast multipole algorithm. In: *Proceedings of the IEEE Antennas and Propagation Society International Symposium*, 2004: 1195–98.

[65] Bogaert I, Peeters J, and Olyslager F. A nondirective plane wave MLFMA stable at low frequencies. *IEEE Trans. Antennas Propag*. 2008;56(12): 3752–67.

[66] Bogaert I and Olyslager F. A low frequency stable plane wave addition theorem. *J. Comput. Phys*. 2009;228(4):1000–16.

[67] Peeters J, Cools K, Bogaert I, *et al*. Embedding Calderon multiplicative preconditioners in multilevel fast multipole algorithms. *IEEE Trans. Antennas Propag*. 2010;58(4):1236–50.

[68] Khalichi B, Ergül Ö, and Ertürk VB. Broadband solutions of potential integral equations with NSPWMLFMA. *IEEE Trans. Antennas Propag*. 2019;67(6):4307–12.

[69] Bogaert I, Pissoort D, and Olyslager F. A normalized plane wave method for 2D Helmholtz problems. *Microw. Opt. Technol. Lett*. 2006;48(2):237–43.

[70] Chen H, Leung KW, and Yung EKN. Fast directional multilevel algorithm for analyzing wave scattering. *IEEE Trans. Antennas Propag*. 2011;59(7):2546–56.

[71] Ergül Ö and Karaosmanoğlu B. Approximate stable diagonalization of the Green's function for low frequencies. *IEEE Antennas Wireless Propag. Lett*. 2014;13:1054–56.

[72] Ergül Ö and Karaosmanoğlu B. Broadband multilevel fast multipole algorithm based on an approximate diagonalization of the Green's function. *IEEE Trans. Antennas Propag.* 2015;63(7):3035–41.
[73] Takrimi M, Ergül Ö, and Ertürk VB. A novel broadband multilevel fast multipole algorithm with incomplete-leaf tree structures for multiscale electromagnetic problems. *IEEE Trans. Antennas Propag.* 2016;64(6):2445–56.
[74] Ergül Ö and Karaosmanoğlu B. Low-frequency fast multipole method based on multiple-precision arithmetic. *IEEE Antennas Wireless Propag. Lett.* 2014;13:975–78.
[75] Kalfa M, Ergül Ö, and Ertürk VB. Error control of multiple-precision MLFMA. *IEEE Trans. Antennas Propag.* 2018;66(10):5651–56.
[76] Vikram M, Huang H, Shanker B, *et al.* A novel wideband FMM for fast integral equation solution of multiscale problems in electromagnetics. *IEEE Trans. Antennas Propag.* 2009;57(7):2094–104.
[77] Melapudi V, Shanker B, Seal S, *et al.* A scalable parallel wideband MLFMA for efficient electromagnetic simulations on large scale clusters. *IEEE Trans. Antennas Propag.* 2011;59(7):2565–77.
[78] Wu L, Zhao Y, Cai Q, *et al.* An adaptive segmented reduced basis method for fast interpolating the wideband scattering of the dielectric–metallic targets. *IEEE Antennas Wireless Propag. Lett.* 2020;19(12):2235–39.
[79] Hughey S, Aktulga HM, Vikram M, *et al.* Parallel wideband MLFMA for analysis of electrically large, nonuniform, multiscale structures. *IEEE Trans. Antennas Propag.* 2019;67(2):1094–107.
[80] Khalichi B, Ergül Ö, Takrimi M, *et al.* Broadband analysis of multiscale electromagnetic problems: novel incomplete-leaf MLFMA for potential integral equations. *IEEE Trans. Antennas Propag.* 2021;69(12):9032–9037.
[81] Wu F, Zhang Y, Oo ZZ, *et al.* Parallel multilevel fast multipole method for solving large-scale problems. *IEEE Antennas Propag. Mag.* 2005;47(4): 110–18.
[82] Gürel L and Ergül Ö. Fast and accurate solutions of extremely large integral-equation problems discretised with tens of millions of unknowns. *Electron. Lett.* 2007;43(9):499–500.
[83] Ergül Ö and Gürel L. Hierarchical parallelisation strategy for multilevel fast multipole algorithm in computational electromagnetics. *Electron. Lett.* 2008;44(1):3–5.
[84] Pan XM and Sheng XQ. A sophisticated parallel MLFMA for scattering by extremely large targets. *IEEE Antennas Propag. Mag.* 2008;50(3):129–38.
[85] Ergül Ö and Gürel L. Efficient parallelization of the multilevel fast multipole algorithm for the solution of large-scale scattering problems. *IEEE Trans. Antennas Propag.* 2008;56(8):2335–45.
[86] Fostier J and Olyslager F. Provably scalable parallel multilevel fast multipole algorithm. *Electron. Lett.* 2008;44(19):1111–13.
[87] Fostier J and Olyslager F. Full-wave electromagnetic scattering at extremely large 2-D objects. *Electron. Lett.* 2009;45(5):245–46.

[88] Ergül Ö and Gürel L. A hierarchical partitioning strategy for an efficient parallelization of the multilevel fast multipole algorithm. *IEEE Trans. Antennas Propag.* 2009;57(6):1740–50.

[89] Taboada JM, Landesa L, Obelleiro F, *et al.* High scalability FMM-FFT electromagnetic solver for supercomputer systems. *IEEE Antennas Propag. Mag.* 2009;51(6):21–8.

[90] Araujo MG, Taboada JM, Obelleiro F, *et al.* Supercomputer aware approach for the solution of challenging electromagnetic problems. *Prog. Electromagnet. Res.* 2010;101:241–56.

[91] Taboada JM, Araujo MG, Bertolo JM, *et al.* MLFMA-FFT parallel algorithm for the solution of large-scale problems in electromagnetics. *Prog. Electromagn. Res.* 2010;105:15–30.

[92] Fostier J and Olyslager F. An open-source implementation for full-wave 2D scattering by million-wavelength-size objects. *IEEE Antennas Propag. Mag.* 2010;52(5):23–34.

[93] Ergül Ö and Gürel L. Rigorous solutions of electromagnetic problems involving hundreds of millions of unknowns. *IEEE Antennas Propag. Mag.* 2011;53(1):18–27.

[94] Pan XM, Pi WC, and Sheng XQ. On OpenMP parallelization of the multilevel fast multipole algorithm. *Prog. Electromagn. Res.* 2011;112:199–213.

[95] Melapudi V, Shanker B, Seal S, *et al.* A scalable parallel wideband MLFMA for efficient electromagnetic simulations on large scale clusters. *IEEE Trans. Antennas Propag.* 2011;59(9):2565–77.

[96] Gürel L and Ergül Ö. Hierarchical parallelization of the multilevel fast multipole algorithm (MLFMA). *IEEE Proc.* 2013;101(2):332–41.

[97] Michiels B, Fostier J, Bogaert I, *et al.* Full-wave simulations of electromagnetic scattering problems with billions of unknowns. *IEEE Trans. Antennas Propag.* 2015;63(2):796–99.

[98] Yang ML, Wu BY, Gao HW, *et al.* Ternary parallelization approach of MLFMA for solving electromagnetic scattering problems with over 10 billion unknowns. *IEEE Trans. Antennas Propag.* 2019;67(11): 6965–78.

[99] Cwikla M, Aronsson J, and Okhmatovski V. Low-frequency MLFMA on graphics processors. *IEEE Antennas Wireless Propag. Lett.* 2010;9: 8–11.

[100] Donepudi KC, Jin JM, and Chew WC. A higher order multilevel fast multipole algorithm for scattering from mixed conducting/dielectric bodies. *IEEE Trans. Antennas Propag.* 2003;51(10):2814–21.

[101] Fostier J and Olyslager F. An asynchronous parallel MLFMA for scattering at multiple dielectric objects. *IEEE Trans. Antennas Propag.* 2008;56(8):2346–55.

[102] Hu J, Lei L, Nie Z, *et al.* Fast solution of electromagnetic scattering from thin dielectric coated PEC by MLFMA and successive overrelaxation iterative technique. *IEEE Microw. Wireless Comp. Lett.* 2009;19(12):762–64.

[103] Tong MS and Chew WC. Multilevel fast multipole acceleration in the Nyström discretization of surface electromagnetic integral equations for composite objects. *IEEE Trans. Antennas Propag.* 2010;58(10): 3411–16.

[104] Tong MS and Chew WC. Fast convergence of fast multipole acceleration using dual basis function in the method of moments for composite structures. *IEEE Trans. Antennas Propag.* 2011;59(7):2741–46.

[105] Ergül Ö. Solutions of large-scale electromagnetics problems involving dielectric objects with the parallel multilevel fast multipole algorithm. *J. Opt. Soc. Am. A.* 2011;28(11):2261–68.

[106] Fostier J, Michiels B, Bogaert I, *et al.* A fast 2-D parallel multilevel fast multipole algorithm solver for oblique plane wave incidence. *Radio Sci.* 2011;46(6006).

[107] Ergül Ö. Parallel implementation of MLFMA for homogeneous objects with various material properties. *Prog. Electromagn. Res.* 2011;121: 505–20.

[108] Ergül Ö. Fast and accurate analysis of homogenized metamaterials with the surface integral equations and the multilevel fast multipole algorithm. *IEEE Antennas Wireless Propag. Lett.* 2011;10:1286–89.

[109] Zha LP, Hu YQ, and Su T. Efficient surface integral equation using hierarchical vector bases for complex EM scattering problems. *IEEE Trans. Antennas Propag.* 2012;60(2):952–57.

[110] Araujo MG, Taboada JM, Rivero J, *et al.* Solution of large-scale plasmonic problems with the multilevel fast multipole algorithm. *Opt. Lett.* 2012;37(3):416–18.

[111] Ergül Ö. Fast and accurate solutions of electromagnetics problems involving lossy dielectric objects with the multilevel fast multipole algorithm. *Eng. Anal. Bound. Elem.* 2012;36(3):423–32.

[112] Ergül Ö. Analysis of composite nanoparticles with surface integral equations and the multilevel fast multipole algorithm. *J. Opt.* 2012;14(062701).

[113] Ergül Ö and Gürel L. Accurate solutions of extremely large integral-equation problems in computational electromagnetics. *IEEE Proc.* 2013;101(2): 342–49.

[114] Ergül Ö and Gürel L. Fast and accurate analysis of large-scale composite structures with the parallel multilevel fast multipole algorithm. *J. Opt. Soc. Am. A.* 2013;30(3):509–17.

[115] Solis DM, Araujo MG, Landesa L, *et al.* MLFMA-MoM for solving the scattering of densely packed plasmonic nanoparticle assemblies. *IEEE Photon. J.* 2015;7(3).

[116] Karaosmanoğlu B, Yılmaz A, Gür UM, *et al.* Solutions of plasmonic structures using the multilevel fast multipole algorithm. *Int. J. RF Microwave Comput.-Aided. Eng.* 2016;26(4):335–41.

[117] Yang M, Wu Y, Ren KF, *et al.* Computation of radiation pressure force exerted on arbitrary shaped homogeneous particles by high-order Bessel vortex beams using MLFMA. *Opt. Exp.* 2016;24(276483).

[118] Kong BB and Sheng XQ. An efficient preconditioning approach for surface MLFMA solution of scattering from multilayer dielectric bodies. *IEEE Antennas Wireless Propag. Lett.* 2017;16:1192–95.

[119] Yu M, Han Y, Cui Z, *et al.* Scattering of a Laguerre–Gaussian beam by complicated shaped biological cells. *J. Opt. Soc. Am. A*. 2018;35(9): 1504–10.

[120] Gonzalez I, Tayebi A, Gomez J, *et al.* Fast analysis of a dual-band reflectarray using two different numerical approaches based on the moment method. *IEEE Trans. Antennas Propag.* 2013;61(4):2333–36.

[121] Yang ML, Gao HW, and Sheng XQ. Parallel domain-decomposition-based algorithm of hybrid FE-BI-MLFMA method for 3-D scattering by large inhomogeneous objects. *IEEE Trans. Antennas Propag.* 2013;61(9):4675–84.

[122] Guan J, Yan S, and Jin JM. An accurate and efficient finite element-boundary integral method with GPU acceleration for 3-D electromagnetic analysis. *IEEE Trans. Antennas Propag.* 2014;62(12):6325–36.

[123] Jiang M, Hu J, Tian M, *et al.* An enhanced preconditioned JMCFIE-DDM for analysis of electromagnetic scattering by composite objects. *IEEE Antennas Wireless Propag. Lett.* 2015;14:1362–65.

[124] Solis DM, Obelleiro F, and Taboada JM. Surface integral equation – domain decomposition scheme for solving multiscale nanoparticle assemblies with repetitions. *IEEE Photon. J.* 2016;8(5).

[125] Guan J, Yan S, and Jin JM. A multi-solver scheme based on combined field integral equations for electromagnetic modeling of highly complex objects. *IEEE Trans. Antennas Propag.* 2017;65(3):1236–47.

[126] Jiang M, Li Y, Rong Z, *et al.* Fast solving scattering from multiple bodies of revolution with arbitrarily metallic-dielectric combinations. *IEEE Trans. Antennas Propag.* 2019;67(7):4748–55.

[127] Li L, He J, Liu Z, *et al.* MLFMA analysis of scattering from multiple targets in the presence of a half-space. *IEEE Trans. Antennas Propag.* 2003;51(4):810–19.

[128] Yang W, Zhao Z, Qi C, *et al.* Iterative hybrid method for electromagnetic scattering from a 3-D object above a 2-D random dielectric rough surface. *Prog. Electromagn. Res.* 2011;117:435–48.

[129] Lu CC. A fast algorithm based on volume integral equation for analysis of arbitrarily shaped dielectric radomes. *IEEE Trans. Antennas Propag.* 2003;51(3):606–12.

[130] Nie XC, Yuan N, Li LW, *et al.* A fast volume-surface integral equation solver for scattering from composite conducting-dielectric objects. *IEEE Trans. Antennas Propag.* 2005;53(2):818–24.

[131] Rawat V and Webb JP. Scattering from dielectric and metallic bodies using a high-order, Nystrom, multilevel fast multipole algorithm. *IEEE Trans. Magn.* 2006;42(4):521–26.

[132] Wang B, He M, Liu J, *et al.* An efficient integral equation/modified surface integration method for analysis of antenna-radome structures in receiving mode. *IEEE Trans. Antennas Propag.* 2014;62(9):4884–89.

[133] Takrimi M, Ergül Ö, and Ertürk VB. Incomplete-leaf multilevel fast multipole algorithm for multiscale penetrable objects formulated with volume integral equations. *IEEE Trans. Antennas Propag.* 2017;65(9):4914–18.

[134] Li X, Lei L, Chen Y, *et al.* Efficient electromagnetic analysis for complex planar thin-layer composite objects by a hybrid method. *IEEE Antennas Wireless Propag. Lett.* 2019;18(9):1706–10.

[135] Wu L, Zhao Y, Cai Q, *et al.* An adaptive segmented reduced basis method for fast interpolating the wideband scattering of the dielectric-metallic targets. *IEEE Antennas Wireless Propag. Lett.* 2020;19(12):2235–39.

[136] İbili H, Koyaz Y, Özmü U, *et al.* Novel SIE implementations for efficient and accurate electromagnetic simulations of zero-index materials. In: *Proceedings of the Photonics & Electromagnetics Research Symposium*, 2019: 3772–80.

[137] Karaosmanoğlu B, Tuygar E, Topçuoğlu U, *et al.* Improving the absorption of solar cells using antenna-inspired cavities. *Microw. Opt. Technol. Lett.* 2019;61(8):1924–30.

[138] Eriş Ö, Ergül Ö. Design and simulation of beam-generating shells with near-zero-index characteristics. In: *Proceedings of the URSI General Assembly and Scientific Symposium*, 2021.

[139] Johnson PB and Christy RW. Optical constants of the noble metals. *Phys. Rev. B.* 1972;6(12):4370–79.

[140] Önol C, Üçüncü A, and Ergül Ö. Efficient multilayer iterative solutions of electromagnetic problems using approximate forms of the multilevel fast multipole algorithm. *IEEE Antennas Wireless Propag. Lett.* 2017;16:3253–56.

[141] Altınoklu A and Ergül Ö. Design, optimization, and analyses of nano-optical couplers consisting of nanocubes to construct efficient nanowire transmission systems. *Prog. Electromagn. Res. C.* 2021;113:13–27.

[142] Haupt RL and Werner DH. *Genetic Algorithms in Electromagnetics*. Hoboken, NJ: Wiley; 2007.

[143] Önol C, Karaosmanoğlu B, and Ergül Ö. Efficient and accurate electromagnetic optimizations based on approximate forms of the multilevel fast multipole algorithm. *IEEE Antennas Wireless Propag. Lett.* 2016;15:1113–15.

[144] Fromm DP, Sundaramurthy A, Schuck PJ, *et al.* Gap-dependent optical coupling of single bowtie nanoantennas resonant in the visible. *Nano Lett.* 2004;4(5):957–61.

[145] Alda J, Rico-Garcia JM, Lopez-Alonso JM, *et al.* Optical antennas for nano-photonic applications. *Nanotechnology*. 2005;16:230–34.

[146] Muhlschlegel P, Eisler HJ, Martin OJF, *et al.* Resonant optical antennas. *Science*. 2005;308:1607–09.

[147] Işıklar G, C. Çetin İ, Algun M, *et al.* Design and analysis of nanoantenna arrays for imaging and sensing applications at optical frequencies. *Adv. Electromagnet.* 2019;8(2):18–27.

[148] Karaosmanoğlu B, Koyaz Y, İbili H, *et al.* Fast and accurate analysis of three-dimensional structures involving near-zero-index materials. In:

Proceedings of the International Conference on Electromagnetics in Advanced Applications, 2019: 1019–1024.

[149] Eriş Ö, İbili H, and Ergül Ö. Low-cost inkjet-printed multiband frequency-selective structures consisting of U-shaped resonators. *Prog. Electromagn. Res. C.* 2020;98:31–44.

[150] Karaosmanoğlu B, Eray H, and Ergül Ö. Full-wave optimization of three-dimensional photonic-crystal structures involving dielectric rods. *J. Opt. Soc. Am. A.* 2018;37(7):1103–13.

[151] Karaova G, Altınoklu A, and Ergül Ö. Full-wave electromagnetic optimisation of corrugated metallic reflectors using a multigrid approach. *Sci. Rep.* 2018;8(1267).

[152] Karaova G, Eriş Ö, and Ergül Ö. Combined potential-field formulation for densely discretized conductors. In: *Proceedings of the IEEE Antennas and Propagation Society International Symposium*, 2021.

[153] Taskinen M and Ylä-Oijala P. Current and charge integral equation formulation. *IEEE Trans. Antennas Propag.* 2006;54(1):58–67.

[154] Chen YP, Jiang L, Qian ZG, *et al.* An augmented electric field integral equation for layered medium Green's function. *IEEE Trans. Antennas Propag.* 2008;59(3):960–68.

[155] Karaosmanoğlu B and Ergül Ö, Acceleration of MLFMA simulations using trimmed tree structures. *IEEE Trans. Antennas Propag.* 2021;69(1):356–65.

Chapter 4
Kernel-independent fast factorization methods for multiscale electromagnetic problems

Mengmeng Li[1]*, Paola Pirinoli*[2]*, Francesca Vipiana*[2] *and Giuseppe Vecchi*[2]

In this chapter, the low-rank factorization methods for real-life multiscale simulations are proposed. The low-rank factorization methods are fully algebraic, the rank-deficient nature is exploited for the coupling matrix blocks produced by two well-separated groups. It is well known, the whole impedance matrix of method of moments (MoM) is full-rank, while the off-diagonal matrix blocks are low-rank. The off-diagonal matrix blocks are produced by the "far" coupling basis functions, which are over sampling than the Nyquist limit [1]. With the low-rank factorization methods, the impedance matrix can be approximated for fast evaluations of matrix–vector products in iterative solutions [2] or fast direct solvers [3,4].

4.1 Introduction

The setup to obtain the low-rank decomposition matrices is generally evaluated with QR decomposition, singular value decomposition (SVD), and adaptive cross approximation (ACA) [5,6]. The low-rank decomposition is purely algebraic, kernel-independent and easily employed for existing MoM codes are important advantages than other fast methods.

For the multiscale problems such as large aircraft platforms installed with fine antenna array and instruments, the coarse meshes are used to capture the large-scale interactions showing high-frequency problems, while the dense meshes are used to capture the geometric details showing low-frequency problems. They are typical mixed frequency problems, the challenges for multiscale simulation are mainly from:

For standard fast multipole method (FMM), it would meet

(1) Low-frequency breakdown would be encountered when the far coupling group size is smaller than 0.2 wavelength for the multipole expansion. Several wide-band fast multipole algorithm methods are proposed to solve the low-frequency problems [7–12].

[1]School of Microelectronics and Integrated Circuits, Nanjing University of Science and Technology, China
[2]Department of Electronics and Telecommunications, Politecnico di Torino, Italy

(2) Dense meshes with much a small electrical size will be appearing. In the standard electric field integral equation (EFIE), the magnitude of vector potential is much smaller than the scalar potential. The produced matrix system will be ill-conditioning and even breaks down due to finite machine precision. The ill-conditioning performance would be further worsened by the nonuniform discretization.

The low-rank factorization methods are one of the most effective solvers for the above challenges. They are low-frequency stable, the hierarchical preconditioner [13,14] and direct solvers [3,4,15,16] are proposed with low-rank factorization respectively, to solve the matrix system from an ill-conditioning matrix system.

4.2 Adaptive cross approximation (ACA) method

The ACA is one of the most popular fast low-rank factorization methods to compress the off-diagonal impedance matrices. It was first introduced with non-oscillating kernels [5], then it was employed to evaluate the EFIE for electromagnetic compatibility simulation [6]. For a low-rank matrix $\mathbf{Z}$ with dimensions $m \times n$, a product $\mathbf{UV}$ can be generated with ACA for approximations, where $\mathbf{U}$ is $m \times r$ and $\mathbf{V}$ is $r \times n$, the time and memory computational cost is proportional to $r(m+n)$ instead of mn. r is the rank for a predefined approximation tolerance ϵ, which is much smaller than m and n.

When the far couplings are evaluated with fast solvers, the MoM discretized linear system can be written as

$$\mathbf{ZI} = (\mathbf{Z}_{\text{near}} + \mathbf{Z}_{\text{far}})\,\mathbf{I} \tag{4.1}$$

where $\mathbf{Z}_{\text{near}}$ is the near-field coupling, it is evaluated with MoM directly. $\mathbf{Z}_{\text{far}}$ is the far-field couplings, it is evaluated with low-rank matrix factorization method, i. e. ACA [6]. For the couplings between groups t and s, the far-field couplings admission condition is

$$R(s,t) \geq 2D_l \tag{4.2}$$

where $R(s,t)$ is the distance between the group center of s and t, D_l is the group size at the peer couplings level l. For low-rank factorization methods, the far-coupling impedance matrix subblock $\mathbf{Z}_{m\times n}$ with dimensions of $m \times n$ can be approximated by as a product of matrices with smaller dimensions

$$\mathbf{Z}_{m\times n} \approx \mathbf{U}_{m\times r} \times \mathbf{V}_{r\times n} \tag{4.3}$$

The low-rank factorization gives a good approximation of the original coupling matrix with a predefined tolerance

$$\mathbf{R}_{m\times n} = \|\mathbf{Z}_{m\times n} - \mathbf{U}_{m\times r}\mathbf{V}_{r\times n}\| \leq \epsilon\,\|\mathbf{Z}_{m\times n}\| \tag{4.4}$$

where $\mathbf{R}$ is the error matrix, $\|.\|$ is the Frobenus norm. The computation cost savings would be achieved for $r \ll min(m,n)$ [6]. Let $\mathbf{I} = [I_1, \ldots I_r]$ and $\mathbf{J} = [J_1, \ldots J_r]$ as the sampled row and column indexes of the original matrix $\mathbf{Z}_{m\times n}$, $\mathbf{u}_k$ and $\mathbf{v}_k$ are the

Algorithm 4.1: ACA

% Initialization:
Set the 1st row index $I_1 = 1$ and set $\tilde{\mathbf{Z}} = 0$
Set the 1st row of the error matrix $\mathbf{R}(I_1, :) = \mathbf{Z}(I_1, :)$
Sample the 1st column index in the first row J_1: $|\mathbf{R}(I_1, J1)| = \max_j(\mathbf{R}(I_1, j))$
$\mathbf{v}_1 = \mathbf{R}(I_1, :)/\mathbf{R}(I_1, J_1)$
Set the 1st column of the error matrix $\mathbf{R}(:, J_1) = \mathbf{Z}(:, J_1)$
$\mathbf{u}_1 = \mathbf{R}(:.J_1)$
$\|\tilde{\mathbf{Z}}^{(1)}\|^2 = \|\tilde{\mathbf{Z}}^{(0)}\|^2 + \|\mathbf{u}_1\|^2\|\mathbf{v}_1\|^2$
Sample 2nd row index I_2: $|\mathbf{R}(I_2, J_1)| = \max_i(\mathbf{R}(i, J_1)), i \neq I_1$
% kth iteration:
Calculate (I_k)th row of the error matrix:$\mathbf{R}(I_k, :) = \tilde{Z}(I_k) - \sum_{l=1}^{k-1}(\mathbf{u}_l)_{I_k}\mathbf{v}_l$
Sample k_{th} column index J_k: $|\mathbf{R}(I_k, J_k)| = \max_j(R(I_k, j)), j \neq J_1, ..., J_{k-1}$
$\mathbf{v}_k = \mathbf{R}(I_k, :)/\mathbf{R}(I_k, J_k)$
Calculate (J_k) th column of the error matrix: $\mathbf{R}(:, J_k) = \mathbf{Z}(I_k, :) - \sum_{l=1}^{k-1}(\mathbf{v}_l)_{J_k}\mathbf{u}_l$
$\mathbf{u}_k = \mathbf{R}(:, J_k)$
$\|\tilde{\mathbf{Z}}^{(k)}\|^2 = \|\tilde{\mathbf{Z}}^{(k-1)}\|^2 + 2\sum_{j=1}^{k-1}|\mathbf{u}_j^T\mathbf{u}_k| \cdot \|\mathbf{u}_k\|^2\|\mathbf{v}_k\|^2$
If $\|\mathbf{u}_k\|\|\mathbf{v}_k\| \leq \epsilon\|\tilde{\mathbf{Z}}^{(k)}\|$, then stop the iterations
Find next row index I_{k+1} : $|\mathbf{R}(I_{k+1}, J_k)| = \max_i(\mathbf{R}(i, J_k)), i \neq I_1, ...I_k$

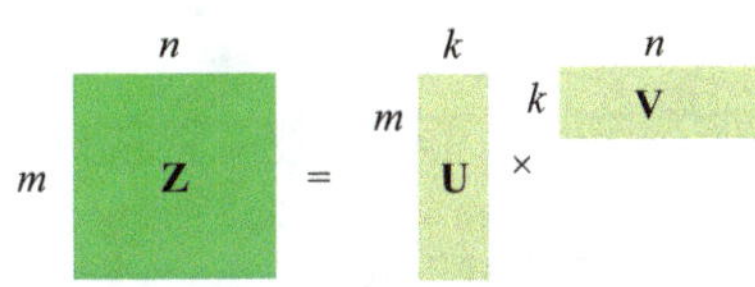

Figure 4.1 Low-rank factorization of the far-coupling impedance matrix between two groups

kth column and row of the low-rank approximation matrices $\mathbf{U}$ and $\mathbf{V}$, $\tilde{\mathbf{Z}}$ is the approximation matrix at kth iteration. The algorithm of ACA for the impedance matrix samplings are show in Algorithm 4.1 [5,6].

For standard low-rank factorization methods, the key differences are the techniques to obtain the low-rank approximation matrices in (4.3). For matrix decomposition algorithm (MDA) [17], equivalent sources are employed to obtain the low-rank approximation, the number of equivalent points is equal to the rank [17]. For ACA [5,6], the dominant columns and rows are sampled automatically with a predefined tolerance. For the UV method [18], the columns and rows are sampled uniformly or randomly. The multiresolution preconditioner is proposed together with the UV method for the modeling of finite size frequency selective surface (FSS) arrays in [19]. The standard low-rank factorization methods are limited to small and medium problems, because the low-rank decompositions in (4.3) are implemented repeatedly for each pair of two coupling groups. The low-rank approximation matrices are constructed level by level even for a multilevel algorithm. The computational complexity is reported as $O(N^{4/3}\log N)$ [5,6] for small and medium problems.

4.3 Multilevel matrix compression method for multiscale problems

4.3.1 Background and theory

Although the gain in efficiency of ACA with respect to direct evaluation of the full impedance matrix is impressive, the rank of the corresponding off-diagonal impedance matrix block grows with the electrical size, initially approximately proportional to the frequency but for an asymptotically large frequency it will grow with the frequency squared [6]. As a consequence, the computational burden of the ACA compression will eventually grow proportionally to $O(N^3)$ and the storage to $O(N^2)$ [6,20].

The low-rank approximation matrices post-compression is usually carried on by application of SVD [21]; the different steps of the process are sketched in Figure 4.2, for the case in which the post-compression is applied to a typical low-rank method. From a mathematical point of view, it consists in writing $\tilde{\mathbf{Z}}_{m\times n}$ as the product of two matrices $\tilde{\mathbf{U}}_{m\times r'}$ and $\tilde{\mathbf{V}}_{r'\times n}$, where r' is smaller than r, i.e.,

$$\tilde{\mathbf{Z}}_{m\times n} = \mathbf{U}_{m\times r} \times \mathbf{V}_{r\times n} = \tilde{\mathbf{U}}_{m\times r'} \times \tilde{\mathbf{V}}_{r'\times n} \tag{4.5}$$

As already noticed in Section 4.2, the matrices $\mathbf{U}$ and $\mathbf{V}$ need to be computed for each pair of coupling groups, and, therefore, the efficiency of the approach results to be lower than that of the FMM [22,23] in which the aggregation and disaggregation operators of a group are defined just once, independently from the far-field interaction groups. For each group, a single aggregation operator for all the interaction groups is obtained through the MultiLevel Matrix Compression Method (MLMCM) [24] while the disaggregation operator is derived via the reciprocal MLMCM (rMLMCM) [25].

In Figure 4.3, the coupling groups in far-field region of group i are shown. Let $[\mathbf{Z}_{ij}]_{m\times n}$ represents the impedance submatrix between i and a far-field group j, being m, n the number of basis functions in groups i and j, respectively. Applying MLMCM, first the submatrix that collects the column vectors $\mathbf{Z}_{i,1}, \mathbf{Z}_{i,2}, \mathbf{Z}_{i,3} \ldots$ representing the interaction between group i and all the far-field groups at peer-level can be obtained: it contains redundant information and, therefore, its rank can be reduced thanks to a modified Gram–Schmidt (MGS) algorithm; in this way, the *receiving compression matrix* $\mathbf{U}_i$ is obtained

$$\mathbf{U}_i = \mathrm{MGS}[\mathbf{Z}_{i,1}, \mathbf{Z}_{i,2}, \mathbf{Z}_{i,3} \ldots] \tag{4.6}$$

and the coupling matrix $\mathbf{Z}_{m\times n}$ can be written as

$$[\mathbf{Z}_{ij}]_{m\times n} = [\mathbf{U}_i]_{m\times r}[\mathbf{D}_{ij}]_{r\times r}[\mathbf{V}_j]_{r\times n} \tag{4.7}$$

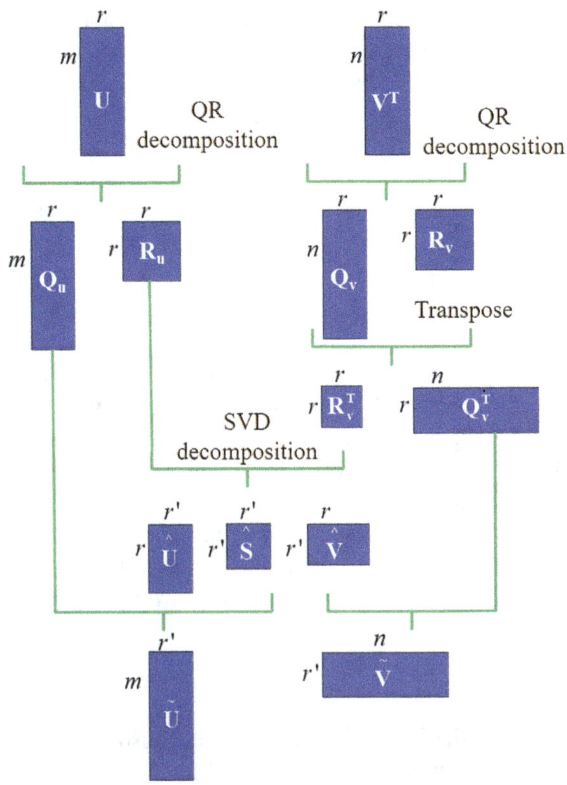

Figure 4.2 SVD post-compression of typical low-rank methods

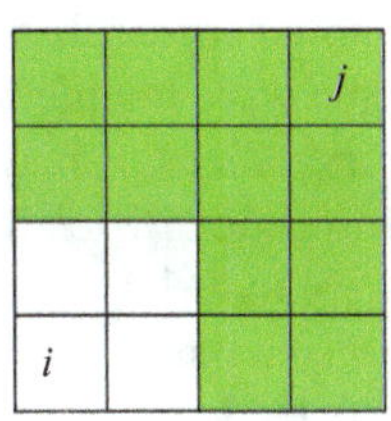

Figure 4.3 Far couplings between groups i and j with MLMCM

where r is the ϵ-rank of $\mathbf{Z}_{ij}$, while $\mathbf{U}_i$, $\mathbf{D}_{ij}$, and $\mathbf{V}_j$ are smaller but dense matrices. Inspired by [26,27], a reciprocal algorithm rMLMCM [25] is developed over [24], where the radiation compression matrix $\mathbf{V}_j$ satisfies

$$\mathbf{V}_j = \mathbf{U}_j^T \tag{4.8}$$

The evaluation of the column vectors in (4.6) is time expensive, but it can be reduced by using an error-controllable procedure, according to which a subset of the column vectors is sampled by ACA [5,6]. By construction, $\mathbf{U}_i$ and $\mathbf{V}_j$ are unitary matrices and therefore the translation matrix $\mathbf{D}_{ij}$ between groups i and j can be written as

$$\mathbf{D}_{ij} = \mathbf{U}_i^\dagger \mathbf{U}_i \mathbf{D}_{i,j} \mathbf{V}_j \mathbf{V}_j^\dagger = \mathbf{U}_i^\dagger \mathbf{Z}_{ij} \mathbf{V}_j^\dagger \tag{4.9}$$

Note that $\mathbf{Z}_{ij}$ in (4.9) is not explicitly computed, but it is instead evaluated with ACA [6]. Since $\mathbf{U}_i^\dagger$ and $\mathbf{V}_j^\dagger$ in (4.9) have dimensions $r \times m$ and $n \times r$ respectively, the translation matrix $[\mathbf{D}_{ij}]_{r\times r}$ in (4.9) has dimensions $(r \times r)$ significantly smaller than those of the original matrix $[\mathbf{Z}_{ij}]_{m\times n}$.

Two different strategies have been introduced for the rMLMCM, suitable for realistic large and multiscale problems. In the first one, a hybridization of the rMLMCM with MLFMA [25] is proposed: the rMLMCM is employed to evaluate the low- and mid-frequency regime where it is most effective due to the dense mesh of the fine structures installed on the large platforms, while the MLFMA [22,23] is used at the high frequencies, where the rMLMCM suffer for the degeneration of its computation performance. A second option consist in further introducing a hierarchical MultiResolution (MR) preconditioner [14,28,29] is in the rMLMCM/MLFMA scheme, to improve the condition number and consequently the convergence of the ill-conditioned matrix.

4.3.2 Accuracy validation

The octree clustering of the rMLMCM/MLFMA is constructed in such a way that the average number of basis functions at the leaf level is about 50. The MR basis has been generated with a grouping of the mesh resulting in an average cell size of approximately 0.125λ at the top level. All the simulations have been carried out using a 64-bit workstation Dell Precision T7400, Intel Xeon CPU E5440 @ 2.88 GHz, 96 GB of RAM. Single-core, double-precision computation is always used.

To test the accuracy of the proposed approach, a plate with dimensions 1 m × 1 m (see Figure 4.4), excited with a plane wave impinging from ($\theta^i = 0°, \phi^i = 0°$), at the frequency of 3 GHz is considered; a smaller square with a side of 0.1 m, highlighted in Figure 4.4 (a), is meshed more densely, to artificially generate a multiscale problem. The mesh length ranges from 8.0e−4 m to 1.6e−2 m, corresponding to 8.0e−3 λ to 1.6e−1 λ, and the resulting number of unknowns is 70,516. The normalized error for the simulated surface current density by doubly hierarchical MoM with respect to full MoM is 0.03.

4.3.3 Computational complexity analysis

For objects with medium electrical size, the computational complexity of the rMLMCM is $O(N\log N)$ [25]. The computational complexity is tested considering a series of cubic cavities as the one shown in the inset in Figure 4.5(a). The edge length of the cube is 2 m while the average mesh size is selected equal to 6e−2λ. As a result, the total number of unknowns is equal to 2,644, 10,495, 42,080, and 168,520, at the considered frequencies of 125 MHz, 250 MHz, 500 MHz, and 1 GHz.

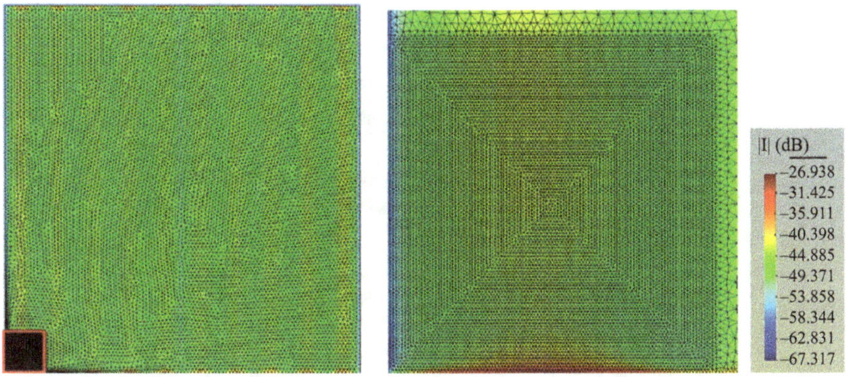

Figure 4.4 Accuracy validation: simulated surface current density distribution [dBA/m] of the square plate with a dense discretization corner at 3 GHz; (a) top view and (b) zoom details of the dense discretization corner. The normalized error for the simulated surface current density by doubly hierarchical MoM with respect to full MoM is 0.03

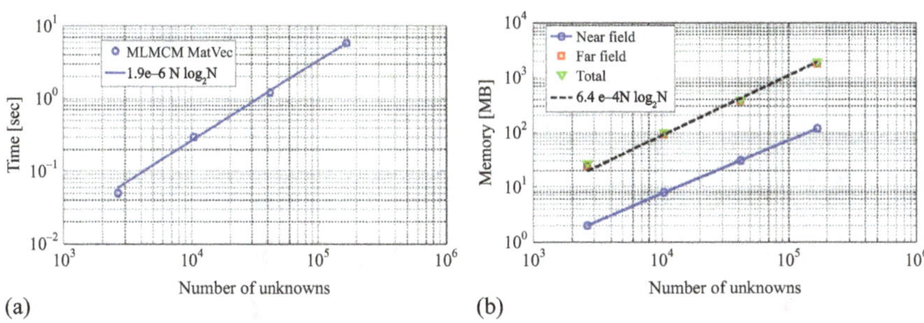

Figure 4.5 Computational complexity test: (a) matrix–vector product (MVP) time and (b) storage requirements

Figure 4.5(a) and (b) shows the computational complexity of a matrix–vector product and the memory requirements of the solver.

4.3.4 Numerical evaluation of the induced fields in a real-life aircraft

The proposed method is finally validated through the analysis of a detailed 3D (morphed) model of a realistic aircraft, the Evektor EV55*. In the model, all the aircraft internal details, such as passenger seats and the instruments board are considered, as shown in Figure 4.6. The aircraft is 14.2 m long, with a wingspan of 16.1 m,

*http://www.evektoraircraft.com/en/aircraft/ev-55-outback/overview

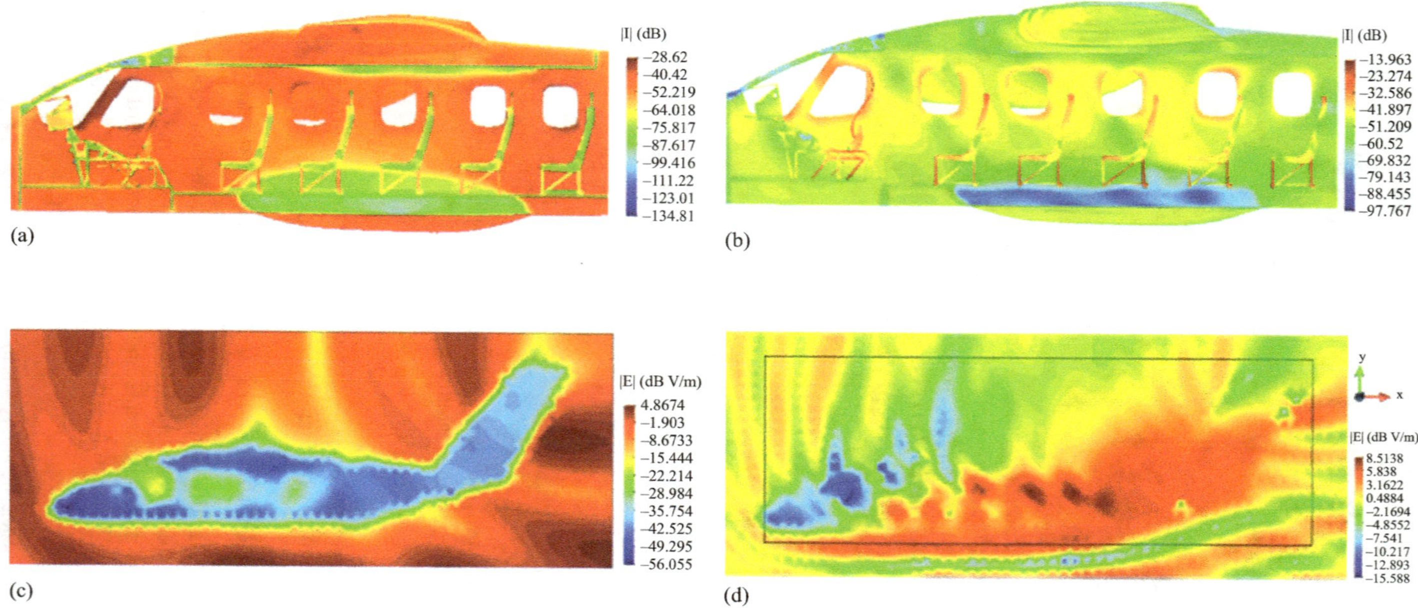

Figure 4.6 Realistic morphed EV55 aircraft: side view of details of surface current density at 75 MHz (a) and 244 MHz (b); electric fields on a symmetry plane of the EV55 aircraft at 75 MHz (c) and 244 MHz (d)

corresponding to 11.5 λ and 13 λ, respectively, at 224 MHz. The mesh size ranges from 4.95e−3 m to 8.5e−2 m, and this means, at the considered frequency, that it varies between 3.6e−3 λ and 6.3e−2 λ. As a consequence, the total number of unknowns is 171,763. The aircraft is illuminated by a plane wave impinging from $(\theta^i = 90^\circ, \phi^i = 225^\circ)$.

The proposed approach is used to evaluate the surface current distribution; single-level rMLMCM and 5-level MLFMA are employed, corresponding to a near-field region of 0.1 λ. The preconditioner is constructed from the near-field region and the skeletons of rMLMCM.

Table 4.1 summarizes the results of the comparison among the performance of the rMLMCM/MLFMA with different preconditioners, i.e., the MR-ILU, ILU, and the diagonal preconditioners. For a mere diagonal preconditioner, convergence cannot be achieved within 2,000 iterations as well as when the ILU preconditioner is employed with a fill-in parameter $p = 1$, since in this case, the ILU decomposition is unstable, as indicated in [13]. When p is increased to 2, the rMLMCM/MLFMA needs approximately 900 iterations to converge, almost the same number that is necessary to reach convergence with the MR-ILU preconditioner (810), but the L and U matrices require 4 times the memory of the near-field matrix and the time for evaluating the decomposition increases of a factor 4 with respect to the case $p = 1$ and it is 62 times larger than the case when MR-ILU [13] is adopted. Finally, the seventh column shows the memory requirements for the different preconditioners: with MR-ILU, only 625 MB are necessary for storing the LU factors, while the memory requirements for the full ILU decomposition are 11 times larger. It is, therefore, possible to conclude that MR-ILU preconditioner guarantees significant savings in total simulation time and memory requirements [25].

Figure 4.6(a) and (b) shows the surface current distributions of the morphed EV55 aircraft at 75 MHz and 244 MHz, respectively. Finally, Figure 4.6(c) and (d) shows the electric field distribution inside the aircraft on a cut in the *xoy* plane.

Table 4.1 Morphed EV55 aircraft simulations: performance comparison of different preconditioning techniques at 244 MHz

	p	**Condest**	**Time for generating L and U (hh: mm: ss)**	**Number of iterations**	**Total simulation time (hh: mm: ss)**	**ILU RAM overhead (MB)**	**Total Memory (GB)**
Diag	–	–	–	–	–	–	5.6
MR-ILU	4	4	00: 03: 56	810	04: 10: 32	625	6.2
ILU	1	NAN	01: 47: 28	–	–	7223	–
ILU	2	8.7E4	04: 09: 17	906	10: 25: 01	14,446	20.0

4.4 Nested equivalence source approximation for low-frequency multiscale problems

As already pointed out in Section 4.3, the low-rank factorization of MLMCM [24,25] is similar to the FMM [22,23], since for each group, only one radiation compression and receiving compression matrices are required to be computed. However, the low-rank factorization has to be computed at each level, with a consequent increasing of the approach computation complexity in the case of large problems. To mitigate this drawback, a nested approximation framework can be constructed, resorting an idea similar to that adopted by MLFMA [22,23] in which the computational complexity is reduced thank to nested factorization denoted as transfer matrices between adjacent levels. In the resulting approach, named nested equivalent source approximation (NESA) at low-to-high frequency [30], the low-rank factorization at all the levels but the leaf one are expressed in terms of that at this finest level, that is the only one computed. More specifically, the idea is that of reusing the radiation and receiving matrices at a level l to represent the matrices at the next, higher level $l-1$, i.e., to write

$$\mathbf{V}_{j^p}^{l-1} = \mathbf{C}^{l-1,l}\mathbf{V}_j^l \tag{4.10}$$

$$\mathbf{U}_{i^p}^{l-1} = \mathbf{U}_i^l\mathbf{B}^{l,l-1} \tag{4.11}$$

where i^p and j^p at level $l-1$ are the parent groups of groups i and j at level l, and $\mathbf{C}^{l-1,l}$ and $\mathbf{B}^{l,l-1}$ are the transfer matrices between them. From 4.4, it emerges that in NESA, to evaluate $\mathbf{V}_j^l$, $\mathbf{U}_j^l$, it is necessary to only compute the radiation and receiving matrices at the leaf level and the transfer matrices between adjacent levels. This recurrence is responsible for the following reduced $O(N)$ complexity scaling [30,31] that becomes the same as the low-frequency MLFMA [8].

4.4.1 Equivalent source distributions for field representation

The NESA is originally inspired by [31] for what concerns the construction of the nested scheme, but it is not a trivial extension from scalar (static) to vector problem. In fact, while in [31], the couplings between Octree groups at different levels are approximated through ACA [5,6], here, the equivalence theorem is applied to evaluate the field outside a closed surface Σ due to sources inside the volume delimited by Σ by substituting these last with equivalent sources located on the surface itself. For doing this, it is necessary to introduce equivalent surfaces on the groups, and *equivalent* sources placed on them, able to reproduce the field radiated by the *actual* sources.

For implementation simplicity, the equivalent sources adapted here are RWGs located on the equivalence surface: as will be discussed in the following, their number affects the factorization accuracy and the efficiency. The adoption of equivalent RWGs is a reminiscence of MLMDA [17,21], while their use and realization is completely different. In fact, here the goal is to build a nested scheme whose recurrence is given by (4.4)–(4.11), and for doing this, *two* concentric equivalence surfaces are introduced for each group, as depicted in Figure 4.7(a): a source surface $\sum_{\tau}^{s}$ around the considered group bounding box, and a "test" surface $\sum_{\sigma}^{s}$, at the interface between near- and far-field regions.

Instead of *individual* sources, source *distribution* are introduced on $\sum_{\tau}^{s}$: they generate on the test surface $\sum_{\sigma}^{s}$ the same field produced by all the RWG basis functions in the considered group and, therefore, they could be interpreted as a change of basis to represent the interactions in terms of the equivalent sources (see Section 4.4.2). These distributions are obtained via a fully algebraic inverse-source process, whose computational cost does not depend on N.

4.4.2 Field representation via equivalent RWG basis functions

Let consider a group of source functions with size S at a specific octree level, as the one shown in Figure 4.7(a), where the two-sphere surfaces, $\sum_{\tau}^{s}$ and $\sum_{\sigma}^{s}$ are also plotted. Both of them are centered in the group center, and they have radius R_τ and $R_\sigma = 3S/2$, respectively. R_τ is fixed as the radius of the minimum sphere encompassing the bounding box, i.e., $R_\tau = \sqrt{3}S/2$, although the choice $R_\tau < \sqrt{3}S/2$ is often more convenient for practical implementations [30]. Spherical surfaces are preferable, when compared with placing the equivalent RWGs on cubes directly [21], since their uniform sampling represents the *optimal* choice [30]. Three quasi-orthogonal RWG basis functions, two of which are tangent basis functions and one is a radial basis function [21] are located in correspondence of each sampling point as depicted in Figure 4.7(b): their directions are $(\hat{r}, \hat{\theta}, \hat{\phi})$, having defined the spherical coordinate system with respect to the center of the group. The edge length of the auxiliary RWGs is taken equal to a fraction of the average meshing size, since in this way, it is possible to reduce the number of integration points on the equivalent RWGs, without loss of accuracy.

4.4.3 Single-level nested matrix compression approximation algorithm

In Figure 4.8, the approximation process for the coupling between groups s and o in the case of a single-level NESA is described. The most significant parameters of the

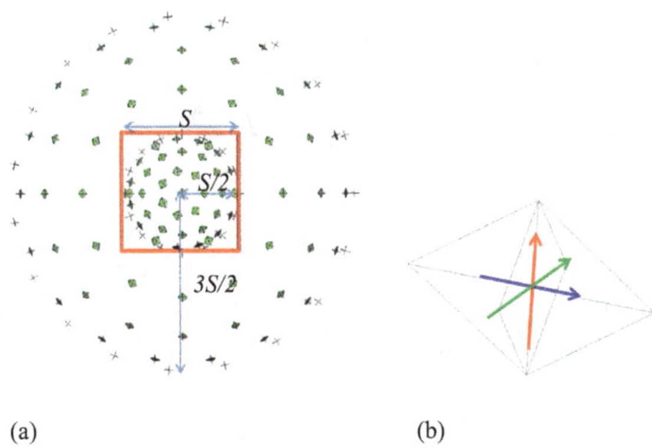

(a) (b)

Figure 4.7 Equivalent RWG basis functions of a group (a) on the internal and external equivalence surface (b) details of three independent RWG basis functions at an equivalent point

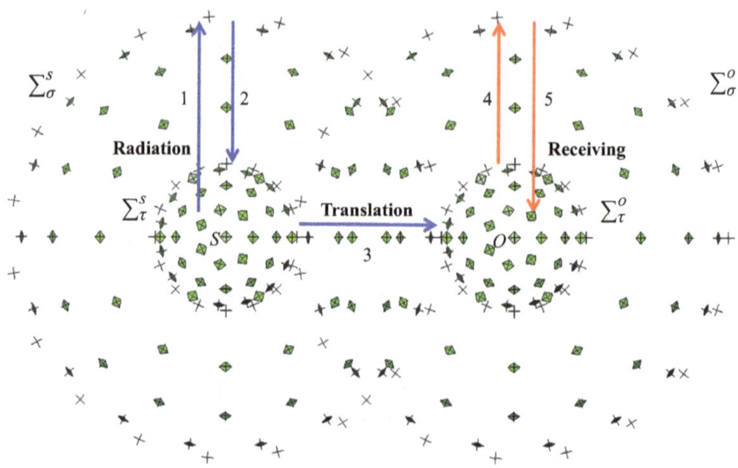

*Figure 4.8 Evaluation of the coupling between two groups at peer level using NESA. The field on the external testing sphere surface is enforced in a weak form by the equivalent and source distributions. Computation of radiation matrix **V** for group s by the forward and backward radiation processes are symbolized by 1 and 2, similarly, 4 and 5 denote the forward and the backward radiation to obtain receiving matrix **U** of group o. The translation between two groups is symbolized by 3*

Table 4.2 Parameter notation in NESA

$\sum_\tau^i$	The *equivalent* sphere of radius R_τ for group i
$\sum_\sigma^i$	The *testing* sphere of radius $R_\sigma = \sqrt{3}S/2$ for group i
$\boldsymbol{\tau}_i$	RWG basis functions on equivalent sphere surface of group i
$\boldsymbol{\sigma}_i$	RWG test functions on testing sphere surface of group i
$\mathbf{Z}_{i,j}$	Impedance sub-matrix between groups i and j
$\mathbf{I}_i$	Current density coefficients for basis function in group i
$\mathbf{E}_i$	Projection of the electric field onto test functions in group i
L	Total number of levels in the Octree

nested algorithm are listed in Table 4.2. Let s be a group of source functions: as shown in Figure 4.8, the goal is to determine the coefficients of the equivalent sources $\boldsymbol{\tau}_s$ on $\sum_\tau^s$ responsible for the same (within a predefined tolerance) field radiated by the actual sources in s on the test surface $\sum_\sigma^s$. These coefficients can be obtained equating (in a weak sense) the fields radiated by the two sets of sources, i.e., by testing both the fields on the testing functions $\boldsymbol{\sigma}_s$ located on $\sum_\sigma^s$

$$\mathbf{Z}_{\sigma_s,s}\mathbf{I}_s = \mathbf{Z}_{\sigma_s,\tau_s}\mathbf{I}_{\tau_s} \tag{4.12}$$

where $\mathbf{I}_s$ and $\mathbf{I}_{\tau_s}$ are the current coefficients of the actual and equivalent RWG basis functions, respectively, and $\mathbf{Z}_{\sigma_s,\tau_s}$ is the matrix whose elements are evaluated testing the radiated fields by RWGs on equivalent and testing surfaces. The coefficients $\mathbf{I}_{\tau_s}$ are, therefore, given by

$$\mathbf{I}_{\tau_s} = \mathbf{Z}^{\dagger}_{\sigma_s,\tau_s}\mathbf{Z}_{\sigma_s,s}\mathbf{I}_s \tag{4.13}$$

where $(\cdot)^{\dagger}$ denotes the pseudo-inverse. Here the pseudo-inverse matrix is obtained by means of a truncated SVD (TSVD), according to which only singular values within a given threshold are kept; the effect of the choice of this threshold is discussed in detail in [30].

By reciprocity, in the case of the observation group o as the one sketched on the right side of Figure 4.8, the fields radiated by sources $\boldsymbol{\sigma}_o$ on $\sum_{\sigma}^{o}$ are tested on functions located on $\sum_{\tau}^{o}$ and onto the original test functions of observation group o

$$\mathbf{E}_{\tau_o} = \mathbf{Z}_{\tau_o,\sigma_o}\mathbf{I}_{\sigma_o} \tag{4.14}$$

$$\mathbf{E}_o = \mathbf{Z}_{o,\sigma_o}\mathbf{I}_{\sigma_o} \tag{4.15}$$

If (4.14) is solved for $\mathbf{I}_{\sigma_o}$, and the result is substituted in (4.15), the filed $\mathbf{E}_o$ can be written as

$$\mathbf{E}_o = \mathbf{Z}_{o,\sigma_o}\mathbf{Z}^{\dagger}_{\tau_o,\sigma_o}\mathbf{E}_{\tau_o} \tag{4.16}$$

Moreover, the field $\mathbf{E}_{\tau_o}$ can be expressed in terms of the currents $\mathbf{I}_{\tau_s}$ in the testing sphere $\sum_{\tau}^{s}$ of source group s

$$\mathbf{E}_{\tau_o} = \mathbf{Z}_{\tau_o,\tau_s}\mathbf{I}_{\tau_s} \tag{4.17}$$

Finally, if (4.17) is substituted in (4.16), and $\mathbf{I}_{\tau_s}$ is expressed in terms of the actual sources $\mathbf{I}_s$ with (4.13), it is possible to write

$$\begin{aligned}\mathbf{E}_o &= \mathbf{Z}_{o,s}\mathbf{I}_s \\ &= \mathbf{Z}_{o,\sigma_o}\mathbf{Z}^{\dagger}_{\tau_o,\sigma_o}\mathbf{Z}_{\tau_o,\tau_s}\mathbf{Z}^{\dagger}_{\sigma_s,\tau_s}\mathbf{Z}_{\sigma_s,s}\mathbf{I}_s\end{aligned} \tag{4.18}$$

Eq. (4.18) represents the single-level NESA low-rank approximation of the impedance matrix $\mathbf{Z}_{o,s}$ produced by groups s and o

$$\begin{aligned}\mathbf{Z}_{o,s} &= \mathbf{Z}_{o,\sigma_o}\mathbf{Z}^{\dagger}_{\tau_o,\sigma_o}\mathbf{Z}_{\tau_o,\tau_s}\mathbf{Z}^{\dagger}_{\sigma_s,\tau_s}\mathbf{Z}_{\sigma_s,s} \\ &= \mathbf{U}_o\mathbf{D}_{o,s}\mathbf{V}_s\end{aligned} \tag{4.19}$$

where $\mathbf{U}_o = \mathbf{Z}_{o,\sigma_o}\mathbf{Z}^{\dagger}_{\tau_o,\sigma_o}$ is the *receiving matrix* of group o, $\mathbf{V}_s = \mathbf{Z}^{\dagger}_{\sigma_s,\tau_s}\mathbf{Z}_{\sigma_s,s}$ is the *radiation matrix* of group s, and $\mathbf{Z}_{\tau_o,\tau_s}$ is the *translation matrix* between groups s and o.

4.4.4 Multilevel NESA

The scheme described in the previous section can be generalized to the case of a multilevel version of the NESA: starting from the top level $l = 1$, the equivalent currents and fields at each level are expressed in terms of the equivalent currents and fields at its child level $(l + 1)$, and, therefore, recursively, in terms the quantities at the leaf level L [30], being L also the number of levels in the scheme. Referring to Figure 4.9, the group s^p is assumed to be the parent of group s, while $\boldsymbol{\tau}_{s^p}$ and $\boldsymbol{\sigma}_{s^p}$ are the equivalent RWGs on the equivalent and testing surface of group s^p. Analogically

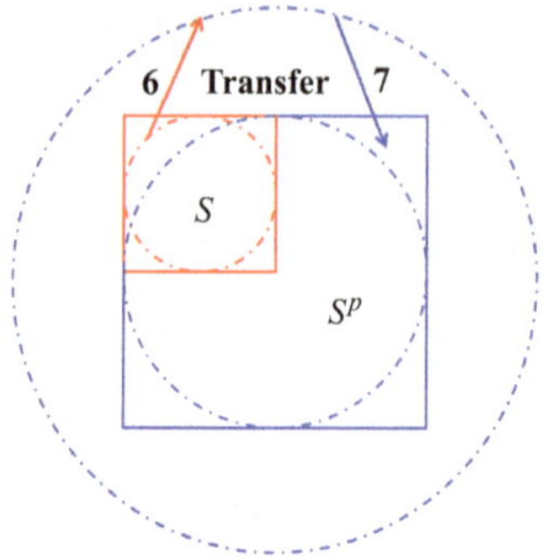

Figure 4.9 Forward and backward radiation noted as 6 and 7 is employed to compute the radiation matrix of parent group s^p by its child groups

to (4.12)–(4.13), it is, therefore, possible to express the equivalent current of group s^p in terms of the current $\mathbf{I}_{\tau_s}$ of its child s

$$\mathbf{I}_{\tau_{s^p}} = \mathbf{Z}^{\dagger}_{\sigma_{s^p},\tau_{s^p}} \mathbf{Z}_{\sigma_{s^p},\tau_s} \mathbf{I}_{\tau_s} \tag{4.20}$$

and the field E_{τ_o} on $\sum_{\tau}^{o}$ in terms of the field $\mathbf{E}_{\tau_{o^p}}$ on $\sum_{\tau}^{o^p}$, where o^p is the parent group of group o, likewise (4.16)

$$\mathbf{E}_{\tau o} = \mathbf{Z}_{\tau_o,\sigma_{o^p}} \mathbf{Z}^{\dagger}_{\tau_{o^p},\sigma_{o^p}} \mathbf{E}_{\tau_{o^p}} \tag{4.21}$$

Let first consider the case in which the nested algorithm just consists in two levels: introducing the *transfer matrix* from parent to child group $\mathbf{B} = \mathbf{Z}_{\tau_o,\sigma_{o^p}} \mathbf{Z}^{\dagger}_{\tau_{o^p},\sigma_{o^p}}$ and the *transfer matrix* from child to parent group $\mathbf{C} = \mathbf{Z}^{\dagger}_{\sigma_{s^p},\tau_{s^p}} \mathbf{Z}_{\sigma_{s^p},\tau_s}$, the approximation of the couplings between groups o and s becomes

$$\mathbf{Z}_{o,s} = \mathbf{U}_o \mathbf{B}_{o,o^p} \mathbf{D}_{o^p,s^p} \mathbf{C}_{s^p,s} \mathbf{V}_s \tag{4.22}$$

where $\mathbf{D}_{o^p,s^p} = \mathbf{Z}_{\tau_{o^p},\sigma_{o^p}}$. The difference between (4.22) and (4.19) consists mainly in the fact that the translation matrix in (4.19) is expanded in terms of:

- a translation matrix $\mathbf{D}_{o^p,s^p}$ between parents of groups s and o;
- two transfer matrices ($\mathbf{B}_{o,o^p}$ and $\mathbf{C}_{s^p,s}$) allowing to ascend/descend the tree.

Finally, the procedure can be extended to the generic multi-level case by recursively expanding the translation matrix in (4.22) up to the level l where the couplings are effectively computed [30]

$$\mathbf{Z}^l_{i,j} = \mathbf{U}^L_i \mathbf{B}^{L,L-1}_i \cdots \mathbf{B}^{l+1,l}_i \mathbf{D}^l_{i,j} \mathbf{C}^{l,l+1}_j \cdots \mathbf{C}^{L-1,L}_j \mathbf{V}^L_j \tag{4.23}$$

where $\mathbf{B}^{l+1,l}_i$ and $\mathbf{C}^{l,l+1}_j$ are the transfer matrices allowing to connect the level l with the $(l-1)$ and vice versa. Note that, in the case of an EFIE problem, the receiving and the radiation matrix of a given group are one the transpose of the other and, therefore, it is sufficient to compute and to store just one of the two. The pseudo-code of the proposed multi-level NESA factorization is shown in Algorithm 4.2.

Algorithm 4.2: Multi-level NESA low-rank approximation

Procedure NESA decompositions ($s, o, \boldsymbol{\tau}_s, \boldsymbol{\tau}_o, \boldsymbol{\sigma}_s, \boldsymbol{\sigma}_o$,
U, D, V, B, C)
% Compute radiation matrices
for all *levels* $i = L : 1 : -1$ **do**
  if $i = L$ **then**
    $\mathbf{V}^L \leftarrow$ compute radiation matrices with (4.13)
  else
    $\mathbf{C}^i \leftarrow$ compute transfer matrices with (4.20)
  end if
end for
% Compute translation matrices
for all *levels* $i = L : 1 : -1$ **do**
  $\mathbf{D}^i \leftarrow$ compute translation matrices with (4.17)
end for
% Compute receiving matrices
for all *levels* $i = L : 1 : -1$ **do**
  if $i = L$ **then**
    $\mathbf{U}^L \leftarrow$ compute receiving matrices with (4.16)
  else
    $\mathbf{B}^i \leftarrow$ compute transfer matrices with (4.21)
  end if
end for

Table 4.3 Parameter notation in the MVP and following

s^l	Non-empty source group s at level l
o^l	Non-empty observation group o at level l
$\mathbf{I}_i$	Subvector of $\mathbf{I}$ restricted to basis functions in group i
$\mathbf{V}_i^l$	Radiation matrix for group i at level l
$\mathbf{B}_i^l$	Transfer matrix for group i at level l
$\boldsymbol{\zeta}_i^l$	Temporary vector in MVP in the radiation process aggregated from the current in group i at level l
$\mathbf{D}_{i,j}^l$	Translation matrix between groups i and j at level l
$\boldsymbol{\xi}_i^l$	Temporary vector in MVP in the translation process translated from group i far interaction list at level l
$\mathbf{U}_i^l$	Receiving matrix for group i at level l
$\mathbf{C}_i^l$	Transfer matrix for group i at level l
$\mathbf{y}_i^l$	Temporary vector in MVP in the receiving process receiving from the field in group i at level l
$\mathbf{y}$	Result of the MVP $\mathbf{y} = \mathbf{ZI}$
N_i	Number of basis functions in group i
M_l	Number of non-empty groups at level l

Algorithm 4.3: MVP with NESA

Procedure MVP(**y**, **I**)
% Far interactions:
% Radiation process
for all *levels* $i = L : 1 : -1$ **do**
  if $i = L$ **then**

$$\boldsymbol{\zeta}^L_{s^L} \leftarrow \boldsymbol{\zeta}^L_{s^L} + \mathbf{V}^L_{s^L} \mathbf{I}_{s^L} \tag{4.24}$$

  else

$$\boldsymbol{\zeta}^i_{s^i} \leftarrow \boldsymbol{\zeta}^i_{s^i} + \mathbf{C}^i_{s^i} \boldsymbol{\zeta}^{i+1}_{s^{i+1}} \tag{4.25}$$

  end if
end for
% Translation process
for all *levels* $i = L : 1 : -1$ **do**

$$\boldsymbol{\xi}^i_{o^i} \leftarrow \boldsymbol{\xi}^i_{o^i} + \mathbf{D}_{o^i,s^i} \boldsymbol{\zeta}^i_{s^i} \tag{4.26}$$

end for
% Receiving process
for all *levels* $i = 1 : L$ **do**
  if $i \neq L$ **then**

$$\boldsymbol{\xi}^{i+1}_{o^{i+1}} \leftarrow \boldsymbol{\xi}^{i+1}_{o^{i+1}} + \mathbf{B}^i_{o^i} \boldsymbol{\xi}^i_{o^i} \tag{4.27}$$

  else

$$\mathbf{y}^L_{o^L} \leftarrow \mathbf{y}^L_{o^L} + \mathbf{U}^L_{o^L} \boldsymbol{\xi}^L_{o^L} \tag{4.28}$$

  end if
end for
% Sum near interactions at the finest level L:

$$\mathbf{y} \leftarrow \mathbf{y}^L + \mathbf{Z}_{near} \mathbf{I} \tag{4.29}$$

Return y
End procedure

4.4.5 Matrix–vector product and computation complexity

In this section, the analysis of the NESA numerical complexity is carried on [30]. In Table 4.3, the main parameters used in the computation of the matrix–vector product (MPV) $\mathbf{y} = \mathbf{ZI}$ and their description are listed, while the pseudo-code of the algorithm is shown in Algorithm 4.3. In the following, both the memory requirements and the computational cost of the MVPs are considered.

Memory storage

At the leaf level, for each group i, it is sufficient to store only one radiation matrix $\mathbf{V}_i$: since it has size $(Q \times N_i)$, it requires a storage QN_i. The total memory requirement for all the $\mathbf{V}_i$ matrices for all the groups at the leaf level is, therefore, equal to $\sum_i QN_i = QN$, being N the total number of unknowns. In addition to the matrices $\mathbf{V}_i$, for each level $l \neq L$, it is necessary to store 8 matrices which describe the possible relations between parents and children, i.e., between two different levels; each of these transfer matrices has size $Q \times Q$, and, therefore, the cost for storing all of them is $8(L-1)Q^2$. Moreover, at each level, it is necessary to store 316 translation matrices, each with size $Q \times Q$ and this corresponds to store in total $316LQ^2$ elements, L being the number of levels. If the average number of basis functions at the leaf level is fixed to a constant value K, the number of non-zero elements in the near-field matrix is approximately bounded by $9KN$ where the factor 9 has been obtained through geometrical considerations on the number of near-field groups for each source group. The overall memory requirements of the NESA is, therefore, given by

$$C_{mem} \approx QN + 8(L-1)Q^2 316LQ^2 + 9KN \tag{4.30}$$

As a consequence of the assumption that the number of functions K per group at the leaf level is constant, the number of levels L grows with the number of unknowns as $O(\log_2 N)$. Furthermore, at low frequency, Q is constant too, and so the memory requirement in (4.30) can be bounded by

$$C_{mem} \approx O(N) + O(\log_2 N) \tag{4.31}$$

Numerical cost of the MPVs

The cost of evaluation the quantity in (4.24) in the radiation process is $O(QN)$, since each product $\mathbf{V}_{s^L}^L \mathbf{I}_{s^L}$ has a cost QN_s^L, and the sum over all the source groups at the leaf level gives $\sum_{s^L} QN_s^L = QN$. The computation of MVPs in (4.25) has a cost equal to $8Q^2 \sum_{l=1}^{L-1} M_l$, because each non-empty group at level $l \neq L$ has at most 8 non-empty child groups. The complexity of the translation process (4.26) is $316Q^2 \sum_{l=1}^{L} M_l$, since the number of far-field groups in the interaction list related to a non-empty group at level l is bounded by 316. By reciprocity, the cost of the receiving process in (4.27)–(4.28) is equal to that just evaluated for the radiation process. Finally, the computational cost of the MVP involving the near-field matrix is related to the number of its non-zero entries that is bounded by $9KN$, as already pointed out.

As it appears from $316Q^2 \sum_{l=1}^{L} M_l$, the complexity of MVPs depends on the scaling properties of $\sum_{l=1}^{L} M_l$. Let the number of non-empty groups at the leaf level be $M_L = N/K$; moving at an higher level $l < L$, the number of non-empty groups is reduced by a factor 4, i.e., $M_{l-1} = M_l/4$; applying recursively this relation, it is possible to express the number of non-empty groups at a generic level l as a function of those at level L $M_l = \frac{M_L}{4^{L-1}}$, and, therefore, the total number of non-empty groups is

$$\sum_{l=1}^{L} M_l = \frac{N}{4^L K} \sum_{l=1}^{L} 4^l = \frac{N}{3K} \left(4 - \frac{1024K^2}{N^2} \right) \approx \frac{4N}{3K} \tag{4.32}$$

Using the above result, an estimation of the overall cost of each MVP is

$$\begin{aligned} C_{MVP} &= 2\left(QN + 8Q^2\sum_{l=1}^{L-1} M_l\right) + 316Q^2\sum_{l=1}^{L} M_l + 9KN \\ &= 2\left(QN + 8Q^2\frac{N}{3K}\right) + 316Q^2\frac{4N}{3K} + 9KN \approx O(N) \end{aligned} \tag{4.33}$$

Looking at 4.30 and 4.33, it is possible to conclude that both the storage requirements and the computational complexity of matrix–vector products theoretically scale linearly with the number of unknowns.

4.4.6 Numerical results

In this section, the results relative to different test cases are presented to confirm the effectiveness of the proposed solver. In all the considered examples, a BiCGStab iterative solver is employed and the convergence criterion is fixed at a threshold of 10^{-4}; the average number of basis functions at the finest level of the Octree is kept constant to a value of 50 and, therefore, the average meshing edge h varies with the wavelength λ. To overcome the low-frequency and dense meshing issues, an MR preconditioner [13,28,32] is applied, as proposed in [32]: it allows to avoid the problems of the convergence and accuracy of the fast solver due to the bad conditioning of the system at very low frequencies. All the simulations have been carried out on a 64-bits workstation Dell Precision T7400, Intel Xeon CPU E5440 @ 2.88 GHz, 96 GB of RAM one-core; double precision computation is always used.

Accuracy

The accuracy of the method is affected by the number Q of equivalent sources but it also depends on other quantities, as the truncation threshold ε_{SVD} adopted in the TSVD is used to evaluate the pseudo-inverse matrix in (4.13). To investigate how the NESA accuracy is related to these different factors, 800 source points $\mathbf{r}_n$ and 800 observation points $\mathbf{r}_m$, randomly distributed in two cubes of edge 0.5 m, far 1 m from each other (from center to center) are considered; the accuracy of the NESA is assessed by its application to the evaluation of the free space scalar Green's function at 300 GHz, for which the analytical solution $G(\mathbf{r}_m, \mathbf{r}_n) = e^{-jk_0|\mathbf{r}_m - \mathbf{r}_n|}/|\mathbf{r}_m - \mathbf{r}_n|$ can be used as a reference. The goodness of the NESA is expressed through the approximation error, defined as $\|\mathbf{G} - \mathbf{G}_{\text{NESA}}\|_2/\|\mathbf{G}\|_2$, where $\mathbf{G}$ and $\mathbf{G}_{\text{NESA}}$ are two column vectors collecting the scalar Green's function between all pairs of source/observation points, evaluated analytically and with NESA, respectively, and $\|\mathbf{x}\|_2$ indicates the ℓ^2 norm of the vector $\mathbf{x}$.

Two different possibilities have been considered for the radius R_τ of the sphere where equivalent sources are located; in the first case, it is the radius $R_\tau = R_0 = \sqrt{3}S/2$ of the minimum sphere enclosing all the actual sources, that is tangent to the vertices of the bounding box cube of side S; as a second choice, the sphere is selected to be internally tangent to the facets of the cube, and therefore $R_\tau = S/2 = 0.57R_0$. Note that in this last case the sphere does not include entirely the box inside which the sources are defined, and, therefore, a strict application of the equivalence theorem would not be possible.

First, the impact of the TSVD threshold is examined, assuming $R_\tau = R_0 = \sqrt{3}S/2$, so that the spherical surface contains all the sources and the equivalence theorem is satisfied. In Figure 4.10, the variation of the approximation error with Q for different value of the TSVD threshold is plotted. Initially, the error decreases when the number Q of equivalent sources increases, independently from the value of ε_{SVD}, while for $Q \geq 300$, the behavior of the different curves changes accordingly to the value of the SVD truncation threshold: when it is between 1^{-4} and 1^{-6} the error stay constant, if $\varepsilon_{SVD} = 1^{-7}$, it continue to decrease up to $Q = 600$ and then it stagnates while this does not happen when $\varepsilon_{SVD} = 1^{-12}$, when all the singular values are retained in the pseudo-inverse. From these results, it emerges that no regularization is necessary, in the sense that keeping all the singular values in the pseudo-inverse has no negative effect. Even though this result seems to be surprising for an inverse source problem, it has to be highlighted that in the considered case, the aim is not to find the sources, but only to evaluate the *field* they radiate on the testing surface and beyond. The ill-conditioning of the field-to-source inverse problem is mainly due to the fact that the radiation operator from R_τ to R_σ behaves as a spatial low-pass filter and, therefore, it reduces the degrees of freedom of the field; on the other hand, this behavior allows us to filter away any noise that may appear in the equivalent sources due to the absence of regularization [33].

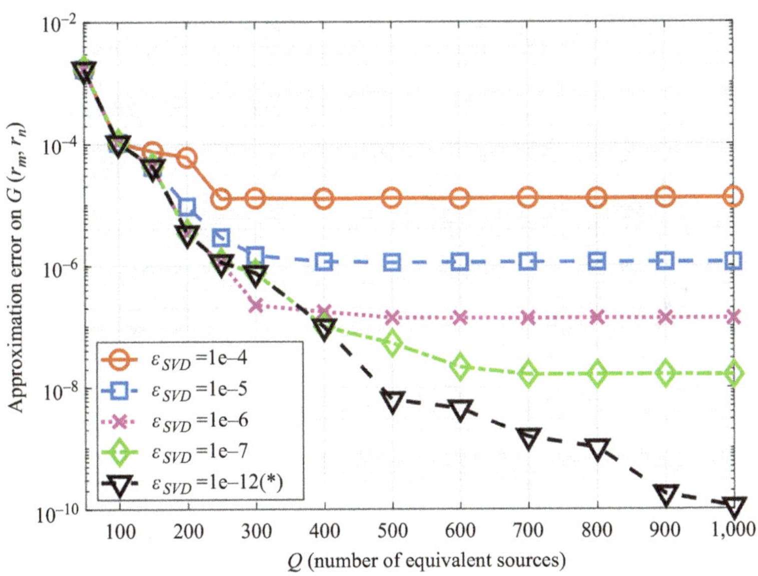

Figure 4.10 Accuracy validation of scalar Green's function with different SVD truncation threshold ε_{SVD}; of external case when $R_\tau = R_0$, R_0 is the radius of the minimum sphere encompassing the bounding box

When the equivalent surface has radius $R_\tau = 0.57R_0$, the sphere is internal to the cube inside which the sources are located. Also in this case, the approximation error has been evaluated as a function of Q and for different values of ε_{SVD}: the resulting curves are plotted in Figure 4.11, where the most significant cases of Figure 4.10 are also reported for comparison. While for higher values of ε_{SVD}, the curves have the same behavior that those for $R_\tau = R_0$, where the error stays constant after a given value of Q, when the threshold is below a critical value it occurs that for $Q > 500$, the accuracy deteriorates and, therefore, the use of regularization could be advantageous. From the comparison between the results relative to the two considered cases, it is possible to deduce that when the spherical surface is internal to the cube including the sources, some field information is lost, and, therefore, regularization is needed to control the error introduced by the approximation that has not to be ascribed to the intrinsic properties of the involved radiation operator. On the other hand, in most of the MoM simulations in cases of practical interest, the accuracy achieved with the "internal" sphere is quite stable with respect to the SVD truncation, with the advantage that it is reached with significantly fewer points [30], as will be proved by the results discussed in the following, relative to the application of NESA to the approximate evaluation of matrix entries.

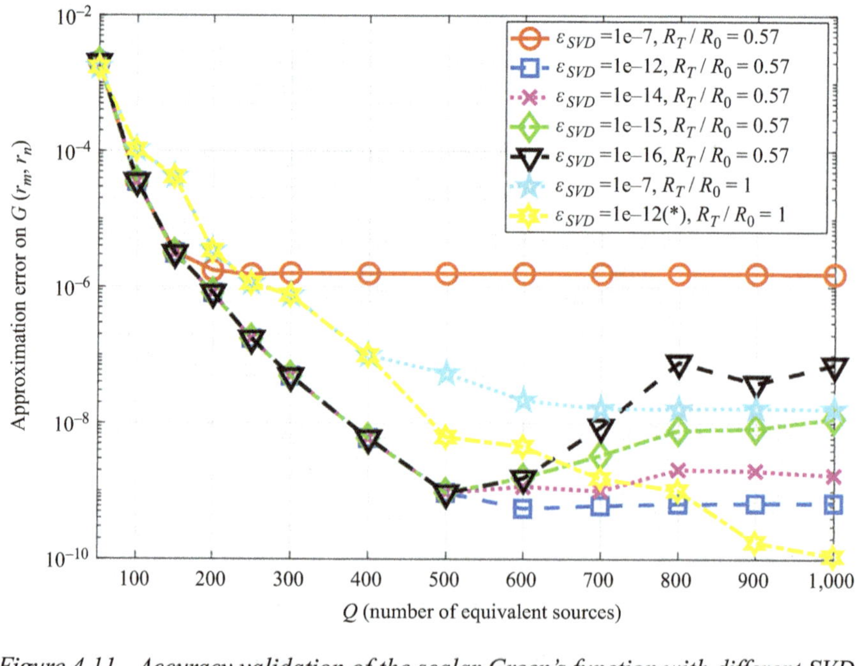

Figure 4.11 Accuracy validation of the scalar Green's function with different SVD truncation threshold ε_{SVD} of internal case when $R_\tau = 0.57R_0$ at 300 MHz

In view of these considerations, in the following, an SVD truncation threshold equal to 1e−12 for the "internal" surface choice, and to 1e–16 for the "external" surface choice will be used.

The further step was that of testing the accuracy of NESA when it is used to approximate the entries of the EFIE operator impedance matrix. For this aim, a sphere with a radius of 2λ has been considered, and discretized with a mesh where 7,188 unknows are defined; from the Octree clustering such mesh, two coupling groups with 256 and 258 RWG basis functions respectively, have then been selected. The entries of the submatrix representing the coupling between the two groups have been evaluated with NESA, following the steps described in (4.12)–(4.17), and adopting different quadrature rules, each one characterized by different numbers m and n of Gaussian points used for the external and the internal triangle, respectively. The obtained submatrices are with that computed using a standard MoM with a very accurate quadrature rule, i.e., with 61 Gaussian points for both the external and the internal. In Figure 4.12, the 2-norm error introduced on the matrix evaluation as a function of Q is plotted, for $R_\tau = R_0$ and $R_\tau = 0.57R_0$ and considering different quadrature rules, each characterized by the label "rule = (m,n)" and by different values for m and n. The accuracy of the matrix evaluation performed with the standard MoM is clearly constant, since it does not depend on Q.

From the curves in Figure 4.12, it appears that the error η_{imp} introduced on the impedance matrix by the approximation depends on both Green's function approximation error η_G and the integral formula precision η_{int}:

$$\eta_{imp} \propto \eta_G + \eta_{int} \tag{4.34}$$

At their turn, the two errors η_G and η_{int} are affected by different parameters: η_G depends only on the number Q of equivalent sources (that also determines, as already discussed, the memory requirements and MVP time of NESA), while η_{int} is determined by the accuracy of the employed quadrature formulas (affecting only the setup time, but not memory and MVP time). Obviously, when the two errors in (4.34) are not comparable, the overall error is dominated by the largest one. Figure 4.12 can, therefore, be interpreted and used as a guideline to choose the parameters of NESA: assuming to select the desired accuracy (that could be, for instance, achievable by standard MoM when integrals are evaluated with a quadrature rule with 3 Gaussian points, as shown in Figure 4.12), it is possible to determine an optimal set of parameters (Q, R_τ, quadrature rule), that allows to minimize the computational cost of the matrix evaluation. If, for instance, the desired accuracy is fixed to be equal to $\backsim 10^{-4}$, the optimal choice consists in taking $Q = 60$, $R_\tau = 0.57R_0$, and a quadrature rule with 3 Gaussian points on triangles.

As a further test case, NESA has been used to evaluate the RCS of a sphere with a radius of 0.5 m, on which a plane wave impinges from the direction ($\theta = 0°, \phi = 0°$); the sphere is discretized with 7188 unknowns and analyzed at the frequency of 3 MHz. The considered number of samples located on the equivalence and testing sphere is 30, corresponding to 90 equivalent RWG functions, since in each sample point, three independent RWGs are defined, as shown in Figure 4.7. Figure 4.13 plots the obtained RCS (in dB) at the considered frequency; as a comparison, the

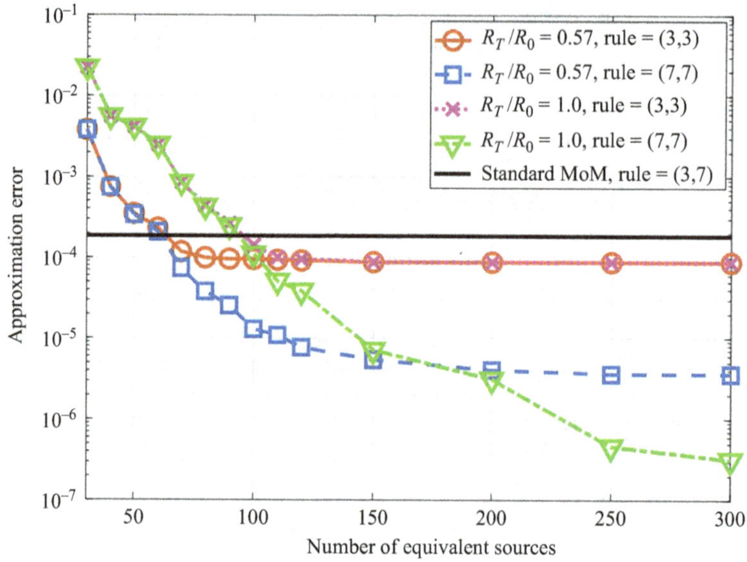

Figure 4.12 NESA approximation of the EFIE matrix entries between two coupling groups with 256 and 258 number of unknowns

RCS evaluated analytically with Mie series is plotted: the agreement between the two result is excellent, and this proves the effectiveness of the approximated approach. It is worth to notice that the choice of taking 30 samples corresponds to the minimum accuracy shown in Figure 4.11: this proves that also a relatively low accuracy can guarantee satisfactory RCS results for simple geometries. In this case, the truncation error in ACA [5,6] is 1e−4.

Finally, to demonstrate the NESA effectiveness also at quasi-resonant frequencies, it has been adopted for computing the RCS at 300 MHz for a sphere with a radius of 2 m (corresponding to 4λ at the working frequency), discretized with 13,246 triangles; the results in Figure 4.14 are compared with those obtained with Mie series: also in this case the agreement between the two results is excellent and this confirms the accuracy of NESA.

Computational complexity

In Section 4.4.5, it was derived that theoretically NESA has a linear computational complexity and memory requirement; to check numerically these results, here is a cube with side 1 m, discretized with 12,672, 50,688, 202,752, and 811,008 RWGs is analyzed. If the ratio between the size mesh and the wavelength, h/λ is kept constant to a value equal to 5.0e−5, the change in the number of basis functions is equivalent to analyze the cube at the frequencies of 37.5 kHz, 75 kHz, 150 kHz, and 300 kHz. In correspondence with these four frequencies, NESA with a different number of level is adopted, ranging from two at 37.5 kHz up to five when the analysis

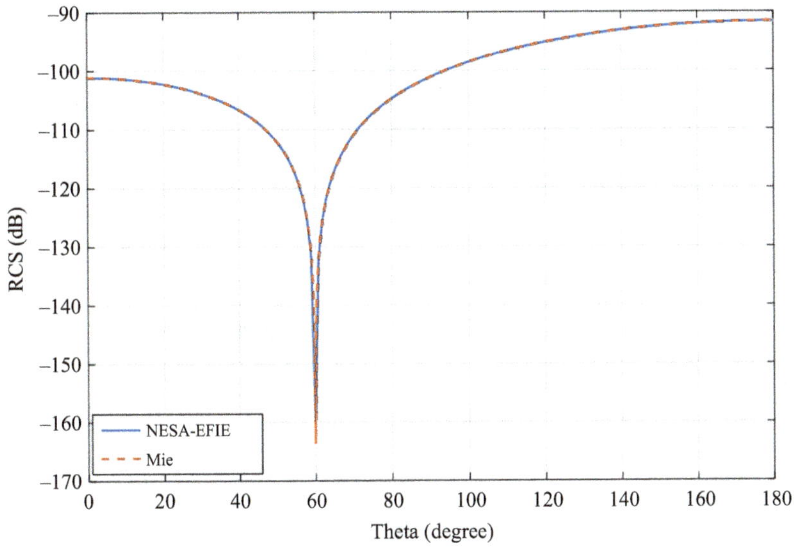

Figure 4.13 Simulated RCS curves of a sphere with a radius of 0.5 m at 3 MHz, the incident direction is $(\theta = 0°, \phi = 0°)$

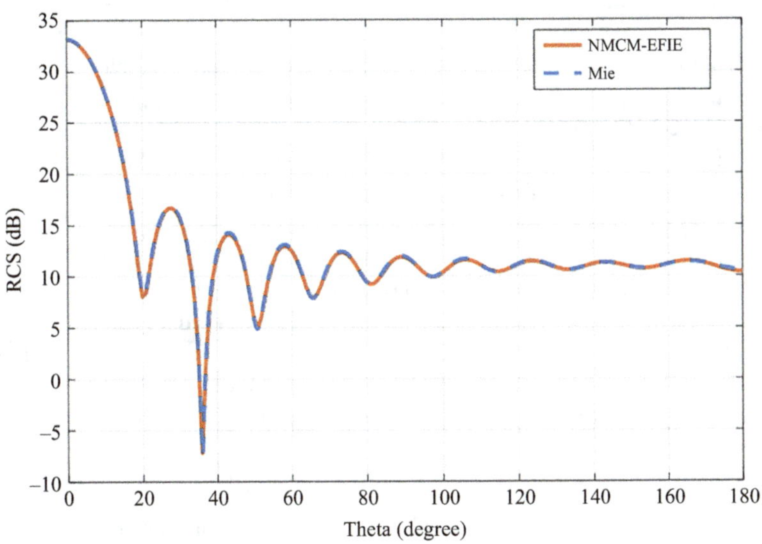

Figure 4.14 Simulated RCS curves of a sphere with a radius of 2 m at 300 MHz, the incident direction is $(\theta = 0°, \phi = 0°)$

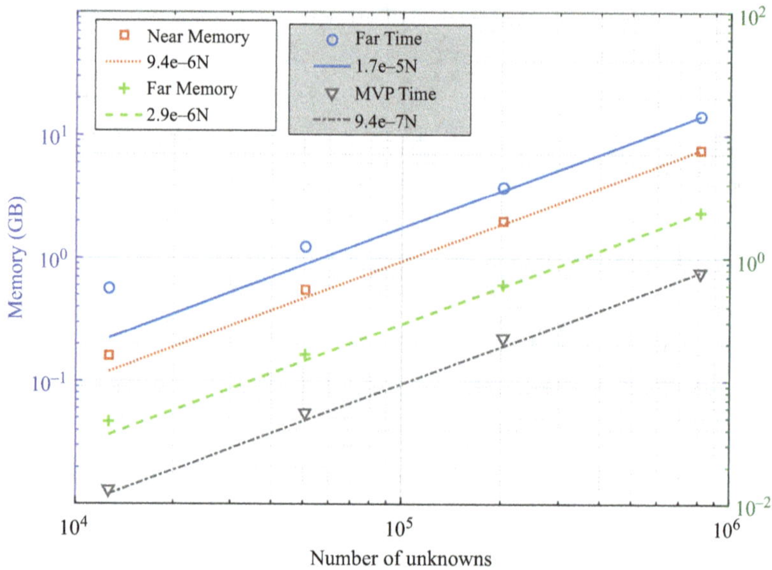

Figure 4.15 Tested time and memory complexity of NESA for the simulation of a cube with side 1 m and fixed $h/\lambda = 1.0e{-5}$

at 300 kHz is performed. In Figure 4.15, the memory requirement and the time necessary to apply the proposed algorithm to the considered case are plotted, versus the number of unknowns: as it clearly appears from the curves in 4.15 the overall complexity of NESA is linear, both for what concerns the memory requirement and the computational time. Note that a linear complexity does not characterized any other standard low-rank method [5,6], that does not employ a nested approximation but are based on the construction of the low-rank approximation level by level. In opposite, the use of the nested approximation and of operators similar to the aggregation and disaggregation operators of the MLFMA, allows to reduce the NESA complexity from $O(N\log N)$ to $O(N)$ [30].

To verify that the proposed NESA still shows a linear scaling even at frequencies where the object size is almost resonant, a further analysis has been performed. For doing this, the cube with an edge equal to 1 m is still analyzed, keeping the ratio h/λ equal to 1.0e−2; the frequency is changed in such a way that the edge of the cube has an electrical length ranging from $\lambda/4$ to 2λ. The curves in Figure 4.16 are linear again and this demonstrates that the proposed algorithm can be effectively applied not only to pure low-frequency problems [30].

NESA validation through the modeling of a real aircraft

As a final example of the NESA capability in modeling high-definition multiscale structures, the proposed approach has been used to simulate a morphed P180 aircraft†

†http://www.piaggioaero.com/#/en/products/p180-avanti-ii/overview

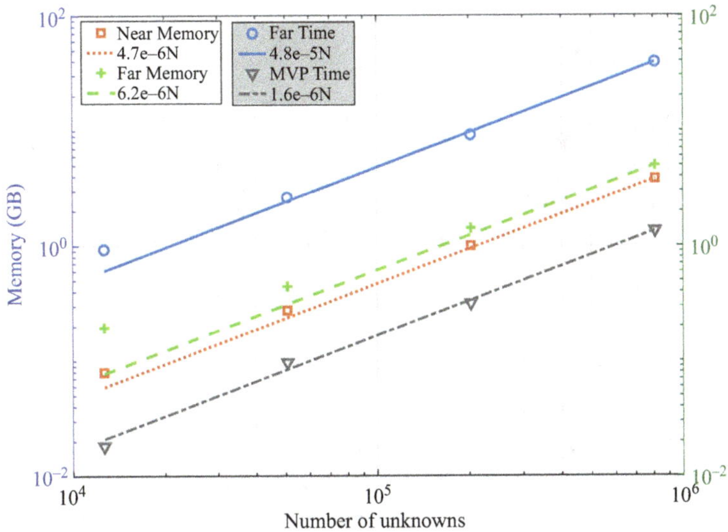

Figure 4.16 Tested time and memory complexity of NESA for the simulation of a cube with side 1 m and fixed $h/\lambda = 1.0e{-2}$

Table 4.4 Simulated memory and time requirement of the proposed NESA for the morphed P180 aircraft model NESA [30] 300 kHz

N	**Near-/far-field memory (GB)**	**Far-field approximation time (mm: ss)**	**Iteration number**	**Total simulation time (hh: mm: ss)**
271,288	6.7/1.1	09: 10	80	02: 03: 01
1,086,083	25.1/3.5	28: 40	138	12: 41: 03

at 300 kHz. Two models with 271,288 and 1,086,083 respectively unknowns have been considered. In the case in which the aircraft is discretized with the coarse mesh, the minimum and maximum h/λ are 2.0e−6 and 7.0e−5, respectively, and a size-level NESA is employed; vice versa, when the dense meshing model is adopted, h/λ ranges from 1.0e−6 to 3.5e−5 and a seven-level NESA is needed. Table 4.4 lists the memory and time requirements for the two cases. Even if the number of unknowns in the second case is almost four times that adopted for the coarser mesh, it appears for the first column of Table 4.4 that memory increases for both the near- and the far-field is lower than a factor 4, and this because of the acceleration techniques described in [30], that allow to compute and store at a high level only part of the radiation, translation, and receiving matrices. In Figure 4.17, it is finally shown the surface current distribution in the case in which the aircraft is modeled with the dense mesh; this current distribution has been obtained with 138 iterations, necessary at the algorithm to converge to a relative residual of 1e−4.

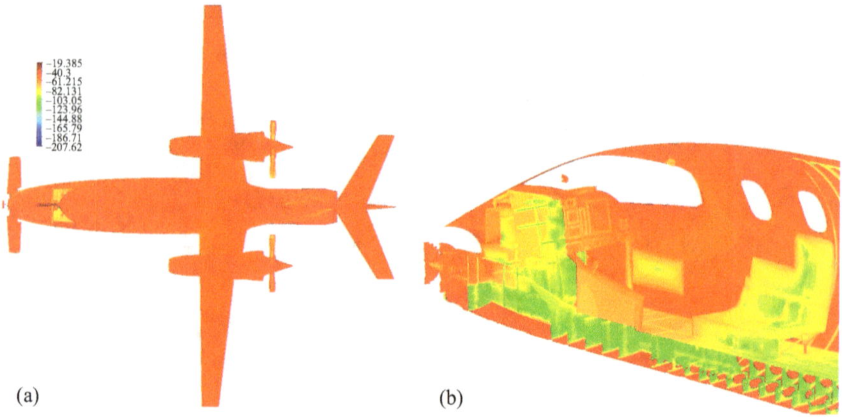

Figure 4.17 *Simulated surface current of the realistic morphed P180 aircraft model with fine meshing at 300 kHz, the incident direction is* $(\theta^i = 90°, \phi^i = 225°)$. *The incident direction of the plane wave is* $(\theta^i = 90°, \phi^i = 225°)$, *(a) top view and (b) details of the inside equipment and seats*

4.5 Wideband nested equivalence source approximation for multiscale problems

As described in Section 4.4, NESA is a multi-level low rank approximation techniques, according to which each level is expressed recursively in terms of its child levels, and consequently in terms of the leaf level. Thank to this scheme, the algorithm has $O(N)$ complexity for static and sufficiently low-frequency problems. In this section, the approach is extended to the solution of large-scale problems. When the coupling in low-frequency regime is evaluated with the algorithm [30], a wideband fast solver with $O(N\log N)$ complexity can be achieved [34].

The proposed method is characterized by the same motivations as other well-known wideband solvers [8–10,25,35–42], but in addition, it has also been developed in such a way as to be a kernel-independent wideband algorithm.

In opposite to what is done in works [31,43], where ACA [5,6] is used to obtain the low-rank approximation, the scheme proposed here consist in defining equivalent sources on surfaces automatically generated and then in enforcing the equivalence, within a prescribed tolerance, between the field radiated by these equivalent sources and by the actual ones on properly defined testing surfaces. A similar idea was exploited in [44] to reduce the number of unknowns of volume integral equations (VIE), by mapping, recursively, volume unknowns onto surfaces.

To extend the use of NESA at high frequency, the partitioning of the interactions in *directions* is employed, to take advantage of the fact that when observation is limited to a specific narrow enough direction, Green's function is smooth, and, therefore, compressible, [42]. In each direction, the rank (number of equivalence sources) does not depend on the group size.

Contrarily to fast directional multilevel algorithms [10,42], which use randomized QR decomposition to approximate Green's function in each direction, here NESA [30] is used to directly compress the matrix entries recursively. The resulting approach is, therefore, named wideband NESA (WNESA) [34]; in the following, the method is described and then its stability and efficiency are validated through its application at the simulation of large real-life multiscale structures.

The aim of WNESA is to extend the nested approximation scheme of [30] to high-frequency couplings, and this means to express radiation and receiving matrices at a given level in terms of those of its child groups, and recursively up to the leaf level. To satisfy the low-rank factorization admission condition of standard low-rank methods [6,21,24,45], in high-frequency regime, the number equivalence sources increases very fast with group size. To bound it, it is necessary to "limit" the direction of observation within *narrow* angles, since the number of directions depends on the observation scale (i.e., group size). Following the ideas presented in [43], and in analogy to (4.4)–(4.11), it is possible to express the relation between the approximation of the coupling between groups t and s in directions d and $-d$ and that of their parent groups t^p and s^p in directions d^p and $-d^p$ as

$$\mathbf{V}_{t^p}^{l,d^p} = \mathbf{C}^{(l,l+1),d^p}\mathbf{V}_t^{l+1,d} \tag{4.35}$$

$$\mathbf{U}_{s^p}^{l,-d^p} = \mathbf{U}_s^{l+1,-d}\mathbf{B}^{(l+1,l),-d^p} \tag{4.36}$$

where $\mathbf{B}^{(l,l-1),-d^p}$ and $\mathbf{C}^{(l-1,l),d^p}$ are the "transfer matrices" while the other parameters are defined in Table 4.5. Equations (4.36)–(4.35) allow to express the radiation and receiving matrices at level l in terms of those of their child level $(l+1)$.

Denoting with D_l the group size at level l, and with D_0 the threshold group size that discriminates between high-frequency and low-frequency couplings, the following three different situations can occur:

1. **High-frequency couplings**: the level l and its child $l+1$ belong both to the high-frequency regime, i.e., $(D_l, D_{l+1}) \geq D_0$;
2. **Interface couplings**: level l belong to the high-frequency regime, while its child $l+1$ is in the low-frequency regime, i.e. $D_l \geq D_0, D_{l+1} < D_0$;
3. **Low-frequency couplings**: both level l and its child $l+1$ belong to the low-frequency regime, i.e., $(D_l, D_{l+1}) < D_0$.

Note that in WNESA [34], the threshold discriminating between high- and low-frequency regimes is chosen to be $D_0 = \lambda$.

4.5.1 *Far-field factorization admissibility conditions*

In this subsection, the admissibility conditions and the "equivalent source" distribution for the three different situations mentioned earlier are discussed. For what concerns the admissibility conditions, they have been fixed following what is done in [10,40,42,43], while the nested approximation is constructed in a way different from the procedures already presented in literature.

Table 4.5 Parameter notation in WNESA

$\sum_{\tau}^{i}$	The *equivalent* sphere of radius R_τ for group i
$\sum_{\sigma^d}^{i}$	The *testing* pyramid surface for group i in direction d
$\boldsymbol{\tau}_i$	RWG basis functions on equivalent sphere surface of group i
$\boldsymbol{\sigma}_i^d$	RWG test functions on testing pyramid surface of group i in direction d
$\mathbf{Z}_{i,j}$	Sub-matrix between groups i and j
$\mathbf{I}_i$	Current density coefficients for basis function in group i
$\mathbf{E}_i$	Projection of the electric field onto test functions in group i
d	Directions number is d
$-d$	Opposite direction of direction d
L	Total number of levels in the Octree
N_d^l	Number of non-zero directions at level l
D_l	Group size (edge of a cube) at level l

In the case of *low-frequency* coupling ($D_l < D_0$), the far coupling admissibility condition is that adopted in traditional rank-based algorithms [5,6,17–19,21], according to which groups t and s are not neighbors if the cubes inside which they are defined do not share any vertex, i.e., if

$$R(s,t) \geq 2D_l \tag{4.37}$$

where $R(s,t)$ is the distance between the centers of groups s and t. If this condition is satisfied, low-frequency couplings are computed using the algorithm [30].

Conversely, for *high-frequency* coupling case, i.e., when $D_l \geq D_0$, the existence of a separated representation of the kernel is assured by the *directional low-rank property* [10,42]. Let us consider a source group s with radius r, and groups t_i is located at a distance $\dfrac{R(s,t_i)}{\lambda} > \left(\dfrac{r}{\lambda}\right)^2$ from s, inside a cone spanning an angle $\frac{\lambda}{r}$, centered in the center of group s; interactions through Helmholtz kernel admit a separable low-rank representation with a desired accuracy, the rank being independent of the radius r. In this case, the coupling admissibility condition between groups s and t is defined as

$$\frac{R(s,t)}{\lambda} \geq \left(\frac{D_l}{\lambda}\right)^2 \tag{4.38}$$

and the directional low-rank property is used to define cones spanning an angle $O(\dfrac{\lambda}{D_l})$. As a result, for the peer far coupling region of groups s and t at level l in high-frequency regime, it is

$$\frac{R(s,t)}{\lambda} \geq \left(\frac{D_l}{\lambda}\right)^2 \tag{4.39a}$$

$$\frac{R(s_p,t_p)}{\lambda} < \left(\frac{D_{l-1}}{\lambda}\right)^2 \tag{4.39b}$$

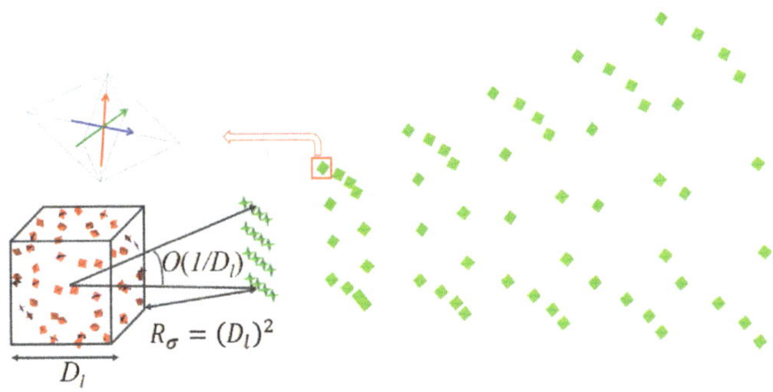

Figure 4.18 High-frequency equivalence process of WNESA: basis functions are placed on the sphere equivalence surface and pyramids testing surface, the testing pyramids are with angle $O(\lambda/D_l)$ to satisfy the far couplings admission condition

where $D_{l-1} = 2D_l$ is the size of the parent group at level $l-1$; 4.39 shows that the far coupling interaction between a source group s and groups t occurs only when these last satisfy the admissibility condition (4.38), while to their parents s^p and t^p do not fulfil (4.38).

Since the unknowns are grouped in cubes rather than in sphere, it is convenient to define directions as the volumes enclosed by square pyramids, whose bases are described in terms of the faces of the Octree cubes. An important advantage of their use is that they allow to define "hierarchical directions," i.e., each direction of a group completely enclosed the directions of its parent groups [42]. As a consequence, if two groups satisfy the admissibility condition in (4.38), also their children fulfil it, allowing to define a nested directional approximation [30,34].

As in NESA [30], also in this case, proper equivalent and testing surfaces are designed, where the equivalent basis functions are defined (see Figure 4.18). However, in opposite to the low-frequency regime [21,30], in the high-frequency regime, the number of equivalent sources Q is independent of the group size only when the directional low-rank property is employed [34]. Moreover, the introduction of equivalent and testing surfaces automatically generates a multiscale family of auxiliary sources, particularly efficient to represent the field in multiscale problems, where the convergence speed is, in this way, significantly enhanced [30].

4.5.2 High-frequency-nested approximation in directions

Figure 4.19 schematically exemplifies the process for the evaluation of the coupling between groups t and s at peer level in the high-frequency regime. The equivalent source distributions, located on surfaces $\Sigma_\tau^{t,s}$, are represented in red, while the testing functions in correspondence with which the field equivalence is enforced are positioned on the surfaces $\Sigma_{\sigma\pm d}^{t,s}$ and are depicted in green. Conventionally, if a group s is in direction d of group t, the opposite direction will be indicated as $-d$, and

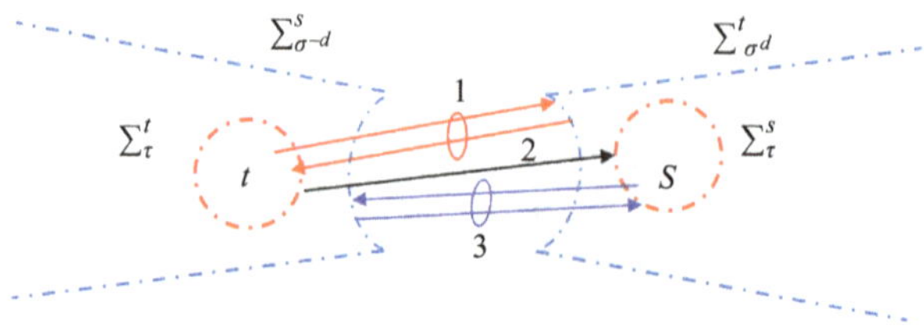

Figure 4.19 Directional far factorization of NESA for two groups t and s at peer level in high-frequency regime. The radiation matrix $\mathbf{V}^d$ *in direction d for groups t symbolized by 1 is evaluated with forward and backward radiations, similarly, the receiving matrix receiving matrix* $\mathbf{U}^{-d}$ *in direction* $-d$ *of group s symbolized by 2 is evaluated, and the peer coupling translation matrix is symbolized by 2*

this means that group t is in direction $-d$ of group s. The equivalent sources $\boldsymbol{\tau}_t$ are obtained by enforcing equivalence in a weak sense between the fields radiated by $\boldsymbol{\tau}_t$ and by the actual sources, on the faces of the wedge enclosing direction d. This operation, labeled with 1 in Figure 4.19, clearly involves a forward radiation operator but also requires to solve an inverse problem, for reconstructing the equivalent sources $\boldsymbol{\tau}_t$ from the fields on $\Sigma^s_{\sigma d}$. Formally, the equivalence between fields can be written as

$$\mathbf{Z}_{\sigma_t^d,t}\mathbf{I}_t = \mathbf{Z}_{\sigma_t^d,\tau_t}\mathbf{I}^d_{\tau_t} \tag{4.40}$$

From the above equation, it is possible to evaluate the vector $\mathbf{I}^d_{\tau_t}$, collecting the coefficients of the equivalent sources $\boldsymbol{\tau}_t$, which radiate the same field as actual sources in the region enclosed in direction d

$$\mathbf{I}^d_{\tau_t} = \mathbf{Z}^\dagger_{\sigma_t^d,\tau_t}\mathbf{Z}_{\sigma_t^d,t}\mathbf{I}_t \tag{4.41}$$

where $(\cdot)^\dagger$ denotes pseudo-inverse, and it is computed resorting to the truncated SVD [30].

By reciprocity, if fields $\mathbf{E}^{-d}_{\tau^s}$ in direction $-d$ tested on Σ^s_τ are known, it is possible to compute the coefficients $\mathbf{I}_{\sigma_s^{-d}}$ of the equivalent sources $\boldsymbol{\sigma}_s^{-d}$ on $\Sigma^s_{\sigma-d}$ radiating the same field $\mathbf{E}^{-d}_{\tau^s}$

$$\mathbf{E}^{-d}_{\tau_s} = \mathbf{Z}_{\tau_s,\sigma_s^{-d}}\mathbf{I}_{\sigma_s^{-d}} \tag{4.42}$$

By solving (4.42) for $I^s_{\sigma-d}$, the field tested on actual testing functions of group s can be expressed as

$$\mathbf{E}^{-d}_s = \mathbf{Z}_{s,\sigma_s^{-d}}\mathbf{Z}^\dagger_{\tau_s,\sigma_s^{-d}}\mathbf{E}^{-d}_{\tau_s}. \tag{4.43}$$

If a *translation matrix* $\mathbf{D}_{s,t}$ collecting the coupling terms between equivalent sources $\boldsymbol{\tau}_t$ and $\boldsymbol{\tau}_s$ is defined as

$$\mathbf{D}_{s,t} = \mathbf{Z}_{\tau_s,\tau_t} \tag{4.44}$$

the fields in group s due to sources in group t can be obtained from (4.42), (4.43), and (4.44)

$$\begin{aligned}\mathbf{E}_s^{-d} &= \mathbf{Z}_{s,t}\mathbf{I}_t = \mathbf{Z}_{s,\sigma_s^{-d}}\mathbf{Z}^{\dagger}_{\tau_s,\sigma_s^{-d}}\mathbf{D}_{s,t}\mathbf{I}^d_{\tau_t} \\ &= \mathbf{Z}_{s,\sigma_s^{-d}}\mathbf{Z}^{\dagger}_{\tau_s,\sigma_s^{-d}}\mathbf{D}_{s,t}\mathbf{Z}^{\dagger}_{\sigma_t^d,\tau_t}\mathbf{Z}_{\sigma_t^d,t}\mathbf{I}_t \end{aligned} \tag{4.45}$$

Equation (4.45) is the single-level WNESA approximation of $\mathbf{Z}_{s,t}$

$$\begin{aligned}\mathbf{Z}_{s,t} &= \mathbf{Z}_{s,\sigma_s^{-d}}\mathbf{Z}^{\dagger}_{\tau_s,\sigma_s^{-d}}\mathbf{D}_{s,t}\mathbf{Z}^{\dagger}_{\sigma_t^d,\tau_t}\mathbf{Z}_{\sigma_t^d,t} \\ &= \mathbf{U}_s^{-d}\mathbf{D}_{s,t}\mathbf{V}_t^d \end{aligned} \tag{4.46}$$

where $\mathbf{U}_s^{-d} = \mathbf{Z}_{s,\sigma_s^{-d}}\mathbf{Z}^{\dagger}_{\tau_s,\sigma_s^{-d}}$ is the *receiving matrix* of the group s in direction $-d$, and $\mathbf{V}_t^d = \mathbf{Z}^{\dagger}_{\sigma_t^d,\tau_t}\mathbf{Z}_{\sigma_t^d,t}$ is the *radiation matrix* of the group t in direction d.

4.5.3 *Multilevel WNESA*

In opposite to standard low-rank methods [5,6,17–19,21], in WNESA, the radiation and receiving matrices at a generic level $l \neq L$ are recursively expressed in terms of those at leaf level L. This is possible thank to the introduction of proper *transfer matrices* allowing to ascend/descend the tree [30,34]. Figure 4.20 illustrates the main idea for the simplest case in which a two-level nested approximation is adopted to evaluate the radiation matrix in the high-frequency regime; according to the directional admission condition in 4.39, the testing surface $\Sigma^{t^p}_{\sigma^{d^p}}$ in direction d^p of group t^p is inside the testing surface $\Sigma^t_{\sigma^d}$ in direction d of its child group t. This implies the possibility of using the radiation matrix in direction d of group t to approximate the coupling terms involving its parent group t^p in direction d^p. In analogy with (4.40)) and (4.41), the coefficients $I^{d^p}_{\tau^p_t}$ of the equivalent sources $\boldsymbol{\tau}_{t^p}$ at the parent level can

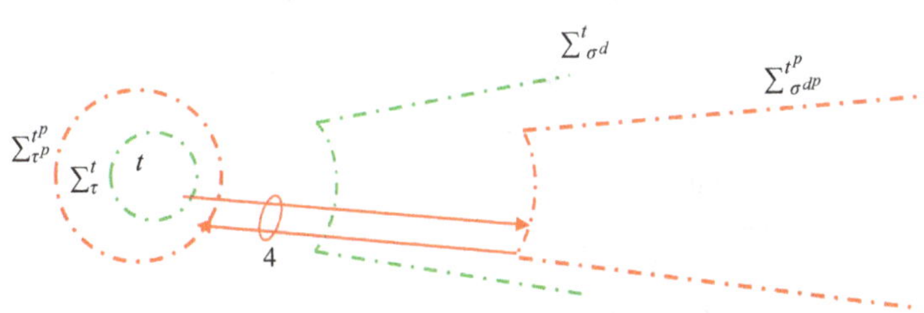

Figure 4.20 Recursive expression of the radiation matrix of parent group t^p by its child group t in WNESA. The directions d^p of parent group are enclosed by its child group, which guarantees the nested factorization in WNESA

be derived by enforcing equivalence between the fields radiated by $\boldsymbol{\tau}_{t^p}$ and $\boldsymbol{\tau}_t$ on the surface $\Sigma^{t^p}_{\sigma^{d^p}}$, indicated with 4 in Figure 4.20:

$$\mathbf{I}^{d^p}_{\tau^p_t} = \mathbf{Z}^{\dagger}_{\sigma^{d^p}_{t^p},\tau_{t^p}} \mathbf{Z}_{\sigma^{d^p}_{t^p},\tau_t} \mathbf{I}_{\tau_{t^d}} \tag{4.47}$$

Expressing fields $\mathbf{E}^{-d}_{\tau_s}$ on Σ^s_τ in terms of $\mathbf{E}^{-d^p}_{\tau_{s^p}}$ on $\Sigma^{s^p}_{\tau^p}$ it yields

$$\mathbf{E}^{-d}_{\tau_s} = \mathbf{Z}_{\tau_s,\sigma^{-d^p}_{s^p}} \mathbf{Z}^{\dagger}_{\tau_{s^p},\sigma^{-d^p}_{s^p}} \mathbf{E}^{-d^p}_{\tau_{s^p}}. \tag{4.48}$$

Finally, if the translation matrix involving groups $\boldsymbol{\tau}_{s^p}$ and $\boldsymbol{\tau}_{t^p}$ at the parent level is defined as

$$\mathbf{D}_{s^p,t^p} = \mathbf{Z}_{\tau_{s^p},\tau_{t^p}} \tag{4.49}$$

the two-level high-frequency WNESA approximation becomes

$$\mathbf{Z}_{s^p,t^p} = \mathbf{U}^{-d}_s \mathbf{B}^{-d^p}_{s,s^p} \mathbf{D}_{s^p,t^p} \mathbf{C}^{d^p}_{t^p,t} \mathbf{V}^d_t \tag{4.50}$$

where the transfer matrices $\mathbf{C}^{d^p}_{t^p,t} = \mathbf{Z}^{\dagger}_{\sigma^{d^p}_{t^p},\tau_{t^p}} \mathbf{Z}_{\sigma^{d^p}_{t^p},\tau_t}$ from child direction d to parent direction d^p, and $\mathbf{B}^{-d^p}_{s,s^p} = \mathbf{Z}_{\tau_s,\sigma^{-d^p}_{s^p}} \mathbf{Z}^{\dagger}_{\tau_{s^p},\sigma^{-d^p}_{s^p}}$ from parent direction $-d^p$ to child direction $-d$ have been introduced.

Equation (4.50) can be easily generalized to the case of an arbitrary number of levels l as

$$\begin{aligned} \mathbf{Z}^l_{s,t} = {} & \mathbf{U}^{L,-d^L}_s \mathbf{B}^{(L,L-1),(-d^L,-d^{L-1})}_s \cdots \\ & \mathbf{B}^{(l+1,l),(-d^l,-d^{l+1})}_s \mathbf{D}^l_{s,t} \mathbf{C}^{(l,l+1),(d^l,d^{l+1})}_t \cdots \\ & \mathbf{C}^{(L-1,L),(d^{L-1},d^L)}_t \mathbf{V}^{L,d^L}_t \end{aligned} \tag{4.51}$$

As mentioned at the beginning of this section, the coupling terms at the bottom of the tree are evaluated as in [30], without exploiting the directional low-rank approximation. If the level at the interface between low- and high-frequency regions is denoted as l_{in}, it is possible to extend (4.51) to a *mixed frequency* scenario as in (4.52).

$$\begin{aligned} \mathbf{Z}^l_{s,t} = {} & \overbrace{\mathbf{U}^L_s \mathbf{B}^{(L,L-1)}_s \cdots}^{\text{low frequency}} \overbrace{\mathbf{B}^{(l_{in}+1,l_{in}),d^{-l_{in}}}_s}^{\text{interface}} \overbrace{\cdots \mathbf{B}^{(l+1,l),(-d^{l+1},-d^l)}_s \mathbf{D}^l_{s,t} \mathbf{C}^{(l,l+1),(d^l,d^{l+1})}_t \cdots}^{\text{high frequency}} \\ & \underbrace{\mathbf{C}^{(l_{in},l_{in}+1),d^{l_{in}}}_t}_{\text{interface}} \underbrace{\cdots \mathbf{C}^{(L-1,L)}_t \mathbf{V}^L_t}_{\text{low frequency}} \end{aligned} \tag{4.52}$$

Note that, when WNESA is applied to a problem described by EFIE, the radiation matrix in a specified direction is the transpose the receiving matrix in the same direction, and, therefore, only one of the them has to be computed and stored. The algorithmic implementation of the WNESA technique described earlier is shown in Algorithm 4.4.

Algorithm 4.4: WNESA low-rank approximation

Initialize an Octree and directions
for $l = L : 1 : -1$ **do**
    for $e = 1 : N_d^l$ **do**
        if $l = L$ **then**
            $\mathbf{V}^{L,d^e} \leftarrow$ radiation matrices with (4.41)
            $\mathbf{U}^{L,d^e} \leftarrow$ receiving matrices with (4.43)
        else
            $\mathbf{C}^{l,d^e} \leftarrow$ transfer matrices with (4.47)
            $\mathbf{B}^{l,d^e} \leftarrow$ transfer matrices with (4.48)
        end if
    end for
    $\mathbf{D}^l \leftarrow$ translation matrices with (4.44) and (4.49)
end for

4.5.4 MVP and computation complexity

The analysis of the algorithm complexity is carried on considering the numerical cost and the storage requirements necessary to compute MVP $\mathbf{y} = \mathbf{ZI}$. In Algorithm 4.5, the pseudocode of the algorithm to compute this product is reported, while the main parameters are listed in Table 4.6. Since it has already been proven that the complexity of the low-frequency regime is $O(N)$ [30,31], in the following only that of the high-frequency regime interactions will be discussed.

It is well known that for a generic 3D problem, described with surface integral equations, the number of unknowns N is proportional to $Ø(S_{max}^2)$, where S_{max} is the maximum electrical size of the object, i.e., the size of the object normalized with respect to wavelength. To study the complexity of the algorithm, it is useful to recall how three important quantities used in the algorithm itself scale:

- the number of nonempty groups at any level l scales as $O(\,(S_{max}/D_l)^2\,)$;
- at any level l, the maximum number of directions is proportional to $O(D_l^2)$ (see Section 4.5.1);
- at any level l, the number of equivalent sources Q in each direction d is constant (see Section 4.5.1).

Taking into account the above considerations, and in the hypothesis that the average number of unknowns K per group at the leaf level is constant, it is possible to demonstrate that the numerical cost of the radiation process at level l, described by lines 3–20 in Algorithm 4.5 is bounded by

$$O\Big(\,(S_{max}/D_L)^2\,D_L^2 KQ\Big) = O\,(N) \quad l = L \tag{4.53a}$$

$$O\Big(\,(S_{max}/D_l)^2\,D_l^2 Q^2\Big) = O\,(N) \quad l = 1\ldots(L-1) \tag{4.53b}$$

Table 4.6 Parameter notation in the MVP and following

s^l	Non-empty source group s at level l
t^l	Non-empty observation group t at level l
$\mathbf{I}_i$	Subvector of **I** restricted to basis functions in group i
$\mathbf{V}_i^{l,d^l}$	Radiation matrix for group i at level l in direction d^l
$\mathbf{B}_i^{l,d^l}$	Transfer matrix for group i at level l in direction d^l
$\boldsymbol{\zeta}_i^{l,d^l}$	Temporary vector in MVP in the radiation process of group i at level l in direction d^l
$\mathbf{D}_{i,j}^l$	Translation matrix between groups i and j at level l
$\boldsymbol{\xi}_i^{l,d^l}$	Temporary vector in MVP in the translation process of group i at level l in direction d^l
$\mathbf{U}_i^{l,d^l}$	Receiving matrix for group i at level l in direction d^l
$\mathbf{C}_i^{l,d^l}$	Transfer matrix for group i at level l in direction d^l
$\mathbf{y}_i^{l,d^l}$	Temporary vector in MVP in the receiving process receiving of group i at level l in direction d^l
$\mathbf{y}$	Result of the MVP $\mathbf{y} = \mathbf{ZI}$
$ch(i)$	Direction number where direction i contained in its child group direction
N_i	Number of basis functions in group i
M_l	Number of non-empty groups at level l
Q	Number of equivalent sources

Since the number of levels is proportional to the logarithm of the object electrical size ($L = O(\log S_{max})$), the overall cost of the radiation process in MVP is $O\left(S_{max}^2 \log S_{max}\right) = O(N \log N)$. The cost of the receiving process (described by lines from 29 to 48 in Algorithm 4.5) is the same, due to reciprocity.

From (4.53a) clearly emerges that radiation patterns $\mathbf{V}_t$ have a linear cost both for fill-in time and memory occupation. Moreover, exploiting symmetry as detailed in [30], the number of transfer matrices that must be computed and stored at each level l is only $8N_d^l$, and, therefore, the memory required for storing all the transfer matrices can be bounded as

$$O\left(8D_l^2 Q^2\right) = O\left(\left(\frac{S_{max}}{2^{l+1}}\right)^2 Q^2\right) \tag{4.54}$$

The evaluation of the partial sum $\sum_{l=1}^{L}\left(\frac{S_{max}}{2^{l+1}}\right)^2 = O(N)$, proves that memory for transfer matrices has a linear complexity.

Let us finally consider the translation process discussed in Section 4.5.1, where the far interaction list at level l includes groups at a distance smaller than $(2D_l)^2$, $2D_l$ being the size of the parent groups at level $(l-1)$. Considering as the starting point

Algorithm 4.5: WNESA Matvec $\mathbf{y} = \mathbf{ZI}$

Procedure WNESA Matvec $(\mathbf{I}, \mathbf{y})$
for $l = L : 1, -1$ **do**
    % *Radiation process*
    if $l = L$ **then**
        if $l > l_{in}$ **then**
            $\boldsymbol{\zeta}_{t^L}^{L} \leftarrow \boldsymbol{\zeta}_{t^L}^{L} + \mathbf{V}_{t^L}^{L}\mathbf{I}_{t^L}$ % *low frequency*
        else
            $\boldsymbol{\zeta}_{t^L}^{L,d^L} \leftarrow \boldsymbol{\zeta}_{t^L}^{L,d^L} + \mathbf{V}_{t^L}^{L,d^L}\mathbf{I}_{t^L}$% *high frequency*
        end if
    else
        if $l > l_{in}$ **then**
            $\boldsymbol{\zeta}_{t^l}^{l} \leftarrow \boldsymbol{\zeta}_{t^l}^{l} + \mathbf{C}_{t^l}^{l}\boldsymbol{\zeta}_{t^{l+1}}^{l+1}$ % *low frequency*
        end if
        if $l = l_{in}$ **then**
            $\boldsymbol{\zeta}_{t^l}^{l,d^l} \leftarrow \boldsymbol{\zeta}_{t^l}^{l,d^l} + \mathbf{C}_{t^l}^{l,d^l}\boldsymbol{\zeta}_{t^{l+1}}^{l+1}$% *interface*
        end if
        if $l < l_{in}$ **then**
            $\boldsymbol{\zeta}_{t^l}^{l,d^l} \leftarrow \boldsymbol{\zeta}_{t^l}^{l,d^l} + \mathbf{C}_{t^l}^{l,d^l}\boldsymbol{\zeta}_{t^{l+1}}^{(l+1),ch(d^l)}$ % *high frequency*
        end if
    end if
    % *Translation process*
    if $l > l_{in}$ **then**
        $\boldsymbol{\xi}_{s^l}^{l} \leftarrow \boldsymbol{\xi}_{t^l}^{l} + \mathbf{D}_{s^l,t^l}\boldsymbol{\zeta}_{t^l}^{l}$ % *low frequency*
    end if
    if $l <= l_{in}$ **then**
        $\boldsymbol{\xi}_{s^l}^{l,d^l} \leftarrow \boldsymbol{\xi}_{t^l}^{l,d^l} + \mathbf{D}_{s^l,t^l}\boldsymbol{\zeta}_{t^l}^{l,-d^l}$ % *high frequency*
    end if
end for
% *Receiving process*
for $l = 1 : L$ **do**
    if $l \neq L$ **then**
        if $l > l_{in}$ **then**
            $\boldsymbol{\xi}_{s^{l+1}}^{l+1} \leftarrow \boldsymbol{\xi}_{s^{l+1}}^{l+1} + \mathbf{B}_{s^l}^{l}\boldsymbol{\xi}_{s^l}^{l}$ % *low frequency*
        end if
        if $l = l_{in}$ **then**
            $\boldsymbol{\xi}_{s^{l+1}}^{l+1} \leftarrow \boldsymbol{\xi}_{s^{l+1}}^{l+1} + \mathbf{B}_{s^l}^{l,d^l}\boldsymbol{\xi}_{s^l}^{l,d^l}$% *interface*
        end if
        if $l < l_{in}$ **then**
            $\boldsymbol{\xi}_{s^{l+1}}^{(l+1),ch(d^l)} \leftarrow \boldsymbol{\xi}_{s^{l+1}}^{l+1} + \mathbf{B}_{s^l}^{l,d^l}\boldsymbol{\xi}_{s^l}^{l,d^l}$% *high frequency*
        end if
    else
        if $l > l_{in}$ **then**
            $\mathbf{y}_{s^L}^{L} \leftarrow \mathbf{y}_{s^L}^{L} + \mathbf{U}_{s^L}^{L}\boldsymbol{\xi}_{s^L}^{L}$ % *low frequency*
        else
            $\mathbf{y}_{s^L}^{L} \leftarrow \mathbf{y}_{s^L}^{L} + \mathbf{U}_{s^L}^{L,d^L}\boldsymbol{\xi}_{s^L}^{L,d^L}$ % *high frequency*
        end if
    end if
end for
$\mathbf{y} \leftarrow \mathbf{y}^{L} + \mathbf{Z}_{near}\mathbf{I}$ % *Sum near interactions*

the admissibility condition of (4.38), it is possible to derive an upper bound for the far interaction list at level l as

$$N_{FIL}^{l} = \left(1 + 2^{L-l}\,3\right)^{2} = O\left(D_l^2\right) N_{FIL}^{l} = 60\left(\frac{D_l}{\lambda}\right)^{2} + 28\left(\frac{D_l}{\lambda}\right) + 3 = O\left(D_l^2\right) \quad (4.55)$$

Since the number of non-empty groups grows as $O\left((S_{max}/D_l)^2\right)$, the cost of the translation process at level l is

$$O\left((S_{max}/D_l)^2\, D_l^2 Q^2\right) = O\,(N) \quad (4.56)$$

and, therefore, summing over $L = O\,(\log N)$ levels, it is finally possible to evaluate the total cost of the translation process, that results to be $O\,(N\log N)$ [34].

4.5.5 Numerical results

To validate the effectiveness of the proposed solver, several test cases have been considered. In the following, h indicates the average mesh edge length, while the average number of basis functions at the finest level is selected to be ~50 for all the considered examples. For all of the examples in the following, the adopted iterative solver is a flexible-GMRES and the maximum number of iterations for the inner solver is fixed to 10. All the simulations have been carried out using on a 64-bits Dell Precision T7400 workstation, Intel Xeon CPU E5440 @ 2.88 GHz, 96 GB of RAM; double precision computation is always used.

Accuracy

As for the NESA, also here the accuracy of WNESA approximation is first tested on the scalar Green's function; with this aim, 500 source points $\mathbf{r}_s$ and 500 observation points $\mathbf{r}_t$ randomly distributed in two cubes satisfying the high-frequency admissibility condition described in Section 4.5.1 are considered. For each pair $(\mathbf{r}_s, \mathbf{r}_t)$, the scalar Green's function $G\,(\mathbf{r}_s, \mathbf{r}_t) = e^{-jk_0|\mathbf{r}_s - \mathbf{r}_t|} / |\mathbf{r}_s - \mathbf{r}_t|$ has been evaluated analytically and with (4.46) and the approximation error of WNESA is defined as $\left|\mathbf{G} - \mathbf{G}_{\mathrm{WNESA}}\right|_2 / |\mathbf{G}|_2$, where $\mathbf{G}$ and $\mathbf{G}_{\mathrm{WNESA}}$ are two column vectors collecting the scalar Green's function between all pairs of source/observation points, evaluated analytically and with WNESA, respectively. In Figure 4.21, plots of the approximation error for group size varying between 1 and 8 λ are shown. As it appears, for a fixed number Q of equivalent sources, the approximation error decreases with the group size, differently from what happens in the other rank-based methods [6,19,21].

A second test consists in evaluating the two-norm error introduced by the approximation on the elements of the EFIE impedance matrix. Two groups containing 636 and 527 RWG basis functions, respectively, and with size 1 λ have been selected. They are obtained from the Octree clustering of a cylinder with diameter 1λ and height 8λ, discretized with 13,168 unknowns. The entries evaluated with standard MoM and with a very accurate quadrature rule (61 Gaussian points on triangles) are taken as a reference. Figure 4.22 shows the approximation error versus the number of equivalent sources Q. The accuracy obtained with $Q = 50$ equivalent sources is

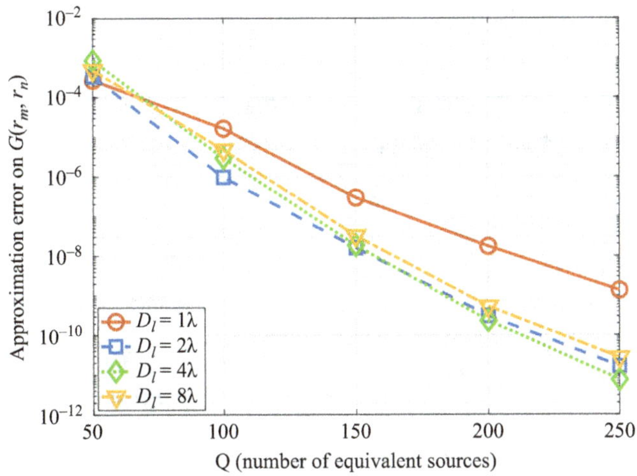

Figure 4.21 WNESA approximation for Green's function matrix $\mathbf{G}_{(\mathbf{r}_s,\mathbf{r}_t)}$, $\mathbf{r}_s$ *and* $\mathbf{r}_t$ *are 500 random distributed points in groups s and t, with different group size of 1 λ, 2 λ, 4 λ, and 8 λ*

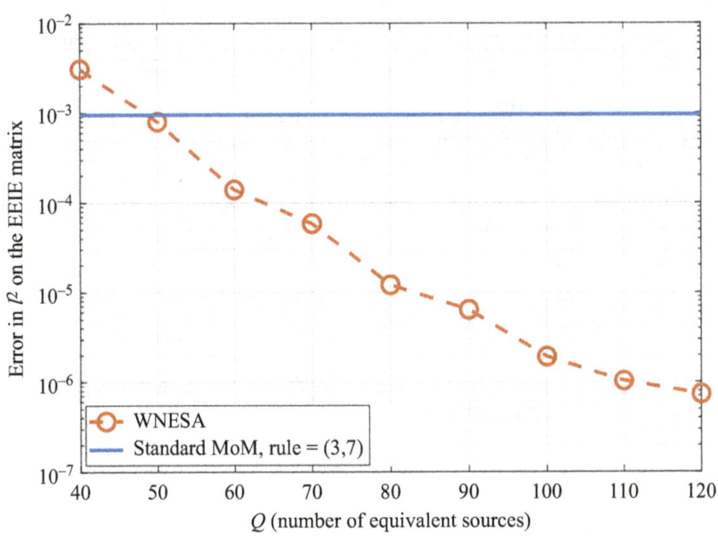

Figure 4.22 WNESA approximation for EFIE entries for two groups with group size 1 λ, and with 636 and 527 number of unknowns

labeled as "Standard MoM, rule = (3,7)," since it represents the typical "goal" accuracy for MoM codes. For this reason, in all the here considered cases, the number of equivalent sources is fixed to $Q = 50$.

Computational complexity

The computational cost of WNESA predicted in Section 4.5.4 is verified numerically, considering several spheres with diameters equal to 8, 16, 32, and 64 λ. The mesh length is fixed to a value satisfying the condition $h/\lambda = 0.15$, and this corresponds to have a number of unknowns equal to 17,808, 71,232, 284,928, and 1,139,712, respectively, for the different spheres. For the spheres with a diameter equal to 8 and 16 λ, a four-level and a four-level WNESA are employed, while for the other two, a five-level scheme is used. In all cases, only two levels are adopted in the low-frequency range. As a worst-case scenario, the code has been forced to consider all the possible directions *d* and not only those including non-empty groups, and, therefore, an upper bound to the actual cost will be evaluated.

In Figure 4.23, the computational cost (right axis) and the memory occupation (left axis) versus the number of unknowns are plotted: the cost of MVP increases as $O(N\log N)$, while that for the setup, as well as the storage requirements, are smaller lower [34]. From the curves in Figure 4.23, it emerges that memory and factorization time for the two smallest spheres, as well as that for those with diameter 32 λ and 64 λ, are almost constant (green and red curves). This is first due to symmetry considerations that allow to evaluate and store the required transfer and translation operators just for a single group at each WNESA level. Moreover, being the number of levels employed in WNESA for analyzing the spheres with a diameter equal to 8 and 16 λ the same, as well as used for the two larger spheres, the cost remains almost constant too, with a small linear increase due to the cost of radiation/receiving patterns at the leaf level. Vice versa, the cost of MVP grows as $O(N\log N)$, since each translator/transfer matrix is multiplied a number of times equal to the number of non-empty groups at each level.

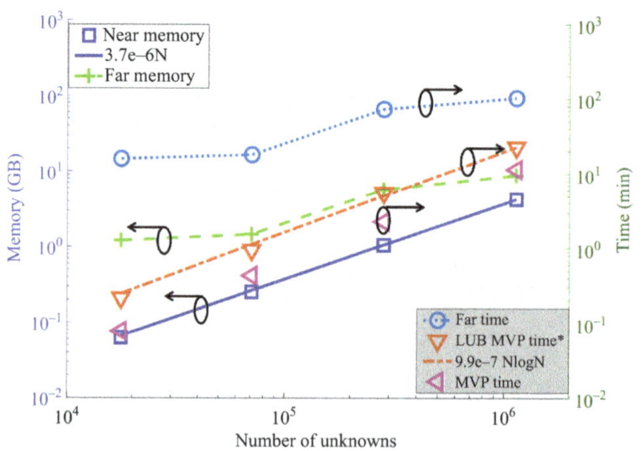

Figure 4.23 Tested time and memory computation complexity of WNESA for the spheres with a fixed $h/\lambda = 0.15$*, the diameters of the spheres are 8, 16, 32, and 64 λ, with number of unknowns of 17,808, 71,232, 284,928, and 1,139,712, correspondingly*

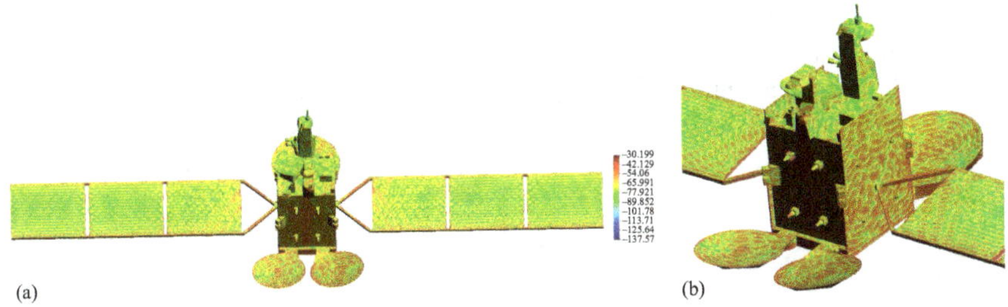

Figure 4.24 Simulated surface current of the 80 λ realistic satellite model with 1,096,225 unknowns with NESA, (a) whole view and (b) details of the body

Finally, also the actual MVP time, i.e., considering only translators in non-empty directions, has been evaluated and plotted in Figure 4.23 as a function of N. Its scaling seems to be "worse" than $O(N\log N)$, that is, the complexity of the MVP computation when all possible directions are taken into account. This proves that MVP time has a cost bounded by $O(N\log N)$ [34].

Application of WNESA to the modeling a realistic multiscale problems

The large and realistic satellite model shown in Figure 4.24 is finally simulated with WNESA. The largest dimension of the satellite is 20 m, corresponding to 80 λ at the considered frequency of 1.2 GHz. To take into account in the model most of the details of the satellite structure, it has been discretized with 1,096,225 unknowns, so that h/λ ranges from 3.5e−3 to 1.9e−1. The incident field has a linear polarization, directed along $\hat{\theta}$ and it is a plane wave impinging on the satellite from the bottom. The scheme adopted to compress the EFIE impedance matrix is characterized by two levels at low-frequency and four levels in the high-frequency regime. The MR-ILU [13,28] preconditioner is employed to improve the convergence performance. The low-rank factorization time and memory are 32 min and 10.6 GB, respectively. A flexible GMRES iterative solution with 10 inner iterations is employed, and only 76 iterations are required to converge with a residual of 1e−3. The overall time for solving the matrix equation describing the problem is 25.1 h, 119 s of which are required for evaluating the MVPs.

The surface current distributions on the satellite are shown in Figure 4.24, and it validates the effectiveness of using the WNESA for realistic multiscale problems.

4.6 Mixed-form nested equivalence source approximation for multiscale problems

In Sections 4.4 and 4.5, equivalence RWGs on a pair of equivalence surfaces have been introduced in the NESA and WNESA to obtain the nested low-rank factorization. The computation accuracy has been tested empirically with respect to the number of equivalence RWGs, and it has been proven that the introduced error is controllable from low-to-high frequency regimes [30,34] also in case of realistic multiscale geometries. However, the number of RWGs is still suboptimal, since it is rather larger, especially at the finest level of the octree, than the minimum rank necessary for obtaining the desired accuracy.

The first objective of mixed-form NESA is, therefore, to find optimal number of skeletons by resorting ACA [5,6]. Even if in [31], the ACA is used to elegantly select skeletons as a subset of those of its child groups, it does not guarantee the sought accuracy for multiscale simulations, as demonstrated in [46]. In [47], the interpolation decomposition (ID) method is applied to scalar problem evaluation: the equivalence points on the auxiliary surfaces are located at the interface between the near and the far region, and the dominant skeletons are selected with the ID compression of the source points against the equivalence points. The technique proposed in [48–50] for the analysis of dense mesh dynamic problems at low frequency, consists in introducing vector equivalence basis functions on the equivalence spheres and then in generating the skeleton bases with random samplings. In the here proposed mixed-form NESA, the spherical distributed auxiliary RWGs [30] are used equivalence vector basis functions, and the skeleton RWGs are selected by compressing the source RWGs against the equivalence RWGs with ACA [5,6] and they are effective in all the far region [46].

The second objective of the mixed-form NESA is to develop a kernel independent, wideband fast solver for multiscale problems more effective than standard wideband fast multipole approaches [8–10,25,35–42]. In opposite to WNESA [34], mixed-form NESA [46] aims to combine the advantages of both the skeleton-based NESA, that is employed at the finest levels, and the standard wideband NESA [34], adopted at the coarser levels; moreover, it also includes the *interface* matrices smoothly mapping skeletons to equivalent sources.

4.6.1 Multiscale sampling for skeletons

The selection of the optimal skeletons for each group is really important for the accuracy of the low-rank approximation. Differently from what is done in [27,47], where the random sampling is used, here the optimal skeleton RWGs $\boldsymbol{\tau}_i$ in each groups are selected with ACA, simply performing a compression of the source group against its far interaction lists once a desired threshold is fixed. As sketched in Figure 4.25, at each level, first an equivalent spherical surface is constructed at the interface between the near and the far regions, then the equivalence points are distributed uniformly on this surface. The radius of the equivalence sphere is selected to be $radius = 3S/2$, where S is the group size. In correspondence with each point, three auxiliary RWG basis functions with different (r, θ, ϕ) directions are placed, and the length of the line is chosen equal to 1/30 of the minimum edge size of the triangles used to discretize the simulated object to obtain fast computation of the integrals involved in the equivalence surface [30]. At the leaf level, ACA is employed to obtain the skeletons by compression of the matrix generated from the RWGs in the source group against the test auxiliary

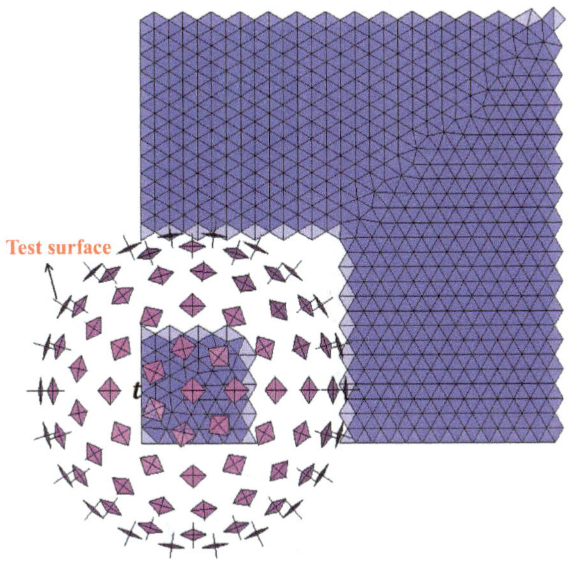

Figure 4.25 Skeletons sampling for group t by ACA, the auxiliary basis functions are placed to represent the couplings from the far region

functions. At the parent level, the skeletons are obtained by compressing the matrix derived from the union of the corresponding child skeletons against the parent group auxiliary RWGs. For the low- and moderate-frequency regime, an exhaustive test campaign has proven that to find the *optimal* dominant RWGs in the source group, a suitable number of equivalence points on the equivalence sphere are 100.

Since at low and mid frequencies, the average number of skeletons at each level is permanent within a predefined tolerance, skeleton NESA is characterized by a liner computation complexity [30,46].

4.6.2 Mixed-form wideband-nested approximation

With respect to standard NESA, the skeleton NESA presents the advantage that it allows to get the correct and small number of dominant RWGs, but, as a counterpart, it is characterized by a higher complexity, due to the computational cost of the ACA samplings and to the memory required to store different transfer matrices between adjacent levels. To compensate for this drawback, in mixed-form WNESA, the skeleton NESA is only adopted for the bottom level evaluations, where the standard NESA has no benefits. In view of this, a transfer matrix from skeleton to equivalence NESA is needed. Referring to Figure 4.26, where the transfer process to map the child skeletons to the parent equivalence sources is sketched, the field radiated by the child skeletons $\boldsymbol{\tau}_t$ in group t is imposed to be equal, in a weak sense, to that radiated by

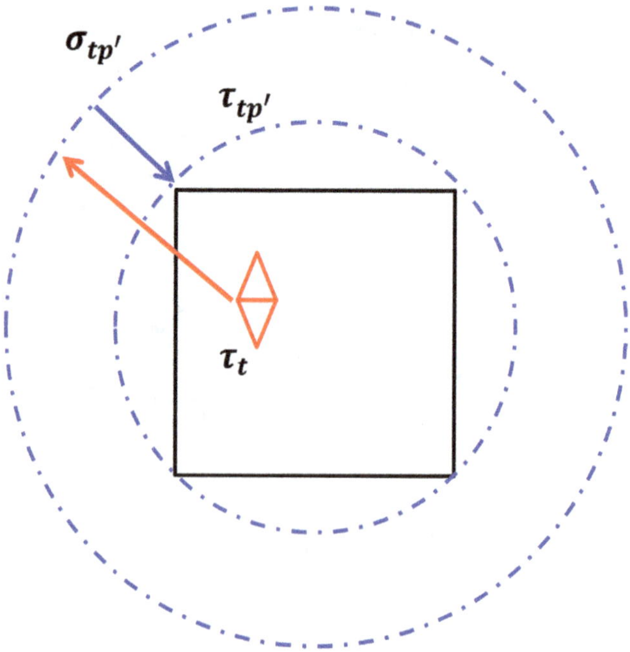

Figure 4.26 *Radiation process in the interface of the mixed-form nested approximation, the radiation matrix of the equivalence RWGs $\boldsymbol{\tau}_{tp'}$ in the parent group is expressed by the child skeletons $\boldsymbol{\tau}_t$*

parent equivalence RWGs $\boldsymbol{\tau}_{tp}$; the transfer matrix from skeleton to equivalence nested approximation [46] could, therefore, be introduced

$$\tilde{\mathbf{B}}_{t,tp} = \mathbf{Z}_{\tau_t,\sigma_{tp}'}\mathbf{Z}^{-1}_{\tau_{tp}',\sigma_{tp}'} \tag{4.57}$$

so that the mixed-form wideband-nested approximation becomes, in analogy to the WNESA approximation in (4.58)

$$\mathbf{Z}^l_{s,t} = \overbrace{\mathbf{U}^L_s\mathbf{B}^{(L,L-1)}_s\ldots}^{\text{Skeleton NESA}}\overbrace{\tilde{\mathbf{B}}^{(l_{se}+1,l_{se})}_s}^{\text{interface}}\overbrace{\ldots\mathbf{B}^{(l+1,l),(-d^{l+1},-d^l)}_s\mathbf{D}^l_{s,t}\mathbf{C}^{(l,l+1),(d^l,d^{l+1})}_t\ldots}^{\text{Standard WNESA}}$$
$$\underbrace{\tilde{\mathbf{C}}^{(l_{se},l_{se}+1)}_t}_{\text{interface}}\underbrace{\ldots\mathbf{C}^{(L-1,L)}_t\mathbf{V}^L_t}_{\text{Skeleton NESA}}. \tag{4.58}$$

where l_{se} denotes the interface level from the skeleton-to-equivalence nested approximation while the parameters in standard NESA and their meaning have already been discussed in [34]. When the mixed-form WNESA is applied to an EFIE problem, radiation and receiving matrices relative to a specific group are one the transpose if the other. Its computational complexity has the same behavior with respect to the problem size, of standard NESA, with a relative smaller coefficient [46]. As a result, the total computational complexity for the proposed technique is $O(N\log N)$ [34,46].

4.6.3 Numerical results

Accuracy

The effectiveness of the proposed mixed-form WNESA has been tested through different cases [46]. Before discussing the results of the simulation carried on, it is, however, necessary to introduce some quantities whose values will be kept the same for all the simulations. The average number of basis functions taken as stopping criterium for the generation of the octree clustering is ~50; the average length of the mesh edge is indicated by h; all the simulations have been carried out serially on a 64-bits Dell workstation with 2.0 GHz CPUs and 512 GB of RAM and double precision computation is always used.

The precision of the skeleton NESA approximation has been first tested by evaluating the EFIE impedance matrix as a function of the threshold used for the ACA sampling; the matrix entries represent the coupling between two groups with 1,283 and 1,037 RWG basis functions, respectively, that discretize a cylinder whose diameter is equal to 1 m and whose height is 4 m. The group size is 1 m and the working frequency is 300 MHz. In Figure 4.27, the 2-norm error is plotted and compared with that relative to the case in which the dense matrix block is computed with a high of accuracy, i.e., using 61 Gaussian points both in the external and in the internal integral. As confirmed by Figure 4.27, the error introduced by the approximation is strongly dependent on the ACA sampling threshold. Since a value of this threshold equal to 1e–4 guarantees the same accuracy as a Gaussian scheme with (3,7) quadrature points in standard MoM implementations, this value will be used in the following examples.

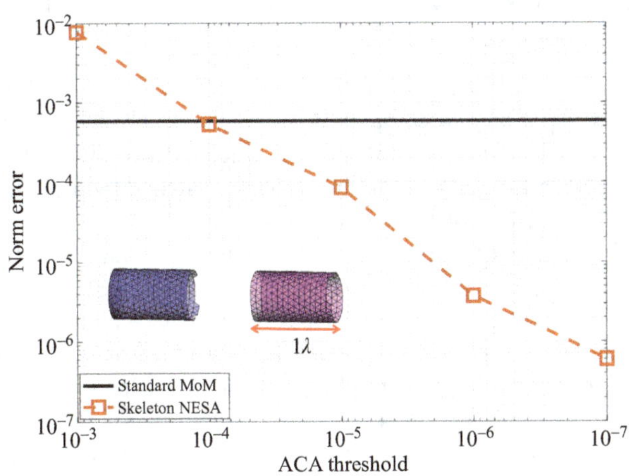

Figure 4.27 Skeleton NESA approximation for the EFIE entries for different ACA thresholds and the number of unknowns of the two groups as inset is 1,283 and 1,037

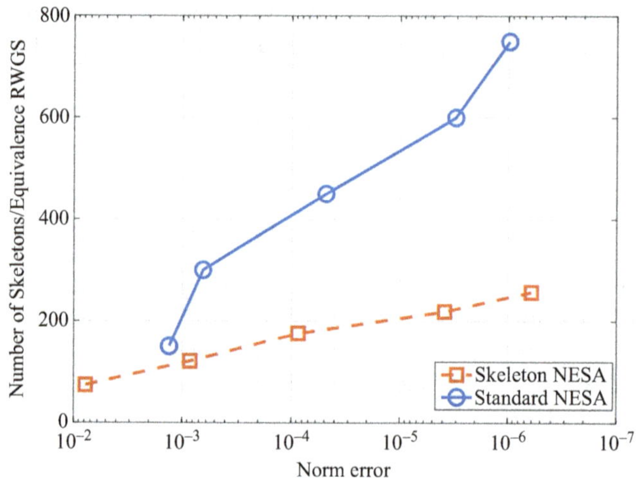

Figure 4.28 Comparison of the number of skeletons and equivalence basis functions for the EFIE entries approximation in Figure 4.27 with skeleton NESA and standard NESA, respectively

In Figure 4.28, the skeleton NESA is compared with the standard one, since it shows the number of skeleton/equivalent basis functions necessary by the two methods to provide a desired accuracy. As expected, skeleton NESA allows to keep the error introduced by approximation at a given value with a much lower average number of skeletons with respect to the number of auxiliary sources needed by standard NESA, and this proves that the strategy in this last technique [30] is indeed suboptimal.

Levels of the skeleton NESA

To discuss the number of levels for the skeleton NESA in the mixed-form WNESA and its effect on the approach complexity, a morphed Porsche 911 car model has been simulated. The car has a size of $4.5 \times 2.7 \times 1.1$ m, it has been discretized with 61,299 unknown (see Figure 4.29(a)), and it is illuminated by a plane wave at 300 GHz, impinging on it from the top.

As stated in Section 4.6.1, the evaluation of the MVP is faster with the skeleton NESA than with the standard one [30], because the number of skeletons necessary to reach a desired accuracy is smaller than that of equivalence RWGs in standard NESA, as confirmed by the results shown in Figure 4.28. Vice versa, skeleton NESA needs more memory and has higher setup time requirements, since it does not allow to exploit symmetry in the far-field couplings decomposition as the standard NESA [30,34]. As a result, a trade-off is necessary in the mixed-form WNESA [46] to balance the advantages of the two methods properly choosing the number of levels in which the skeleton-based NESA is applied rather than the standard one.

In Table 4.7, the variation of the memory requirements and computational time with the number of levels in which the skeleton or standard NESA are used is listed.

Table 4.7 Memory and time requirements for the simulations of a morphed Porsche 911 car model with zero-, one-, two-, and four- skeleton NESA for the mixed-form WNESA [46], respectively

Skeleton/standard NESA levels	Far memory (MB)	Far time (mm:ss)	MVP time (ss)
0/4	602	01:05	2.1
1/3	1,849	06:32	1.2
2/2	2,577	08:09	0.9
4/0	3,091	09:19	0.6

Note that in all the cases, the total number of levels is four, but they are shared in different ways between the two approaches. In particular, the first row refers to the case in which standard NESA is used at all the levels, while in the last row, the opposite situation is considered. In the second and third columns of the table, the (far) memory and (far) time requirements for the far-field nested approximation are reported, while in the last column the MVP time is listed. As expected, increasing the number of levels in correspondence with which the skeleton NESA is used, the MVP time is reduced, but at the cost of the increase of both the far memory and time requirements. From several experimental tests, it emerges that a good balance in the mixed-form WNESA [46] is reached with a two-level skeleton NESA at the bottom level.

In Figure 4.29(b), the surface currents distribution on the morphed Porsche 911 car is shown: it has been evaluated with four-level mixed-form WNESA, two for the skeleton NESA and two for the standard NESA one.

Wideband computational complexity

First, the computational complexity of the mixed-form WNESA at low-frequency regime is checked, considering a sphere with a diameter equal to 8 m, and fixing $h/\lambda = 1.5e-3$. The structure has been simulated at the four frequencies of 3, 6, 12, and 24 MHz. Since each of them corresponds to a different value of λ, the number of unknowns for the four cases is different, equal to 23,703, 94,812, 379,248, and 1,516,992, respectively. Also the number of levels adopted in the mixed-form WNESA changes from one frequency to the other, being equal to two at 3 MHz and then increasing up to five at 24 MHz. As discussed in Section 4.6.3, the number of levels for skeleton NESA is two. The numerical complexity of the approach in the considered low-frequency region is reported in Figure 4.30: the curves relative to the computational cost and memory requirement prove that both of them increase linearly with the number of unknowns [46].

Second, the computational complexity at high frequencies is evaluated. The same sphere used above is considered, at the frequencies of 150 MHz, 300 MHz, 600 MHz,

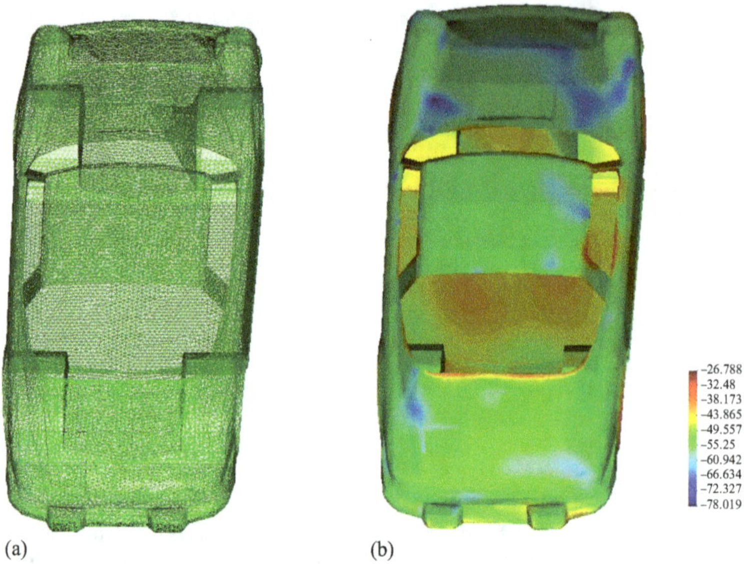

Figure 4.29 Simulated surface current of a morphed Porsche 911 car model: (a) mesh cells and (b) surface currents

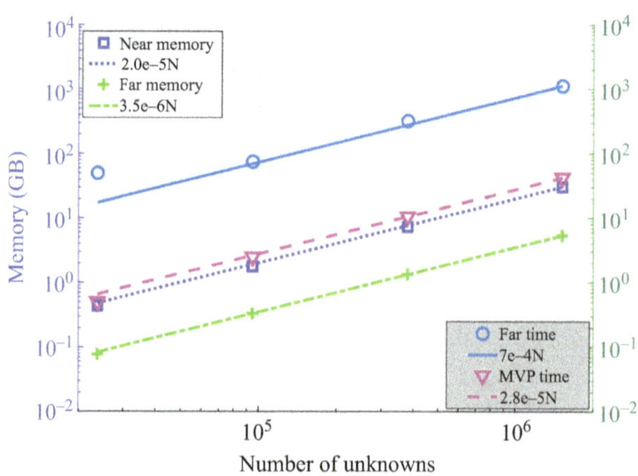

Figure 4.30 Mixed-form WNESA time and memory complexity test at low-frequency with fixed $h/\lambda = 1.5e-3$, the spheres are with number of unknowns of 23,703, 94,812, 379,248, and 1,516,992

and 1.2 GHz: in their correspondence, the electrical size of the sphere becomes equal to 4, 8, 16, and 32 λ, respectively, and consequently also the number of unknowns needed for its discretization, still keeping constant h/λ increases. As shown in Figure 4.31, where both the computational cost and the memory requirements are

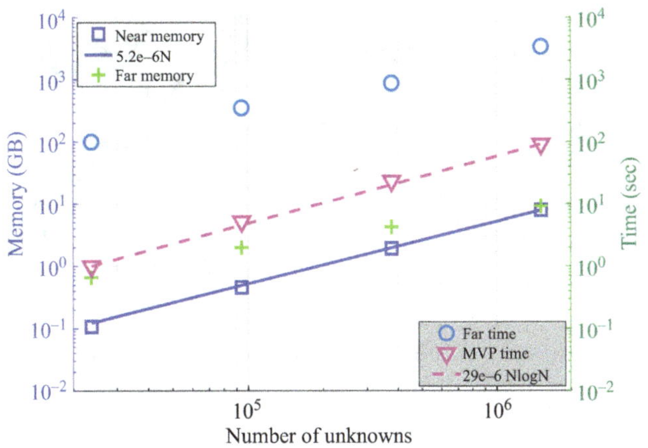

Figure 4.31 Mixed-form WNESA time and memory complexity test at high frequency with fixed $h/\lambda = 0.085$

plotted versus the number of unknown, in this case, the complexity for MVP is $O(N\log N)$ [46].

Wideband simulations of realistic aircraft

As a final test, wideband simulations of a morphed EV55 aircraft‡ at 75 MHz and 1 GHz have been performed, to verify the effectiveness of the mixed-form WNESA when applied to realistic multiscale problems. The aircraft is 14.2 m long and has a wingspan of 16.1 m; these sizes correspond to 3.5 λ and 4.0 λ at 75 MHz, and to 47.3 λ and 53.7 λ at 1 GHz, respectively. The total number of unknowns used for discretizing the airplane is 695,629.

At 75 MHz, the mesh size h/λ ranges from 1.1e−2 to 6.1e−4. First, a six-level skeleton NESA has been compared with a six-level mixed-form WNESA, employing two-level skeleton NESA and four-level standard WNESA. The average number of unknowns at the finest level has been fixed to be 52, while the number of iterations necessary to the FGMRES solver (10 inner iterations) using the MR preconditioner [32] to converge with 1e−4 residual is 18. The required far memory and time required by skeleton NESA result to be equal to 2.2 GB and 3.1 h, respectively, while the MVP time is 5.6 s, and the total solution time is 16.8 min. For what concerns the mixed-form WNESA, the far memory and time are, respectively, equal to 1.2 GB and 25 min, the MVP time is 5.7 s, the total simulation time from start to the end is 1.9 h, while for the WNESA, it is 2.1 h [34].

At the high-frequency regime (1 GHz), if the number of unknowns is kept constant, the mesh size h/λ ranges from 1.4e−1 to 7.9e−3. Also in this case, a six-level mixed-form WNESA (two-level skeleton NESA and four-level standard

‡[Online Available]: http://www.evektoraircraft.com/en/aircraft/ev-55-outback/overview.

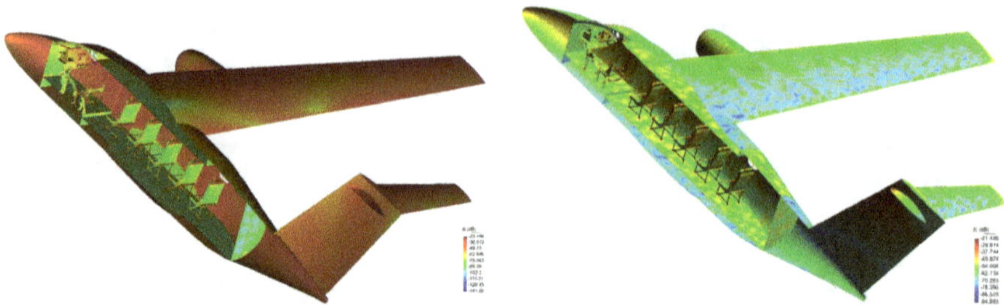

(a) (b)

Figure 4.32 Mixed-form WNESA simulations for the surface currents of a morphed realistic EV55 aircraft with 695,629 unknowns: (a) 75 MHz and (b) 1 GHz

WNESA [30,34]) is considered. The results of the performed simulations show that the required far memory and time are in this case equal to 5.3 GB and 54 min, respectively, the MVP time is 33 s while the FGMRES solver with the MR-ILU preconditioner [13] needs 60 iterations to reach 1e−4 residual and the overall simulation time is 16.4 h.

In Figure 4.32, the details of surface current over the EV55 aircraft obtained through these simulations at the two considered frequencies are shown; these results prove the validity and the accuracy of the mixed-form WNESA for wideband simulation of complex multiscale problems.

4.7 Conclusion and prospect

In this chapter, the kernel-independent fast factorization for realistic multiscale simulations are reviewed. First, the ACA, that is a kernel-independent method, is introduced, pointing out its limitations to low- and medium-frequency problems and the high-computation resources required for both the low-rank decomposition set-up and MVP process. As an alternative, the single-level FMM is discussed, showing that it could achieve an $O(N\log N)$ computational complexity with a post-compression technique for low- and medium-frequency problems. Finally, the NESA and its wideband version of WNESA and mixed-form WNESA are presented. They perform like the MLFMA, but unlike the standard kernel-independent methods, the vector equivalence sources and transfer matrices between adjacent levels are introduced. Optimal computational complexities of $O(N)$ and $O(N\log N)$ are achieved for low- and high-frequency problems. The efficiency of all the proposed methods is validated through the simulations of morphed realistic models provided the collaborative partners Evektor, Piaggio, IDS, and European Space Agency (ESA).

As mentioned, the proposed approaches are not limited to the EFIE, application of WNESA for penetrable problems simulations are demonstrated in [51]. It would be preferred fast solvers for the penetrable problems with high contrasts, where much dense or non-conformal meshes might be involved [52–55]. Another proper topic is to develop fast direct solvers with the proposed kernel-independent methods, the popular format of $\mathcal{H}$-matrix, $\mathcal{H}^2$-matrix, hierarchical off-diagonal low-rank systems (HODLR) [56–59], hierarchically semiseparable (HSS) [60,61] are able to be employed as references. Kernel-independent fast solvers and fast direct solvers are one of the most competitive choices for high-efficiency realistic multiscale simulations.

Acknowledgments

This work was supported in part by the Natural Science Foundation of China (NSFC) of 61890541 and 32261133623, and 62222108, the Fundamental Research Funds for the Central Universities of 30921011101, and the "INSIGHT—An innovative microwave sensing system for the evaluation and monitoring of food quality and safety" (CUP E13C23000180005), a joint research project within the Executive Program of Scientific and Technological Cooperation between Italy and China (period 2023–2025), funded by NSFC and the Italian Ministry of Foreign Affairs and International Cooperation (MAECI). The authors would also like to thank Prof. Dazhi Ding and Prof. Zhenhong Fan with the School of Microelectronics (School of Integrated Circuits) in Nanjing University of Science and Technology, Dr. Matteo Alessandro Francavilla originally with the Antenna and EMC Laboratory (LACE) in Politecnico di Torino for the contributions in developing the methods proposed here.

References

[1] Bucci OM and Franceschetti G. On the degrees of freedom of scattered fields. *IEEE Trans Antennas Propag*. 1989;37(7):918–926.

[2] Saad Y. *Iterative Methods for Sparse Linear Systems*. 2nd ed. Boston, MA: PWS Pub. Co.; 2003.

[3] Heldring A, Rius JM, Tamayo JM, *et al*. Fast direct solution of method of moments linear system. *IEEE Trans Antennas Propag*. 2007;55(3):3220–3228.

[4] Shaeffer J. Direct solve of electrically large integral equations for problem sizes to 1 M unknowns. *IEEE Trans Antennas Propag*. 2008;56(8):2306–2313.

[5] Bebendorf M. Approximation of boundary element matrices. *Numer Math*. 2000;86(4):565–589.

[6] Zhao K, Vouvakis MN, and Lee JF. The adaptive cross approximation algorithm for accelerated method of moments computations of EMC problems. *IEEE Trans Electromagn Compat*. 2005;47(4):763–773.

[7] Jiang LJ and Chew WC. Low-frequency fast inhomogeneous plane-wave algorithm (LF-FIPWA). *Microw Opt Technol Lett*. 2004;40(2):117–122.

[8] Jiang LJ and Chew WC. A mixed-form fast multipole algorithm. *IEEE Trans Antennas Propag*. 2005;53(12):4145–4156.

[9] Vikram M, Huang H, Shanker B, *et al.* A novel wideband FMM for fast integral equation solution of multiscale problems in electromagnetics. *IEEE Trans Antennas Propag*. 2009;57(7):2094–2104.
[10] Chen H, Leung KW, and Yung EKN. Fast directional multilevel algorithm for analyzing wave scattering. *IEEE Trans Antennas Propag*. 2011;59(7): 2546–2556.
[11] Ergül Ö, BK. Broadband multilevel fast multipole algorithm based on an approximate diagonalization of the Green's function. *IEEE Trans Antennas Propag*. 2015;63(7):3035–3041.
[12] Kalfa M, Ergül Ö, and Ertürk VB. Error control of multiple-precision MLFMA. *IEEE Trans Antennas Propag*. 2018;66(10):5651–5656.
[13] Vipiana F, Francavilla MA, and Vecchi G. EFIE modeling of high-definition multiscale structures. *IEEE Trans Antennas Propag*. 2010;58(7):2362–2374.
[14] Andriulli FP, Vipiana F, and Vecchi G. Hierarchical bases for nonhierarchic 3-D triangular meshes. *IEEE Trans Antennas Propag*. 2008;56(8):2288–2297.
[15] Heldring A, Rius JM, Tamayo JM, *et al.* Multiscale compressed block decomposition for fast direct solution of method of moments linear system. *IEEE Trans Antennas Propag*. 2011;59(2):526–536.
[16] Heldring A, Tamayo JM, Ubeda E, *et al.* Accelerated direct solution of the method-of-moments linear system. *Proc IEEE*. 2013;101(2):364–371.
[17] Michielssen E and Boag A. A multilevel matrix decomposition algorithm for analyzing scattering from large structures. *IEEE Trans Antennas Propag*. 1996;44(8):1086–1093.
[18] Tsang L, Li Q, Xu P, *et al.* Wave scattering with UV multilevel partitioning method: 2. Three-dimensional problem of nonpenetrable surface scattering. *Radio Sci*. 2004;39(5):1–11.
[19] Li M, Ding JJ, Ding DZ, *et al.* Multiresolution preconditioned multilevel UV method for analysis of planar layered finite frequency selective surface. *Microw Opt Tech Lett*. 2010;52(7):1530–1536.
[20] Li M, Ding D, Heldring A, *et al.* Low-rank matrix factorization method for multiscale simulations: a review. *IEEE Open J Antennas Propag*. 2021;2: 286–301.
[21] Rius JM, Parrón J, Heldring A, *et al.* Fast iterative solution of integral equations with method of moments and matrix decomposition algorithm—singular value decomposition. *IEEE Trans Antennas Propag*. 2008;56(8):2314–2324.
[22] Greengard L and Rokhlin V. A fast algorithm for particle simulations. *J Comput Phys*. 1987;73(2):325–348.
[23] Song J, Lu CC, and Chew WC. Multilevel fast multipole algorithm for electromagnetic scattering by large complex objects. *IEEE Trans Antennas Propag*. 1997;45(10):1488–1493.
[24] Li M, Li C, Ong CJ, *et al.* A novel multilevel matrix compression method for analysis of electromagnetic scattering from PEC targets. *IEEE Trans Antennas Propag*. 2012;60(3):1390–1399.
[25] Li M, Francavilla MA, Vipiana F, *et al.* A doubly hierarchical MoM for high-fidelity modeling of multiscale structures. *IEEE Trans Electromagn Compat*. 2014;56(5):1103–1111.

[26] Martinsson PG. A fast randomized algorithm for computing a hierarchically semiseparable representation of a matrix. *SIAM J Matrix Anal Appl.* 2011;32(4):1251–1274.

[27] Wei JG, Peng Z, and Lee JF. A fast direct matrix solver for surface integral equation methods for electromagnetic wave scattering from non-penetrable targets. *Radio Sci.* 2012;47(05):1–9.

[28] Francavilla M, Vipiana F, Vecchi G, *et al.* Hierarchical fast MoM solver for the modeling of large multiscale wire-surface structures. *IEEE Antennas Wireless Propag Lett.* 2012;11:1378–1381.

[29] Vipiana F, Polemi A, Maci S, *et al.* A mesh-adapted closed-form regular kernel for 3D singular integral equations. *IEEE Trans Antennas Propag.* 2008;56(6):1687–1698.

[30] Li M, Francavilla MA, Vipiana F, *et al.* Nested equivalent source approximation for the modeling of multiscale structures. *IEEE Trans Antennas Propag.* 2014;62(7):3664–3678.

[31] Bebendorf M and Venn R. Constructing nested bases approximations from the entries of non-local operators. *Numer Math.* 2012;121(4):609–635.

[32] Bautista ME, Francavilla MA, Vipiana F, *et al.* A hierarchical fast solver for EFIE-MoM analysis of multiscale structures at very low frequencies. *IEEE Trans Antennas Propag.* 2014;62(3):1523–1528.

[33] Quijano JLA and Vecchi G. Field and source equivalence in source reconstruction on 3D surfaces. *Progn Electromagnet Res.* 2010;103:67–100.

[34] Li M, Francavilla MA, Chen R, *et al.* Wideband fast kernel-independent modeling of large multiscale structures via nested equivalent source approximation. *IEEE Trans Antennas Propag.* 2015;63(5):2122–2134.

[35] Cheng H, Crutchfield WY, Gimbutas Z, *et al.* A wideband fast multipole method for the Helmholtz equation in three dimensions. *J Comput Phys.* 2006;216(1):300–325.

[36] Bogaert I, Peeters J, and Olyslager F. A nondirective plane wave MLFMA stable at low frequencies. *IEEE Trans Antennas Propag.* 2008;56(12):3752–3767.

[37] Melapudi V, Shanker B, Seal S, *et al.* A scalable parallel wideband MLFMA for efficient electromagnetic simulations on large scale clusters. *IEEE Trans Antennas Propag.* 2011;59(7):2565–2577.

[38] Baczewski AD, Dault DL, and Shanker B. Accelerated Cartesian expansions for the rapid solution of periodic multiscale problems. *IEEE Trans Antennas Propag.* 2012;60(9):4281–4290.

[39] Schobert DT and Eibert TF. Fast integral equation solution by multilevel Green's function interpolation combined with multilevel fast multipole method. *IEEE Trans Antennas Propag.* 2012;60(9):4458–4463.

[40] Messner M, Schanz M, and Darve E. Fast directional multilevel summation for oscillatory kernels based on Chebyshev interpolation. *J Comput Phys.* 2012;231(4):1175–1196.

[41] Ying L, Biros G, and Zorin D. A kernel-independent adaptive fast multipole algorithm in two and three dimensions. *J Comp Phys.* 2004;196(2):591–626.

[42] Engquist B and Ying L. Fast directional multilevel algorithms for oscillatory kernels. *SIAM J Sci Comput.* 2007;29(4):1710–1737.

[43] Bebendorf M, Kuske C, and Venn R. Wideband nested cross approximation for Helmholtz problems. *Numer Math*. 2015;130(1):1–34.

[44] Chew WC and Lu CC. The use of Huygens' equivalence principle for solving the volume integral equation of scattering. *IEEE Trans Antennas Propag*. 1993;41(7):897–904.

[45] Tamayo JM, Heldring A, and Rius JM. Multilevel adaptive cross approximation (MLACA). *IEEE Trans Antennas Propag*. 2011;59(12):4600–4608.

[46] Li M, Francavilla MA, Ding D, *et al.* Mixed-form nested approximation for wideband multiscale simulations. *IEEE Trans Antennas Propag*. 2018;66(11):6128–6136.

[47] Martinsson PG and Rokhlin V. A fast direct solver for boundary integral equations in two dimensions. *J Comput Phys*. 2005;205(1):1–23.

[48] Wei JG, Peng Z, and Lee JF. Multiscale electromagnetic computations using a hierarchical multilevel fast multipole algorithm. *Radio Sci*. 2014;49(11): 1022–1040.

[49] Bautista MAE, Francavilla MA, Martinsson PG, *et al.* $O(N)$ nested skeletonization scheme for the analysis of multiscale structures using the method of moments. *IEEE J Multiscale Multiphys Comput Tech*. 2016;1:139–150.

[50] Pan XM, Wei JG, Peng Z, *et al.* A fast algorithm for multiscale electromagnetic problems using interpolative decomposition and multilevel fast multipole algorithm. *Radio Sci*. 2012;47(01):1–11.

[51] Li M, Francavilla MA, Chen R, *et al.* Nested equivalent source approximation for the modeling of penetrable bodies. *IEEE Trans Antennas Propag*. 2016;65(2):954–959.

[52] Peng Z, Lim KH, and Lee JF. A discontinuous Galerkin surface integral equation method for electromagnetic wave scattering from nonpenetrable targets. *IEEE Trans Antennas Propag*. 2013;61(7):3617–3628.

[53] Ubeda E, Rius JM, and Heldring A. Nonconforming discretization of the electric-field integral equation for closed perfectly conducting objects. *IEEE Trans Antennas Propag*. 2014;62(8):4171–4186.

[54] Bautista MAE, Vipiana F, Francavilla MA, *et al.* A nonconformal domain decomposition scheme for the analysis of multiscale structures. *IEEE Trans Antennas Propag*. 2015;63(8):3548–3560.

[55] Li M, Ding D, Li J, *et al.* Nested equivalence source approximation with adaptive group size for multiscale simulations. *Eng Anal Bound Elem*. 2017;83:188–194.

[56] Ambikasaran S and Darve E. An $O(N\log N)$ fast direct solver for partial hierarchically semi-separable matrices. *J Sci Comput*. 2013;57(3):477–501.

[57] Chen X, Gu C, Li Z, *et al.* Accelerated direct solution of electromagnetic scattering via characteristic basis function method with Sherman-Morrison-Woodbury formula-based algorithm. *IEEE Trans Antennas Propag*. 2016;64(10):4482–4486.

[58] Wang K, Li M, Ding D, *et al.* A parallelizable direct solution of integral equation methods for electromagnetic analysis. *Eng Anal Bound Elem*. 2017;85: 158–164.

[59] Rong Z, Jiang M, Chen Y, *et al.* Fast direct solution of integral equations with modified HODLR structure for analyzing electromagnetic scattering problems. *IEEE Trans Antennas Propag*. 2019;67(5):3288–3296.

[60] Xia J, Chandrasekaran S, Gu M, *et al.* Fast algorithms for hierarchically semiseparable matrices. *Numer Linear Algebra Appl*. 2010;17(6):953–976.

[61] Ma M and Jiao D. Non-leaf-level algorithms in structure preserving hss matrix inversion in exact arithmetic. In: *2019 IEEE International Conference on Computational Electromagnetics (ICCEM)*. IEEE; 2019. p. 1–2.

Chapter 5
Domain decomposition method (DDM)

Víctor Martín[1], Hong-Wei Gao[2], Diego M. Solís[1], José M. Taboada[1] and Zhen Peng[3]

This chapter concerns the use of domain decomposition (DD) methods for the surface integral equation (SIE)-based solution of time-harmonic electromagnetic wave problems. DD methods have attracted significant attention for solving partial differential equations [1–16]. These methods are appealing due to their ability to obtain effective, efficient preconditioned iterative solution algorithms. They are also attractive because of their inherently parallel nature, an important consideration in keeping with current trends in computer architecture.

In the literature of computational mathematics, DD methods have also been extended to SIE methods [17–19]. Several noteworthy works include boundary element tearing and interconnecting methods [20], local and global multi-trace formulations [21–23], the Nitsche-based DD method for hypersingular integral equations [24], the mortar boundary element method [25], and the overlapping DD preconditioner for Laplacian problems [26]. In computational electromagnetics, overlapping integral equation DD methods have been discussed in [27–29]. A volume-based SIE DD method has been proposed to compute EM wave scattering from large multi-scale PEC objects in [30,31].

In this chapter, we focus our discussion on two classes of geometry-based SIE DD methods aiming to address the geometric complexity and material complexity in real-life EM engineering applications. The first class of methods leverages recent advances in the surface-based discontinuous Galerkin (DG) formulation [32,33]. The continuity of the currents across the domain boundaries is directly enforced through the interior penalty DG formulation. As a result, they do not require modifications on the original CAD object nor do they introduce artificial interfaces and auxiliary unknowns as in [31]. Moreover, there is no need to construct an additional contour subdomain at subdomain contour boundaries. The DD matrix equation resulting from the boundary

[1]Department of Computer and Communications Technology, University of Extremadura, Spain
[2]School of Cyberspace Science and Technology, Beijing Institute of Technology, China
[3]Electromagnetics Lab and Center for Computational Electromagnetics, Department of Electrical, Computer Engineering University of Illinois, USA

element discretization is solved with an additive Schwarz preconditioner, which offers high parallel efficiency on distributed memory HPC platforms.

The second class of SIE DD method is based on the tear-and-interconnect approach [34,35]. Similar to the DG formulation, the transmission conditions are imposed along the tearing contours between subdomains, without the need for the insertion of artificial surfaces to close the open subdomains that may arise from the splitting stage. This brings the ability to deal with both open and closed bodies. The main difference compared to the DG scheme is that the transmission conditions through the tear contours are imposed by employing a domain enlargement technique. In addition, the tear-and-interconnect DDM can be easily integrated with existing SIE implementation.

From a simulation-based engineering perspective, both classes of methods decompose an arbitrarily complicated virtual prototype into a collection of components directly based on the design description embodied in CAD models. Each component has its own geometric representation and is represented as one sub-domain. Individual sub-domains are allowed to choose the solution strategy the best adapted to the local EM characteristics and geometry features. Thus, it enables us to rapidly generate high-fidelity models of complex geometries and material properties.

The chapter is organized as follows. We first introduce the mathematical ingredients of the DG DDM for SIE (DG-DDM-SIE)-based solution of EM scattering from nonpenetrable PEC objects. Next, the DG-DDM-SIE approach is extended to the homogeneous penetrable objects. We will then illustrate the flexibility and capability of DG-DDM-SIE for piecewise homogeneous (composite) objects. Finally, the tear-and-interconnect DD method is presented for the solution of large-scale and multi-scale targets.

5.1 Discontinuous Galerkin DD method for PEC objects

5.1.1 Introduction to discontinuous Galerkin method

SIEs are often solved via the Galerkin method, which is based on a variational formulation in suitable trial and testing function spaces. It starts by representing the unknown quantities (currents and/or fields) using a set of trial functions, testing the integral equation based on the principle of duality pairing and solving the resulting discrete system. Both trial and testing functions are often defined in terms of interpolatory polynomials on a simplicial discretization of the target's surface. Noteworthy examples in this category are the $H(\mathrm{curl})$-conforming [36,37] and $H(\mathrm{div})$-conforming vector functions [38–40]. These vector functions are required to possess a certain degree of continuity, and/or smoothness across adjacent element boundaries. Therefore, the discretizations are commonly required to be conformal.

Generating conformal discretizations for real-life system-level simulations is far from trivial, as the complexity of modern engineering applications increases at a fast pace. Moreover, engineering analysis and design are not separate endeavors. In most cases, many iterative loops are involved in the engineering design process. When conformal discretizations are employed, a redesigned module may result in remeshing the entire system. Lastly, most of the conformal boundary element spaces are tightly associated with the underlying discretization. Mixing different types of basis functions, employing different discretizations, and/or incorporating the underlying physics to construct special basis functions within local regions pose great challenges.

DG methods have attracted considerable attention for discretizing partial differential equations (PDEs) [41]. They have been applied to the finite element-based solution of the time-dependent Maxwell's equations [42–45] and time-harmonic wave problems [46,47]. Since the tangential continuities of the fields and boundary conditions are enforced weakly through the Galerkin testings [48], the methods naturally fit into the framework of FEM and are adapted to heterogeneous media and finite non-conformal discretizations.

Recently, the DG method has also been investigated for SIE methods. After the first demonstration of EM scattering from nonpenetrable PEC objects [32], the DG SIE methodology has been extended to impedance boundary condition (IBC) objects [49], homogeneous penetrable objects [50], and piecewise homogeneous objects [51, 52]. The main benefit of the DG SIE method is that it allows square-integrable vector function spaces to be employed for both trial and testing functions. Therefore, it supports various types and shapes of elements, non-conformal discretizations, and non-uniform orders of approximation.

This section will introduce a SIE DD method based on the interior penalty (IP) DG formulation, which is termed as DG-DDM-SIE. The continuity of the electric surface current across the boundary between adjacent subdomains is enforced by a skew-symmetric IP DG formulation. For the solution of the linear system of equations resulting from boundary element discretization, a nonoverlapping additive Schwarz preconditioner is constructed and examined.

5.1.2 SIE formulation

We focus on the solution of the time-harmonic EM scattering problem from a non-penetrable PEC object Ω with its exterior boundary S, as illustrated in Figure 5.1(a). The exterior background Ω^{ext} is assumed to be a free space. For such a problem, the SIE method will transform it into the solution of an equivalent electric current $\boldsymbol{J}$ on the boundary S of the object according to the equivalence principle, which is shown in Figure 5.1(b). In Figure 5.1, $\boldsymbol{E}^{\text{i}}$ and $\boldsymbol{H}^{\text{i}}$ denote the incident electric and magnetic fields, respectively. They have an $e^{\imath\omega t}$ time dependence with the angular frequency operation $\omega = 2\pi f$ and the imaginary unit $\imath \equiv \sqrt{-1}$. Moreover, the permittivity and the permeability of free space are denoted by ε_0 and μ_0, respectively. Besides, the free-space wave number is denoted by $k_0 = \omega\sqrt{\mu_0\varepsilon_0}$, and the free-space intrinsic impedance is denoted by $\eta_0 = \sqrt{\mu_0/\varepsilon_0}$.

According to EM theories, we have the following boundary conditions on PEC surface S: the tangential electric field is zero, e.g., $\boldsymbol{E}_{\text{tan}} = \hat{\boldsymbol{n}} \times \boldsymbol{E} \times \hat{\boldsymbol{n}} = 0$; the twisted tangential magnetic field is $\boldsymbol{J}$, e.g. $\hat{\boldsymbol{n}} \times \boldsymbol{H} = \boldsymbol{J}$. It notes that $\boldsymbol{E}$ and $\boldsymbol{H}$ are the total

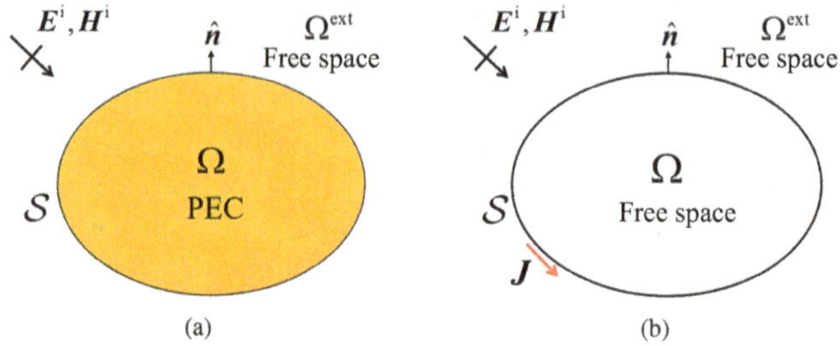

Figure 5.1 Illustration of EM scattering problem from a nonpenetrable PEC object. (a) Original problem. (b) Equivalent problem.

electric and magnetic fields. Based on the above boundary conditions, we can derive the electric field integral equation (EFIE) as

$$\mathscr{L}(\bar{\boldsymbol{J}}; S)(\boldsymbol{r}) + \boldsymbol{E}^{i}_{tan}(\boldsymbol{r}) = 0 \qquad \boldsymbol{r} \in S \tag{5.1}$$

and the magnetic field integral equation (MFIE) as

$$\hat{\boldsymbol{n}} \times \mathscr{K}(\bar{\boldsymbol{J}}; S)(\boldsymbol{r}) - \frac{1}{2}\bar{\boldsymbol{J}}(\boldsymbol{r}) + \eta_0 \hat{\boldsymbol{n}} \times \boldsymbol{H}^{i}(\boldsymbol{r}) = 0 \qquad \boldsymbol{r} \in S \tag{5.2}$$

where $\bar{\boldsymbol{J}} = \eta_0 \boldsymbol{J}$ is the scaled electric current. $\mathscr{L}$ and $\mathscr{K}$ are two integro-differential operators, defined as

$$\mathscr{L}(\boldsymbol{X}; S)(\boldsymbol{r}) := -\imath k_0 \Psi_A(\boldsymbol{X}; S)(\boldsymbol{r}) + \frac{1}{\imath k_0} \nabla \Psi_F(\nabla_\tau \cdot \boldsymbol{X}; S)(\boldsymbol{r}) \tag{5.3}$$

$$\mathscr{K}(\boldsymbol{X}; S)(\boldsymbol{r}) := p.v.[\nabla \times \Psi_A(\boldsymbol{X}; S)(\boldsymbol{r})] \tag{5.4}$$

where Ψ_A and Ψ_F are the single-layer vector and scalar potential, defined by:

$$\Psi_A(\boldsymbol{X}; S)(\boldsymbol{r}) := \int_S \boldsymbol{X}(\boldsymbol{r}') G(\boldsymbol{r}, \boldsymbol{r}') d\boldsymbol{r}' \tag{5.5}$$

$$\Psi_F(\rho; S)(\boldsymbol{r}) := \int_S \rho(\boldsymbol{r}') G(\boldsymbol{r}, \boldsymbol{r}') d\boldsymbol{r}' \tag{5.6}$$

with the free-space Green's function $G(\boldsymbol{r}, \boldsymbol{r}') := e^{-\imath k_0 |\boldsymbol{r} - \boldsymbol{r}'|}/4\pi|\boldsymbol{r} - \boldsymbol{r}'|$. $\boldsymbol{r}$ and $\boldsymbol{r}'$ denote the observation point and the source point, respectively. Moreover, $p.v.$ in (5.4) stands for the Cauchy principal value.

5.1.3 Domain partitioning and basis function space

The first step of DDMs is to decompose the entire computational domain into several small subdomains. For the discontinuous Galerkin (DG) DDM, the surface S is decomposed into several nonoverlapping open sub-surfaces (subdomains), as shown in Figure 5.2. For simplicity and without loss of generality, the number of subdomains is considered as $M = 3$ such that $S = S_1 \cup S_2 \cup S_3$. Each subdomain is a part of the

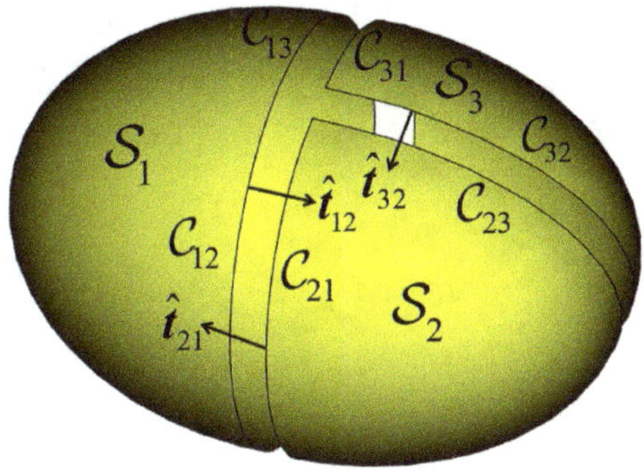

Figure 5.2 Notations for surface decomposition

original surface S. The boundary of subdomain S_i is denoted by C_i, and $\hat{\boldsymbol{t}}_i$ is the corresponding exterior unit normal on C_i tangent to S_i. Besides, the contour boundaries between two adjacent subdomains S_i and S_j are denoted as C_{ij} w.r.t. S_i and C_{ji} w.r.t. S_j. Further, associated with each subdomain contour C_{ij}, we define a unit normal $\hat{\boldsymbol{t}}_{ij}$, which points from subdomain S_i toward subdomain S_j.

Under this setting, the electric current can be written as $\boldsymbol{J} = \bigoplus_{i=1}^{M} \boldsymbol{J}_i$. In the discontinuous Garlerkin DDM, the trial function space $\boldsymbol{u}_i$ for each subdomain S_i is constructed independently, where $\boldsymbol{u}_i(\boldsymbol{r}) \in \mathbf{H}^{-1/2}(\text{div}_\tau, S_i)$ is the local approximation within each subdomain S_i. The space $\boldsymbol{H}^{-1/2}(\text{div}_\tau, S_i)$ is defined as

$$\boldsymbol{H}^{-1/2}(\text{div}_\tau, S_i) := \left\{ \boldsymbol{u} \in \boldsymbol{H}^{-1/2}(S_i) | \nabla_\tau \cdot \boldsymbol{u} \in \mathrm{H}^{-1/2}(S_i) \right\} \tag{5.7}$$

Namely, the requirement for $\boldsymbol{u}_i(\boldsymbol{r})$ is that both the function and its surface divergence have finite energy in S_i. Thus, it allows us to choose trial basis functions whose in-plane normal components are continuous within each subdomain S_i but can be discontinuous across subdomain boundaries. Thereby, we may employ the well-known Rao–Wilton–Glisson (RWG) basis functions [38] with extensions to discretize the trial function spaces. As shown in Figure 5.3, each subdomain S_i is first meshed by triangles independently. Associating with the edges located in the interior of S_i, the traditional RWG basis functions are defined. For the edges located at the contours C_i, only half of RWG basis functions are used because one edge is just supported by one triangle. These basis functions can be called half-RWG functions or monopolar-RWG functions [53]. In Figure 5.3, l_i^n denotes the length of the nth edge of subdomain S_i, and A_i^{n+} denotes the area of a triangle T_i^{n+}.

Although the construction of trial function space for each subdomain could be independent, one should introduce additional conditions to enforce the continuity of the electric current across contours between different subdomains. To simplify the notions for the following discussion, we introduce the following jump operator at contours between subdomains:

$$[[\boldsymbol{u}]]_{ij} = \hat{\boldsymbol{t}}_{ij} \cdot \boldsymbol{u}_i - \hat{\boldsymbol{t}}_{ij} \cdot \boldsymbol{u}_j \quad \text{at} \quad C_{ij} \tag{5.8}$$

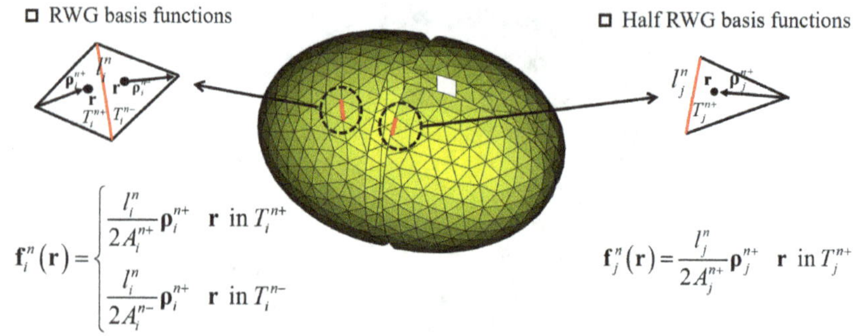

Figure 5.3 Different RWG basis functions for discretizing trial function spaces

The vector and scalar inner products are defined by $<\boldsymbol{x},\boldsymbol{y}>_{S_i}= \int_{S_i}\boldsymbol{x}\cdot\boldsymbol{y}ds$ and $< x,y >_{S_i}= \int_{S_i} xyds$, respectively.

5.1.4 Interior penalty formulation

In this section, we present the variational formulation of DG-DDM-SIE for the decomposed problem shown in Figure 5.2 based on the IP method [48], whose basic concept is to weakly enforce the interelement continuity through penalization terms. We begin with the derivation of variational weak formulation for the EFIE equation. For each subdomain $S_i \in S$, we test (5.1) with a test function $\boldsymbol{v}_i \in \boldsymbol{H}^{-1/2}(\mathrm{div}_\tau, S_i)$, which gives

$$\left\langle \boldsymbol{v}_i, \sum_{j=1}^{M} \mathscr{L}(\bar{\boldsymbol{J}}_j; S_j) \right\rangle_{S_i} = -\left\langle \boldsymbol{v}_i, \boldsymbol{E}^{\mathrm{i}}_{\tan} \right\rangle_{S_i} \tag{5.9}$$

whose detailed expansion is

$$\imath k_0 \left\langle \boldsymbol{v}_i, \sum_{j=1}^{M} \Psi_A(\bar{\boldsymbol{J}}_j; S_j) \right\rangle_{S_i} - \frac{1}{\imath k_0} \left\langle \boldsymbol{v}_i, \sum_{j=1}^{M} \nabla \Psi_F(\nabla'_\tau \cdot \bar{\boldsymbol{J}}_j; S_j) \right\rangle_{S_i} = \left\langle \boldsymbol{v}_i, \boldsymbol{E}^{\mathrm{i}}_{\tan} \right\rangle_{S_i} \tag{5.10}$$

Then, we apply a surface Green's formula to the second summand to shift the surface gradient onto the testing function, which introduces a boundary term involving a line integral along subdomain boundary C_i. We arrive at the following expression:

$$\begin{aligned} &\imath k_0 \left\langle \boldsymbol{v}_i, \sum_{j=1}^{M} \Psi_A(\bar{\boldsymbol{J}}_j; S_j) \right\rangle_{S_i} + \frac{1}{\imath k_0} \left\langle \nabla \cdot \boldsymbol{v}_i, \sum_{j=1}^{M} \Psi_F(\nabla'_\tau \cdot \bar{\boldsymbol{J}}_j; S_j) \right\rangle_{S_i} \\ &- \frac{1}{\imath k_0} \left\langle \boldsymbol{v}_i \cdot \hat{\boldsymbol{t}}_i, \sum_{j=1}^{M} \Psi_F(\nabla'_\tau \cdot \bar{\boldsymbol{J}}_j; S_j) \right\rangle_{C_i} = \left\langle \boldsymbol{v}_i, \boldsymbol{E}^{\mathrm{i}}_{\tan} \right\rangle_{S_i} \end{aligned} \tag{5.11}$$

Summing over all subdomains and using the continuity of $\int_{S_j} \nabla \cdot \bar{\boldsymbol{J}}_j d\boldsymbol{r}'$, we are able to rewrite the boundary term in (5.11) through normal jumps of $\boldsymbol{v}_i$ using (5.8). Explicitly, we have

$$-\frac{1}{\imath k_0}\left\langle \boldsymbol{v}_i \cdot \hat{\boldsymbol{t}}_i, \sum_{j=1}^{M} \Psi_F(\nabla'_\tau \cdot \bar{\boldsymbol{J}}_j; S_j)\right\rangle_{C_i} = -\frac{1}{2\imath k_0}\sum_{C_{ij}\in C}\left\langle [[\boldsymbol{v}]]_{ij}, \sum_{j=1}^{M} \Psi_F(\nabla'_\tau \cdot \bar{\boldsymbol{J}}_j; S_j)\right\rangle_{C_{ij}} \tag{5.12}$$

To this, one could add an antisymmetric boundary IP term involving the normal jump of $\bar{\boldsymbol{J}}$ as

$$\frac{1}{2\imath k_0}\sum_{C_{ij}\in C}\left\langle [[\bar{\boldsymbol{J}}]]_{ij}, \sum_{j=1}^{M} \Psi_F(\nabla'_\tau \cdot \boldsymbol{v}_j; S_j)\right\rangle_{C_{ij}} = 0 \tag{5.13}$$

We note that this boundary IP term will vanish for the exact solution and, thus, does not compromise the consistency of the formulation. However, it is necessary to recover the ellipticity of the weak formulation [33].

Finally, since it is our intention to search for the solution in a space that allows in-plane normal discontinuity across subdomain boundaries, we add a penalty term to enforce continuity weakly across the boundary of subdomains, defined by

$$\left\langle [[\boldsymbol{v}]]_{ij}, [[\bar{\boldsymbol{J}}]]_{ij}\right\rangle_{C_{ij}} = 0 \tag{5.14}$$

The above suggests a variational weak formulation for the EFIE equation, which can be formally stated as: Find $\boldsymbol{u} = \bigoplus_{i=1}^{M} \boldsymbol{u}_i,\ \boldsymbol{u}_i \in \boldsymbol{H}^{-1/2}(\mathrm{div}_\tau, S_i)$, such that

$$\mathrm{a}\,(\boldsymbol{v}, \boldsymbol{u}) = \left\langle \boldsymbol{v}, \boldsymbol{E}^{\mathrm{i}}_{\tan}\right\rangle_{\bigoplus_{i=1}^{M} S_i} \tag{5.15}$$

with $\forall \boldsymbol{v} = \bigoplus_{i=1}^{M} \boldsymbol{v}_i,\ \boldsymbol{v}_i \in \boldsymbol{H}^{-1/2}(\mathrm{div}_\tau, S_i)$. Besides, the bilinear form is defined as $\mathrm{a}\,(\boldsymbol{v}, \boldsymbol{u}) = \mathrm{a_S}\,(\boldsymbol{v}, \boldsymbol{u}) + \mathrm{a_C}\,(\boldsymbol{v}, \boldsymbol{u}) + \mathrm{a_P}\,(\boldsymbol{v}, \boldsymbol{u})$.

The first term, $\mathrm{a_S}\,(\boldsymbol{v}, \boldsymbol{u})$, denotes the contribution from the standard EFIE bilinear form localized to the subdomains:

$$\mathrm{a_S}\,(\boldsymbol{v}, \boldsymbol{u}) := \imath k_0 \sum_{i=1}^{M}\left\langle \boldsymbol{v}_i, \sum_{j=1}^{M} \Psi_A(\boldsymbol{u}_j; S_j)\right\rangle_{S_i} + \frac{1}{\imath k_0}\sum_{i=1}^{M}\left\langle \nabla \cdot \boldsymbol{v}_i, \sum_{j=1}^{M} \Psi_F(\nabla'_\tau \cdot \boldsymbol{u}_j; S_j)\right\rangle_{S_i} \tag{5.16}$$

The second term, $\mathrm{a_C}(\boldsymbol{v}, \boldsymbol{u})$, involves the following line integrals along contours:

$$\mathrm{a_C}(\boldsymbol{v}, \boldsymbol{u}) := -\frac{1}{2\imath k_0} \sum_{C_{ij} \in C} \left\langle [[\boldsymbol{v}]]_{ij}, \sum_{j=1}^{M} \Psi_F(\nabla'_\tau \cdot \boldsymbol{u}_j; S_j) \right\rangle_{C_{ij}} + \frac{1}{2\imath k_0} \sum_{C_{ij} \in C} \left\langle [[\boldsymbol{u}]]_{ij}, \sum_{j=1}^{M} \Psi_F(\nabla'_\tau \cdot \boldsymbol{v}_j; S_j) \right\rangle_{C_{ij}} \tag{5.17}$$

We remark that $[[\mathbf{v}]]_{ij}$ can be viewed as virtual charges that accumulate on the contour boundaries (due to the normal discontinuity of the trial currents). Following the proper duality pairing, the virtual charges should be paired with the electric potential to ensure that there is no measurable electric potential energy in the weak form. This indeed holds true for the consistency term in (5.17).

The last term, $\mathrm{a_P}(\boldsymbol{v}, \boldsymbol{u})$, is the IP stabilization term, defined by

$$\mathrm{a_P}(\boldsymbol{v}, \boldsymbol{u}) := \frac{\beta}{k_0} \sum_{C_{ij} \in C} \left\langle [[\boldsymbol{v}]]_{ij}, [[\bar{\boldsymbol{J}}]]_{ij} \right\rangle_{C_{ij}} \tag{5.18}$$

which is introduced to stabilize the weak formulation. In recent work on DG-BEM, the stabilization parameter, β, is chosen inversely proportional to the characteristic element size h. It leads to $\beta = \zeta h^{-1}$, in which ζ is a small positive number independent of h. Numerical experiments suggest that $\beta = |\log \bar{h}|/10$ is a good choice, where $\bar{h}$ is the average element size over the entire discretization.

Similarly, the weak formulation for MFIE is to find

$$\mathrm{b}(\boldsymbol{v}, \boldsymbol{u}) = \left\langle \boldsymbol{v}, \hat{\boldsymbol{n}} \times \boldsymbol{H}^{\mathrm{i}} \right\rangle_{\bigoplus_{i=1}^{M} S_i} \tag{5.19}$$

where the bilinear form is defined by

$$\mathrm{b}(\boldsymbol{v}, \boldsymbol{u}) = \frac{1}{2} \sum_{i=1}^{M} \langle \boldsymbol{v}_i, \boldsymbol{u}_i \rangle_{S_i} - \sum_{i=1}^{M} \left\langle \boldsymbol{v}_i, \hat{\boldsymbol{n}}_i \times \sum_{j=1}^{M} \mathscr{K}(\boldsymbol{u}_j; S_j) \right\rangle_{S_i} \tag{5.20}$$

It remarks that no consistency terms are required here.

Finally, to avoid spurious resonances, the EFIE and MFIE weak equations are added linearly which results in the CFIE weak equation:

$$\alpha \mathrm{a}(v, \boldsymbol{u}) + (1-\alpha)\mathrm{b}(\boldsymbol{v}, \boldsymbol{u}) = \alpha \left\langle \boldsymbol{v}, \boldsymbol{E}^{\mathrm{i}}_{\mathrm{tan}} \right\rangle_{\bigoplus_{i=1}^{M} S_i} + (1-\alpha) \left\langle \boldsymbol{v}, \hat{\boldsymbol{n}} \times \boldsymbol{H}^{\mathrm{i}} \right\rangle_{\bigoplus_{i=1}^{M} S_i} \tag{5.21}$$

The coupling parameter α is generally chosen to be $\frac{1}{2}$.

5.1.5 Matrix equation and preconditioner

Discretizing the testing function space $\boldsymbol{v}$ and the trial function space $\boldsymbol{u}$ by the aforementioned RWG and half-RWG basis functions, the CFIE weak equation (5.21) will be converted into a linear matrix equation. For the case of three subdomains, the linear matrix equation is written as:

$$\begin{bmatrix} \boldsymbol{Z}_{11} & \boldsymbol{Z}_{12} & \boldsymbol{Z}_{13} \\ \boldsymbol{Z}_{21} & \boldsymbol{Z}_{22} & \boldsymbol{Z}_{23} \\ \boldsymbol{Z}_{31} & \boldsymbol{Z}_{32} & \boldsymbol{Z}_{33} \end{bmatrix} \begin{Bmatrix} \boldsymbol{x}_1 \\ \boldsymbol{x}_2 \\ \boldsymbol{x}_3 \end{Bmatrix} = \begin{Bmatrix} \boldsymbol{b}_1 \\ \boldsymbol{b}_2 \\ \boldsymbol{b}_3 \end{Bmatrix} \tag{5.22}$$

where $\boldsymbol{Z}_{ii}$ denotes the CFIE impedance matrix for subdomain S_i, and $\boldsymbol{Z}_{ij}$ is the coupling matrix between subdomains S_i and S_j. In addition, $\boldsymbol{x}_i$ denotes the vector containing the unknown coefficients of expanding basis functions for $\bar{\boldsymbol{J}}_i$, and $\boldsymbol{b}_i$ denotes the source vector for the subdomain S_i due to the incident electric and magnetic fields.

To further improve the conditioning number of the matrix equation (5.22), we construct an effective preconditioner $\boldsymbol{P}^{-1}$ through the diagonal submatrices, explicitly written as:

$$\boldsymbol{P}^{-1} = \begin{bmatrix} \boldsymbol{Z}_{11}^{-1} & 0 & 0 \\ 0 & \boldsymbol{Z}_{22}^{-1} & 0 \\ 0 & 0 & \boldsymbol{Z}_{33}^{-1} \end{bmatrix} \tag{5.23}$$

This preconditioner is called an additive Schwarz preconditioner or a diagonal block preconditioner. Applying $\boldsymbol{P}^{-1}$ to (5.22), we will obtain the final preconditioned matrix equation as:

$$\begin{bmatrix} \boldsymbol{I}_1 & \boldsymbol{Z}_{11}^{-1}\boldsymbol{Z}_{12} & \boldsymbol{Z}_{11}^{-1}\boldsymbol{Z}_{13} \\ \boldsymbol{Z}_{22}^{-1}\boldsymbol{Z}_{21} & \boldsymbol{I}_2 & \boldsymbol{Z}_{22}^{-1}\boldsymbol{Z}_{23} \\ \boldsymbol{Z}_{33}^{-1}\boldsymbol{Z}_{31} & \boldsymbol{Z}_{33}^{-1}\boldsymbol{Z}_{32} & \boldsymbol{I}_3 \end{bmatrix} \begin{Bmatrix} \boldsymbol{x}_1 \\ \boldsymbol{x}_2 \\ \boldsymbol{x}_3 \end{Bmatrix} = \begin{Bmatrix} \boldsymbol{Z}_{11}^{-1}\mathrm{b}_1 \\ \boldsymbol{Z}_{22}^{-1}\mathrm{b}_2 \\ \boldsymbol{Z}_{33}^{-1}\mathrm{b}_3 \end{Bmatrix} \tag{5.24}$$

5.1.6 Iterative solution of preconditioned matrix equation

The preconditioned matrix equation (5.24) will be solved with a Krylov subspace method, e.g., such as GCR [54] and GMRES [55]. The multilevel fast multipole algorithm (MLFMA) [56,57] can be employed to accelerate the dense matrix–vector multiplication. It is clear that the application of the preconditioner only requires the inversion of each diagonal block CFIE matrix. In practical computations, these CFIE matrices can be either factorized in a pre-processing step or solved via another preconditioned Krylov method at each DDM iteration (inner-loop iteration). Moreover, in many applications, there might exist different types of repetitions and periodicities within the computational domain, where the partitions can be wrought to exploit local repetitions.

5.1.6.1 Discussion of the computational complexity

When a Krylov subspace iterative method is applied to solve equation (5.24), the iteration count of preconditioned systems increases logarithmically on the number of sub-domains and the electrical size of the object, $K \propto O(\log M)$. Moreover, the application of the preconditioner requires the inversion of each sub-domain matrix on the diagonal. If we keep the size of the sub-domain to be constant, the number of sub-domain increases linearly as the number of degrees of freedom (DOFs), $M \propto O(N)$. The computational costs for applying the preconditioner exhibit linear computational complexity. Therefore, the method provides a very effective preconditioning scheme for very large-scale EM problems.

Assuming a fast direct solver is applied for the factorization of subdomain matrices and MLFMA is employed for the matrix–vector multiplication, the overall computational complexity for the iterative solution of (5.24) can be expressed as:

$$\underbrace{O(\log M)}_{iterations}\left[\underbrace{O\left(M\left(\frac{N}{M}\right)^{1.5}\right)}_{preconditioning}+\underbrace{O(N\log N)}_{coupling}\right]\xrightarrow{M\propto O(N)} O\left(N\,(\log N)^2\right) \quad (5.25)$$

It can be concluded from (5.25) that the method presents a quasi-linear computational complexity for high-frequency EM wave problems.

5.1.6.2 Discussion of the parallel implementation

To fully exploit the recent success of multi-core processors and massively parallel-distributed memory supercomputers, we can develop a hybrid MPI/OpenMP implementation of the proposed algorithms. The computation in solving the preconditioned DD matrix equation can be divided into two parts: (*i*) the application of the additive Schwarz preconditioner involving local sub-domain solutions; and (*ii*) the coupling among sub-domains corresponding to off-diagonal matrices.

The application of the additive Schwarz preconditioner requires the solution of individual sub-domain problems. One advantage of the proposed preconditioner is the ability to solve all the sub-domain problems simultaneously in each Krylov iteration. In the parallel implementation, we may employ a task-based parallelism for sub-domain solutions. Namely, those sub-domain solutions are considered independent tasks in an MPI programming model. Individual MPI processes will execute different tasks simultaneously. The parallelization within each task is attained using OpenMP, which exploits the fast memory access in the shared-memory multi-core processors. Therefore, each sub-domain is allowed to choose its own appropriate sub-domain solver based on local EM wave characteristics and geometrical features.

The second part of the parallel computing is the coupling among multiple sub-domains. To achieve low computational costs, we may adopt separate representations based on the relative spatial positions of two sub-domains. For readers who are interested in this subject, a detailed discussion is provided in [58,59].

5.1.7 Numerical experiments

In this section, numerical experiments are presented to validate the analysis and evaluate the performance of the proposed DG-DDM-SIE method. We first validate the solution accuracy of the proposed work. It is followed by the convergence analysis. Finally, we present the parallel strong scaling and weak scaling experiments.

5.1.7.1 Validation of the solution accuracy

We first apply the DG-DDM-SIE to analyze the monostatic EM scattering problem from a NASA metallic almond [60]. Figure 5.4 shows the geometry of the almond

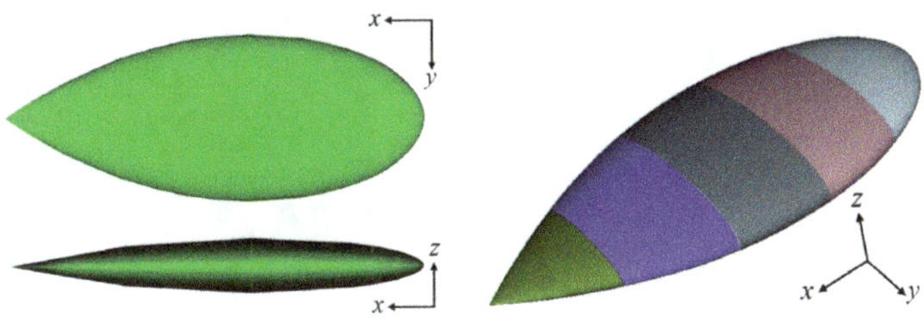

(a) (b)

Figure 5.4 The geometry and domain partition of the NASA almond. (a) Front and side view of the almond and (b) regular five subdomains.

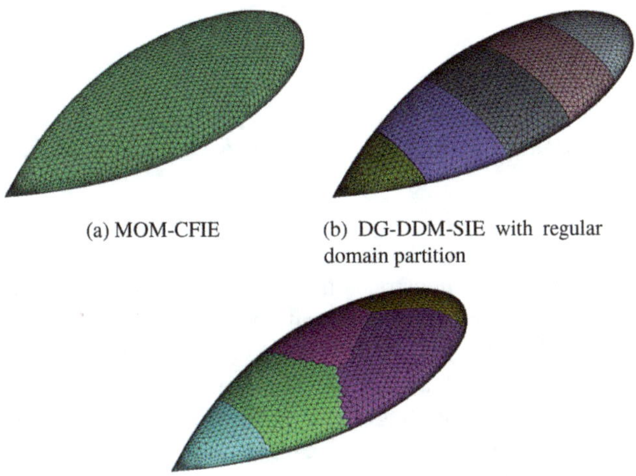

(a) MOM-CFIE

(b) DG-DDM-SIE with regular domain partition

(c) DG-DDM-SIE with irregular domain partition

Figure 5.5 The triangular meshes and domain partitionings used for different numerical schemes

with a length of 252.37 mm. It is illuminated by plane waves at 7 GHz coming from $\theta = 90°, \phi = 0° : 1° : 180°$. Three numerical methods are considered in this study: the standard CFIE method with boundary element discretization, the DG-DDM-SIE with regular subdomain partitioning, and the DG-DDM-SIE with irregular subdomain partitioning using the algorithmic graph partitioning algorithm, METIS [61]. The illustration of the discretizations and domain partitionings is given in Figure 5.5.

The calculated monostatic RCS using the three methods are plotted in Figure 5.6. As a reference, the measurements published in [60] are also included. It can be seen the simulation results from CFIE and DG-DDM-SIE all agree very well with

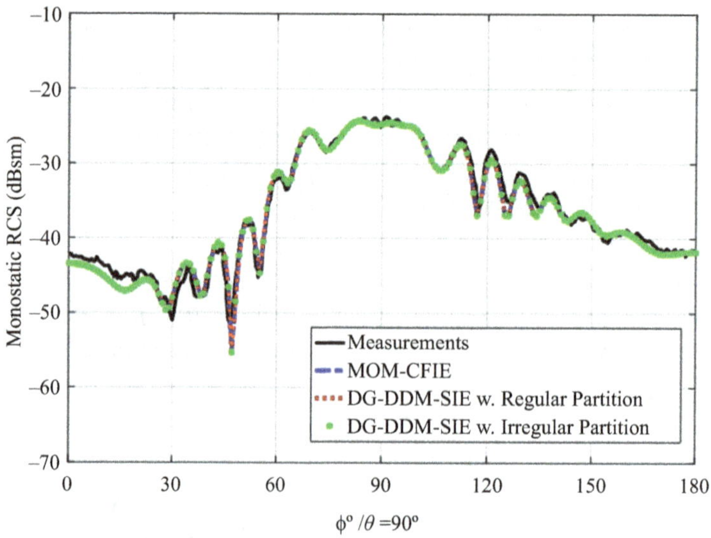

Figure 5.6 Monostatic RCS from different methods for the metallic NASA almond at 7 GHz

the measurements, which demonstrates the accuracy of the simulation. Moreover, Figure 5.7 shows the induced electric current obtained by the three methods for the incident direction of $\theta = 90°, \phi = 30°$. We observe that the currents from MOM and DG-DDM-SIE agree very well, and the currents from DG-DDM-SIE are continuous and smooth for both regular and irregular partitions. From the above results, one can conclude that the DG-DDM-SIE shows the same level of accuracy as the standard CFIE with boundary element discretization.

5.1.7.2 Convergence study

Next, we investigate the convergence in iterative matrix solutions of (5.24) with respect to the problem size using the DG-DDM-SIE solver. Namely, we keep the size of each sub-domain constant and increase the number of sub-domains for higher operating frequencies. The studies performed here demonstrate the potential of the proposed method to solve very large-scale problems. A PEC cube of dimensions 1 m by 1 m by 1 m is used and the performance of the method is evaluated at five different frequencies, 0.6 GHz, 1.2 GHz, 2.4 GHz, 4.8 GHz, and 9.6 GHz. The electrical size of the cube ranges from 2 λ to 32 λ, and the number of sub-domains increases from 2 to 512, correspondingly. The iterative solver convergences for various simulations are plotted in Figure 5.8. We observe that the convergence behavior is quite insensitive to the increase in the number of sub-domains. Regarding to the cases of 2 and 512 sub-domains, the iteration counts increase from 4 to 7 to reach a relative residual 10^{-2}, and from 24 to 35 to reach a relative residual 10^{-6}. Figure 5.9 gives electric current distributions at 9.6 GHz using the proposed work with 512 sub-domains. We notice

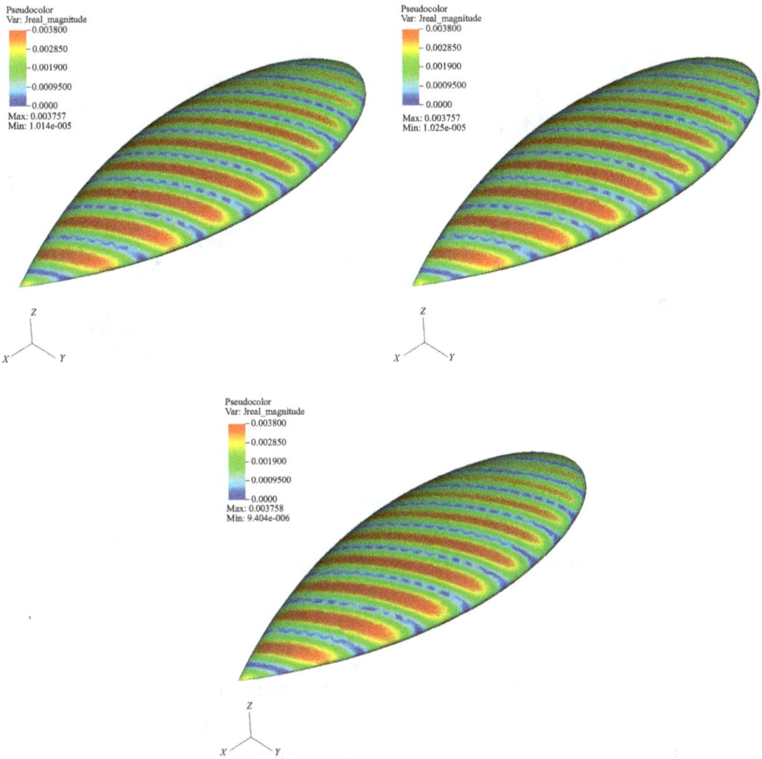

Figure 5.7 *The electric current distribution obtained from different methods for the metallic almond under the plane wave at 7 GHz illuminating from* $\theta = 90°, \phi = 30°$.

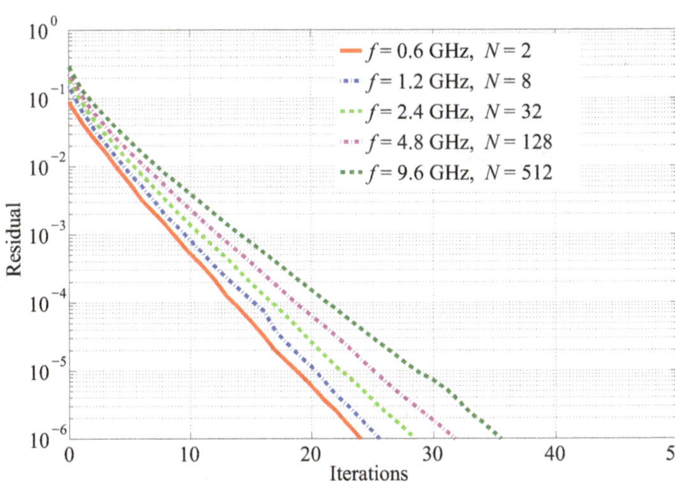

Figure 5.8 *Iterative solver convergence with respect to the operating frequency*

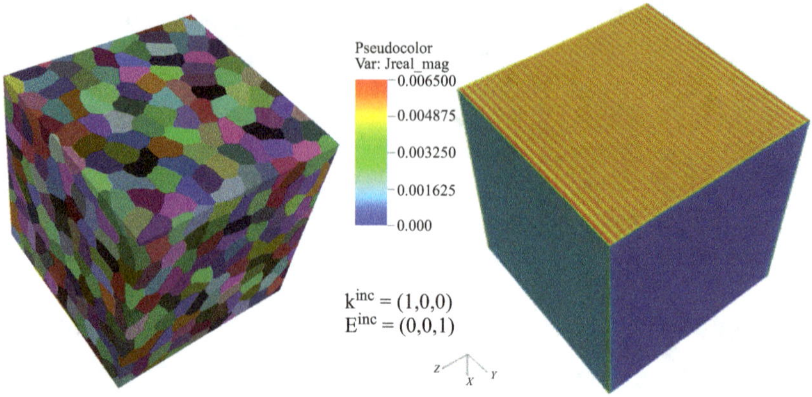

Figure 5.9 Electric current on a PEC cube at 9.6 GHz using 512 sub-domains

that the normal continuity of electric current is enforced correctly at the surface of the PEC cube.

5.1.7.3 Parallel efficiency study

We consider a plane-wave scattering from a mockup unmanned aerial vehicle (UAV) made of non-penetrable PEC materials. The dimensions of the geometry are approximately 14.4 m wide, 8.2 m long, and 2.5 m high. The plane-wave incidents upon the UAV from the nose (radome) direction. The electric field is perpendicular to the wing span. In the weak scalability experiment, the performance of the method is evaluated at four different frequencies, 7 GHz, 10 GHz, 14 GHz, and 20 GHz. The electrical size of the UAV ranges from 336 λ to 960 λ. The average mesh size is kept fixed at $h = \lambda/12$ for individual operating frequencies. The surface discretization at each frequency is then partitioned into a number of sub-domains with a nearly constant number of DOFs, 2 million, per sub-domain. The number of sub-domains increases from 5 sub-domains at 7 GHz to 40 sub-domains at 20 GHz, correspondingly. The resulting partitions are displayed in Figure 5.10.

In the parallel computation, each sub-domain is assigned one MPI process in which 32 OpenMP threads are used. The parallelization efficiency relative to 32 cores is plotted in Figure 5.11. It is evident that the efficiency is close to 90% for all the computations. Moreover, the iteration counts to reach a relative residual of 10–2 only increase from 5 to 10 when comparing the cases of 7 GHz with 8.75 million DOFs and 20 GHz with 84.3 million DOFs. The promising iterative solver convergence in the proposed work is one of the crucial factors to reach scalability for very large-scale EM simulations.

Next, in the strong scaling experiment, we examine the solution time required for a fixed problem size by increasing the number of cores. To do so, we consider EM scattering from the UAV at 10 GHz and increase the number of sub-domain from 4 to 28. In parallel computing, each sub-domain is assigned one MPI process in which

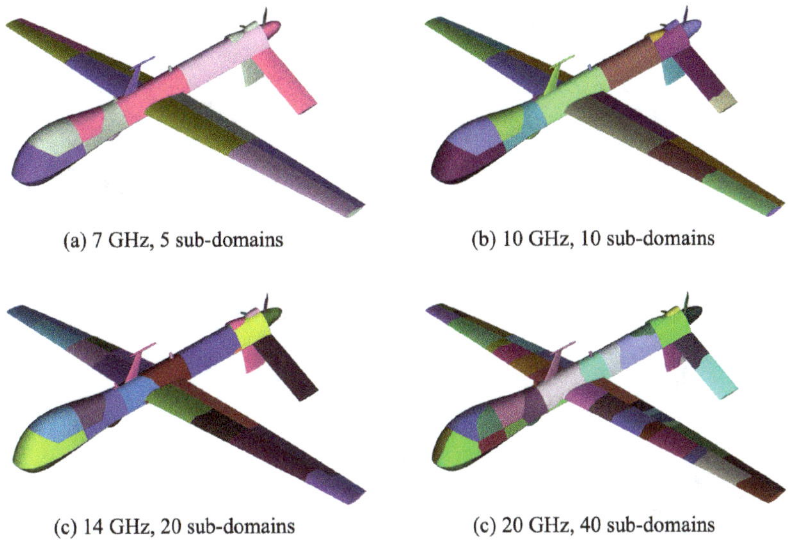

Figure 5.10 Domain partitioning for UAV simulation

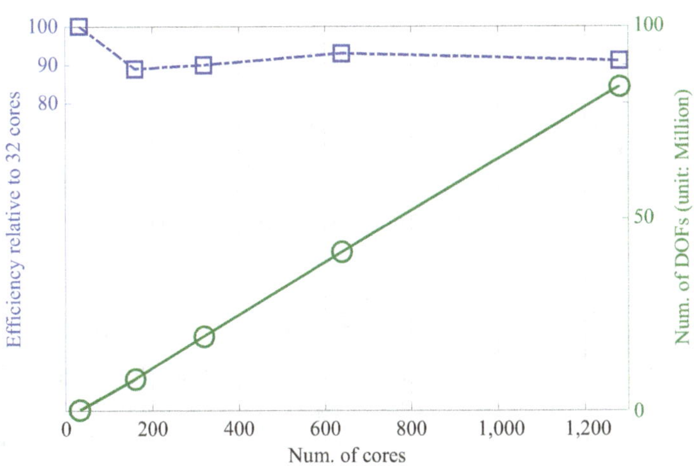

Figure 5.11 Parallelization efficiency in various simulations

32 OpenMP threads are used. Thus, the total computing cores increase from 128 to 896. The timings of various simulations are given in Figure 5.12. We observe only a slight growth of iteration numbers with respect to the increase in the number of sub-domains. The speedup in run time is almost linear. At peak performance using 896 cores, the speedup relative to 128 cores is approximately a 6.5 times decrease in simulation time.

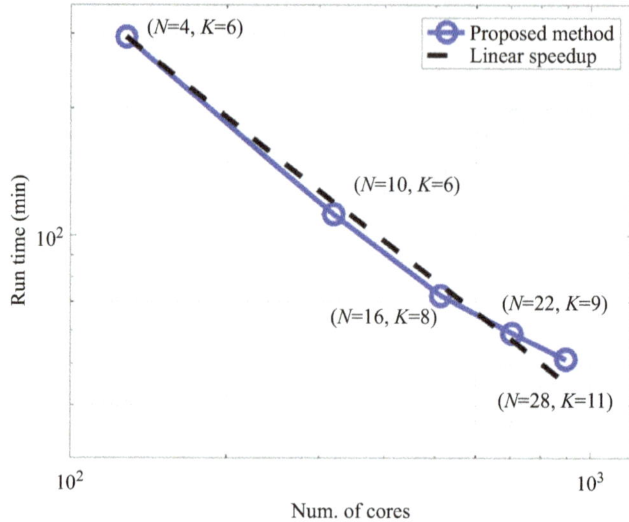

Figure 5.12 Parallel performance tests with respect to strong scalability, N: number of MPI processes (sub-domains) and K: number of iterations

5.2 DG DD method for penetrable objects

The previously introduced DG-DDM-SIE formulation for non-penetrable PEC objects can be extended to the analysis of penetrable objects. In this section, we will first present the DG-DDM-SIE formulation for a homogeneous penetrable object, then discuss the analysis of piecewise homogeneous composite objects including junctions between different materials.

5.2.1 DG-DDM-SIE for homogeneous objects

The SIE method has been an appealing full-wave approach for solving EM problems of homogeneous penetrable objects. By considering the boundary conditions for the electric and magnetic fields separately, one can derive four different integral equations, namely, the tangential electric-field integral equation (T-EFIE), the normal electric-field integral equation (N-EFIE), the tangential magnetic-field integral equation (T-MFIE), and the normal magnetic-field integral equation (N-MFIE). Moreover, these equations derived from the inner and outer regions can be combined in various ways resulting in different equations systems, such as Poggio–Miller–Chang–Harrington–Wu–Tsai(PMCHWT) equations [62], M*ü*ller equations [63], JMCFIE equations [64], etc. Among them, JMCFIE equations are derived by combined-field integral equations (CFIEs) for inner and outer regions by linearly combing EFIE and MFIE. Numerical experiments have demonstrated that JMCFIE can avoid internal resonances and provides better convergence than other equations. For electrically large and multi-scale electromagnetic wave problems, we shall need an effective and reliable preconditioning strategy to assure fast convergence in iterative matrix solutions. In this subsection, we will present the DG DD preconditioning for the JMCFIE-based solution of homogeneous objects.

5.2.1.1 JMCFIE formulation

As shown in Figure 5.13, a closed surface S is decomposed into several open surface subdomains. Here, we take three subdomains as an example. On each subdomain S_i, there will be an electric current $\boldsymbol{J}_i$ and a magnetic current $\boldsymbol{M}_i$. The outward unit normal of S_i is denoted as $\hat{\boldsymbol{n}}_i$. The identification of contour boundaries between different surface sub-domains is the same as that used in the DG-DDM-SIE for PEC objects. Moreover, the exterior region of the homogeneous object is denoted as R_1 and the interior region is denoted as R_2. The permittivity and permeability of region $R_l (l = 1, 2)$ are ε_l and μ_l respectively.

Under such a setting, JMCFIEs may be derived for each subdomain associating with R_1 and R_2. For S_i, the J-CFIE with respect to the exterior region R_1 is established by combining T-EFIE and N-MFIE, written as:

$$-\frac{(1-\alpha)}{2}\boldsymbol{J}_i(\boldsymbol{r}) + \sum_{j=1}^{3}\mathscr{C}_1^{\alpha}(\boldsymbol{J}_j; S_j) - \frac{\alpha}{2}\eta_1^{-1}\hat{\boldsymbol{n}}_i \times \boldsymbol{M}_i(\boldsymbol{r}) + \eta_1^{-1}\sum_{j=1}^{3}\hat{\boldsymbol{n}}_i \times \mathscr{C}_1^{1-\alpha}(\boldsymbol{M}_j; S_j)$$
$$= -\alpha\eta_1^{-1}\boldsymbol{E}^{\mathrm{i}}_{\tan}(\boldsymbol{r}) - (1-\alpha)\hat{\boldsymbol{n}}_i \times \mathbf{H}^{\mathrm{i}}(\boldsymbol{r}) \quad \boldsymbol{r} \in S_i \tag{5.26}$$

and M-CFIE w.r.t. R_1 is established by combining N-EFIE and T-MFIE, written as

$$-\frac{\alpha}{2}\eta_1\hat{\boldsymbol{n}}_i \times \boldsymbol{J}_i(\boldsymbol{r}) + \eta_1\sum_{j=1}^{3}\hat{\boldsymbol{n}}_i \times \mathscr{C}_1^{1-\alpha}(\boldsymbol{J}_j; S_j) + \frac{1-\alpha}{2}\boldsymbol{M}_i(\boldsymbol{r}) - \sum_{j=1}^{3}\mathscr{C}_1^{\alpha}(\boldsymbol{M}_j; S_j)$$
$$= -(1-\alpha)\hat{\boldsymbol{n}}_i \times \boldsymbol{E}^{\mathrm{i}}(\boldsymbol{r}) + \alpha\eta_1\boldsymbol{H}^{\mathrm{i}}_{\tan}(\boldsymbol{r}) \quad \boldsymbol{r} \in S_i \tag{5.27}$$

Analogously, the J-CFIE and M-CFIE associating with the interior region R_2 are written as

$$\frac{(1-\alpha)}{2}\boldsymbol{J}_i(\boldsymbol{r}) - \sum_{j=1}^{3}\mathscr{C}_2^{\alpha}(\boldsymbol{J}_j; S_j) - \frac{\alpha}{2}\eta_2^{-1}\hat{\boldsymbol{n}}_i \times \boldsymbol{M}_i(\boldsymbol{r}) + \eta_2^{-1}\sum_{j=1}^{3}\hat{\boldsymbol{n}}_i \times \mathscr{C}_2^{1-\alpha}(\boldsymbol{M}_j; S_j) = 0 \tag{5.28}$$

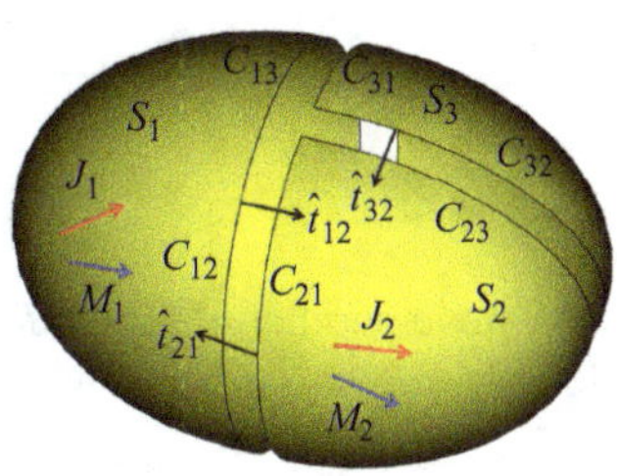

Figure 5.13 DD for the exterior surface of a homogeneous object

$$-\frac{\alpha}{2}\eta_2\hat{\boldsymbol{n}}_i \times \boldsymbol{J}_i(\boldsymbol{r}) + \eta_2\sum_{j=1}^{3}\hat{\boldsymbol{n}}_i \times \mathscr{C}_2^{1-\alpha}(\boldsymbol{J}_j;S_j) - \frac{(1-\alpha)}{2}\boldsymbol{M}_i(\boldsymbol{r}) + \sum_{j=1}^{3}\mathscr{C}_2^{\alpha}(\boldsymbol{M}_j;S_j) = 0 \tag{5.29}$$

In the above equations, $\mathscr{C}_l^{\alpha}, l = 1,2$ is an integral operator of linear combination of EFIE operator $\mathscr{L}$ and MFIE operator $\mathscr{K}$, which is written as

$$\mathscr{C}_l^{\alpha}(\boldsymbol{X};S) = \alpha\mathscr{L}_l(\boldsymbol{X};S) + s_l(1-\alpha)\hat{\boldsymbol{n}} \times \mathscr{K}_l(\boldsymbol{X};S) \tag{5.30}$$

where s_l is defined by

$$s_l = \begin{cases} +1 & l = 1 \\ -1 & l = 2 \end{cases} \tag{5.31}$$

Subsequently, the combination of J-CFIE$_i^1$ (5.26) and J-CFIE$_i^2$ (5.28) results into the final J-CFIE for S_i as

$$\text{J-CFIE}_i^1 - \text{J-CFIE}_i^2 \tag{5.32}$$

and combining M-CFIE$_i^1$(5.27) and M-CFIE$_i^2$(5.29) results into the final M-CFIE for S_i as

$$-\text{M-CFIE}_i^1 + \text{M-CFIE}_i^2 \tag{5.33}$$

5.2.1.2 Interior penalty formulations

In addition to (5.32) and (5.33), one also needs to introduce suitable formulations at the contours between subdomains to guarantee the continuity of currents across different subdomains. Similar to the PEC cases, we will extend the antisymmetric boundary IP term (5.12) and penalty term (5.14) for both electric currents $\boldsymbol{J}_i, i = 1,3$ and magnetic currents $\boldsymbol{M}_i, i = 1,3$. Moreover, we should consider the contributions from both the interior region and the exterior region.

Specifically, the IP formulations for electric currents are written as

$$\sum_{l=1}^{2}\left\{\frac{1}{2\imath k_l}\sum_{C_{ij}\in C}\left\langle[[\boldsymbol{J}]]_{ij}, \sum_{j=1}^{M}\Psi_F^l(\nabla_\tau' \cdot \boldsymbol{v}_j;S_j)\right\rangle_{C_{ij}} + \frac{\beta}{k_l}\sum_{C_{ij}\in C}\langle[[\boldsymbol{v}]]_{ij}, [[\boldsymbol{J}]]_{ij}\rangle_{C_{ij}}\right\} = 0 \tag{5.34}$$

where $k_l = \omega\sqrt{\varepsilon_l\mu_l}$ denotes the wave number of region R_l. Ψ_F^l is the scalar potential in homogeneous region R_l, defined by:

$$\Psi_F^l(\rho;S)(\mathbf{r}) := \int_S \rho(\mathbf{r}')G_l(\mathbf{r},\mathbf{r}')d\mathbf{r}' \tag{5.35}$$

with the homogeneous Green's function $G_l(\mathbf{r},\mathbf{r}') := e^{-\imath k_l|\mathbf{r}-\mathbf{r}'|}/4\pi|\mathbf{r}-\mathbf{r}'|$. If we replace $\boldsymbol{J}$ in (5.34) with $\boldsymbol{M}$, it can easily obtain the IP formulation for magnetic currents:

$$\sum_{l=1}^{2}\left\{\frac{1}{2\imath k_l}\sum_{C_{ij}\in C}\left\langle [[\boldsymbol{M}]]_{ij}, \sum_{j=1}^{M}\Psi_F^l(\nabla'_\tau\cdot\boldsymbol{v}_j;S_j)\right\rangle_{C_{ij}} + \frac{\beta}{k_l}\sum_{C_{ij}\in C}\langle [[\boldsymbol{v}]]_{ij}, [[\boldsymbol{M}]]_{ij}\rangle_{C_{ij}}\right\} = 0 \tag{5.36}$$

5.2.1.3 Matrix equation and solution

Next, each subdomain surface is discretized by triangular meshes. The electric current $\boldsymbol{J}_i$ and the magnetic current $\boldsymbol{M}_i$ of subdomain S_i are expanded in a sum of RWG basis functions $\boldsymbol{f}_i^q$ in the form

$$\boldsymbol{J}_i = \sum_{q=1}^{N_i} J_i^q \boldsymbol{f}_i^q \tag{5.37}$$

$$\boldsymbol{M}_i = \sum_{q=1}^{N_i} M_i^q \boldsymbol{f}_i^q \tag{5.38}$$

where J_i^q and M_i^q are the unknown expansion complex coefficients, and N_i is the number of basis functions for S_i which is the same to the number of edges. As shown in Figure 5.3, $\boldsymbol{f}_i^q$ is a full RWG basis function when its supporting edge is located inside the surface, while $\boldsymbol{f}_i^q$ is a half RWG basis function when its supporting edge is located at boundary contours.

By submitting (5.37) and (5.38) into (5.32), (5.33),(5.34), and (5.34) and applying the Galerkin testing procedure, a system of linear matrix equations can be derived from the integral equations and IP formulations. Explicitly, the matrix equation can be written as:

$$\begin{bmatrix} \tilde{\boldsymbol{Z}}_{11} & \tilde{\boldsymbol{Z}}_{12} & \tilde{\boldsymbol{Z}}_{13} \\ \tilde{\boldsymbol{Z}}_{21} & \tilde{\boldsymbol{Z}}_{22} & \tilde{\boldsymbol{Z}}_{23} \\ \tilde{\boldsymbol{Z}}_{31} & \tilde{\boldsymbol{Z}}_{32} & \tilde{\boldsymbol{Z}}_{33} \end{bmatrix} \begin{Bmatrix} \tilde{\boldsymbol{x}}_1 \\ \tilde{\boldsymbol{x}}_2 \\ \tilde{\boldsymbol{x}}_3 \end{Bmatrix} = \begin{Bmatrix} \tilde{\boldsymbol{b}}_1 \\ \tilde{\boldsymbol{b}}_2 \\ \tilde{\boldsymbol{b}}_3 \end{Bmatrix} \tag{5.39}$$

where $\tilde{\boldsymbol{Z}}_{ii}$ is the self-impedance matrix for subdomain S_i, and $\tilde{\boldsymbol{Z}}_{ij}$ is the coupling matrix between subdomains S_i and S_j. $\tilde{\boldsymbol{x}}_i$ is a vector containing $2N_i$ unknown coefficients of basis functions expanding $\boldsymbol{J}_i$ and $\boldsymbol{M}_i$. Besides, $\tilde{\boldsymbol{b}}_i$ is a RHS vector for S_i. Obviously, the form of (5.39) for homogeneous objects is the same as the form of (5.22) for PEC objects. However, the evaluation of elements in block submatrices and RHS vectors of (5.39) is more complicated.

Each block submatrix $\tilde{\boldsymbol{Z}}_{ij}$ can be expanded into the following format:

$$\tilde{\boldsymbol{Z}}_{ij} = \begin{bmatrix} \tilde{\boldsymbol{Z}}_{ij}^{\mathrm{JJ}} & \tilde{\boldsymbol{Z}}_{ij}^{\mathrm{JM}} \\ \tilde{\boldsymbol{Z}}_{ij}^{\mathrm{MJ}} & \tilde{\boldsymbol{Z}}_{ij}^{\mathrm{MM}} \end{bmatrix} \tag{5.40}$$

with

$$\begin{aligned}\tilde{\boldsymbol{Z}}_{ij}^{\mathrm{JJ}}(m,n) &= \left\langle \boldsymbol{f}_m, \mathscr{C}_1^{\alpha}(\boldsymbol{f}_n;T_n) + \mathscr{C}_2^{\alpha}(\boldsymbol{f}_n;T_n)\right\rangle_{T_m} - (1-\alpha)\left\langle \boldsymbol{f}_m, \boldsymbol{f}_n\right\rangle_{T_m} \\ &+ \left\langle \frac{1}{\imath k_1}\Psi_F^1(\nabla_\tau'\cdot\boldsymbol{f}_m;T_m) + \frac{1}{\imath k_2}\Psi_F^2(\nabla_\tau'\cdot\boldsymbol{f}_m;T_m), \hat{\boldsymbol{t}}_j\cdot\boldsymbol{f}_n\right\rangle_{C_j} \\ &+ \left(\frac{\beta}{k_1}+\frac{\beta}{k_2}\right)\left\langle \hat{\boldsymbol{t}}_m\cdot\boldsymbol{f}_m, \hat{\boldsymbol{t}}_m\cdot\boldsymbol{f}_n\right\rangle_{C_i}\end{aligned} \tag{5.41}$$

$$\begin{aligned}\tilde{\boldsymbol{Z}}_{ij}^{\mathrm{JM}}(m,n) &= \left\langle \boldsymbol{f}_m, \eta_1^{-1}\hat{\boldsymbol{n}}_m\times\mathscr{C}_1^{1-\alpha}(\boldsymbol{f}_n;T_n) - \eta_2^{-1}\hat{\boldsymbol{n}}_m\times\mathscr{C}_2^{1-\alpha}(\boldsymbol{f}_n;T_n)\right\rangle_{T_m} \\ &- \frac{\alpha}{2}\left\langle \boldsymbol{f}_m, (\eta_1^{-1}-\eta_2^{-1})\hat{\boldsymbol{n}}_m\times\boldsymbol{f}_n\right\rangle_{T_m}\end{aligned} \tag{5.42}$$

$$\begin{aligned}\tilde{\boldsymbol{Z}}_{ij}^{\mathrm{MJ}}(m,n) &= \left\langle \boldsymbol{f}_m, -\eta_1\hat{\boldsymbol{n}}_m\times\mathscr{C}_1^{1-\alpha}(\boldsymbol{f}_n;T_n) + \eta_2\hat{\boldsymbol{n}}_m\times\mathscr{C}_2^{1-\alpha}(\boldsymbol{f}_n;T_n)\right\rangle_{T_m} \\ &- \frac{\alpha}{2}\left\langle \boldsymbol{f}_m, (-\eta_1+\eta_2)\hat{\boldsymbol{n}}_m\times\boldsymbol{f}_n\right\rangle_{T_m}\end{aligned} \tag{5.43}$$

$$\begin{aligned}\tilde{\boldsymbol{Z}}_{ij}^{\mathrm{MM}}(m,n) &= \left\langle \boldsymbol{f}_m, \mathscr{C}_1^{\alpha}(\boldsymbol{f}_n;T_n) + \mathscr{C}_2^{\alpha}(\boldsymbol{f}_n;T_n)\right\rangle_{T_m} - (1-\alpha)\left\langle \boldsymbol{f}_m, \boldsymbol{f}_n\right\rangle_{T_m} \\ &+ \left\langle \frac{1}{\imath k_1}\Psi_F^1(\nabla_\tau'\cdot\boldsymbol{f}_m;T_m) + \frac{1}{\imath k_2}\Psi_F^2(\nabla_\tau'\cdot\boldsymbol{f}_m;T_m), \hat{\boldsymbol{t}}_j\cdot\boldsymbol{f}_n\right\rangle_{C_j} \\ &+ \left(\frac{\beta}{k_1}+\frac{\beta}{k_2}\right)\left\langle \hat{\boldsymbol{t}}_m\cdot\boldsymbol{f}_m, \hat{\boldsymbol{t}}_m\cdot\boldsymbol{f}_n\right\rangle_{C_i}\end{aligned} \tag{5.44}$$

where T_m denotes the subdomain in which function $\boldsymbol{f}_m$ is defined. Notice that $\tilde{\boldsymbol{Z}}_{ij}^{\mathrm{MM}}$ is same with $\tilde{\boldsymbol{Z}}_{ij}^{\mathrm{JJ}}$. Thus, one can take advantage of this repetition to save computational resources. In addition, $\tilde{\boldsymbol{b}}_i$ contains two terms, written as

$$\tilde{\boldsymbol{b}}_i = \begin{Bmatrix} \tilde{\boldsymbol{b}}_i^{\mathrm{J}} \\ \tilde{\boldsymbol{b}}_i^{\mathrm{M}} \end{Bmatrix} \tag{5.45}$$

with

$$\tilde{\boldsymbol{b}}_i^{\mathrm{J}}(m) = -\alpha\eta_1^{-1}\left\langle \boldsymbol{f}_m, \boldsymbol{E}_{\tan}^{\mathrm{i}}\right\rangle_{T_m} - (1-\alpha)\left\langle \boldsymbol{f}_m, \hat{\boldsymbol{n}}\times\boldsymbol{H}^{\mathrm{i}}\right\rangle_{T_m} \tag{5.46}$$

$$\tilde{\boldsymbol{b}}_i^{\mathrm{M}}(m) = (1-\alpha)\left\langle \boldsymbol{f}_m, \hat{\boldsymbol{n}}\times\boldsymbol{E}_{\tan}^{\mathrm{i}}\right\rangle_{T_m} - \alpha\eta_1\left\langle \boldsymbol{f}_m, \boldsymbol{H}^{\mathrm{i}}\right\rangle_{T_m} \tag{5.47}$$

Based on (5.39), we can construct an additive Schwarz preconditioner $\tilde{\boldsymbol{P}}^{-1}$, which is explicitly written as

$$\tilde{\boldsymbol{P}}^{-1} = \begin{bmatrix} \tilde{\boldsymbol{Z}}_{11}^{-1} & 0 & 0 \\ 0 & \tilde{\boldsymbol{Z}}_{22}^{-1} & 0 \\ 0 & 0 & \tilde{\boldsymbol{Z}}_{33}^{-1} \end{bmatrix} \tag{5.48}$$

With the left preconditioning by $\tilde{\boldsymbol{P}}^{-1}$, the final matrix equation to be solved is given as

$$\begin{bmatrix} \tilde{\boldsymbol{I}}_1 & \tilde{\boldsymbol{Z}}_{11}^{-1}\tilde{\boldsymbol{Z}}_{12} & \tilde{\boldsymbol{Z}}_{11}^{-1}\tilde{\boldsymbol{Z}}_{13} \\ \tilde{\boldsymbol{Z}}_{22}^{-1}\tilde{\boldsymbol{Z}}_{21} & \tilde{\boldsymbol{I}}_2 & \tilde{\boldsymbol{Z}}_{22}^{-1}\tilde{\boldsymbol{Z}}_{23} \\ \tilde{\boldsymbol{Z}}_{33}^{-1}\tilde{\boldsymbol{Z}}_{31} & \tilde{\boldsymbol{Z}}_{33}^{-1}\tilde{\boldsymbol{Z}}_{32} & \tilde{\boldsymbol{I}}_3 \end{bmatrix} \begin{Bmatrix} \tilde{\boldsymbol{x}}_1 \\ \tilde{\boldsymbol{x}}_2 \\ \tilde{\boldsymbol{x}}_3 \end{Bmatrix} = \begin{Bmatrix} \tilde{\boldsymbol{Z}}_{11}^{-1}\tilde{\boldsymbol{b}}_1 \\ \tilde{\boldsymbol{Z}}_{22}^{-1}\tilde{\boldsymbol{b}}_2 \\ \tilde{\boldsymbol{Z}}_{33}^{-1}\tilde{\boldsymbol{b}}_3 \end{Bmatrix} \tag{5.49}$$

5.2.1.4 Numerical experiments

We consider the EM scattering from a homogeneous dielectric almond object. The geometry is identical to the PEC "NASA almond" of characteristic length 252.37 mm. The dielectric almond is made of a low-loss resin material with a relative permittivity of $\varepsilon_r = 2.9 - j0.082$ at 7 GHz [65]. The incident angles are from $\theta = 0°$ to $\theta = 180°$ with a step of 1° while fixing ϕ at 90°. In analogy to the PEC case study, we have considered three numerical methods, i.e. the standard JMCFIE method with boundary element discretization, the DG-DDM-SIE with regular subdomain partitioning (five subdomains), and the DG-DDM-SIE with irregular subdomain partitioning (five subdomains). The illustration of the discretizations and domain partitionings is given in Figure 5.14.

Figure 5.15(a) shows the computed VV-polarized monostatic RCS, and Figure 5.15(b) shows the computed HH-polarized monostatic RCS. Moreover, the results from measurement data [65] are included as a comparison. It can be seen that results from DG-DDM-SIE match those from JMCFIE very well, and the simulated results are in agreement with those from the measurement. The surface magnetic currents

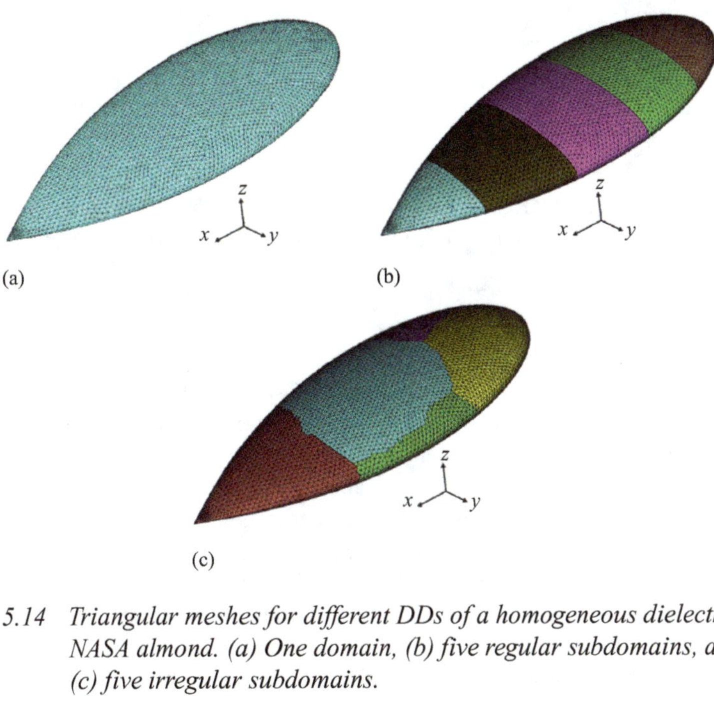

Figure 5.14 Triangular meshes for different DDs of a homogeneous dielectric NASA almond. (a) One domain, (b) five regular subdomains, and (c) five irregular subdomains.

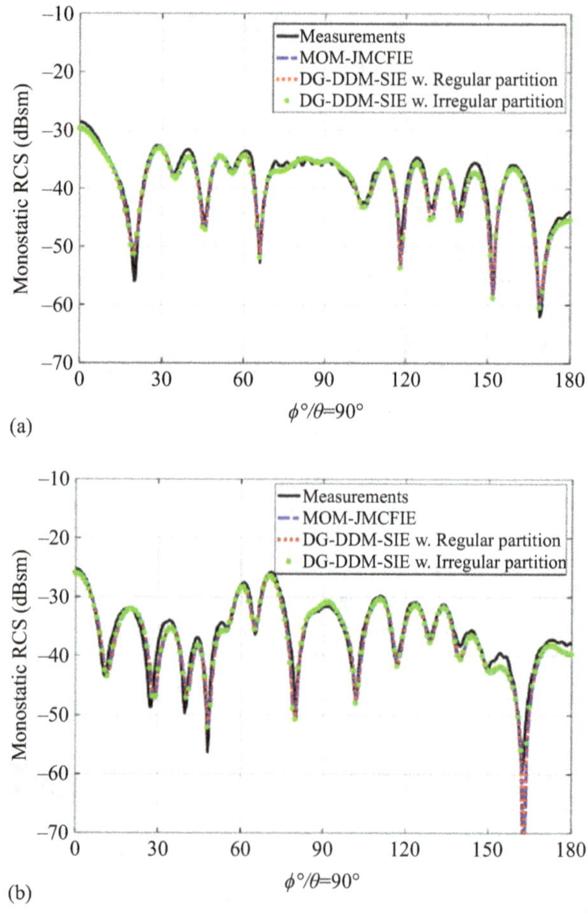

Figure 5.15 Monostatic RCS obtained by different methods at xy-plane from a dielectric almond at 7.0 GHz. (a) VV-polarized and (b) HH-polarized.

obtained by JMCFIE and DG-DDM-SIE are plotted in Figure 5.16, which demonstrates that the surface magnetic currents from DG-DDM-SIE have a good agreement with that form MOM and there is no discontinuity even though the boundaries of the subdomains are irregular. Validated by Figures 5.15 and 5.16, the DG-DDM-SIE based on JMCFIE is correct and has high accuracy.

Finally, we compare the iterative solver convergence for the PEC almond and dielectric almond. The same operating frequency at 7 GHz, plane wave incident angle, and polarization are considered. Figure 5.17 gives the convergence histories for regular domain partitioning and irregular domain partitioning. We observe rapid convergence for all the study cases. A moderate increase in iteration numbers is seen for irregular domain partitioning for the dielectric almond object.

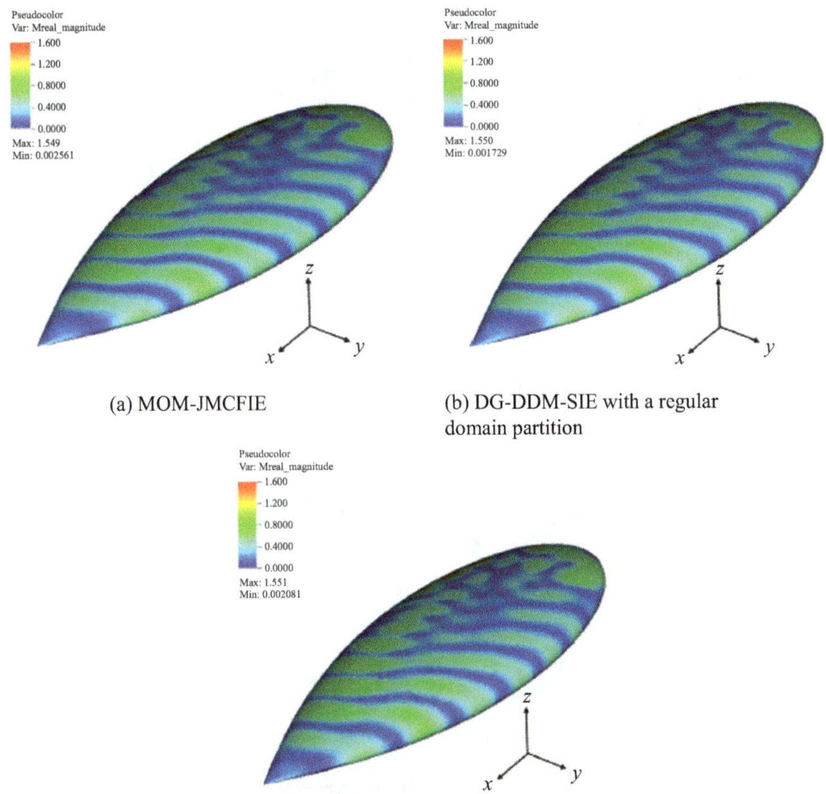

(a) MOM-JMCFIE

(b) DG-DDM-SIE with a regular domain partition

(c) DG-DDM-SIE with an irregular domain partition

Figure 5.16 The magnetic current distribution obtained from different SIE methods using different DD schemes for a dielectric almond under the illumination of $\theta = 90^\circ, \phi = 30^\circ$ *plane wave at 7 GHz.*

5.2.2 DG-DDM-SIE for piecewise homogeneous objects

We have shown previously how the use of DG provides improved flexibility for CAD designing and meshing complex problems involving multiscale features. But an additional advantage of this same strategy for the analysis of piecewise homogeneous problems is that it completely avoids the awkward multi-material junction problem [52,66,67]. The combination of the boundary conditions implicit in the JMCFIE formulation and the interior penalties [32]—locally applied in the different regions conforming to the junction—are enough to weakly enforce the normal current continuity, without requiring further specific treatment/action by the user in these areas.

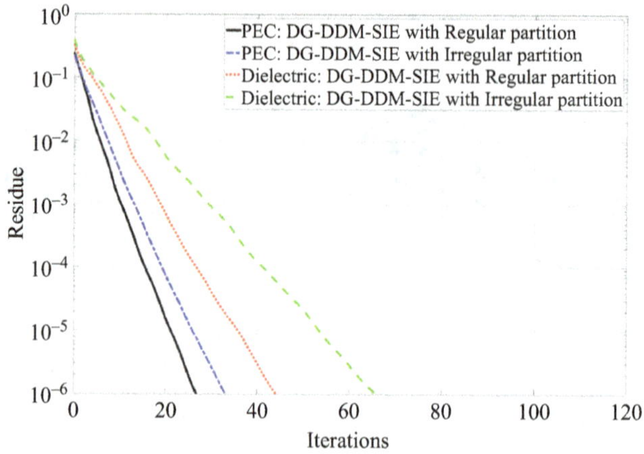

Figure 5.17 Iterative solver convergence for PEC and dielectric almonds under the illumination of $\theta = 90°, \phi = 30°$ plane wave at 7 GHz

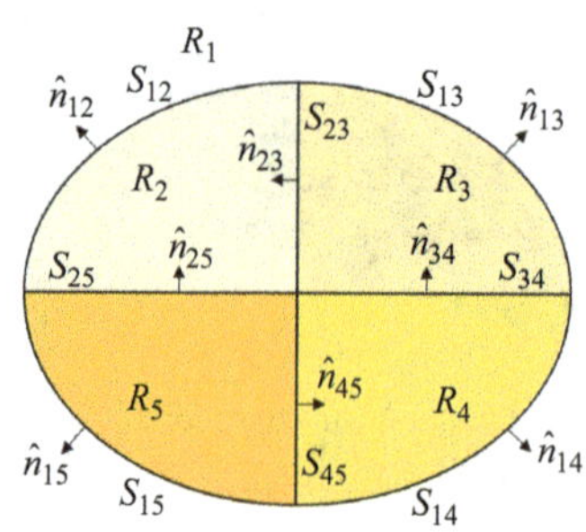

Figure 5.18 Notations for SIE formulation on piecewise objects

5.2.2.1 Review integral equations

We focus on the electromagnetic (EM) solution of piecewise homogeneous penetrable objects in a homogeneous (unbounded) medium. The JMCFIE formulation [68] previously derived for a single dielectric object is extended here for the case of piecewise homogeneous bodies including multiple material junctions, as illustrated in Figure 5.18. Each homogeneous region is denoted by R_l. The material properties of each region corresponding to the penetrable objects or the background are defined by the permittivity ε_l and the permeability μ_l. Let us denote with S_{kl} (or S_{lk}) the interface between regions R_k and R_l. $\hat{\boldsymbol{n}}_{kl}$ denotes the unit vector normal to S_{kl} and pointing toward R_k, hence, $\hat{\boldsymbol{n}}_{kl} = -\hat{\boldsymbol{n}}_{lk}$. In region R_k, the JMCFIEs can be derived for the equivalent electric and magnetic currents $\boldsymbol{J}_{kl}(\boldsymbol{r})$ and $\boldsymbol{M}_{kl}(\boldsymbol{r})$ on the interfaces S_{kl} that surround R_k, which are related to the fields on this region through $\boldsymbol{J}_{kl}(\boldsymbol{r}) = \hat{\boldsymbol{n}}_{kl} \times \boldsymbol{H}_k(\boldsymbol{r})$

and $\boldsymbol{M}_{kl}(\boldsymbol{r}) = -\hat{\boldsymbol{n}}_{kl} \times \boldsymbol{E}_k(\boldsymbol{r})$, respectively. Applying the equivalence principle to the total electric and magnetic fields $\boldsymbol{E}_k(\boldsymbol{r})$ and $\boldsymbol{H}_k(\boldsymbol{r})$ in R_k, we obtain the J-CFIE equation in S_{kl} through region R_k as

$$
\begin{aligned}
&-\frac{(1-\alpha)}{2}\boldsymbol{J}_{kl}(\boldsymbol{r}) + \sum_{l' \in G_k} \mathscr{C}_k^{\alpha}(\boldsymbol{J}_{kl'}; S_{kl'}) \\
&-\frac{\alpha}{2}\eta_k^{-1}\hat{\boldsymbol{n}}_{kl} \times \boldsymbol{M}_{kl}(\boldsymbol{r}) + \eta_k^{-1}\hat{\boldsymbol{n}}_{kl} \times \sum_{l' \in G_k} \mathscr{C}_k^{(1-\alpha)}(\boldsymbol{M}_{kl'}; S_{kl'}) \\
&= -\alpha\eta_k^{-1}\boldsymbol{E}^i_{k,\tan}(\boldsymbol{r}) - (1-\alpha)\hat{\boldsymbol{n}}_{kl} \times \boldsymbol{H}^i_k(\boldsymbol{r}) \quad \boldsymbol{r} \in S_{kl}
\end{aligned}
\tag{5.50}
$$

and the M-CFIE equation in S_{kl} through R_k as

$$
\begin{aligned}
&-\frac{\alpha}{2}\eta_k\hat{\boldsymbol{n}}_{kl} \times \boldsymbol{J}_{kl}(\boldsymbol{r}) + \eta_k\hat{\boldsymbol{n}}_{kl} \times \sum_{l' \in G_k} \mathscr{C}_k^{(1-\alpha)}(\boldsymbol{J}_{kl'}; S_{kl'}) \\
&+\frac{(1-\alpha)}{2}\boldsymbol{M}_{kl}(\boldsymbol{r}) - \sum_{l' \in G_k} \mathscr{C}_k^{\alpha}(\boldsymbol{M}_{kl'}; S_{kl'}) \\
&= -(1-\alpha)\hat{\boldsymbol{n}}_{kl} \times \boldsymbol{E}^i_k(\boldsymbol{r}) + \alpha\eta_k\boldsymbol{H}^i_{k,\tan}(\boldsymbol{r}) \quad \boldsymbol{r} \in S_{kl}
\end{aligned}
\tag{5.51}
$$

Similarly, J-CFIE and M-CFIE in S_{kl} through region R_l can be written as

$$
\begin{aligned}
&-\frac{(1-\alpha)}{2}\boldsymbol{J}_{lk}(\boldsymbol{r}) + \sum_{l'' \in G_l} \mathscr{C}_l^{\alpha}(\boldsymbol{J}_{ll''}; S_{ll''}) \\
&-\frac{\alpha}{2}\eta_l^{-1}\hat{\boldsymbol{n}}_{lk} \times \boldsymbol{M}_{lk}(\boldsymbol{r}) + \eta_l^{-1}\hat{\boldsymbol{n}}_{lk} \times \sum_{l'' \in G_l} \mathscr{C}_l^{(1-\alpha)}(\boldsymbol{M}_{ll''}; S_{ll''}) \\
&= -\alpha\eta_l^{-1}\boldsymbol{E}^i_{l,\tan}(\boldsymbol{r}) - (1-\alpha)\hat{\boldsymbol{n}}_{lk} \times \boldsymbol{H}^i_l(\boldsymbol{r}) \quad \boldsymbol{r} \in S_{kl}
\end{aligned}
\tag{5.52}
$$

$$
\begin{aligned}
&-\frac{\alpha}{2}\eta_l\hat{\boldsymbol{n}}_{lk} \times \boldsymbol{J}_{lk}(\boldsymbol{r}) + \eta_l\hat{\boldsymbol{n}}_{lk} \times \sum_{l'' \in G_l} \mathscr{C}_l^{(1-\alpha)}(\boldsymbol{J}_{ll''}; S_{ll''}) \\
&+\frac{(1-\alpha)}{2}\boldsymbol{M}_{lk}(\boldsymbol{r}) - \sum_{l'' \in G_l} \mathscr{C}_l^{\alpha}(\boldsymbol{M}_{ll''}; S_{ll''}) \\
&= -(1-\alpha)\hat{\boldsymbol{n}}_{lk} \times \boldsymbol{E}^i_l(\boldsymbol{r}) + \alpha\eta_l\boldsymbol{H}^i_{l,\tan}(\boldsymbol{r}) \quad \boldsymbol{r} \in S_{kl}
\end{aligned}
\tag{5.53}
$$

In the above equations, $\mathscr{C}_l^{\alpha}$ is the integral operator defined in (5.30), G_l denotes the set of indices of the regions surrounding region R_l, and the sums in equations (5.50) to (5.53) account for all the interfaces $S_{kl'}$ that surround region R_k, which includes surface S_{kl}, and the interfaces $S_{ll''}$ that surround region R_l, which includes surface $S_{lk} = S_{kl}$. Note that here we must set $s_l = 1$ in (5.30), since we are explicitly doubling the electric and magnetic currents and the normal vector to have a local set within each region, satisfying $\boldsymbol{J}_{lk} = -\boldsymbol{J}_{kl}$, $\boldsymbol{M}_{lk} = -\boldsymbol{M}_{kl}$, and $\hat{\boldsymbol{n}}_{lk} = -\hat{\boldsymbol{n}}_{kl}$. $\boldsymbol{E}^{\mathrm{i}}_l$ and $\boldsymbol{H}^{\mathrm{i}}_l$ denote the incident fields due to the sources in R_l. These are the fields excited by the sources when all other objects (including the background) are replaced by the constitutive parameters of the region R_l. If there are no sources in R_l, these incident fields vanish.

Subsequently, combining equations J-CFIE^k and J-CFIE^l results in the final J-CFIE for interface S_{kl}, while combining M-CFIE^k and M-CFIE^l leads to the final M-CFIE for this interface:

$$\text{J-CFIE}^k - \text{J-CFIE}^l \tag{5.54}$$

$$-\text{M-CFIE}^k + \text{M-CFIE}^l \tag{5.55}$$

5.2.2.2 Matrix equation and solution

To obtain the equivalent currents $\boldsymbol{J}_{kl}$ and $\boldsymbol{M}_{kl}$ on the surfaces S_{kl}, we first expand the currents on the interfaces in a weighted sum of known basis functions $\mathbf{f}_n$ as

$$\boldsymbol{J}_{kl} = \sum_n J_n \mathbf{f}_n \quad \text{in } S_{kl},\ k < l \tag{5.56}$$

$$\boldsymbol{M}_{kl} = \sum_n M_n \mathbf{f}_n \quad \text{in } S_{kl},\ k < l \tag{5.57}$$

where J_n and M_n are the unknown expansion complex coefficients. Substituting (5.56) and (5.57) into (5.54) and (5.55) and applying the Galerkin testing procedure, a system of linear equations is derived from the integral equations, which can be expressed as a dense matrix as

$$\begin{bmatrix} [Z_{kl,kl}] & [Z_{kl,pq}] & \cdots & [Z_{kl,rs}] \\ [Z_{pq,kl}] & [Z_{pq,pq}] & \cdots & [Z_{pq,rs}] \\ \vdots & \vdots & \ddots & \vdots \\ [Z_{rs,kl}] & [Z_{rs,pq}] & \cdots & [Z_{rs,rs}] \end{bmatrix} \begin{bmatrix} [x_{kl}] \\ [x_{pq}] \\ \vdots \\ [x_{rs}] \end{bmatrix} = \begin{bmatrix} [b_{kl}] \\ [b_{pq}] \\ \vdots \\ [x_{rs}] \end{bmatrix} \tag{5.58}$$

where each block of the impedance matrix is given by

$$[Z_{kl,k'l'}] = \begin{bmatrix} \left[Z^{JJ}_{kl,k'l'} \right] & \left[Z^{JM}_{kl,k'l'} \right] \\ \left[Z^{MJ}_{kl,k'l'} \right] & \left[Z^{MM}_{kl,k'l'} \right] \end{bmatrix} \tag{5.59}$$

with

$$Z^{JJ}_{kl,kl}(m,n) = \left\langle \boldsymbol{f}_m, \mathscr{C}^{\alpha}_k(\boldsymbol{f}_n, T_n) + \mathscr{C}^{\alpha}_l(\boldsymbol{f}_n, T_n) \right\rangle_{T_m} - (1-\alpha) \left\langle \boldsymbol{f}_m, \boldsymbol{f}_n \right\rangle_{T_m} \tag{5.60}$$

$$\begin{aligned} Z^{JM}_{kl,kl}(m,n) = &\left\langle \boldsymbol{f}_m, \eta_k^{-1} \hat{\boldsymbol{n}}_m \times \mathscr{C}^{(1-\alpha)}_k(\boldsymbol{f}_n, T_n) - \eta_l^{-1} \hat{\boldsymbol{n}}_m \times \mathscr{C}^{(1-\alpha)}_l(\boldsymbol{f}_n, T_n) \right\rangle_{T_m} \\ &- \frac{\alpha}{2} \left\langle \boldsymbol{f}_m, (\eta_k^{-1} - \eta_l^{-1}) \hat{\boldsymbol{n}}_m \times \boldsymbol{f}_n \right\rangle_{T_m} \end{aligned} \tag{5.61}$$

$$\begin{aligned} Z^{MJ}_{kl,kl}(m,n) = &\left\langle \boldsymbol{f}_m, -\eta_k \hat{\boldsymbol{n}}_m \times \mathscr{C}^{(1-\alpha)}_k(\boldsymbol{f}_n, T_n) + \eta_l \hat{\boldsymbol{n}}_m \times \mathscr{C}^{(1-\alpha)}_l(\boldsymbol{f}_n, T_n) \right\rangle_{T_m} \\ &- \frac{\alpha}{2} \left\langle \boldsymbol{f}_m, (-\eta_k + \eta_l) \hat{\boldsymbol{n}}_m \times \boldsymbol{f}_n \right\rangle_{T_m} \end{aligned} \tag{5.62}$$

$$Z^{MM}_{kl,kl}(m,n) = \left\langle \boldsymbol{f}_m, \mathscr{C}^{\alpha}_k(\boldsymbol{f}_n, T_n) + \mathscr{C}^{\alpha}_l(\boldsymbol{f}_n, T_n) \right\rangle_{T_m} - (1-\alpha) \left\langle \boldsymbol{f}_m, \boldsymbol{f}_n \right\rangle_{T_m} \tag{5.63}$$

for the self-coupling of surface S_{kl}, where $k < l$ is assumed,

$$Z^{JJ}_{kl,kl'}(m,n) = s_{kl'} \langle \boldsymbol{f}_m, \mathscr{C}^{\alpha}_k(\boldsymbol{f}_n, T_n) \rangle_{T_m} \tag{5.64}$$

$$Z^{JM}_{kl,kl'}(m,n) = s_{kl'} \left\langle \boldsymbol{f}_m, \eta_k^{-1} \hat{\boldsymbol{n}}_m \times \mathscr{C}^{(1-\alpha)}_k(\boldsymbol{f}_n, T_n) \right\rangle_{T_m} \tag{5.65}$$

$$Z^{MJ}_{kl,kl'}(m,n) = s_{kl'} \left\langle \boldsymbol{f}_m, -\eta_k \hat{\boldsymbol{n}}_m \times \mathscr{C}^{(1-\alpha)}_k(\boldsymbol{f}_n, T_n) \right\rangle_{T_m} \tag{5.66}$$

$$Z^{MM}_{kl,kl'}(m,n) = s_{kl'} \langle \boldsymbol{f}_m, \mathscr{C}^{\alpha}_k(\boldsymbol{f}_n, T_n) \rangle_{T_m} \tag{5.67}$$

for the mutual coupling between surfaces S_{kl} and $S_{kl'}$ sharing region R_k,

$$Z^{JJ}_{kl,ll''}(m,n) = s_{ll''} \langle \boldsymbol{f}_m, \mathscr{C}^{\alpha}_l(\boldsymbol{f}_n, T_n) \rangle_{T_m} \tag{5.68}$$

$$Z^{JM}_{kl,ll''}(m,n) = s_{ll''} \left\langle \boldsymbol{f}_m, -\eta_l^{-1} \hat{\boldsymbol{n}}_m \times \mathscr{C}^{(1-\alpha)}_l(\boldsymbol{f}_n, T_n) \right\rangle_{T_m} \tag{5.69}$$

$$Z^{MJ}_{kl,ll''}(m,n) = s_{ll''} \left\langle \boldsymbol{f}_m, \eta_l \hat{\boldsymbol{n}}_m \times \mathscr{C}^{(1-\alpha)}_l(\boldsymbol{f}_n, T_n) \right\rangle_{T_m} \tag{5.70}$$

$$Z^{MM}_{kl,ll''}(m,n) = s_{ll''} \langle \boldsymbol{f}_m, \mathscr{C}^{\alpha}_l(\boldsymbol{f}_n, T_n) \rangle_{T_m} \tag{5.71}$$

for the mutual coupling between surfaces S_{kl} and $S_{ll''}$ sharing region R_l, and zero otherwise. The internal sign s_l in (5.30) is set to $+1$ for $\mathscr{C}^{\alpha}_k$ and $\mathscr{C}^{(1-\alpha)}_k$, and to -1 in $\mathscr{C}^{\alpha}_l$ and $\mathscr{C}^{(1-\alpha)}_l$. $[x_{kl}] = [J_n\ M_n]^T$ (with T denoting matrix transposition) is the vector with the unknown coefficients expanding $\boldsymbol{J}_{kl}$ and $\boldsymbol{M}_{kl}$, and $[b_{kl}]$ are the excitation vectors that can written as

$$[b_{kl}] = \begin{bmatrix} \begin{bmatrix} b^J_{kl} \\ b^M_{kl} \end{bmatrix} \end{bmatrix} \tag{5.72}$$

with

$$\begin{aligned} b^J_{kl}(m) = & -\alpha\eta_k^{-1} \langle \boldsymbol{f}_m, \boldsymbol{E}^{\mathrm{i}}_k \rangle_{T_m} - (1-\alpha) \langle \boldsymbol{f}_m, \hat{\boldsymbol{n}}_m \times \boldsymbol{H}^{\mathrm{i}}_k \rangle_{T_m} \\ & +\alpha\eta_l^{-1} \langle \boldsymbol{f}_m, \boldsymbol{E}^{\mathrm{i}}_l \rangle_{T_m} - (1-\alpha) \langle \boldsymbol{f}_m, \hat{\boldsymbol{n}}_m \times \boldsymbol{H}^{\mathrm{i}}_l \rangle_{T_m} \end{aligned} \tag{5.73}$$

and

$$\begin{aligned} b^M_{kl}(m) = & (1-\alpha) \langle \boldsymbol{f}_m, \hat{\boldsymbol{n}}_m \times \boldsymbol{E}^{\mathrm{i}}_k \rangle_{T_m} - \alpha\eta_k \langle \boldsymbol{f}_m, \boldsymbol{H}^{\mathrm{i}}_k \rangle_{T_m} \\ & +(1-\alpha) \langle \boldsymbol{f}_m, \hat{\boldsymbol{n}}_m \times \boldsymbol{E}^{\mathrm{i}}_l \rangle_{T_m} + \alpha\eta_l \langle \boldsymbol{f}_m, \boldsymbol{E}^{\mathrm{i}}_l \rangle_{T_m} \end{aligned} \tag{5.74}$$

Also, $s_{kl'}$ in (5.64) to (5.67) is defined by

$$s_{kl'} = \begin{cases} +1 & k < l' \\ -1 & k > l' \end{cases} \tag{5.75}$$

and $s_{ll''}$ in (5.68) to (5.71) is defined similarly. The introduction of these signs implicitly ensures that $\boldsymbol{J}_{kl'} = -\boldsymbol{J}_{l'k}$, $\boldsymbol{M}_{kl'} = -\boldsymbol{M}_{l'k}$, which enforce the boundary conditions at the dielectric interfaces.

Similar to the previous subsection, each interface S_{kl} can be further decomposed into several open surface subdomains $\mathscr{S}_i$, in which conformal discretizations are applied, while allowing nonconformal meshes across the tear lines and junctions

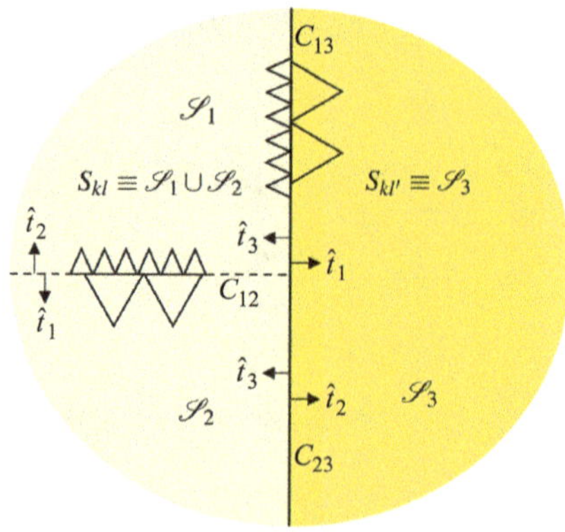

Figure 5.19 Decomposition of the boundary surfaces and interfaces of the piecewise object and notations for DG

between different subdomain surfaces. This decomposition and the corresponding notation are indicated in Figure 5.19. In the example of this figure, the interface S_{kl} between regions R_k and R_l is decomposed into two nonoverlapping surface pieces as $S_{kl} = \mathscr{S}_1 \cup \mathscr{S}_2$, yielding the tear contour C_{12}. Interface $S_{kl'}$ separating regions R_k and $R_{l'}$ is modeled by surface $\mathscr{S}_3$. Nonconformal meshes are allowed in the tear contour C_{12} and multi-material junction contours C_{13} and C_{23}. Thus, without loss of generality, we can write

$$S_{kl} = \bigcup_i \mathscr{S}_i \tag{5.76}$$

and the equivalent currents at interfaces, S_{kl} ($\boldsymbol{J}_{kl}$, $\boldsymbol{M}_{kl}$), can be particularized for subdomain surfaces S_i, denoted as $\boldsymbol{J}_i$ and $\boldsymbol{M}_i$ respectively. When the supporting edge is located inside the surface, the integral equation on each surface $\mathscr{S}_i$ is discretized by applying equations (5.60) to (5.71), (5.73), and (5.74) using full-RWG basis functions [69], while half-RWG (or monopolar-RWG) basis functions [32,70] are used for the supporting edges located at the tear contours between two surfaces $\mathscr{S}_i$ and $\mathscr{S}_j$. To guarantee the continuity of electric and magnetic currents across these contours, the IP of (5.34) and (5.36) is submitted into (5.60) to (5.74), thus expanding the formulation analogously to (5.40) to (5.47).

The Schwarz preconditioner [71,72] $\left[P^{-1}\right]$ of (5.48) can be then constructed in terms of the diagonal blocks corresponding to the subdomain surfaces $\mathscr{S}_i$, and applied to derive a left or right preconditioned system.

5.2.2.3 Multimaterial junctions

One of the most challenges in the analysis of composite objects using the SIE methodology is the treatment of junctions where three or more different materials meet. The imposition of normal current continuity across the entire junction becomes particularly tedious in those cases, requiring the explicit enforcement of the boundary conditions to combine and remove extra unknowns and equations at junctions [68].

Using the DG-DDM-SIE approach, the multi-material junction problem can be straightforwardly solved without any particular restriction or differentiated procedure and, importantly, without concern for mesh conformity across junction lines. Figure 5.20 shows an example of junction between four different regions R_l, $l = 2..5$ in an unbounded background R_1, posing four internal nonoverlapping interfaces S_{23}, S_{34}, S_{45}, and S_{25}. Let us assume that these interfaces are meshed separately, resulting in four meshed open surfaces $\mathscr{S}_i$, respectively, $i = 1...4$. Four contour lines arise at the junctions between pairs of surfaces sharing a region, denoted by C_{14}, C_{12}, C_{23}, and C_{34} in the figure. The pairs of straight arrows entering and leaving the junction represent half-RWG basis functions supporting an unknown X_i, which stands for the electric (or magnetic) current $\boldsymbol{J}_i$ ($\boldsymbol{M}_i$) entering or leaving the junction on surface $\mathscr{S}_i$. These coefficients account for the currents on both sides of this surface, which are equal but flow in opposite directions according to the boundary conditions implicitly imposed by JMCFIE on $\mathscr{S}_i$ (this is denoted by the continuous circular arrows in Figure 5.20). Consequently, four independent unknowns arise for the junction current. Normal current continuity is then weakly restored by applying the IP to the four junction contours (independently defined within their respective regions), setting up four independent equations to equalize current flow. This is denoted as $IP^l_{C_{ii'}}$ and circular dashed arrows in the figure.

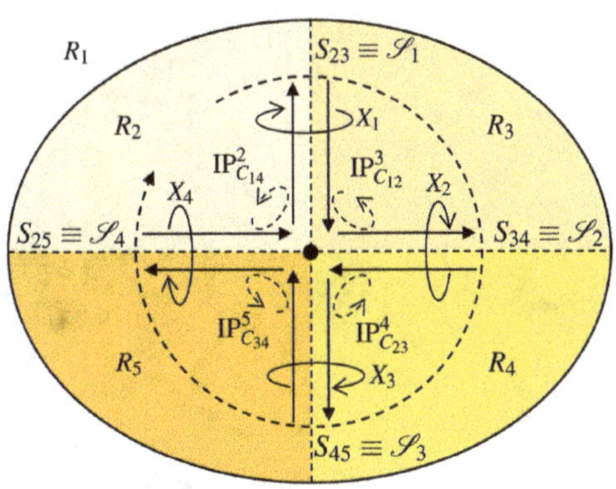

Figure 5.20 Multi-material junction between the four regions of the example in Figure 5.18. Straight dashed lines denote interfaces between regions. The center point denotes the four overlapping contour lines between the surfaces that meet at the junction. Solid curved arrows denote boundary conditions. Small curved dashed arrows denote IP conditions. The large dashed circular arrow denotes weakly enforced continuity of the normal current across the junction.

5.2.2.4 Numerical examples

A numerical experiment is presented in this section to demonstrate the flexibility of the DG-DDM-SIE formulation in solving composite objects of different materials. It consists of a composite dielectric cone with a diameter of 2 m and 4 m height, in vacuum (background region R_1 with $\varepsilon_{r1} = 1.0$). The cone is made up of nine regions R_l, $l = 2 \ldots 10$, with relative permittivity constants growing correlatively from $\varepsilon_{r2} = 2.0$ to $\varepsilon_{r10} = 10.0$.

The geometry is depicted in Figure 5.21. The cone is divided into three main sections along the z-axis. The two lower sections are subdivided into four regions each (R_2 to R_5 and R_6 to R_9). The upper section is made of a homogeneous material (R_{10}). This yields 29 boundary surfaces and interfaces between regions. The boundary surface of the last upper region decomposed into four subdomain surfaces, resulting in a total of 32 nonoverlapping surfaces. These subdomain surfaces are independently tessellated and assembled back into the entire geometry. This process results in different nonconformal tear contours: junction contours at the intersections between three or four surfaces, separating three or four different regions; and tear contours between different surface subdomains assembling a given interface between two homogeneous regions. Nonconformal meshes are found both in the tear and junction contour lines,

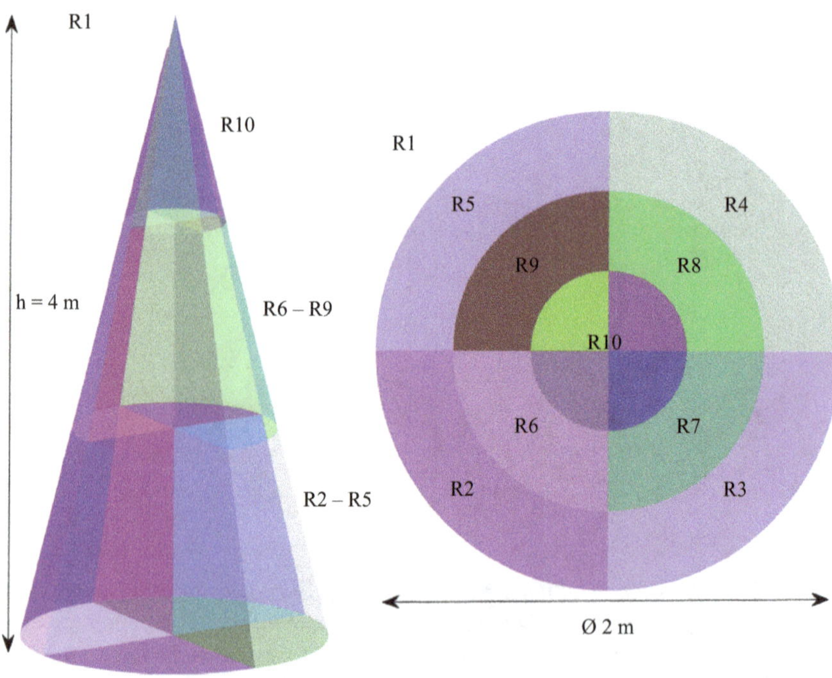

Figure 5.21 Cone-shaped composite object made up of nine different regions with dielectric permittivity growing correlatively from $\varepsilon_{r2} = 2.0$ *to* $\varepsilon_{r10} = 10.0$*, in vacuum background*

which in some cases are curved. A total of 69,294 unknowns are applied to model the problem: 66,436 RWG within the conformal surfaces and 2,858 half-RWG along the nonconformal junction and tear contours. The excitation is an $\hat{x}$-polarized plane wave propagating in the $-\hat{z}$ direction at a frequency of 300 MHz.

Figure 5.22 shows the number of iterations needed with this problem to converge to a residual error below 10^{-6}, versus β. The DG-DDM-SIE method is compared with the conventional approach, using conformal RWG basis functions and MLFMA (RWG-MLFMA). In both approaches, results without preconditioning and with Jacobi (J), block-Jacobi (BJ), and incomplete LU (ILU) preconditioners [71] are shown. Highly accurate results are obtained within a reasonable number of iterations despite the use of nonconformal meshes and the high contrast of the media involved in this case, provided that an appropriate penalty factor in the range from $\beta = 0.5$ to $\beta = 1.5$ is chosen (these are otherwise typical values).

The equivalent electric and magnetic currents induced at the external boundary surfaces of the cube are shown in Figure 5.23 for the nonconformal DG-DDM-SIE and the reference RWG-MLFMA solutions. The reference is solved using conformal triangles on both sides of the junctions and tear contours, which means that those RWG functions defined in the different regions which share a given junction edge are combined under a single unknown, thus imposing the normal current continuity through the junction (in this case, the so-called multi-region basis functions of [73] were applied to facilitate this solution). Looking at this figure, it can be observed that both the electric and magnetic current distributions posed by DG-DDM-SIE perfectly match the reference solution, without the presence of any discontinuity or artifact around the junction contours.

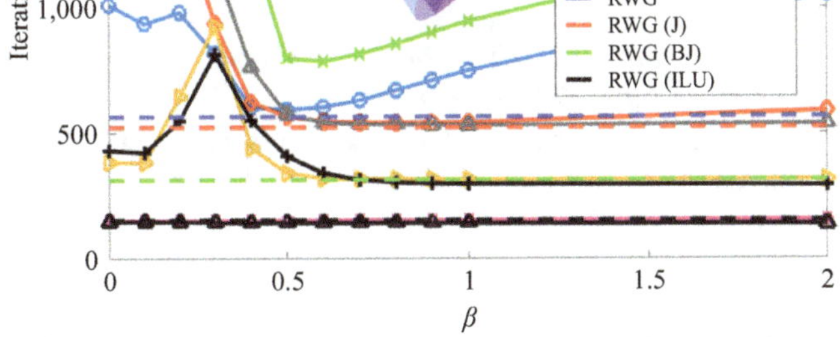

Figure 5.22 Iteration count as a function of β to reach a relative residual error below 10^{-6} for the composite object of Figure 5.21

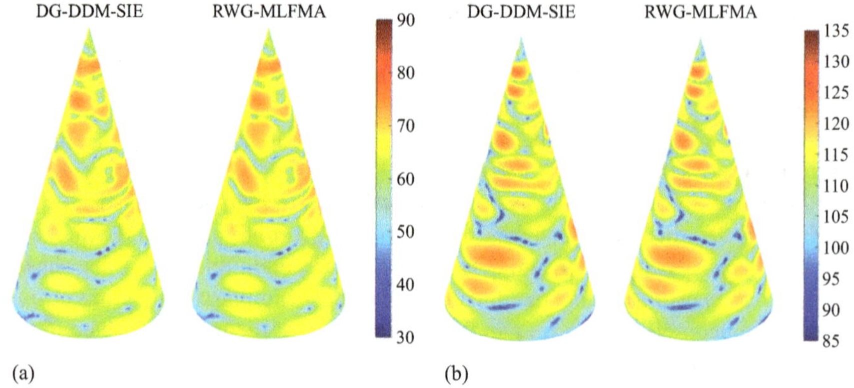

Figure 5.23 Equivalent (a) electric and (b) magnetic current distributions on the external boundary surfaces of the composite object of Figure 5.21 under plane wave excitation at a frequency of 300 MHz calculated by DG-DDM-SIE formulation (left) and reference RWG-MLFMA solution (right)

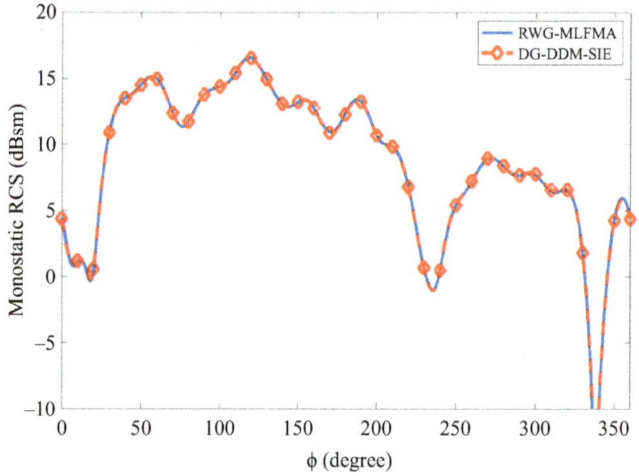

Figure 5.24 $\hat{\theta}\hat{\theta}$-monostatic RCS of the cone-shaped composite object of Figure 5.21 at 300 MHz in the xy-plane, calculated by DG-DDM-SIE formulation and compared with the conformal RWG-MLFMA reference solution

Figure 5.24 shows the $\hat{\theta}\hat{\theta}$-monostatic RCS of the composite cone in the $\theta = 90°$ plane, calculated via DG-DDM-SIE, compared with the reference result obtained via conventional RWG-MLFMA. A very good agreement is observed between the results provided by both methodologies.

5.3 Tear-and-interconnect DDM

A different class of SIE DD method [29,33,74,75] is based on the tear-and-interconnect approach. This class has some advantages that might make it appealing at first glance. Similar to the DG approach, the transmission conditions are imposed along the tearing contours between subdomains, without requiring the addition of artificial surfaces to close the open subdomains that may arise from the splitting stage. This brings the ability to deal with both open and closed bodies. However, instead of using $\mathbf{L}^2$ half-RWG basis functions, the transmission conditions to ensure the near-field current continuity through the tear contours are imposed by employing a domain enlargement technique [29,34,35]. This has the disadvantage of requiring conformal meshes between subdomains (which is not a serious drawback considering the capabilities of current CAD and meshing software). In return, it avoids dealing with perturbations due to charge accumulation in non-conformally meshed tear contours, posing a compact and well-conditioned formulation. Interestingly, this choice eases the incorporation of robust DDM schemes into any existing SIE implementation.

This section will introduce a DD method of SIE based on the tear-and-interconnect approach for the solution of large-scale and multi-scale piecewise homogeneous objects.

5.3.1 Preconditioner formulation

Conceptually, the DDM can be understood as a left or right additive Schwarz preconditioner for the solution of the matrix system. Assuming a partition of the geometry into a collection of surface subdomains $\mathscr{S}_i$, as in Figure 5.19, the Schwarz preconditioner $\left[P^{-1}\right]$ of (5.48) can be constructed and the system matrix equation is left- or right-preconditioned throughout the local solutions of the individual subdomains $\mathscr{S}_i$, as

$$\left[P^{-1}\right] \cdot [Z] \cdot [x] = \left[P^{-1}\right] \cdot [b] \tag{5.77}$$

in the case of left preconditioning, and

$$[Z] \cdot \left[P^{-1}\right] \cdot ([P] \cdot [x]) = [b] \tag{5.78}$$

for right preconditioning.

Each $\left[P^{-1}\right]$ diagonal block conceptually denotes the inverse of the respective impedance matrix block $[Z_{ii}]$ governing the local problem in $\mathscr{S}_i$. Although formally stated in (5.48) with the inverses of the block matrices, they can be solved by the method considered most appropriate in each case. Hence, the subdomain problems can be written as

$$[Z_{ii}] \cdot [\tilde{x}_i] = \left[\tilde{b}_i\right] \tag{5.79}$$

where $[\tilde{x}_i]$ is the local system solution and $\left[\tilde{b}_i\right]$ is the RHS vector.

The subdomain matrix block $[Z_{ii}]$ can be formally obtained from the full impedance matrix $[Z]$ as:

$$[Z_{ii}] = [R_i] \cdot [Z] \cdot \left[R_i^T\right] \tag{5.80}$$

where $[R_i]$ is the restriction matrix mapping the unknown vector to the unknown sub-vector corresponding to subdomain $\mathscr{S}_i$ (i.e., $[x_i] = [R_i] \cdot [x]$). Reciprocally, $\left[R_i^T\right]$ is the prolongation matrix that extends the sub-vector $[x_i]$ to the whole domain. The block diagonal preconditioner can then be built as

$$[P] = \sum_i \left[R_i^T\right] \cdot \left[Z^{-1}\right] \cdot [R_i] \tag{5.81}$$

Achieving good convergence with the above preconditioner demands that normal current continuity is satisfied between the subdomains in contact, and for this transmission conditions must be applied. Using the tear-and-interconnect approach, these conditions are achieved by enlarging the subdomains to incorporate the near-field current flowing across the tearing contours. The enlargement is done by including "flaps" [29] of a quarter to a half-wavelength width belonging to the adjacent touching subdomains. A quarter to a half-wavelength width reveals to be sufficient for the near fields to be able to cancel the potentials due to charge accumulation at the open edges. Consequently, instead of solving equation (5.79), the following equation is solved for the augmented subdomains:

$$\left[Z'_{ii}\right] \cdot \left[\tilde{x}'_i\right] = \left[\tilde{b}'_i\right] \tag{5.82}$$

where $\left[Z'_{ii}\right] = \left[R'_i\right] \cdot [Z] \cdot \left[R_i'^T\right]$ is the impedance matrix block of the augmented subdomain S'_i, $\left[\tilde{x}'_i\right] = \left[R'_i\right] \cdot [\tilde{x}]$, $\left[\tilde{b}'_i\right] = \left[R'_i\right] \cdot \left[\tilde{b}\right]$, and $\left[R'_i\right]$ and $\left[R_i'^T\right]$ are the restrictions and prolongation matrices for the augmented subdomain $\mathscr{S}'_i$, respectively.

The solutions of the augmented systems are then restricted back to the original restricted subdomains as

$$[\tilde{x}_i] = [R_i] \cdot \left[R_i'^T\right] \cdot \left[\tilde{x}'_i\right] \tag{5.83}$$

and assembled together as

$$\sum_{i=1} \left[R_i^T\right] \cdot [\tilde{x}_i] \tag{5.84}$$

The generation of the augmented subdomains can be efficiently accomplished without explicitly building the restriction/prolongation matrices, as follows:

- **Subdomain generation:** The CAD models of the different subdomains $\mathscr{S}_i$ are generated as independent CAD models composed of two components: (i) a reduced subdomain, which is the portion that does not touch or overlap with any adjacent subdomains; (ii) one or more inner flaps surfaces, extending from the reduced subdomain to the boundary of the adjacent subdomain(s). The union of the reduced subdomain and its inner flap(s) constitute the actual (restricted) subdomain $\mathscr{S}_i$. The union of the restricted subdomain with the inner flap(s) of the adjacent subdomain(s) constitutes the augmented subdomain $\mathscr{S}'_i$. (The inner flaps actually constitute the outer flaps of the adjacent subdomains, being part of the assembly of their respective augmented subdomains.)
- **Conformal meshing:** The DG approach could be still applied for the different tear contours between reduced and augmented subdomains. Notwithstanding, one

can rely on the use of conformal meshes and RWG basis functions to ensure the current flow through contacting subdomains. To accomplish this, the complete set of subdomains (with their separated components) should be assembled and meshed together. Current CAD packages often support this workflow, allowing different entities to be grouped and (conformally) meshed together, still providing separate access to the sub-meshes of the individual entities.

- **DDM assembly:** The previous process gives a set of meshed surfaces for the different subdomains. The main geometry is built by orderly assembling this collection of subdomain surfaces, which poses a global topological data set that includes geometry meshes and indexing lists to the different restricted and augmented subdomain elements. The required restriction and prolongation of operations can be straightforwardly obtained from this topological set: the restriction operator consists of picking out the vector elements belonging to a specific subdomain (either restricted or augmented), while the prolongation operator corresponds to putting the elements back in their particular positions of the global structure. Additionally, the RWG basis functions supporting the current flow through the tear contours of adjacent surfaces are created at this stage and assigned to only one of the two contacting surfaces in each case.

5.3.2 A note on parallelization

DDM is a method prone to parallelization in distributed or mixed-memory computers. Perhaps one of the most efficient parallelization approaches for mixed-memory computers is to use MPI to assign subdomains to processes and processes to compute nodes, along with thread parallelization within compute nodes. Looking for a good load balance, the assignment of subdomains to MPI processes must be made according to the available computational resources. The internal (fast) solvers running locally can be parallelized using the OpenMP standard or a similar approach, by distributing local computations across threads. This results in optimal shared-memory computing performance.

The outlined parallelization scheme is advantageous from the point of view of resource management in mixed-memory computers. Since the DDM preconditioner drastically reduces the outer (global) iteration count, most of the computation is then performed locally, on shared-memory compute nodes, without any need for inter-process communication. The (time-consuming) global inter-node communications associated with global iterations, required to calculate the mutual couplings between subdomains, are reduced in the same proportion. Considering that such global communications are often the biggest bottleneck in solving extremely large-scale problems in distributed environments, DDM brings a major competitive advantage compared to parallel distributed versions of conventional fast solvers.

5.3.3 Numerical examples

Two realistic challenging examples of practical interest are shown in this section to demonstrate the effectiveness of the tear-and-interconnect DDM-SIE approach to solving real-life problems. First, an example of electromagnetic compatibility and interference (EMC/EMI) engineering is considered, involving a variety of complex antennas on board a large platform. Second, the simulation of an F-22 model aircraft that includes composite antennas and noticeable multiscale character is presented.

5.3.3.1 EMC study of a modern ship

A realistic application example is presented in this section, consisting of the evaluation of the electromagnetic coupling between the antennas belonging to different radio systems installed on board a ship [34]. The S_1 system is intended for air-to-ship communications operating in the VHF band. It consists of a transceiver connected via a multi-coupler to four patch antennas built into the intermediate level of the main mast. The antennas are designed to work simultaneously providing a 360° coverage. The S_2 system operates in the V/UHF band and is intended for line-of-sight communications. It consists of a transceiver connected to two omnidirectional vertically polarized coaxial dipole antennas [76] located at the main yardarm of the aft mast, in port/starboard arrangement. The third system, S_3, is composed of two transceivers connected to respective UHF omnidirectional vertically polarized sleeve dipole antennas [76,77]. These antennas are located on the upper yardarm of the aft mast, in port/starboard arrangement. The S_4 system is a V/UHF system for broadband surface naval applications and operates via four omnidirectional, vertically polarized bicone antennas located at the aft mast.

The dimensions of the realistic ship are around 140 m long, 20 m wide, and 40 m high ($140\,\lambda \times 20\,\lambda \times 40\,\lambda$ at the working wavelength λ corresponding to a frequency of 300 MHz). A total of 13, 841, 560 unknowns are required for the entire surface. The resulting mesh shows a largely disparate mesh size, from around $\lambda/10$ on the smooth surfaces of the geometry to over $\lambda/1,600$ in regions containing finely detailed antennas [78]. This highlights the marked multi-scale nature of this problem.

According to this multi-scale character, the geometry is decomposed into $M = 23$ subdomains, as shown in Figures 5.25 to 5.27:

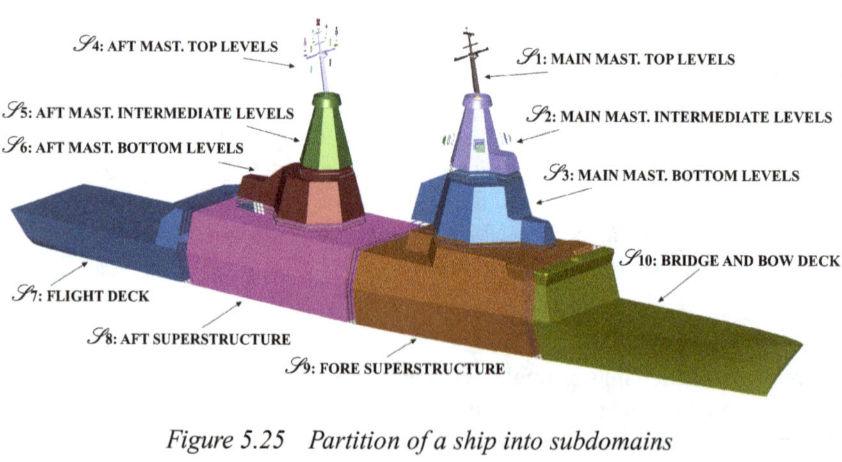

Figure 5.25 Partition of a ship into subdomains

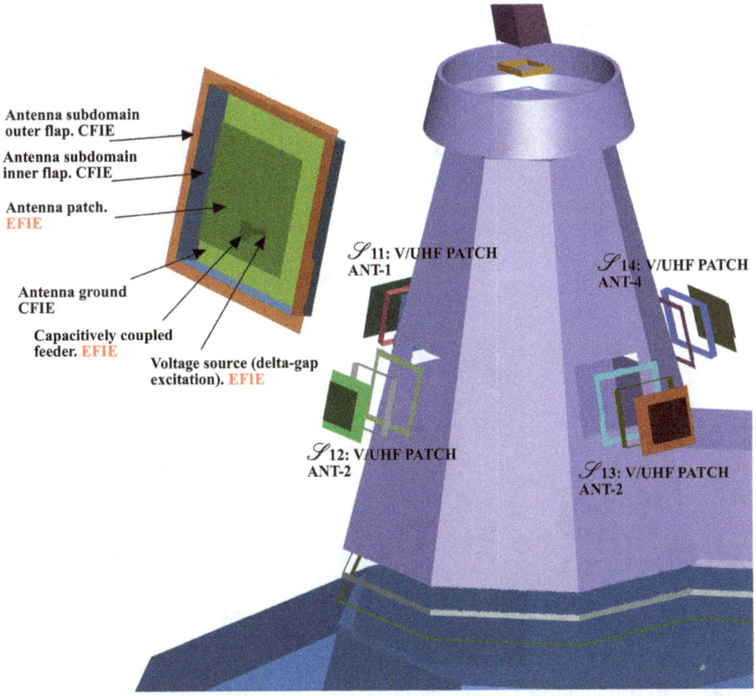

Figure 5.26 Arrangement of the antennas on the main (forward) mast and partition into subdomains. The inner flaps of each touching domain (outer flaps of the adjacent domains) are shown.

- 10 large subdomains ($\mathscr{S}_1$ to $\mathscr{S}_{10}$) that describe the structural parts of the ship (see Figure 5.25). These subdomains are solved using independent four-level MLFMA [56,57,79–81] solvers.
- 13 electrically small subdomains ($\mathscr{S}_{11}$ to $\mathscr{S}_{23}$) including the 13 antennas onboard (Figures 5.26 and 5.27): $\mathscr{S}_{11}$ to $\mathscr{S}_{14}$ include the antennas A-1 to A-4 of system S_1, $\mathscr{S}_{15}$ and $\mathscr{S}_{16}$ include the antennas A-5 and A-6 of S_2, $\mathscr{S}_{17}$ and $\mathscr{S}_{18}$ include the antennas A-7 and A-8 of S_3, and $\mathscr{S}_{19}$ to $\mathscr{S}_{23}$ include the antennas A-9 to A-13 of S_4. All these (smaller) subdomains are solved via factorization of the MoM matrix.

The antenna subdomains include the proper inner and outer flaps to enable the formation of restricted and augmented subdomains.

The excitation of the problem consists of delta-gap voltage sources placed at the feed terminals [82]. The terminals are formed by the RWG bases defined in the perimeter of the contact between the surface of the feeder wire and the ground surface, as illustrated in the right insets of Figures 5.25 and 5.26. The entire problem is modeled as PEC. The EFIE formulation ($\alpha = 1$) is used for open surfaces (open parts

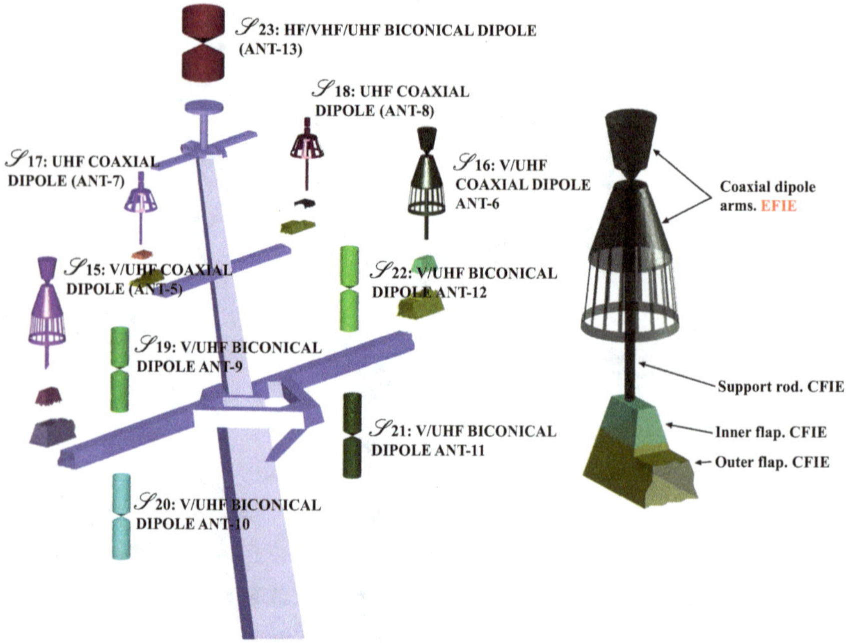

Figure 5.27 Arrangement of the antennas on the aft mast and partition into subdomains. The inner flaps of each touching domain (outer flaps of the adjacent domains) are shown.

of the antennas as well as the feed points). CFIE formulation ($\alpha = 0.5$) is used for the rest of the surfaces. Remarkably, although the application of DDM causes closed surfaces to open due to partitioning into subdomains (e.g., with holes corresponding to the connection of antenna subdomains to the superstructure), these apparently open surfaces are actually part of closed surfaces in the original problem (indeed they are closed back by the incoming radiation coming from all the other subdomains in the right-hand side of the local solvers). Therefore, CFIE can be perfectly applied to these surfaces. The global MVP, accounting for the mutual coupling between subdomains at the external Krylov stage, is accelerated via MLFMA-FFT. A residual error of 10^{-6} is chosen as GMRES [71] stopping criterion.

To conduct a comprehensive EMC study in a real engineering case like this, the simulations should be extended from HF to X-band or above. In this academic case study, however, the numerical simulations have been constrained to a frequency sweep from 100 MHz to 550 MHz, according to the different operating bands of the systems involved in the example. The results are limited only to a simulated frequency, 300 MHz, except for the S parameters S_{11} and S_{12}, which are shown as a function of the frequency that covers the usual frequency band for communications in V/UHF.

Figure 5.28 compares the GMRES residual error of the global preconditioned system compared with that of the original matrix system. The curves are shown

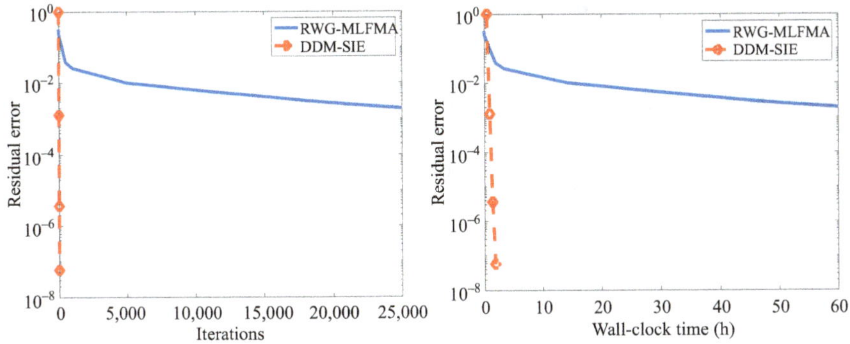

Figure 5.28 Convergence performance of DG-DDM-SIE compared to conventional MLFMA (left) in number of iterations, (right) in time

Figure 5.29 Real part of the equivalent electric current on the ship boundary surfaces calculated by DG-DDM-SIE

versus the number of iterations and the wall-clock time taken to reach a certain convergence threshold (the last is a better figure of merit). An excellent convergence is observed for the DDM approach, which takes 4 outer Krylov iterations and 1.8 h of simulation to converge to a residual error of $5 \cdot 10^{-8}$. This contrasts with the more than 25,000 iterations and the 60 h spent by conventional MLFMA to reach a residual error of $2 \cdot 10^{-3}$. This is a good example of how the concurrence of large-scale realistic platforms that include locally excited radiating elements and fine mesh details, resulting in disparate mesh sizes, slows down the convergence of conventional (not intended for multiscale) fast solvers. In light of the above, it is evident that the rapid convergence of DDM to highly accurate solutions provides a great competitive advantage for solving real-life problems with great interest to the industry, where in addition to the usual requirements in terms of accuracy, there are there are often tight due dates.

Figures 5.29 and 5.30 show the equivalent current distribution in the superstructure. The S_{12} parameters accounting for the coupling between antennas are computed from these currents. As an example, the mutual couplings between antenna A-3 and the other onboard antennas sharing an operating band are shown in Figure 5.31.

Figure 5.30 Detailed view of the current distribution on the masts around the antennas

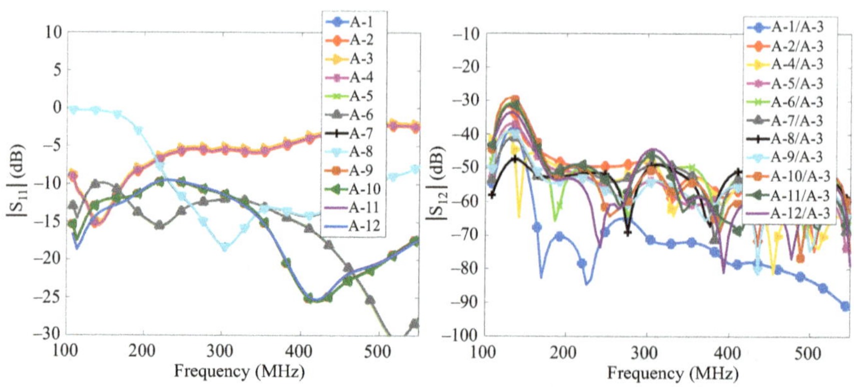

Figure 5.31 S-parameters: (left) S11 parameter; (b) mutual coupling (S12) for the transmitting patch antenna A-3 in the frequency range from 108 to 550 MHz

The computational statistics corresponding to this example can be looked up in Table 5.1, where the following information is shown for each augmented subdomain: number of unknowns, memory consumption, setup time (which is consumed only once per frequency), and the solving time (computed as the average of the solving times throughout all the external iterations).

Table 5.1 Computational simulation parameters of the vessel example

Domain	$\mathcal{S}_1$	$\mathcal{S}_2$	$\mathcal{S}_3$	$\mathcal{S}_4$
Unknowns	64,575	480,887	992,569	64,314
Memory (GB)	1.4	9.6	19.2	1.3
Setup time (s)	8.28	38.66	73.34	8.6
Solving time (s)	9.22	71.7	74.6	9.08
Domain	$\mathcal{S}_5$	$\mathcal{S}_6$	$\mathcal{S}_7$	$\mathcal{S}_8$
Unknowns	439,633	698,046	2,771,909	2,820,127
Memory (GB)	8.8	13.4	50.5	51.7
Setup time (s)	35.66	51.01	193.66	196.89
Solving time (s)	63.66	51.28	459.97	264.29
Domain	$\mathcal{S}_9$	$\mathcal{S}_{10}$	$\mathcal{S}_{11}$	$\mathcal{S}_{12}$
Unknowns	2,881,272	2,770,883	12,309	12,309
Memory (GB)	52.9	51.9	5.5	5.5
Setup time (s)	200.91	193.66	92.9	92.9
Solving time (s)	299.12	459.97	0.23	0.23
Domain	$\mathcal{S}_{13}$	$\mathcal{S}_{14}$	$\mathcal{S}_{15}$	$\mathcal{S}_{16}$
Unknowns	12,309	12,309	6,804	6,804
Memory (GB)	5.5	5.5	1.4	1.4
Setup time (s)	92.9	92.9	40.8	40.8
Solving time (s)	0.23	0.23	0.11	0.11
Domain	$\mathcal{S}_{17}$	$\mathcal{S}_{18}$	$\mathcal{S}_{19}$	$\mathcal{S}_{20}$
Unknowns	4,320	4,320	2,498	2,498
Memory (GB)	0.6	0.6	0.2	0.2
Setup time (s)	18.43	18.43	3.72	3.72
Solving time (s)	0.05	0.05	0.03	0.03
Domain	$\mathcal{S}_{21}$	$\mathcal{S}_{22}$	$\mathcal{S}_{23}$	
Unknowns	2,498	2,498	2,384	
Memory (GB)	0.2	0.2	0.1	
Setup time (s)	3.72	3.72	3.38	
Solving time (s)	0.03	0.03	0.02	

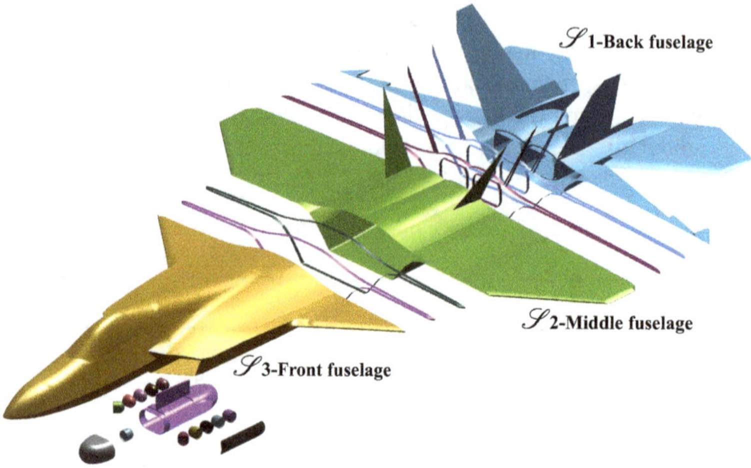

Figure 5.32 Partition of an F-22 aircraft into subdomains

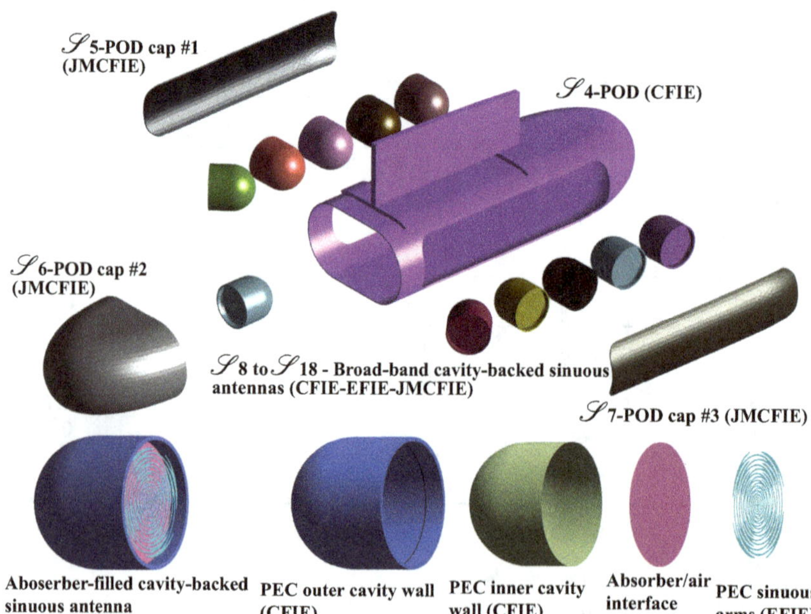

Figure 5.33 Detailed view of POD mounted antennas, and single antenna subdomain at the bottom

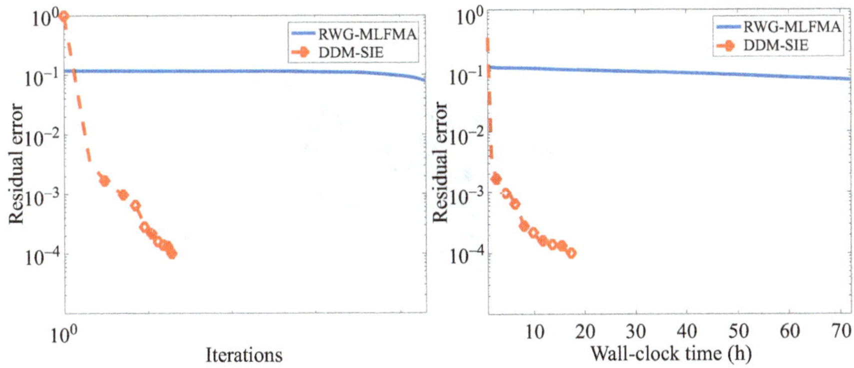

Figure 5.34 Convergence performance of DDM-SIE compared to conventional MLFMA (left) in a number of iterations, (right) in time

5.3.3.2 Tactical fighter aircraft

The following example consists of the solution of an F-22 aircraft with onboard sensors. The geometry of the problem is shown in Figure 5.32. The total number of unknowns is 5,766,388, with the discretization size varying from $\lambda/10$ to $\lambda/360$, being λ the wavelength corresponding to 1 GHz. The F-22 fuselage, including the engine air intake cavities, is split into 3 large PEC-touching subdomains, as shown in Figure 5.33. The respective inner and outer flaps are included to apply the near-field transmission conditions. The partition does not respond to the different subsystems of the fuselage, but it simply divides it into three approximately equal parts for load balancing. A partition by subsystems could be more convenient, although it is not imperative. The sensor systems consist of 11 absorbent-loaded cavity-backed sinuous antennas (subdomains $\mathscr{S}_8$ to $\mathscr{S}_{18}$) assembled inside a PEC/dielectric avionics POD, which is located in the underbelly of the aircraft. The POD is built up of four pieces: the central PEC body to hold antennas ($\mathscr{S}_4$), and three dielectric covers ($\mathscr{S}_5$ to $\mathscr{S}_7$). The sensors operate at 1 GHz and are similar to those described in [83,84]. Their structure is detailed at the bottom of Figure 5.33. Each one is an antenna consisting of a complex multi-scale composite structure including four sinuous PEC arms of about 3 mm width ($\lambda/100$). The sinuous arms describe a complex pattern including tiny details, demanding a careful generation of the triangular mesh. The sinuous arms are embedded on a PEC cavity filled with absorbing material to avoid back radiation, providing circularly polarized transmissions in the forward direction in the UHF band. The axis dimensions of one antenna are 30 × 30 × 30 cm (λxλxλ approx.). A residual error of 10^{-4} was prescribed to stop the GMRES algorithm.

The comparative analysis of the convergence performance of the SIE-DD approach and MLFMA is shown in Figure 5.34. Only 20 iterations of the outer GMRES and 18 h of simulation were required with the SIE-DD approach to converge to a residual error of $8.5 \cdot 10^{-5}$, below the prescribed threshold. The MLFMA

Figure 5.35 *Real part of the equivalent electric current on the aircraft boundary surfaces calculated by DDM-SIE*

Table 5.2 *Computational simulation parameters of the F22 example*

Domain	$\mathscr{S}_1$	$\mathscr{S}_2$	$\mathscr{S}_3$	$\mathscr{S}_4$
Unknowns	1,904,096	1,677,374	1,746,978	255,084
Memory (GB)	258.6	162.8	189.4	44.1
Setup time (s)	5,380	33,136	3,851	1,188
Solving time (s)	6084	665	708	184
Domain	$\mathscr{S}_5$	$\mathscr{S}_6$	$\mathscr{S}_7$	$\mathscr{S}_7$
Unknowns	54,360	109,872	54,504	19,874
Memory (GB)	10.5	22.2	10.6	12.0
Setup time (s)	350	885	350	262
Solving time (s)	13	43	14	0.5
Domain	$\mathscr{S}_9$	$\mathscr{S}_{10}$	$\mathscr{S}_{11}$	$\mathscr{S}_{12}$
Unknowns	19,874	19,874	19,874	19,874
Memory (GB)	12.0	12.0	12.0	12.0
Setup time (s)	262	262	262	262
Solving time (s)	0.5	0.5	0.5	0.5
Domain	$\mathscr{S}_{13}$	$\mathscr{S}_{14}$	$\mathscr{S}_{15}$	$\mathscr{S}_{16}$
Unknowns	19,874	19,874	19,874	19,874
Memory (GB)	12.0	12.0	12.0	12.0
Setup time (s)	262	262	262	262
Solving time (s)	0.5	0.5	0.5	0.5
Domain	$\mathscr{S}_{17}$	$\mathscr{S}_{18}$		
Unknowns	19,874	19,874		
Memory (GB)	12.0	12.0		
Setup time (s)	262	262		
Solving time (s)	0.5	0.5		

simulation, however, took more than 20,000 external iterations and 72 h to achieve a residual error of $7 \cdot 10^{-3}$. The challenging multi-scale and multi-material features of this problem make it difficult for the global problem to converge if a proper preconditioning strategy is not employed. Moreover, the possibility offered by the SIE-DD approach of using multiple formulations within a subdomain (see Figure 5.33) has special positive implications in the convergence when there are composite subdomains, as in this example.

The computational parameters of the simulation corresponding to this problem are gathered in Table 5.2. The approach exploits the fact that the single antenna subdomains D_8 to D_{18} are geometrically identical, except for rotation and/or translation movements, by reusing the MoM impedance matrix calculations of the first antenna subdomain for all of them. The convergence records of Figure 5.34, together with the data of Table 5.2, point out again that the SIE-DD implementation allows to take on real-life projects involving convergence difficulties, high requirements of computational resources, and time constraints. Finally, Figure 5.35 shows the equivalent surface current distribution over the aircraft surface.

References

[1] Després B, Joly P, and Roberts JE. A domain decomposition method for the harmonic Maxwell equations. In: *Iterative methods in linear algebra (Brussels, 1991)*. Amsterdam: North-Holland; 1992. p. 475–484.

[2] Stupfel B. A hybrid finite element and integral equation domain decomposition method for the solution of the 3-D scattering problem. *Journal of Computational Physics*. 2001;172:451–471.

[3] Toselli A and Widlund O. *Domain Decomposition Methods—Algorithms and Theory*. Berlin: Springer; 2005.

[4] Bendali A, Boubendir Y, and Fares M. A FETI-like domain decomposition method for coupling finite elements and boundary elements in large-size problems of acoustic scattering. *Computers and Structures*. 2007;85(9): 526–535.

[5] Lee SC, Vouvakis MN, and Lee JF. A non-overlapping domain decomposition method with non-matching grids for modeling large finite antenna arrays. *Journal of Computational Physics*. 2005;203(1):1–21.

[6] Zhao K, Rawat V, Lee SC, *et al.* A domain decomposition method with nonconformal meshes for finite periodic and semi-periodic structures. *IEEE Transactions on Antennas and Propagation*. 2007;55(9):2559–2570.

[7] Li YJ and Jin JM. A new dual-primal domain decomposition approach for finite element simulation of 3-D large-scale electromagnetic problems. *IEEE Transactions on Antennas and Propagation*. 2007;55(10):2803–2810.

[8] Rawat V and Lee JF. Non-overlapping domain decomposition method with second order transmission condition for the time-harmonic Maxwell's equations. *SIAM Journal on Scientific Computing*. 2010;32:3584–3603.

[9] Peng Z, Rawat V, and Lee JF. One way domain decomposition method with second order transmission conditions for solving electromagnetic wave problems. *Journal of Computational Physics*. 2010;229:1181–1197.

[10] Peng Z and Lee JF. Non-conformal domain decomposition method with second-order transmission conditions for time-harmonic electromagnetics. *Journal of Computational Physics*. 2010;229(16):5615–5629.

[11] Peng Z and Lee JF. Non-conformal domain decomposition method with mixed true second order transmission condition for solving large finite antenna arrays. *IEEE Transactions on Antennas and Propagation*. 2011;59(5):1638–1651.

[12] Peng Z and Lee JF. A scalable non-overlapping and non-conformal domain decomposition method for solving time-harmonic Maxwell Equations in $\mathbb{R}^3$. *SIAM Journal of Scientific Computing*. 2012;34(3):A1266–A1295.

[13] Xue MF and Jin JM. A preconditioned dual-primal finite element tearing and interconnecting method for solving three-dimensional time-harmonic Maxwell's equations. *Journal of Computational Physics*. 2014;274(0): 920–935.

[14] Xue MF and Jin JM. A hybrid conformal/nonconformal domain decomposition method for multi-region electromagnetic modeling. *IEEE Transactions on Antennas and Propagation*. 2014;62(4):2009–2021.

[15] Dolean V, Gander MJ, Lanteri S, *et al.* Effective transmission conditions for domain decomposition methods applied to the time-harmonic curl-curl Maxwell's equations. *Journal of Computational Physics*. 2015;280: 232–247.

[16] Peng Z, Shao Y, Gao HW, *et al.* High-fidelity, high-performance computational algorithms for intrasystem electromagnetic interference analysis of IC and electronics. *IEEE Transactions on Components, Packaging and Manufacturing Technology*. 2017;7(5):653–668.

[17] Maischak M, Stephan EP, and Tran T. Multiplicative Schwarz algorithms for the Galerkin boundary element method. *SIAM Journal of Numerical Analysis*. 2000;38(3):1243–1268.

[18] Hsiao GC, Khoromskij B, and Wendland WL. Preconditioning for boundary element methods in domain decomposition. *Engineering Analysis with Boundary Elements*. 2001;25(4-5):323–338.

[19] Steinbach O and Windisch M. Stable boundary element domain decomposition methods for the Helmholtz equation. *Numerical Mathematics*. 2011;118(1):171–195.

[20] Langer U, Of G, Steinbach O, *et al.* Inexact data-sparse boundary element tearing and interconnecting methods. *SIAM Journal of Scientific Computing*. 2007;29(1):290–314.

[21] Claeys X and Hiptmair R. Multi-trace boundary integral formulation for acoustic scattering by composite structures. *Pure Applied Mathematics*. 2013;66(8):1163–1201.

[22] Claeys X and Hiptmair R. Electromagnetic scattering at composite objects: A novel multi-trace boundary integral formulation. *Mathematics Modelling and Numerical Analysis*. 2012;46(6):1421–1445.

[23] Hiptmair R, Jerez-Hanckes C, Lee JF, *et al.* Domain decomposition for boundary integral equations via local multi-trace formulations. In: *Seminar for Applied Mathematics*, ETH Zürich; 2013. 2013-08.

[24] Chouly F and Heuer N. A Nitsche-based domain decomposition method for hypersingular integral equations. *Numerische Mathematik*. 2012;121(4): 705–729. Available from: http://dx.doi.org/10.1007/s00211-012-0451-2.

[25] Healey M and Heuer N. Mortar boundary elements. *SIAM Journal of Numerical Analysis*. 2010;48(4):1395–1418.

[26] Heuer N and Stephan E. An overlapping domain decomposition preconditioner for high order BEM with anisotropic elements. *Advances in Computational Mathematics*. 2003;19(1–3):211–230.

[27] Li WD, Hong W, and Zhou HX. An IE-ODDM-MLFMA scheme with DILU preconditioner for analysis of electromagnetic scattering from large complex objects. *IEEE Transactions on Antennas and Propagation*. 2008;56(5):1368–1380.

[28] Wiedenmann O and Eibert TF. A domain decomposition method for boundary integral equations using transmission condition based on the near-zone coupling. *IEEE Transactions on Antennas and Propagation*. 2014;62:4105–4114.

[29] Echeverri Bautista MA, Vipiana F, Francavilla MA, *et al.* A nonconformal domain decomposition scheme for the analysis of multiscale structures. *IEEE Transactions on Antennas and Propagation*. 2015;63(8):3548–3560.

[30] Peng Z, Wang XC, and Lee JF. Integral equation based domain decomposition method for solving electromagnetic wave scattering from non-penetrable objects. *IEEE Transactions on Antennas and Propagation*. 2011;59(9):3328–3338.

[31] Peng Z, Lim KH, and Lee JF. Non-conformal domain decomposition methods for solving large multi-scale electromagnetic scattering problems. *Proceedings of IEEE*. 2013;101(2):298–319.

[32] Peng Z, Lim KH, and Lee JF. A discontinuous Galerkin surface integral equation method for electromagnetic wave scattering from nonpenetrable targets. *IEEE Transactions on Antennas and Propagation*. 2013;61(7):3617–3628.

[33] Peng Z, Hiptmair R, Shao Y, *et al.* Domain decomposition preconditioning for surface integral equations in solving challenging electromagnetic scattering problems. *IEEE Transactions on Antennas and Propagation*. 2016;64(1): 210–223.

[34] Solís DM, Martín VF, Araújo MG, *et al.* Accurate EMC engineering on realistic platforms using an integral equation domain decomposition approach. *IEEE Transactions on Antennas and Propagation*. 2020;68(4):3002–3015.

[35] Martín VF, Larios D, Solís DM, *et al.* Tear-and-interconnect domain decomposition scheme for solving multiscale composite penetrable objects. *IEEE Access*. 2020;8:107345–107352.

[36] Nédélec JC. Mixed finite elements in $\mathbb{R}^3$. *Numerische Mathematik*. 1980;35(3):315–341.

[37] Barton ML and Cendes ZJ. New vector finite elements for three-dimensional magnetic field computation. *Journal of Applied Physics*. 1987;61:3919–3921.

[38] Rao SM, Wilton DR, and Glisson AW. Electromagnetic scattering by surfaces of arbitrary shape. *IEEE Transactions on Antennas and Propagation*. 1982;30(3):409–418.

[39] Graglia RD, Wilton DR, and Peterson AF. Higher order interpolatory vector bases for computational electromagnetics. *IEEE Transactions on Antennas and Propagation*. 1997;45(3):329–342.

[40] Demkowicz L and Buffa A. H^1, $H(\text{curl})$ and $H(\text{div})$-conforming projection-based interpolation in three dimensions Quasi-optimal p-interpolation estimates. *Computer Methods in Applied Mechanics and Engineering*. 2005;194:267–296.

[41] Bernardo Cockburn CWS and Karniadakis GE. *Discontinuous Galerkin Methods: Theory Computation and Applications*. Berlin, Heidelberg: Springer; 2000.

[42] Lu T, Zhang P, and Cai W. Discontinuous Galerkin methods for dispersive and lossy Maxwell's equations and PML boundary conditions. *Journal of Computational Physics*. 2004;200(2):549–580. Available from: https://www.sciencedirect.com/science/article/pii/S0021999104001792.

[43] Lu T, Cai W, and Zhang P. Discontinuous galerkin time-domain method for GPR simulation in dispersive media. *IEEE Transactions on Geoscience and Remote Sensing*. 2005;43(1):72–80.

[44] Xiao T and Liu QH. Three-dimensional unstructured-grid discontinuous Galerkin method for Maxwell's equations with well-posed perfectly matched layer. *Microwave and Optical Technology Letters*. 2005;46(5):459–463. Available from: https://onlinelibrary.wiley.com/doi/abs/10.1002/mop.21016.

[45] Gedney SD, Young JC, Kramer TC, *et al*. A discontinuous Galerkin finite element time-domain method modeling of dispersive media. *IEEE Transactions on Antennas and Propagation*. 2012;60(4):1969–1977.

[46] Buffa A, Houston P, and Perugia I. Discontinuous Galerkin computation of the Maxwell eigenvalues on simplicial meshes. *Journal of Computational and Applied Mathematics*. 2007;204(2):317–333. Special Issue: The Seventh International Conference on Mathematical and Numerical Aspects of Waves. Available from: https://www.sciencedirect.com/science/article/pii/S0377042706003608.

[47] Houston P, Perugia I, and Schtzau D. An a posteriori error indicator for discontinuous Galerkin discretizations of H(curl)-elliptic partial differential equations. *IMA Journal of Numerical Analysis*. 2007;27(1):122–150. Available from: https://doi.org/10.1093/imanum/drl012.

[48] Houston P, Perugia I, Schneebeli A, *et al*. Interior penalty method for the indefinite time-harmonic Maxwell equations. *Numerische Mathematik*. 2005; 100:485–518.

[49] Gao HW, Yang ML, and Sheng XQ. A new SDIE based on CFIE for electromagnetic scattering From IBC objects. *IEEE Transactions on Antennas and Propagation*. 2020;68(1):388–399.

[50] Kong BB and Sheng XQ. A discontinuous Galerkin surface integral equation method for scattering from multiscale homogeneous objects. *IEEE Transactions on Antennas and Propagation*. 2018;66(4):1937–1946.

[51] Martin VF, Larios D, Taboada JM, *et al.* DG-JMCFIE formulation for the simulation of composite objects. In: *2021 International Applied Computational Electromagnetics Society Symposium (ACES)*; 2021. p. 1–4.

[52] Martín VF, Landesa L, Obelleiro F, *et al.* A discontinuous Galerkin combined field integral equation formulation for electromagnetic modeling of piecewise homogeneous objects of arbitrary shape. *IEEE Transactions on Antennas and Propagation*. 2022;70(1):487–498.

[53] Ubeda E and Rius JM. Novel monopolar MFIE MoM-discretization for the scattering analysis of small objects. *IEEE Transactions on Antennas and Propagation*. 2006;54(1):50–57.

[54] Jackson CP and Robinson PC. A numerical study of various algorithms related to the preconditioned conjugate gradient method. *International Journal for Numerical Methods in Engineering*. 1985;21(7):1315–1338.

[55] Saad Y and Schultz MH. GMRES: a generalized minimal residual algorithm for solving nonsymmetric linear systems. *SIAM Journal on Scientific and Statistical Computing*. 1986;7(3):856–869.

[56] Song JM and Chew WC. Multilevel fast-multipole algorithm for solving combined field integral equations of electromagnetic scattering. *Microwave and Optical Technology Letters*. 1995;10(1):14–19.

[57] Chew WC, Jin JM, Michielssen E, *et al. Fast and Efficient Algorithms in Computational Electromagnetics*. Norwood: Artech House; 2001.

[58] MacKie-Mason B, Greenwood A, and Peng Z. Adaptive and parallel surface integral equation solvers for very large-scale electromagnetic modeling and simulation. *Progress in Electromagnetics Research*. 2015;154: 143–162.

[59] MacKie-Mason B, Shao Y, Greenwood A, *et al.* Supercomputing-enabled first-principles analysis of radio wave propagation in urban environments. *IEEE Transactions on Antennas and Propagation*. 2018;66(12):6606–6617.

[60] Kelley JT, Chamulak DA, Courtney CC, *et al.* Rye Canyon radar cross-section measurements of benchmark Almond targets [EM Programmer's Notebook]. *IEEE Antennas and Propagation Magazine*. 2020;62(1):96–106.

[61] Karypis G and Kumar V. A fast and high quality multilevel scheme for partitioning irregular graphs. *SIAM Journal on Scientific Computing*. 1999;20(1): 359–392.

[62] Sheng XQ, Jin JM, Song J, *et al.* Solution of combined-field integral equation using multilevel fast multipole algorithm for scattering by homogeneous bodies. *IEEE Transactions on Antennas and Propagation*. 1998;46(11): 1718–1726.

[63] Müller C. *Foundations of the Mathematical Theory of Electromagnetic Waves*. Springer, Berlin, Heidelberg; 1969.

[64] Yla-Oijala P and Taskinen M. Application of combined field integral equation for electromagnetic scattering by dielectric and composite objects. *IEEE Transactions on Antennas and Propagation*. 2005;53(3):1168–1173.

[65] Kelley JT, Yilmaz AE, Chamulak DA, *et al.* Measurements of non-metallic targets for the Austin RCS benchmark suite. In: *2019 Antenna Measurement Techniques Association Symposium (AMTA)*; 2019. p. 1–6.

[66] Martín VF, Solís DM, Araújo MG, *et al.* A discontinuous Galerkin integral equation approach for electromagnetic modeling of realistic and complex radiating systems. *IEEE Transactions on Antennas and Propagation*. 2023;99: 1–1.
[67] Martín VF, Solís DM, Jericó D, *et al.* Discontinuous Galerkin integral equation method for light scattering from complex nanoparticle assemblies. *Optics Express*. 2023;31(2):1034–1048. Available from: https://opg.optica.org/oe/abstract.cfm?URI=oe-31-2-1034.
[68] Ylä-Oijala P, Taskinen M, and Järvenpää S. Surface integral equation formulations for solving electromagnetic scattering problems with iterative methods. *Radio Science*. 2005;40(6):RS6002.
[69] Rao S, Wilton D, and Glisson A. Electromagnetic scattering by surfaces of arbitrary shape. *IEEE Transactions on Antennas and Propagation*. 1982;30(3):409–418.
[70] Ubeda E and Rius JM. Novel monopolar MFIE MoM-discretization for the scattering analysis of small objects. *IEEE Transactions on Antennas and Propagation*. 2006;54(1):50–57.
[71] Saad Y. *Iterative Methods for Sparse Linear Systems*. PWS Publishing Company, a division of International Thomsom Publishing Inc., Boston, MA; 1996.
[72] Stupfel B. A fast-domain decomposition method for the solution of electromagnetic scattering by large objects. *IEEE Transactions on Antennas and Propagation*. 1996;44(10):1375–1385.
[73] Solís DM, Taboada JM, and Basteiro FO. Surface integral equation-method of moments with multiregion basis functions applied to plasmonics. *IEEE Transactions on Antennas and Propagation*. 2015;63(5):2141–2152.
[74] Gao HW, Peng Z, and Sheng XQ. A geometry-aware domain decomposition preconditioning for hybrid finite element-boundary integral method. *IEEE Transactions on Antennas and Propagation*. 2017;65(4):1875–1885.
[75] Jia PH, Lei L, Hu J, *et al.* Twofold domain decomposition method for the analysis of multiscale composite structures. *IEEE Transactions on Antennas and Propagation*. 2019;67(9):6090–6103.
[76] Burberry RA. VHF and UHF antennas. *IEE Electromagnetic Waves Series*, No. 35. Peter Peregrinus Ltd. On behalf of the Institution of Electrical Engineers; 1992.
[77] Stutzman WL and Thiele GA. *Antenna Theory and Design*. 3rd ed., John Wiley & Sons, Hoboken; 2012.
[78] Leugner D and Bruns HD. Modeling antenna feeds by electric and magnetic current sheets in conjunction with the method of moments. In: *2007 2nd International ITG Conference on Antennas*; 2007. p. 100–104.
[79] Taboada JM, Araujo MG, Bertolo JM, *et al.* MLFMA-FFT parallel algorithm for the solution of large-scale problems in electromagnetics. *Progress in Electromagnetics Research*. 2010;105:15–30.
[80] Ergül O and Gürel L. Rigorous solutions of electromagnetic problems involving hundreds of millions of unknowns. *IEEE Antennas and Propagation Magazine*. 2011;53(1):18–27.

[81] Taboada JM, Araujo MG, Basteiro FO, *et al.* MLFMA-FFT parallel algorithm for the solution of extremely large problems in electromagnetics. *Proceedings of the IEEE*. 2013;101(2):350–363.

[82] Gibson WC. *The Method of Moments in Electromagnetics*. Chapman & Hall/CRC, Taylor & Francis Group, UK; 2008.

[83] Cloete JH and Sickel T. The planar dual-polarized cavity backed sinuous antenna – a design summary. In: *2012 IEEE-APS Topical Conference on Antennas and Propagation in Wireless Communications (APWC)*; 2012. p. 1169–1172.

[84] Sammeta R and Filipovic DS. Improved efficiency lens-loaded cavity-backed transmit sinuous antenna. *IEEE Transactions on Antennas and Propagation*. 2014;62(12):6000–6009.

Chapter 6
Multi-resolution preconditioner

Francesca Vipiana[1], Victor F. Martin[2] and Jose M. Taboada[2]

The purpose of this chapter is to provide the main guidelines for an efficient implementation of the multi-resolution (MR) preconditioner for the electromagnetic (EM) analysis of perfect electric conductor (PEC) structures of arbitrary 3-D shape via the method of moments (MoM) applied to the electric field integral equation (EFIE) and to the combined field integral equation (CFIE).

The chapter is structured in four main parts. First, the generation of the MR basis functions as a linear combination of the standard basis functions is described. Second, the generation of a multi-level set of meshes, starting from the usual mesh, is reported: the MR functions are defined on each level of the generated set of meshes. These two parts are essential to implement the proposed preconditioner. Then, the third part is dedicated to the insertion of the MR preconditioner into the solution algorithm, together with the description of some implementation tricks. Finally, numerical results, where the MR preconditioner is applied to complex realistic 3-D structures, are reported and commented.

The expected property of the MR preconditioner is an improvement of the convergence rates of iterative solvers, with a limited computational cost for its generation and application. The proposed preconditioner can be applied to realistic structures with arbitrary topological complexity.

6.1 Preliminaries

6.1.1 Introduction and scope

This chapter is based on the preconditioning method introduced in [1] for the efficient and accurate EM modeling of multi-scale structures. In the analysis of multi-scale structures, the solution shows multiple scales of variation because the structures can be electrically large but, at the same time, they possess details much smaller than the working wavelength. The consequence, of the MoM solution to the EM problem, is an ill-conditioning of the linear system that impacts strongly on the solution cost and

[1]Wavision Research Group, Department of Electronics and Telecommunications, Politecnico di Torino, Italy
[2]Departamento de Tecnologias de Computadores y Comunicaciones, Universidad de Extremadura, Spain

accuracy. The key idea in the MR preconditioner is to keep in the basis functions, implemented to discretize the problem, the different scales of variation of the solution.

Multi-resolution and wavelets revolutionized signal processing. Almost ever since, researchers in computational electromagnetics have tried to inject the attractive properties of the wavelet representation into the integral equation method. However, the application to general 3-D vector surface problems remained a significant challenge, since wavelets had been devised for 1-D or separable domains, and applied to 3-D scalar problems (in the framework of computer graphics). The approach described in this chapter has enabled the use of wavelet methods to the electromagnetic analysis. An important aspect of this research is that the main impact of wavelets in integral equations is different than expected from a direct extrapolation of image compression; MR can act as an efficient preconditioner, especially for fine-meshed problems, that are of special relevance in antenna and circuit EM analysis.

To build the MR basis functions, the initial mesh cells are grouped subsequently, giving rise to a nested set of meshes formed by polygonal cells. Then, the MR basis functions generation procedure is based on the decomposition of the solution space into a solenoidal part and a non-solenoidal remainder, which intrinsically avoids the so-called low-frequency breakdown [2–8]. The obtained hierarchic basis functions can act as effective preconditioners in the analysis of multi-scale structures, characterized by dense meshes. In [9], it has been shown that the MR basis functions have different spatial supports, on the various mesh levels, leading to a spectral resolution. This property allows to "sample" the background Green's function at different spatial frequencies, which arises on a peaked matrix diagonal, and to regularize the problem via a simple diagonal preconditioning.

In this chapter, the MR preconditioner is presented for the analysis of PEC structures of arbitrary 3-D shape via the MoM applied to the EFIE and CFIE. Because of the way it is constructed, it will also be called a *hierarchical* preconditioner; the two expressions have to be considered equivalent in the following. The proposed MR preconditioner is constructed as a linear combination of the basis functions usually employed in the MoM-based codes, called piecewise linear (PWL) basis functions for wires discretized via segments, Rao–Wilton–Glisson (RWG) basis functions for surfaces discretized via triangles, and junction basis functions for modelling the connections between surfaces and wires [10–14]. Moreover, in the case of non-conformal triangular meshes, the MR basis functions are expressed as a linear combination of the Multi-Branch RWG (MB-RWG) basis functions that guarantee normal current continuity and no line charges accumulation along the non-conformal mesh edges [15].

6.1.2 Basis functions

In the structure under analysis, the surfaces are discretized into (planar) triangles S_n^B, and the wires into line segments S_n^W. The unknown surface current on S_B can be represented as

$$\boldsymbol{J}(\boldsymbol{r}) \approx \sum_{n=1}^{N_B} I_n^B \boldsymbol{f}_n^B(\boldsymbol{r}) + \sum_{n=1}^{N_{MB}} I_n^{MB} \boldsymbol{f}_n^{MB}(\boldsymbol{r}) + \sum_{n=1}^{N_J} I_n^J \boldsymbol{f}_n^J(\boldsymbol{r}) \qquad \boldsymbol{r} \in S_B \tag{6.1}$$

and the axial current on the wires S_W is given by

$$\boldsymbol{I}(\boldsymbol{r}) \approx \sum_{n=1}^{N_W} I_n^W \boldsymbol{f}_n^W(\boldsymbol{r}) + \sum_{n=1}^{N_J} I_n^J \boldsymbol{f}_n^J(\boldsymbol{r}) \qquad \boldsymbol{r} \in S_W \tag{6.2}$$

where $\boldsymbol{f}_n^B$ are the RWG basis functions, $\boldsymbol{f}_n^{MB}$ are the MB-RWG basis functions, $\boldsymbol{f}_n^J$ are the junction basis functions, and $\boldsymbol{f}_n^W$ are the PWL basis functions [10–15]. The total number of unknowns is $N = N_B + N_{MB} + N_W + N_J$, where N_B and N_{MB} are the number of surface current unknowns discredited with RWG and MB-RWG basis functions, respectively, N_W is the number of wire current unknowns, and N_J is the number of junction current unknowns.

The RWG and PWL basis functions are expressed as

$$\boldsymbol{f}_n^\gamma(\boldsymbol{r}) = \begin{cases} \dfrac{\boldsymbol{\rho}_n^\pm}{h_n^{\gamma\pm}} & \boldsymbol{r} \in S_n^{\gamma\pm} \\ 0 & \text{otherwise} \end{cases} \tag{6.3}$$

where $S_n^{\gamma\pm}$, with $\gamma = B, W$, is the $\pm$ triangle or segment attached to the nth edge or node of the surface or wire, respectively, $h_n^{\gamma\pm}$ is the height (length) of $S_n^{\gamma\pm}$ relative to the nth edge (node) of S_γ, and $\boldsymbol{\rho}_n^\pm$ is $\pm$ the vector from the free node of $S_n^{\gamma\pm}$ to $\boldsymbol{r}$, as shown in Figure 6.1(a) and (b). The corresponding surface divergence is given by

$$\nabla_S \cdot \boldsymbol{f}_n^\gamma(\boldsymbol{r}) = \begin{cases} \pm\dfrac{\sigma_\gamma}{h_n^{\gamma\pm}} & \boldsymbol{r} \in S_n^{\gamma\pm} \\ 0 & \text{otherwise} \end{cases} \tag{6.4}$$

where $\sigma_B = 2$, and $\sigma_W = 1$ $(\gamma = B, W)$.

The MB-RWG basis functions are expressed as

$$\boldsymbol{f}_n^{MB}(\boldsymbol{r}) = \begin{cases} \dfrac{\boldsymbol{\rho}_n^+}{h_n^{B+}} & \boldsymbol{r} \in S_n^{B+} \\ \dfrac{\boldsymbol{\rho}_{n,l}^-}{h_{n,l}^{B-}} & \boldsymbol{r} \in S_{n,l}^{B-} \\ 0 & \text{otherwise} \end{cases} \tag{6.5}$$

where the double subscript n, l refers to the lth negative triangle $S_{n,l}^{B-}$ that has an edge partially superimposed to the edge of the positive triangle S_n^{B+} where the nth MB-RWG basis function is defined; $\boldsymbol{\rho}_{n,l}^-$ is minus the vector from the free node of

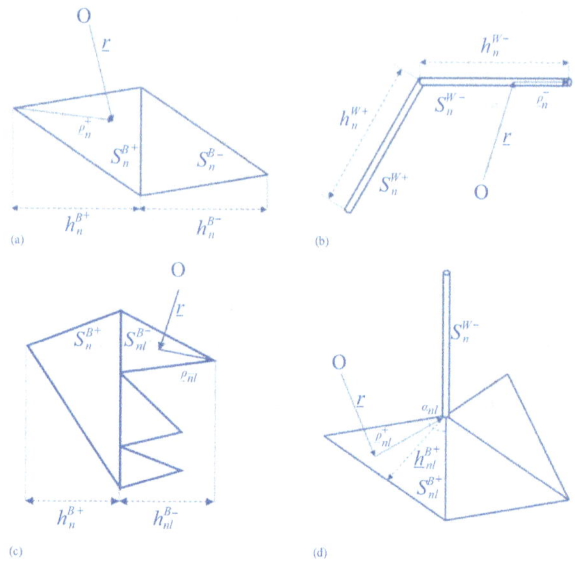

Figure 6.1 Geometrical definition domain and associated parameters for (a) RWG basis function, (b) PWL basis function, (c) MB-RWG basis function, and (d) junction basis function

$S_{n,l}^{B-}$ to $\boldsymbol{r}$, and $h_{n,l}^{B-}$ is the height of $S_{n,l}^{B-}$ from its free node, as shown in Figure 6.1(c). The corresponding surface divergence is given by

$$\nabla_S \cdot \boldsymbol{f}_n^{MB}(\boldsymbol{r}) = \begin{cases} +\dfrac{2}{h_n^{B+}} & \boldsymbol{r} \in S_n^{B+} \\ -\dfrac{2}{h_{n,l}^{B-}} & \boldsymbol{r} \in S_{n,l}^{B-} \\ 0 & \text{otherwise.} \end{cases} \tag{6.6}$$

The junction basis function is equal to

$$\boldsymbol{f}_n^J(\boldsymbol{r}) = \begin{cases} K_{n,l}\left[1 - \left(\dfrac{h_{n,l}^{B+}}{\boldsymbol{\rho}_{n,l}^{+} \cdot \widehat{\boldsymbol{h}}_{n,l}^{B+}}\right)^2\right] \boldsymbol{f}_{n,l}^{B}(\boldsymbol{r}) & \boldsymbol{r} \in S_{n,l}^{B+} \\ \boldsymbol{f}_n^{W}(\boldsymbol{r}) & \boldsymbol{r} \in S_n^{W-} \\ 0 & \text{otherwise} \end{cases} \tag{6.7}$$

where the double subscript n, l refers to the lth triangle, called the "junction triangle" in the following, attached to the vertex where the nth junction is defined. The vector from $\boldsymbol{r}$ to the node of $S_{n,l}^{B+}$ coincident with the nth junction vertex is $\boldsymbol{\rho}_{n,l}^{+}$, and $\widehat{h}_{n,l}^{B+}$ is the unit vector from the nth junction vertex and along the height of $S_{n,l}^{B+}$, as shown in Figure 6.1(d). Finally, $K_{n,l} = \alpha_{n,l}/\left(\ell_{n,l}\alpha_n^t\right)$, where $\ell_{n,l}$ is the edge of the triangle $S_{n,l}^{B+}$ opposite the nth junction vertex, $\alpha_{n,l}$ is the angle between the two edges of $S_{n,l}^{B+}$ common to the nth junction vertex, called the "junction vertex angle," and α_n^t is the sum of all the junction vertex angles at the nth junction. The surface divergence of a junction basis function is given by

$$\nabla_S \cdot \boldsymbol{f}_n^J(\boldsymbol{r}) = \begin{cases} \dfrac{2K_{n,l}}{h_{n,l}^{B+}} & \boldsymbol{r} \in S_{n,l}^{B+} \\ -\dfrac{1}{h_n^{W-}} & \boldsymbol{r} \in S_n^{W-} \\ 0 & \text{otherwise} \end{cases} \tag{6.8}$$

6.1.3 *MoM linear system*

By substituting (6.1) and (6.2) into the considered integral equation and testing it with the functions $\boldsymbol{f}_n^{\gamma}$, where $\gamma = B, MB, W$ and J, and $n = 1, \ldots, N_{\gamma}$, we obtain the following linear system in the matrix format

$$[Z] \cdot [I] = [V] \tag{6.9}$$

where

$$[Z] = \begin{bmatrix} [Z^{B,B}] & [Z^{B,MB}] & [Z^{B,W}] & [Z^{B,J}] \\ [Z^{MB,B}] & [Z^{MB,MB}] & [Z^{MB,W}] & [Z^{MB,J}] \\ [Z^{W,B}] & [Z^{W,MB}] & [Z^{W,W}] & [Z^{W,J}] \\ [Z^{J,B}] & [Z^{J,MB}] & [Z^{J,W}] & [Z^{J,J}] \end{bmatrix} \tag{6.10}$$

$$[I] = \begin{bmatrix} [I^B] \\ [I^{MB}] \\ [I^W] \\ [I^J] \end{bmatrix} \tag{6.11}$$

$$[V] = \begin{bmatrix} [V^B] \\ [V^{MB}] \\ [V^W] \\ [V^J] \end{bmatrix} \tag{6.12}$$

The dimension of the system matrix [Z] is $N \times N$, with $N = N_B + N_{MB} + N_W + N_J$, and $[I^\gamma]$, with $\gamma = B$, *MB*, *W* and *J*, is a column vector with dimension $N_\gamma \times 1$ that collects the corresponding coefficients I_n^γ of (6.1) and (6.2).

In the case of the EFIE, each matrix element is defined as

$$Z_{m,n}^{\mathrm{EFIE},\alpha,\beta} = \left\langle \boldsymbol{f}_m^\alpha, \boldsymbol{E}_{tan}^s \left\{ \boldsymbol{f}_n^\beta \right\} \right\rangle \qquad \alpha, \beta = B, MB, W, J \tag{6.13}$$

where $\langle\ ,\ \rangle$ expresses the reaction-type inner product between two vector functions, and $\boldsymbol{E}_{tan}^s$ is the electric field, tangent to the PEC body, radiated by the basis function $\boldsymbol{f}_n^\beta$. Then, each element of the $N \times 1$ vector [V] is equal to

$$V_m^{\mathrm{EFIE},\gamma} = -\left\langle \boldsymbol{f}_m^\gamma, \boldsymbol{E}_{tan}^i \right\rangle \qquad \gamma = B, MB, W, J \tag{6.14}$$

where $\boldsymbol{E}_{tan}^i$ is the incident electric field, tangent to the PEC body.

In the case of the MFIE, which can be applied to the body's closed parts, each matrix element is defined as

$$Z_{m,n}^{\mathrm{MFIE},\alpha,\beta} = \left\langle \boldsymbol{f}_m^\alpha, \hat{n} \times \boldsymbol{H}^s \left\{ \boldsymbol{f}_n^\beta \right\} \right\rangle \qquad \alpha, \beta = B, MB \tag{6.15}$$

where $\boldsymbol{H}^s$ is the magnetic field radiated by the basis function $\boldsymbol{f}_n^\beta$, and $\hat{n}$ is the outer unit normal of the considered PEC closed body. Each element of [V] is equal to

$$V_m^{\mathrm{MFIE},\gamma} = -\left\langle \boldsymbol{f}_m^\gamma, \hat{n} \times \boldsymbol{H}^i \right\rangle \qquad \gamma = B, MB \tag{6.16}$$

where $\boldsymbol{H}^i$ is the incident magnetic field.

Finally, the CFIE linear system can be obtained as:

$$\left\{ \eta \left[Z^{\mathrm{EFIE}} \right] + (1-\eta) \left[Z^{\mathrm{MFIE}} \right] \right\} \cdot [I] = \left\{ \eta \left[V^{\mathrm{EFIE}} \right] + (1-\eta) \left[V^{\mathrm{MFIE}} \right] \right\} \tag{6.17}$$

where $0 < \eta < 1$ is the weight controlling the contribution of the EFIE and MFIE equations, which is selected to 0.5 in the numerical results reported at the end the chapter.

6.1.4 *Multi-resolution strategy*

The MR preconditioning of the MoM system is logically and algorithmically divided in two main phases.

First, there is the generation of the MR basis functions, as detailed in Sections 6.2 and 6.3. Since the MR basis functions are expressed as a linear combination of standard basis functions (i.e., RWG, MB-RWG, PWL, and junction basis functions), the output of this phase is a basis-change matrix that contains the coefficients of the MR basis functions and allows to (algebraically) transform the MoM system matrix into the MR–MoM system matrix.

Then, there is the construction of the MR-MoM matrix, applying the obtained basis-change matrix to the original MoM system matrix, as described in Section 6.4.

6.2 Basis functions generation

For a structure discretized with N unknowns, the basis functions of the initial level-0 mesh are indicated with $\boldsymbol{f}_i^0$; these functions are equal to the standard functions defined on triangles and segments, as detailed in Section 6.1.2, divided by the corresponding common edge ℓ_j

$$\{f_i^0(\boldsymbol{r})\} = \bigcup_{\gamma=B,MB,W,J} \left\{\frac{f_j^\gamma(\boldsymbol{r})}{\ell_j}\right\} \qquad i = 1,\ldots,N \qquad j = 1,\ldots,N_\gamma \tag{6.18}$$

where ℓ_j is equal to 1 in the case of PWL ($\gamma = W$) or junction basis functions ($\gamma = J$), and equal to the corresponding length of the common edge of the positive triangle in the case of RWG and MB-RWG basis functions ($\gamma = B, MB$). With the previous notation, the solution current $\widetilde{\boldsymbol{J}}$ is written as

$$\widetilde{\boldsymbol{J}}(\boldsymbol{r}) \cong \sum_{i=1}^{N} \alpha_i f_i^0(\boldsymbol{r}). \tag{6.19}$$

where in comparison to (6.1) and (6.2) in Section 6.1.2,

$$\widetilde{\boldsymbol{J}}(\boldsymbol{r}) = \begin{cases} \boldsymbol{J}(\boldsymbol{r}) \text{ if } \boldsymbol{r} \in S_B \\ \boldsymbol{I}(\boldsymbol{r}) \text{ if } \boldsymbol{r} \in S_W \end{cases} \tag{6.20}$$

and

$$\alpha_i = \begin{cases} I_n^B \cdot \ell_B & \text{for } i = 1,\ldots,N_B \\ & \text{and } n = 1,\ldots,N_B \\ I_n^{MB} \cdot \ell_{MB} & \text{for } i = (N_B+1),\ldots,(N_B+N_{MB}) \\ & \text{and } n = 1,\ldots,N_{MB} \\ I_n^W \cdot \ell_W & \text{for } i = (N_B+N_{MB}+1),\ldots,(N_B+N_{MB}+N_W) \\ & \text{and } n = 1,\ldots,N_W \\ I_n^J \cdot \ell_J & \text{for } i = (N_B+N_{MB}+N_W+1),\ldots,N \\ & \text{and } n = 1,\ldots,N_J \end{cases} \tag{6.21}$$

We recall that the total number of unknowns is equal to $N = N_B + N_{MB} + N_W + N_J$.

The hierarchical basis functions are denoted with $\boldsymbol{w}_j, j = 1, ..., N$, in the following and are organized in levels as

$$\{\boldsymbol{w}_j\} = \bigcup_{l=0}^{L} \{\boldsymbol{w}_i^l\} \qquad j = 1,\ldots,N \qquad i = 1,\ldots,N_w^l, \tag{6.22}$$

where $N = \sum_{l=0}^{L} N_w^l$. Each level hierarchical basis functions are defined on the corresponding level mesh constituted by non-simplex (polygonal) cells, in general. Hence, the basis is the collection of all the hierarchical functions of all levels. The number of hierarchical functions at each level, N_w^l, is related to the grouping scheme (see Section 6.3), applied to generate the different level meshes and to the initial mesh; in general, $N_w^l > N_w^{l+1}$. The different levels meshes are built by grouping the cells (triangles or segments) of the initial mesh, indicated with level $l = 0$, and each level-l cell is included in one cell of the level-($l + 1$) mesh only. The cell grouping algorithm is discussed in detail in Section 6.3. All the generated basis functions can be written as a linear combination of the functions of the initial level-0 mesh (6.18), i.e.,

$$\boldsymbol{w}_j(\boldsymbol{r}) = \sum_{i=1}^{N} T_{ij} \boldsymbol{f}_i^0(\boldsymbol{r}) \tag{6.23}$$

and the basis change is obtained algebraically by a "change-of-basis" matrix $[T]$ with dimension $N \times N$ and whose elements are the coefficients T_{ij} defined in (6.23). The matrix $[T]$ is nonsingular and square, because the standard basis and the hierarchical one are constituted by the same number of basis functions and describe the same space. Hence, the solution current $\tilde{\boldsymbol{J}}$ can be written as a linear combination of the hierarchical basis,

$$\tilde{\boldsymbol{J}}(\boldsymbol{r}) \cong \sum_{i=1}^{N} \beta_i \boldsymbol{w}_i(\boldsymbol{r}), \tag{6.24}$$

and the matrix $[T]$ connects the coefficients in (6.24) to the ones in (6.19) as

$$\alpha_i = \sum_{j=1}^{N} T_{i,j} \beta_j. \tag{6.25}$$

In the following, Section 6.2.1 shows how the standard basis functions can be generalized on non-simplex mesh cells, and Section 6.2.2 describes how generating the hierarchical basis functions for each level mesh starting from the generalized basis functions. The hierarchical basis functions are separated into solenoidal and non-solenoidal parts [1,16–18].

6.2.1 Generalized basis functions

The initial step is the building of a set of basis functions on each level mesh. These functions are a generalization of the standard basis functions (RWG, MB-RWG, PWL, and junction basis functions) and are called in the following "generalized" basis functions. The standard basis functions are defined on pairs/groups of triangles or on pairs of segments or on a group of triangles all connected to a segment, while the generalized basis functions are always defined on pairs of non-simplex (polygonal) cells [1]. The generalized basis functions of a level-l mesh are expressed as a linear combination of the generalized basis functions of the previous level-$(l-1)$ mesh. Then, via a recursive application of this inter-mesh relation, all generalized basis functions can be described as a linear combination of the initial (level-0) mesh basis functions that are the standard ones. The proposed procedure in building the generalized functions is an extension to non-simplex mesh cells of the generation process for multi-level RWG basis functions, presented in [19,20] for the case of hierarchic triangular cells only.

On each level-l mesh, with $l > 0$, a set of generalized basis functions $\boldsymbol{f}_i^l$ is generated, with $i = 1, \ldots, N^l$ and N^l equal to the total number of linearly independent functions at that level. Each level-l generalized basis function, $\boldsymbol{f}_i^l$, is defined on a pair of adjacent level-l generalized cells, $C_i^{l,+}$ and $C_i^{l,-}$. All $\boldsymbol{f}_i^l$ basis functions are described as a linear combination of the level-($l-1$) functions as

$$\boldsymbol{f}_i^l(\boldsymbol{r}) = \sum_{n=1}^{N_i^{l-1}} f_{n,i}^l \boldsymbol{f}_{\mu(n)}^{l-1}(\boldsymbol{r}) \qquad i = 1, \ldots, N^l \quad l = 1, \ldots, L \qquad \mu(n) \in \mathscr{I}_i^l \tag{6.26}$$

and

$$\mathscr{I}_i^l = \left\{ j = j_1, j_2, \ldots, j_n \ldots, j_{N_i^{l-1}} : \text{support}\left(\boldsymbol{f}_j^{l-1}\right) \in C_i^{l,+} \bigcup C_i^{l,-} \right\} \tag{6.27}$$

where N_i^{l-1} is the number of level-($l-1$) functions defined strictly inside the support, $C_i^{l,+} \bigcup C_i^{l,-}$, of the considered $\boldsymbol{f}_i^l$ function. In (6.26), we use a local index n on the support ($C_i^{l,+} \bigcup C_i^{l,-}$) of the considered $\boldsymbol{f}_i^l$ function, and a global index $\mu(n)$ that uniquely identifies a function on the level-($l-1$) mesh. To illustrate (6.26) and (6.27) and better describe the correspondence between global and local numbering of the considered level-($l-1$) generalized functions, we refer to the example reported in Figure 6.2. The boundaries of the two cells, $C_i^{l,+}$ and $C_i^{l,-}$, where the $\boldsymbol{f}_i^l$ function is defined, are highlighted with thick solid black lines; the $\boldsymbol{f}_{\mu(n)}^{l-1}$ functions, forming the considered $\boldsymbol{f}_i^l$, are indicated with yellow arrows, and the corresponding definition cells are shown with a group of triangles with the same color. Then (6.26), explicitly referred to the example of Figure 6.2, can be written as

$$\begin{aligned}\boldsymbol{f}_i^2 &= f_{1,i}^2 \boldsymbol{f}_{10}^1 + f_{2,i}^2 \boldsymbol{f}_5^1 + f_{3,i}^2 \boldsymbol{f}_1^1 + f_{4,i}^2 \boldsymbol{f}_{11}^1 + f_{5,i}^2 \boldsymbol{f}_{16}^1 + \\ &\quad f_{6,i}^2 \boldsymbol{f}_{14}^1 + f_{7,i}^2 \boldsymbol{f}_{12}^1 + f_{8,i}^2 \boldsymbol{f}_{13}^1\end{aligned} \tag{6.28}$$

The coefficient indices refer to the local numbering in Figure 6.2(b), instead the level-1 generalized function indices correspond to the global numbering in Figure 6.2(a). In the reported example $l = 2$, $N_i^{l-1} = 8$, and $N^l = 54$.

The coefficients $f_{n,i}^l$ in (6.26) have to be found to build, on all the levels meshes, basis functions that are a generalization of the standard basis functions. To derive the coefficients $f_{n,i}^l$, the surface divergence operator $\nabla_s \cdot$ is applied to (6.26), and, then, the obtained equation is projected on the level-($l-1$) cells, within the $\boldsymbol{f}_i^l$ definition domain. Hence, the following linear system is obtained:

$$\left[Q_i^l\right]\left[f_i^l\right] = \left[q_i^l\right] \qquad i = 1, \ldots, N^l \qquad l = 1, \ldots, L \tag{6.29}$$

where

$$\left[f_i^l\right] = \left[f_{1,i}^l, f_{2,i}^l, \ldots, f_{N_i^{l-1},i}^l\right]^T, \tag{6.30}$$

$$\left[Q_i^l\right]_{m,n} = \langle p_m^{l-1}, \nabla_s \cdot \boldsymbol{f}_{\mu(n)}^{l-1} \rangle, \qquad m = 1, \ldots, M_i^{l-1}, \quad n = 1, \ldots, N_i^{l-1}, \tag{6.31}$$

and

$$\left[q_i^l\right]_m = \langle p_m^{l-1}, \nabla_s \cdot \boldsymbol{f}_i^l \rangle, \qquad m = 1, \ldots, M_i^{l-1}, \tag{6.32}$$

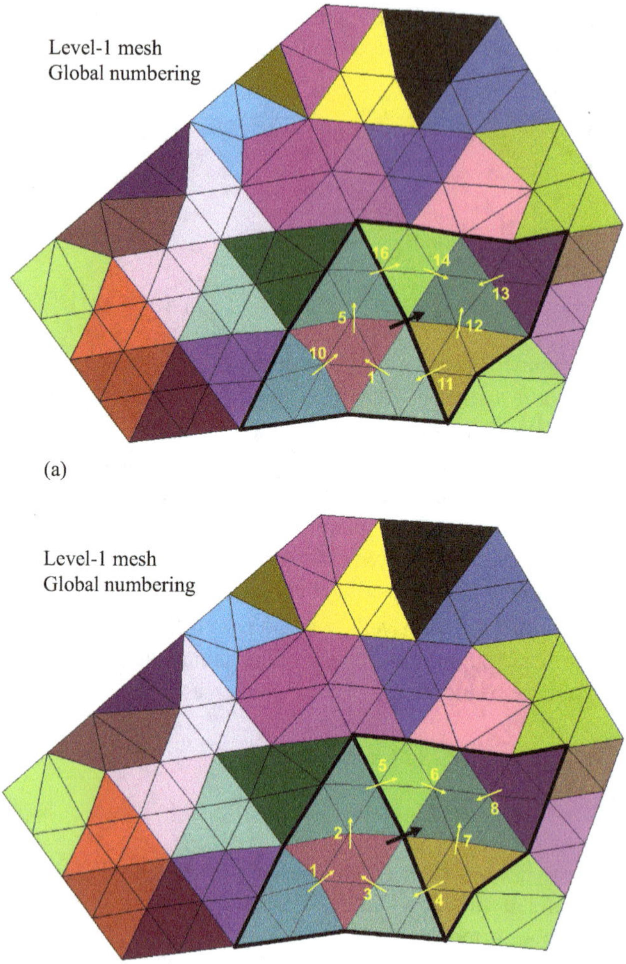

Figure 6.2 Example of level-1 mesh with highlighted the level-1 generalized functions that form a level-2 generalized function. The cells of the level-1 mesh correspond to the triangle groups with the same color. (a) Global numbering of the considered level-1 generalized functions; (b) local numbering of the considered level-1 generalized functions.

where M_i^{l-1} is the number of level-($l-1$) cells inside the definition domain of the considered $\boldsymbol{f}_i^l$ function, $C_i^{l,+} \bigcup C_i^{l,-}$, and p_m^{l-1} is a pulse function, equal to one inside the corresponding level-($l-1$) cell, C_m^{l-1}, and zero outside. In the example reported in Figure 6.2, $M_i^{l-1} = 8$. In the following, the matrix $\left[Q_i^l\right]$ is called "charge" matrix [5].

As shown in (6.31) and (6.32), the generalized basis functions are built from their surface divergence. The divergence of a RWG or MB-RWG basis function, divided by its common edge (6.18), or of the portion of a junction basis function defined on the surface is proportional to $\frac{1}{A}$, where A is the area of the considered triangle within the function domain. Instead, the divergence of a PWL basis function or of the portion of a junction basis function defined on the wire is proportional to $\frac{1}{h}$, where h is the length of the considered segment within its domain. If we consider a generalized function $\boldsymbol{f}_i^l$, the cells within its definition domain can cover, at the same time, both surfaces and wires; hence, we need to generalize the definition of area, applicable to both surfaces and wires. The key step is to associate an area to segments equal to $(2\,\pi\,a\,h)$, where a is the segment radius and h its length. With this choice, the area A_j^l of a generic level-l cell C_j^l can be defined as

$$A_j^l = \sum_{m=1}^{M_j^{l,0}} A_m^0 \tag{6.33}$$

where $M_j^{l,0}$ is the total number of level-0 cells inside the considered cell, and A_m^0 is the area of corresponding mth level-0 cell, C_m^0. If C_m^0 is a triangle, A_m^0 is equal to the triangle area; if, instead, C_m^0 is a segment, A_m^0 is equal to the segment length multiplied by $2\,\pi$ its radius, as described earlier. Once defined the generalized cells' areas (6.33), the divergence of a generalized function $\boldsymbol{f}_i^l$, with $l > 0$, is equal to

$$\nabla_s \cdot \boldsymbol{f}_i^l\,(\boldsymbol{r}) = \begin{cases} \pm\frac{1}{A_i^{l,\pm}} & \boldsymbol{r} \in C_i^{l,\pm} \\ 0 & \text{otherwise} \end{cases} \tag{6.34}$$

where $A_i^{l,\pm}$ is the area of the corresponding generalized cell $C_i^{l,\pm}$ of its domain. For $l = 0$, we have

$$\nabla_s \cdot \boldsymbol{f}_i^0\,(\boldsymbol{r}) = \begin{cases} +\frac{K_{i,j}}{A_{i,j}^{0,+}} & \boldsymbol{r} \in C_{i,j}^{0,+} \\ -\frac{1}{A_{i,k}^{0,-}} & \boldsymbol{r} \in C_{i,k}^{0,-} \\ 0 & \text{otherwise} \end{cases} \tag{6.35}$$

where if $\boldsymbol{f}_i^0$ is a RWG or a PWL function, $j = 1$, $k = 1$, $K_{i,1} = 1$, and $A_{i,1}^{0,\pm}$ are the areas of the corresponding $C_{i,1}^{0,\pm}$ cells, triangles, or segments, respectively. If $\boldsymbol{f}_i^0$ is a MB-RWG function, $j = 1$ and $K_{i,1} = 1$, as in the case of RWG and PWL basis functions, but $k = 1, \ldots, M_i^{0,-}$, where $M_i^{0,-}$ is the number of negative triangles $C_{i,k}^{0,-}$, with area $A_{i,k}^{0,-}$, with an edge in common with the corresponding positive triangle $C_{i,1}^{0,+}$ (6.6). If, instead, $\boldsymbol{f}_i^0$ is a junction basis function, $k = 1, j = 1, \ldots, M_i^{0,+}$, where $M_i^{0,+}$ is the number of triangles $C_{i,j}^{0,+}$, with area $A_{i,j}^{0,+}$, around the considered junction vertex, and $K_{i,j}$ is equal to the ratio between the corresponding junction vertex angle and the sum of all the junction vertex angles at the considered junction (6.8).

Substituting (6.34) into (6.31), for $l > 1$, we obtain

$$\left[Q_i^l\right]_{m,n} = \begin{cases} \pm 1 & \text{if } C_m^{l-1} \equiv C_n^{l-1,\pm} \\ 0 & \text{otherwise} \end{cases} \tag{6.36}$$

For $l = 1$, substituting (6.35) into (6.31), we have

$$\left[Q_i^1\right]_{m,n} = \begin{cases} +K_{n,j} & \text{if } C_m^0 \equiv C_{n,j}^{0,+} \\ -1 & \text{if } C_m^0 \equiv C_{n,k}^{0,-} \\ 0 & \text{otherwise} \end{cases} \tag{6.37}$$

where $j = 1$, $k = 1$, and $K_{n,1} = 1$, if the function $\boldsymbol{f}_n^0$ is a RWG or a PWL function. In the case of a MB-RWG function, $j = 1$, $K_{n,1} = 1$ and $k = 1, \ldots, M_n^{0,-}$. If, instead, $\boldsymbol{f}_n^0$ is a junction basis function, $k = 1, j = 1, \ldots, M_n^{0,+}$, and $K_{n,j}$ is equal to the ratio between the corresponding junction vertex angle and the sum of all the junction vertex angles at the considered junction, as in (6.35). Finally, substituting (6.34) into (6.32), we obtain

$$\left[q_i^l\right]_m = \pm \frac{A_m^{l-1}}{A_i^{l,\pm}} \qquad \text{with } l > 0. \tag{6.38}$$

As shown in [1], the charge matrix $\left[Q_i^l\right]$ is rank deficient, so we define a modified charge matrix $\widetilde{\left[Q_i^l\right]}$, obtained by deleting an arbitrary row of $\left[Q_i^l\right]$ [5], and a modified right-hand side (RHS) vector $\widetilde{\left[q_i^l\right]}$, obtained by deleting the corresponding row in $\left[q_i^l\right]$. The $\left(N_i^{l-1} - M_i^{l-1} + 1\right)$ missing equations of (6.29) correspond to the solenoidal null space of $\left[Q_i^l\right]$ and can be found through the singular value decomposition (SVD) of that matrix. As a result, the final system for the coefficients $\left[f_i^l\right]$ of the generalized function $\boldsymbol{f}_i^l$ is obtained by adding the condition that solenoidal currents do not contribute to the considered generalized function, i.e.,

$$\begin{bmatrix} \widetilde{\left[Q_i^l\right]} \\ \left[U_i^l\right]^T \end{bmatrix} \left[f_i^l\right] = \begin{bmatrix} \widetilde{\left[q_i^l\right]} \\ [0] \end{bmatrix} \tag{6.39}$$

where $\left[U_i^l\right]$ is the set of right singular vectors in the null space of $\left[Q_i^l\right]$, and $[0]$ is a null vector of dimension $\left(N_i^{l-1} - M_i^{l-1} + 1\right)$. Solving the system (6.39), the coefficients $f_{n,i}^l$ in (6.26) are found.

A level-l generalized function $\boldsymbol{f}_i^l$ can be expressed also as a linear combination of the level-0 basis functions $\boldsymbol{f}_n^0$,

$$\boldsymbol{f}_i^l(\boldsymbol{r}) = \sum_{n=1}^{N} f_{n,i}^{l,0}\, \boldsymbol{f}_n^0(\boldsymbol{r}) \qquad l = 1, \ldots, L \tag{6.40}$$

where $\boldsymbol{f}_n^0$, through (6.18), are immediately related to the original RWG, MB-RWG, PWL, and junction basis functions. To derive the coefficients

$$\left[f_i^{l,0}\right] = \left[f_{1,i}^{l,0}, f_{2,i}^{l,0}, \ldots, f_{N,i}^{l,0}\right]^T, \tag{6.41}$$

we define an $N^{l-1} \times 1$ vector $[\check{f}_i^l]$ adding appropriate zeros in $[f_i^l]$ (6.30), to refer the local coefficients in (6.30) to the global numbering of the level-($l-1$) functions. Applying (6.26) recursively, and considering the previous definitions, we obtain the coefficients $f_{n,i}^{l,0}$ in (6.40):

$$\left[f_i^{l,0}\right] = [f^1] \cdot [f^2] \cdot \ldots \cdot [f^{l-1}] \cdot [\check{f}_i^l], \tag{6.42}$$

where

$$[f^j] = \left[[\check{f}_1^j], [\check{f}_2^j], \ldots, [\check{f}_{N^j}^j]\right] \qquad j = 1, \ldots, (l-1). \tag{6.43}$$

To better describe the matrix–vector product in (6.42), Figure 6.3 is a graphical representation of it. The correspondence between the local coefficients in $[f_i^l]$ and the global numbering in $[\check{f}_i^l]$ is shown in Figure 6.4 for the example reported in Figure 6.2.

Finally to give a clear idea of the shape of the generalized basis functions obtained following the previous procedure, Figures 6.5–6.7 show examples of generalized functions, defined surfaces, and wires, respectively.

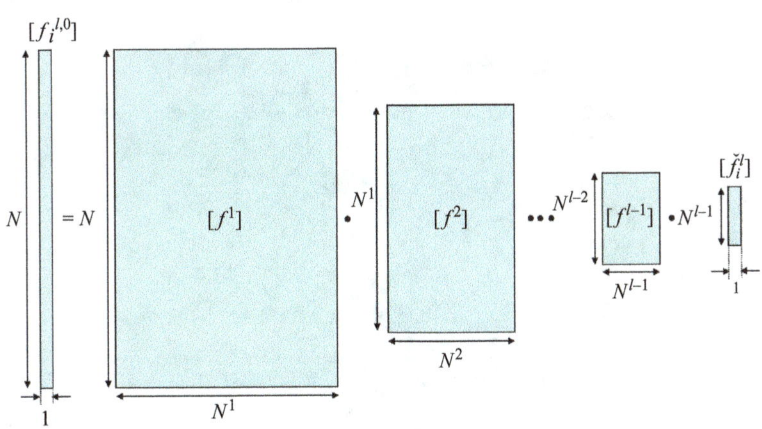

Figure 6.3 Graphical representation of the matrix–vector product in (6.42) to obtain the vector $\left[f_i^{l,0}\right]$

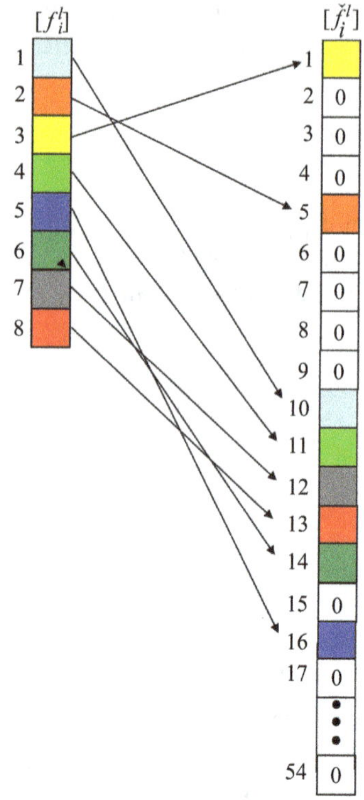

Figure 6.4 *Correspondence between the local coefficients in* $[f_i^l]$ *and the global numbering in* $[\check{f}_i^l]$ *for the example in Figure 6.2*

Figure 6.5 *Example of a level-2 generalized function defined on a pair of surface non-simplex cells*

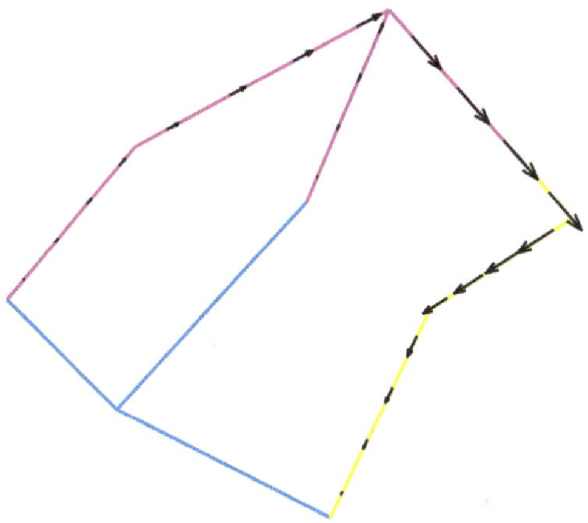

Figure 6.6 Example of a level-2 generalized functions defined on wires; the boundaries between the three level-2 cells are highlighted with different colors

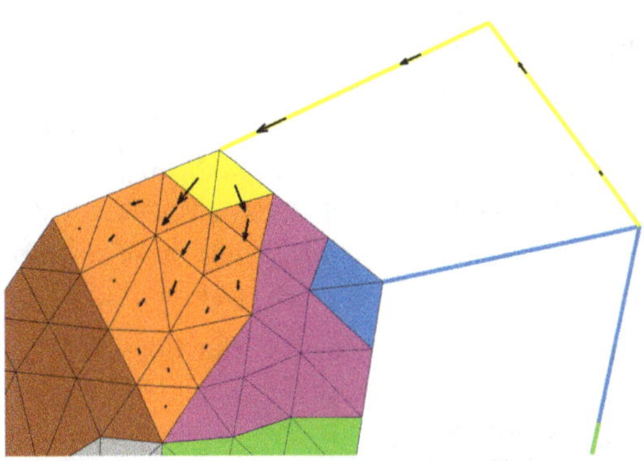

Figure 6.7 Example of a level-2 generalized functions defined on surfaces and wires with junctions. The boundaries between the level-2 cells are highlighted with different colors.

6.2.2 Multi-resolution basis functions

The generalized basis functions (see Section 6.2.1) will now be employed to generate the MR (hierarchical) basis functions.

As mentioned at the beginning of the chapter, the proposed MR basis functions divide the current into a solenoidal part and a non-solenoidal reminder, and are organized into levels, where each level-l group of functions is defined on the corresponding mesh level. A generic level-l MR function $\boldsymbol{w}_i^l$ can be written as a linear combination of the same level generalized functions (see Section 6.2.1):

$$\boldsymbol{w}_i^l(\boldsymbol{r}) = \sum_{k=1}^{K_i^l} T_{k,i}^l \boldsymbol{f}_{\mu(k)}^l(\boldsymbol{r}) \qquad l = 0, \ldots, L \qquad i = 1, \ldots, N_w^l, \tag{6.44}$$

where K_i^l is the number of $\boldsymbol{f}_{\mu(k)}^l$ functions defined inside the considered level-l MR function support, and N_w^l is the number of level-l MR functions. As in (6.26), also here a local index k is used on the support of the considered $\boldsymbol{w}_i^l$ function, and a global index $\mu(k)$ to uniquely identify a level-l generalized basis function.

To generate the proposed MR basis functions set, the first step is to build the charge matrices, as defined in (6.31), corresponding to each group of level-l cells inside the *same* level-$(l+1)$ jth cell, C_j^{l+1}:

$$\left[Q_j^l\right]_{h,k} = \langle p_h^l, \nabla_s \cdot \boldsymbol{f}_{\mu(k)}^l \rangle, \qquad h = 1, \ldots, H_j^l, \quad k = 1, \ldots, K_j^l, \tag{6.45}$$

where H_j^l and K_j^l are the number of level-l cells and generalized functions, respectively, inside C_j^{l+1}. Note that in (6.45) $\nabla_s \cdot \boldsymbol{f}_{\mu(k)}^l$ is evaluated as defined in (6.34) and (6.35). Figure 6.8 reports, on the left, an example of C_j^{l+1} cell that contains four level-l cells and three level-l generalized functions: so in this case, the dimension of the corresponding charge matrix $\left[Q_j^l\right]$ is 4×3.

The corresponding MR level-l basis functions are obtained directly from the SVD of the charge matrices $\left[Q_j^l\right]$. Each set of coefficients $T_{k,i}^l$ (6.44), with $k = 1, \ldots, K_i^l$, corresponds to a right singular vector of the matrix (6.45). Note that in this operation, we employ *all* singular vectors; those associated with non-zero singular values will generate non-solenoidal functions, while those corresponding to the null-space generate the solenoidal ones. From each $\left[Q_j^l\right]$ matrix, K_j^l level-l MR functions are generated, so the total number of actual $\boldsymbol{w}_i^l$ functions is $\sum_{j=1}^{M^{l+1}} K_j^l$, where M^{l+1} is the total number of level-$(l+1)$ cells. In the example of Figure 6.8, the generated MR functions are reported on the right: in this case, only non-solenoidal functions are obtained through the SVD of the corresponding charge matrix, because there are not loop paths internal to the considered C_j^{l+1} cell; this is automatically detected by the SVD of the charge matrix that in this case does not find any zero singular value.

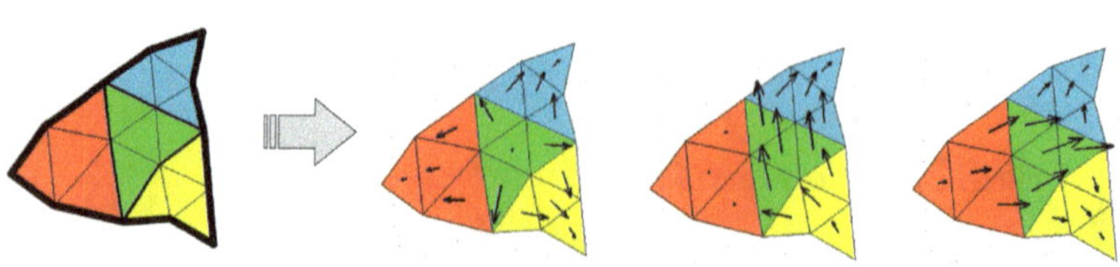

Figure 6.8 Left: example of C_j^{l+1} cell containing 4 level-l cells. Right: corresponding MR functions.

The obtained set of basis functions $\boldsymbol{w}_i^l$ is not yet complete in its solenoidal part, because $\sum_{j=1}^{M^{l+1}} K_j^l < N_w^l$ (this is easily evident in the case of a hierarchic, diadic mesh, as detailed [19]). To complete the MR level-l basis functions, the divergence-free functions, with support across each *pair* of adjacent level-($l+1$) cells, must be added.

By employing the SVD on the charge matrix of the cells included in every pair of neighboring level-($l+1$) cells, we choose the singular vectors $\left[U_n^l\right]$ from the null-space. Nevertheless, this null-space encompasses the subspace formed by the solenoidal functions produced independently for each cell before. Consequently, the obtained singular vectors $\left[U_n^l\right]$ (through the aforementioned SVD procedure) are subsequently rendered orthogonal to the level-l solenoidal functions, which are defined within each individual cell separately $\left[T_s^l\right]$ [16]:

$$\left[\widetilde{U_n^l}\right] = \left[U_n^l\right] - \sum_{s=1}^{S_n^l} c_{s,n}^l \left[U_n^l\right] \tag{6.46}$$

$$c_{s,n}^l = \left[T_s^l\right]^T \cdot \left[U_n^l\right] \tag{6.47}$$

where S_n^l is the number of considered solenoidal functions (generated previously). In the end, the set of vectors $\left[\widetilde{U_n^l}\right]$ is rank deficient, thereby necessitating the extraction of linearly independent vectors associated with the solenoidal functions spanning the pair of neighboring ($l+1$) cells under consideration. This extraction is achieved by employing SVD decomposition on $\left[\widetilde{U_n^l}\right]$ and choosing the right singular vectors corresponding to non-zero singular values. To clarify the previous procedure for the generation of the MR solenoidal basis, we refer to the example in Figure 6.9. The two adjacent level-($l+1$) cells considered in this example are indicated in pink and in light blue, respectively, with thick yellow boundaries. The level-l cells inside each level-($l+1$) cell are indicated with thick yellow boundaries, and the inner nodes are highlighted with yellow spots. Figure 6.9(a)–(c) shows the level-l MR solenoidal basis functions inside the pink level-($l+1$) cell obtained through the SVD of the corresponding 6×8 charge matrix, and Figure 6.9(d) shows the MR solenoidal basis function inside the light blue level-($l+1$) cell. Finally, Figure 6.9(e) reports the level-l MR solenoidal basis function defined across the two level-($l+1$) considered cells, that is generated with the previous procedure, and it is linear independent from the other MR functions defined inside the orange and light blue cells.

A generic MR function $\boldsymbol{w}_i^l$ of level-l is expressed as a linear combination of the generalized basis functions of the same level, as in (6.44); in turn, all generalized functions of all levels can be expressed as a linear combinations of the original conventional functions (of level-0), as clear from (6.40). As a result, a generic level-l MR function can be also described as a linear combination of the level-0 basis functions,

$$\boldsymbol{w}_i^l(\boldsymbol{r}) = \sum_{n=1}^{N} T_{n,i}^{l,0} \boldsymbol{f}_n^0(\boldsymbol{r}) \qquad l = 0, \ldots, L \qquad i = 1, \ldots, N_w^l \tag{6.48}$$

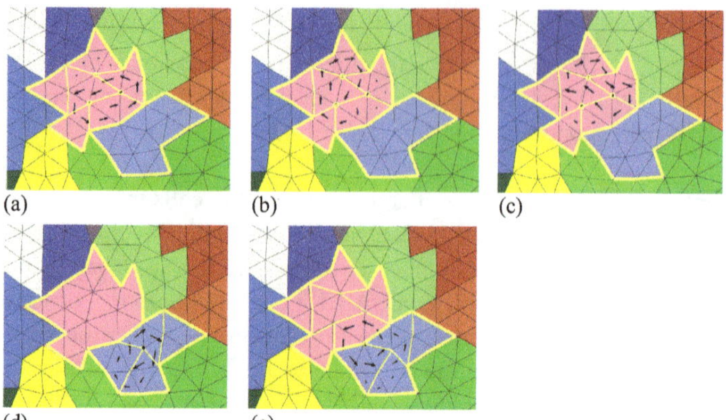

Figure 6.9 (a)–(c) MR level-l solenoidal functions inside the orange level-($l+1$) cell; (d) MR level-l solenoidal function inside the light blue level-($l+1$) cell; (e) MR level-l solenoidal function across the two considered level-($l+1$) cells

and the coefficients $T_{n,i}^{l,0}$ represent the basis change from conventional to MR basis functions. This basis change transformation allows to cast the MoM in the MR basis in a direct manner.

The explicit expression of the change-of-basis transformation can be obtained by substituting equation (6.40) into (6.44), whence it can be easily derived that

$$\left[T_i^{l,0}\right] = \left[T_{1,i}^{l,0}, T_{2,i}^{l,0}, \ldots, T_{N,i}^{l,0}\right]^T \tag{6.49}$$

are equal to

$$\left[T_i^{l,0}\right] = \left[f^1\right] \cdot \left[f^2\right] \cdot \ldots \cdot \left[f^l\right] \cdot \left[\check{T}_i^l\right], \tag{6.50}$$

where $\left[\check{T}_i^l\right]$ is an $N^l \times 1$ column vector that collects the coefficients of the same MR function $\boldsymbol{w}_i^l$ described as a linear combination of the level-l generalized functions (6.44). The row index of the coefficients in $\left[\check{T}_i^l\right]$ corresponds to the global numbering of the level-l generalized functions.

Finally, we can collect the coefficients of all the N_w^l level-l MR functions as

$$\left[T^l\right] = \left[\left[T_1^{l,0}\right], \left[T_2^{l,0}\right], \ldots, \left[T_{N_w^l}^{l,0}\right]\right] \tag{6.51}$$

which allows to obtain the matrix $[T]$ that accounts for the global *change-of-basis* from conventional to MR bases as

$$[T] = \left[\left[T^0\right], \left[T^1\right], \ldots, \left[T^L\right]\right], \tag{6.52}$$

The change-of-basis matrix $[T]$ is a highly sparse matrix of dimension $N \times N$, and each column corresponds to an MR function, constructed as a linear combination of the original basis functions of the problem.

In Figure 6.10, for a simple triangular mesh, the number of generated hierarchical functions is reported level by level, together with the corresponding level mesh. It is

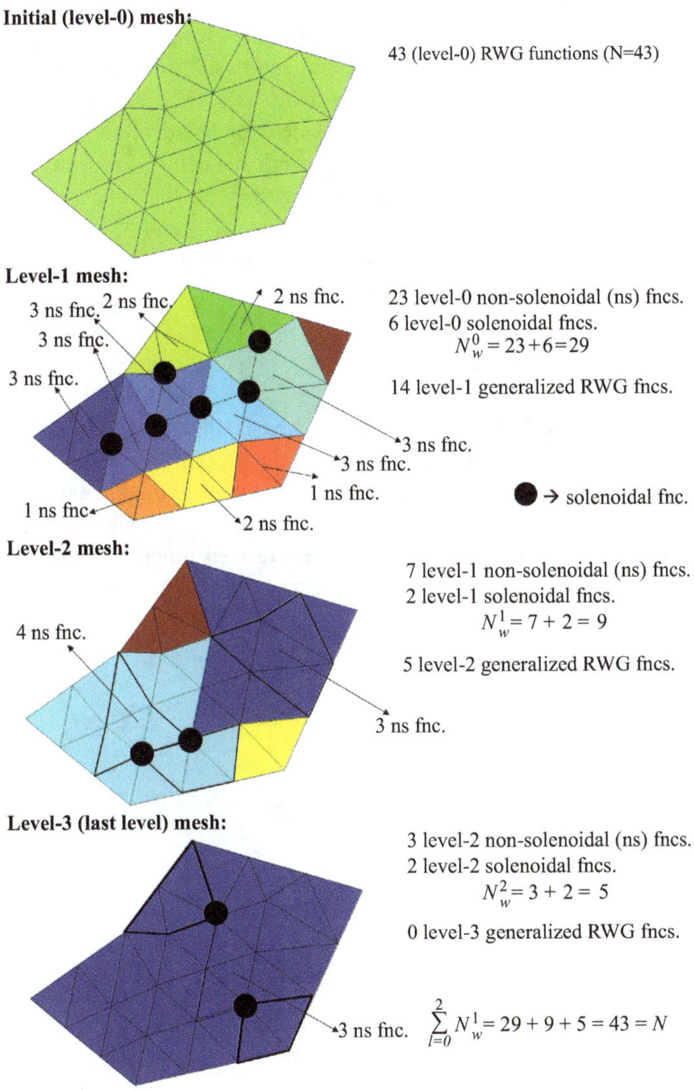

Figure 6.10 Analysis of the number of generated hierarchical functions and generalized functions

evident that the total number of hierarchical functions is equal to the total number of RWG functions defined on the initial (standard) mesh.

6.2.3 PEC ground plane handling

The generation of the generalized basis functions (Section 6.2.1) and of the MR basis functions (Section 6.2.2) can be easily extended to case of a structure connected to an infinite PEC ground plane. The only difference, with respect to the previously described scheme, involves the cells connected to the ground plane.

About the generation of the (half) generalized basis functions, the change is in the definition of the corresponding charge matrices (6.31). On each level-l mesh, for each $C_i^{l,+}$ cell connected to the ground plane, we considered an image cell $C_i^{l,-}$, equal to $C_i^{l,+}$, where the same functions $\boldsymbol{f}_{\mu(n)}^{l-1}$, inside the cell $C_i^{l,+}$, are defined, simply with opposite direction. Hence, as shown in Figure 6.11, the corresponding charge matrix $[Q_i^l]$ has dimension $(2M_i^{l-1}) \times N_i^{(l-1)}$, where M_i^{l-1} and $N_i^{(l-1)}$ are the number of level-$(l-1)$ cells and level-$(l-1)$ generalized functions, respectively, inside the $C_i^{l,+}$, and a generic element of $[Q_i^l]$ is equal to

$$[Q_i^l]_{(m+M_i^{l-1}),n} = -[Q_i^l]_{m,n} \qquad m = 1,\ldots,M_i^{l-1} \tag{6.53}$$

Analogously, in the construction of the MR basis, the difference, with respect to the scheme described in Section 6.2.2, is only in the definition of the charge matrices for the generation of the MR solenoidal function across two adjacent cells. For the cells attached to the ground plane, the corresponding charge matrices are generated exactly as previously described for the half-generalized functions, and as shown in Figure 6.11.

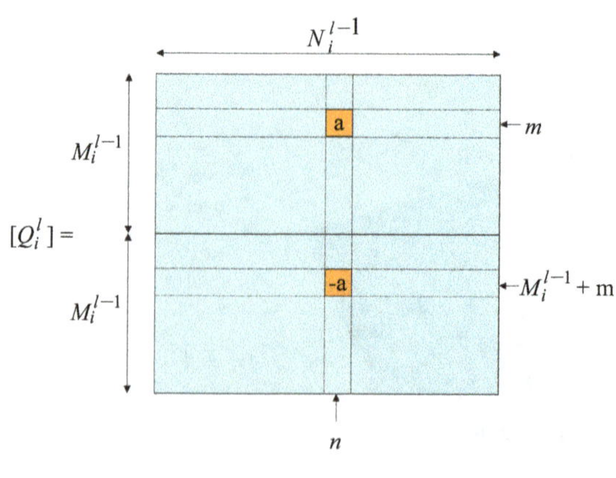

Figure 6.11 Level-l charge matrix corresponding to a level-l cell connected to the ground plane

6.2.4 Basis for electrical sizes beyond the resonance region

The fundamental concept, for effectively analyzing structures that have electrical dimensions exceeding the resonant region, involves employing the hierarchical decomposition, as explained in Section 6.2.2, on the densely discretized surface portions. Subsequently, the generalized functions (described in Section 6.2.1) are utilized within the macro-cells, which have dimensions comparable to the operating wavelength λ. By implementing this approach, the dense mesh can be efficiently handled using the hierarchical functions, while the generalized functions perform effectively on the coarse cells [21,22].

The algorithm for grouping cells, which is extensively explained in the subsequent Section 6.3, is utilized on the original triangular mesh. A stopping criterion is employed based on the size of the resulting macro-cells. Consequently, if a cell reaches the maximum allowable size, its aggregation is prohibited, and it remains unchanged in all subsequent meshes of the subsequent levels. The final level-$(L+1)$ mesh is attained when no more cells can be connected, and it corresponds to the coarsest level mesh.

The generation of the basis occurs in two stages. Initially, on the meshes of all levels l except for the coarsest one, $l \leq L$, the solenoidal and non-solenoidal subspaces are generated precisely as outlined in Section 6.2.2. However, the generalized functions generated on the level-$(L+1)$ mesh are directly the basis functions for the coarsest level. So the change-of-basis matrix $[T]$ is divided in two parts:

$$[T] = \left[[T_{MR}], \left[f^{L+1,0}\right]\right] \tag{6.54}$$

where $[T_{MR}]$ corresponds to (6.52) and collects the set of hierarchical functions defined on the detail level-l meshes with $l \leq L$, and $\left[f^{L+1,0}\right]$ is the set of generalized functions defined on the level-$(L+1)$ mesh (6.41). We observe that the total number of hierarchical function $N_{MR} = \sum_{l=0}^{L} N_w^l$ plus the number of level-$(L+1)$ generalized functions N^{L+1} is equal to the total number of functions defined on the initial mesh, that is:

$$N = N_{MR} + N^{L+1}. \tag{6.55}$$

6.2.5 Algorithm flow chart and computational complexity

To summarize the scheme described in the previous sections (Sections 6.2.1 and 6.2.2), Figure 6.12 reports the flow chart of the proposed algorithm for the generation of the MR basis.

The only requested input is the mesh of the discretized structure, that includes vertices, cells (segments and triangles), and functions (RWG, MB-RWG, PWL, and junction basis functions). It is important to emphasize that the algorithm proposed for generating hierarchical functions is purely algebraic and does not rely on any

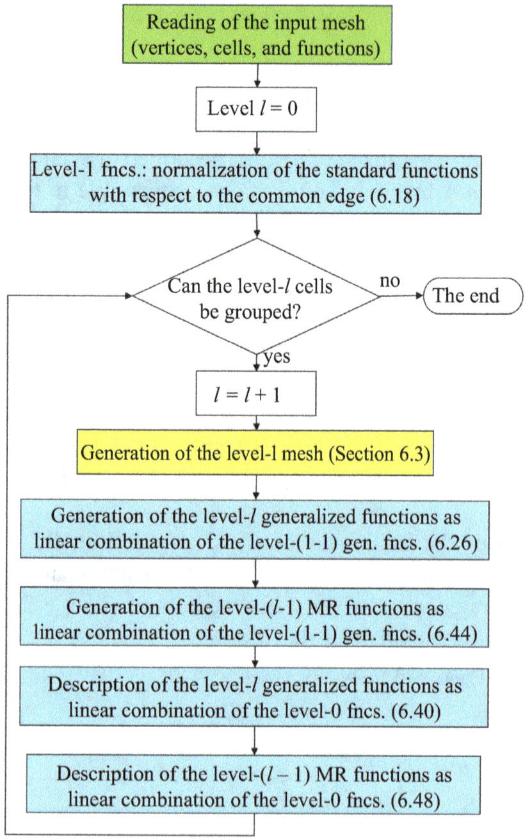

Figure 6.12 Flow chart scheme of the MR basis generation scheme

knowledge of the structural topology, such as the presence of holes or handles. Furthermore, its complexity remains unaffected by the topological complexity of the structure.

The algorithm described in the preceding sections for generating the proposed MR functions exhibits a complexity of $O\,(N \log N)$, where N represents the total number of unknowns and $\log N$ corresponds to the number of levels. This complexity is evident in the sparsity of the change-of-basis matrix, $[T]$, and in the CPU time required to generate it, both of which are proportional to $N \log N$, as demonstrated in Figure 6.13 for the case of a sphere. The $N \log N$ complexity is achieved by performing local operations at each level-l mesh. It is worth noting that the dimension of the involved charge matrices (6.31), (6.45), where an SVD is employed, is consistently small and *independent* of the total number of unknowns N.

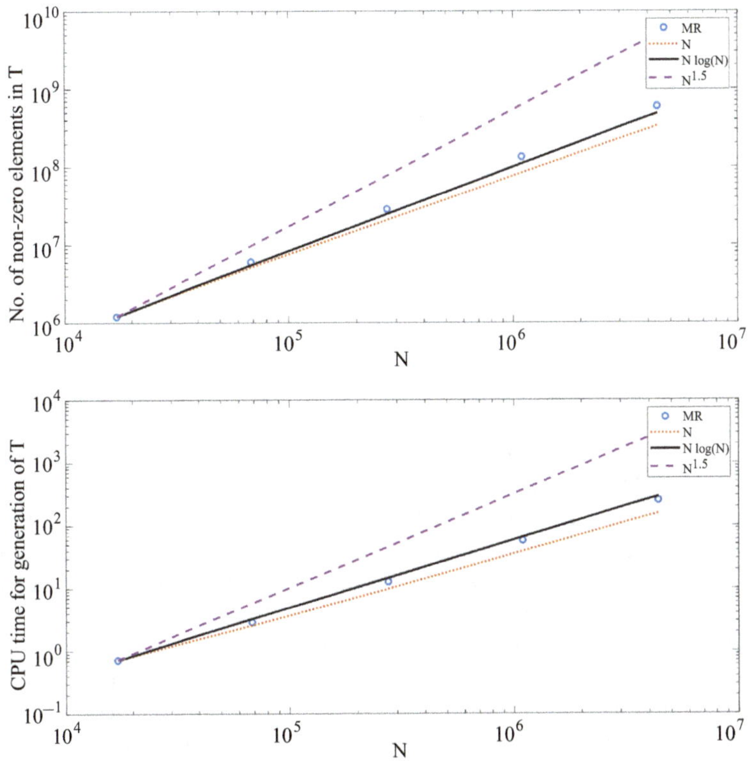

Figure 6.13 Basis-change matrix [T]. Top: no. of non-zero elements versus number of unknowns N; bottom: generation CPU time (seconds) versus no. of unknowns N. Case of a sphere. CPU time relevant to 2xAMD EPYC 7H12@2.6 GHz (using one core only) and 2 TB of RAM memory.

6.3 Generation of a hierarchical family of meshes

6.3.1 Cells grouping strategy

The proposed basis, described in detail in Section 6.2, incorporates a hierarchical structure. Each level within this structure corresponds to specific functions and is associated with a corresponding mesh. Consequently, it is necessary to generate a series of hierarchical meshes, starting with the initial (standard) mesh denoted as M^0 and referred to as the level-0 mesh. The level-0 mesh serves as the starting point, wherein the surfaces are discretized using planar triangles, and the wires are represented by line segments. Additionally, we assume that each triangle can only be linked to a single segment. If more segments are connected to the same triangle, the initial mesh is modified subdividing the considered triangles into two triangles, as

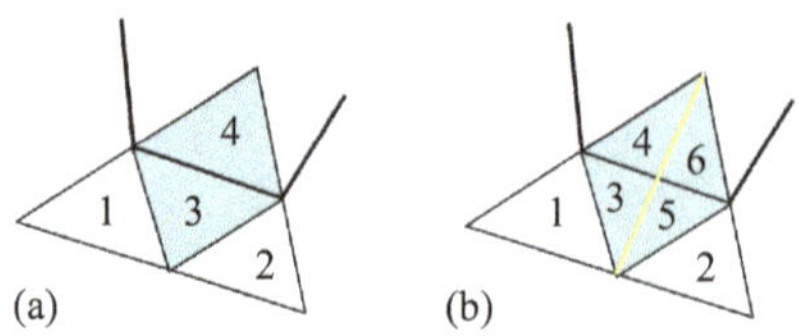

Figure 6.14 (a) Example of a mesh with two segments connected to the same triangle; (b) triangle splitting to have no more than one segment connected to the same triangle

shown in Figure 6.14(b). Moreover, the initial mesh can be also non-conformal with triangles whose edges are in common with two or more other triangles, as shown in Figure 6.1(c).

The meshes M^l for levels $l > 0$ consist of clusters of neighboring cells from the initial mesh M^0, known as "macro-cells." These macro-cells are typically non-simplex cells that can encompass surfaces and wires. The process begins with the level-0 mesh (initial mesh) and continues to construct subsequent levels, such as level-1 and beyond. The final level $(L + 1)$ is attained either when the corresponding $M^{(L+1)}$ mesh comprises a single macro-cell, formed by combining all level-0 cells, or when the level-$(L + 1)$ cells have reached the maximum allowed size.

The grouping algorithm, described in the following, to generate the different levels meshes has been studied to properly handle complex topologies and markedly non-uniform meshes. The initial stage involves an arbitrary standard mesh, M^0, which lacks any constraints on mesh properties or structural topology. This mesh is typically employed for analyzing the structure with the desired level of accuracy. The objective remains focused on achieving an analogous result to the dyadic subdivision of triangular cells as described in [19,20]. However, since we begin with an existing mesh, the process of grouping cells must proceed in a reverse order. The underlying concept of the proposed algorithm is to allow the cells at level l to "grow" on the surfaces and wires that comprise the structure, transitioning them into cells at level $(l + 1)$.

In the following, the algorithm to obtain the (generalized) mesh of level $(l + 1)$ by proper aggregation of the (macro) cells of level l is described. The procedure is the same for all levels, except for the cases of a level-0 triangle connected to a segment (a triangle where a junction basis function is defined, called in the following junction triangle) and a level-0 triangle connected to more than one triangle through the same edge (non-conformal mesh where a MB-RWG basis is defined).

When dealing with a junction basis function, it is essential to combine all the junction triangles that are connected to the same vertex into a single macro-cell. Similarly, the segment associated with the junction is merged with its connected segments, as depicted in the example illustrated in Figure 6.15. This merging process, where the triangles surrounding the junction-vertex form a level-1 cell, proves beneficial in generating basis functions defined on the level-1 mesh (as described in Section 6.2.1) where, consequently, the constructed functions are consistently defined only on pairs of adjacent cells.

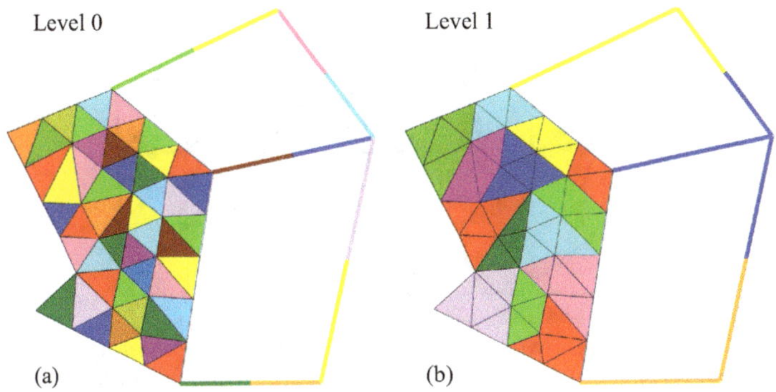

*Figure 6.15 Example of cells grouping; left: level-*0 *mesh, right: level-*1 *mesh*

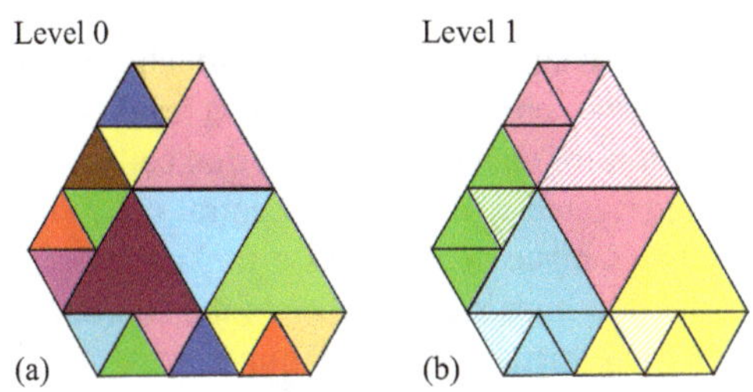

*Figure 6.16 Example of cells grouping; left: level-*0 *mesh, right: level-*1 *mesh*

In the case of a non-conformal triangular mesh, it is necessary to merge all negative triangles associated with the same MB-RWG basis function into a level-1 macro-cell. This ensures that each level-1 generalized basis function can be expressed as a linear combination of complete level-0 RWG and/or MB-RWG functions. Additionally, all triangles with a vertex that belongs to the internal nodes of the corresponding MB-RWG are also grouped within the same level-1 macro-cell. These constraints guarantee charge conservation at each level. Figure 6.16 illustrates an example of cell grouping from level-0 to level-1 in the context of a non-conformal mesh. All the triangles with the same color of Figure 6.16(b) correspond to the same level-1 macro-cell; the cell filled with a downward diagonal lines pattern is the "central" cell of the grouping as detailed below.

In the following, our terminology will not anymore distinguish between standard cells (triangles and segments) of the finest mesh $l = 0$ and the macro cells of the following (coarser) levels $l = 1, .., (L + 1)$. Moreover, we underline the importance of generating non-overlapping coverings for the meshed structure during the grouping process. This means that each cell at level-l should belong exclusively to a single cell at level-$(l + 1)$. Similarly, at any given level, each triangle or segment should be assigned to one and only one cell of that level.

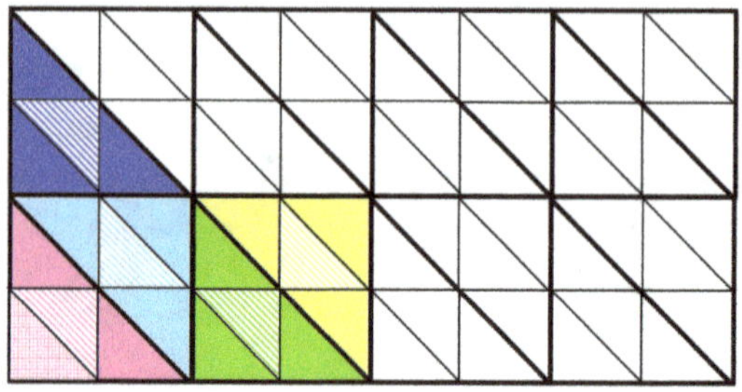

Figure 6.17 Example of a dyadic (hierarchic) mesh via a bottom-up grouping. Triangles with thick solid black edges (and the same inner color): following level cells; triangles marked with downward diagonal lines: central cells; triangle marked with a square grid: seed cell.

Essentially, the process of aggregation involves identifying and selecting "good" cells as central cells, around which the cells of the next level are formed. These central cells serve as the foundation for attaching neighboring cells. It is crucial to note that each cell can only be aggregated with a single central cell. Therefore, the order in which the aggregation takes place necessitates a ranking system. Furthermore, the chosen ranking is expected to impact the quality of the resulting grouping.

To establish quantitative criteria for selecting central cells, it is important to outline the qualitative objectives of the overall grouping algorithm. As mentioned earlier, the main objective is to replicate the behavior of a hierarchical mesh obtained through sequential dyadic splitting of the triangles in the initial coarse mesh (as shown in Figure 6.17). The following guidelines apply to all levels of the grouping.

1. *Maximize cell aggregation*: strive to maximize the aggregation of cells, similar to the dyadic and hierarchical mesh where all three cells adjacent to a central cell are aggregated.
2. *Avoid multiply-connected cells*: prevent the formation of multiply-connected cells, such as ring-like cells that enclose another cell.
3. *Avoid "sterile" cells*: minimize the presence of small cells among larger cells, as it would be challenging to merge them effectively in subsequent grouping steps.

While it is desirable to avoid "holes" in the mesh, the presence of multiply connected cells can be addressed by the proposed hierarchical basis.

When describing the grouping algorithm, it is beneficial to introduce specific terminology and metrics. Throughout the aggregation process, cells are categorized as either "used" if they have already been attached to another cell or "free" if they have not yet been attached. Cells that share a common generalized edge are referred to as "connected." The ranking of cells is based on the integer distance metric d_{ij} as follows:

- $d_{ij} = 1$ if the macro cells i and j are connected.
- $d_{ij} = 2$ if the macro cells i and j are not directly connected but share at least one cell, k, that is connected to both i and j.

The ranking of each free cell i incorporates two parameters:

- F_i^1 that is the number of neighboring free cells j at a distance $d_{ij} = 1$ (this parameter indicates how many cells can be aggregated with cell i if it is chosen as a central cell, aligning with the qualitative criterion 1 mentioned earlier);
- F_i^2 that is the number of used non-central cells j at a distance $d_{ij} = 2$ (F_i^2 is utilized to manage the occurrence of holes within the grouping process).

By considering F_i^1 and F_i^2, the algorithm can effectively control the aggregation of cells and prevent the formation of holes in the grouping.

6.3.2 Cells ranking and aggregation

The grouping algorithm for aggregating cells from level-l mesh to the coarser level-($l+1$) is based on the iterative "rank and aggregate" (RA) procedure, which can be summarized as follows:

1. rank all free cells based on a specified criterion (to be described later) and select the central cell;
2. aggregate all free cells that are connected to the selected central cell;
3. update the ranking, considering that the aggregation process alters the distribution of free cells.

The RA procedure is fully determined once the ranking algorithm is defined. However, special consideration should be given to the first application of RA when dealing with open or partially open structures. In such cases, it is crucial to prevent cells on the borders of the open parts from remaining "sterile" (isolated). Since open or partially open structures are common, the general algorithm that accounts for this is we outlined in the following. Within the algorithm, the ranking criteria only change between the first application of RA and subsequent iterations. To make this explicit, we introduce the application order index, p, to label each application. The initial application is denoted as $p = 0$.

6.3.2.1 RA initial application

To visually determine the starting point for aggregation in structures with open parts, it is preferable to begin from a corner cell, if available, or from a cell located on the border. To facilitate this, we assign a "periphery index" to each level-0 cell, taking into account the number of functions defined on it (see Table 6.1). The cell with the highest periphery index is designated as the "seed" cell. For the initial step ($p = 0$) of the grouping algorithm, the central cell is selected based on the following criteria:

- it should be connected to the seed cell;
- it should have the maximum F^1 value (indicating the highest number of connected cells).

Importantly, the grouping algorithm does not require prior knowledge of whether the structure is closed or open. In the case of an open structure, a random cell is chosen as the central cell. For subsequent levels ($l > 0$), the periphery index of each cell is

Table 6.1 Periphery index for a level-0 cell

	Periphery index	
No. of functions	**Triangle**	**Segment**
0	0	0
1	2	1
2	1	0
≥ 3	0	0

simply the sum of the periphery indices of the previous level-($l-1$) cells, thereby forming the periphery index of the corresponding level-l cell.

6.3.2.2 RA subsequent applications

The central cell is selected as the free cell with the highest F^2 value. If multiple cells have the same F^2 value, the one with the highest F^1 is chosen. The RA algorithm is then iteratively applied until there are no more free cells remaining.

The grouping algorithm is demonstrated using the simple mesh depicted in Figure 6.18. For a more comprehensive understanding of the entire process, a visual representation is provided in the complex example shown in Figure 6.19. Figure 6.18 displays the level-1 mesh (on the left), which results from grouping the initial triangular mesh, and the level-2 mesh (on the right), formed by aggregating cells from the level-1 mesh. In the level-1 mesh, cell number 29 is selected as the seed for generating the level-2 mesh. The first cell of the level-2 mesh (cell number 1) is created by aggregating the central cell number 22 with all the connected cells (i.e., numbered

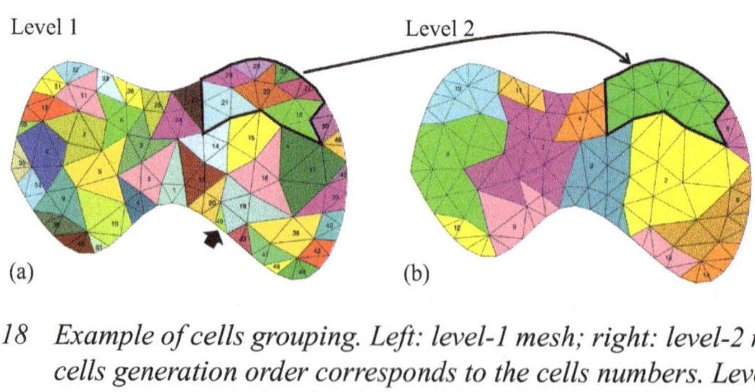

Figure 6.18 Example of cells grouping. Left: level-1 mesh; right: level-2 mesh. The cells generation order corresponds to the cells numbers. Level-1 cell no. 29: seed cell; level-1 cell no. 22: central cell of the level-2 cell no. 1.

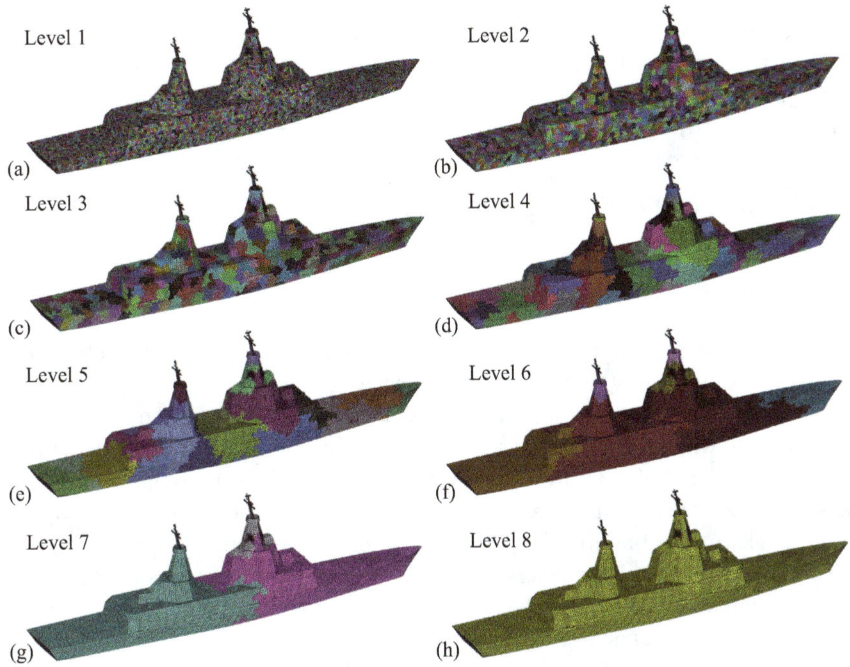

*Figure 6.19 Cell grouping algorithm; (a) level-*1 *mesh, (b) level-*2 *mesh, (c) level-*3 *mesh, (d) level-*4 *mesh, (e) level-*5 *mesh, (f) level-*6 *mesh, (g) level-*7 *mesh, and (h) level-*8 *mesh*

18, 27, 37, 29, 28, 21), including the seed cell. The second level-2 cell (cell number 2) is generated with central cell number 16, incorporating its connected cells. This pattern continues for the other level-2 cells, with the index in the figure denoting the order of generation.

6.3.3 Cells grouping refinement

After completing the grouping process at each level, a refinement step is performed on all the generated cells to address any imbalances caused by cells that are significantly smaller than their neighboring cells (i.e., the cells they are connected to). If such a situation arises, the smaller cell is merged with one of its larger connected cells. An illustration of this correction procedure is evident in the example depicted in Figure 6.18. It can be observed that cell number 49 in the level-1 mesh (highlighted with an arrow) is included within cell number 3 of the level-2 mesh, even though it is not directly connected to the central cell (cell number 13) at level 1. This inclusion is the result of a previous refinement operation conducted at the level-1 stage.

To check if a cell is "small" with respect to the neighboring cells, the area and the perimeter of the surface part, and the length of the wire part of each cell are evaluated. Then, the area, the perimeter, and the length of each cell are compared with the corresponding parameter of all the adjacent cells. If the cell under test has all these parameters x times lower than all the adjacent cells, it is a small cell and is merged with one of the adjacent cells. The x value is an user choice: a suggested range is $x \in [3,10]$.

6.3.4 Maximum cell size grouping limiting

The cell grouping algorithm, described in the previous sections, can be limited to a maximum cell size, with typical range $\Delta_L = \lambda/8 - \lambda/4$, where λ is the working wavelength.

A way to implement the grouping limiting is to enclose the meshed object in a prismatic box, uniformly partitioned into cartesian cells with cell side Δ_L, as shown in Figure 6.20. Hence, each mesh cell (triangle or segment) is associated to a cartesian cell, checking which cartesian cell contains the corresponding barycenter, and marking it with the label of the Cartesian cell. Finally, to avoid small groups of cells with the same label, the area, the perimeter, and the length of each group are compared with the maximum allowed value, Δ_L^2, $4\Delta_L$, and Δ_L, respectively. If the cell group under test has all these parameters x times lower than the corresponding limit, it is a small group and is merged with one of the adjacent groups. The x value is an user choice: a suggested range is $x \in [3,10]$.

Once all the cells of the initial mesh are labeled, the grouping algorithm is applied as described in the previous sections with only one difference: two (macro) cells can be grouped together only if they have the same label. The cell labeling is easily translated level by level considering that each generic macro cell has the same label of the cells that are forming it. The grouping algorithm stops when no more cells can be connected together: this corresponds to the last coarsest level $(L + 1)$.

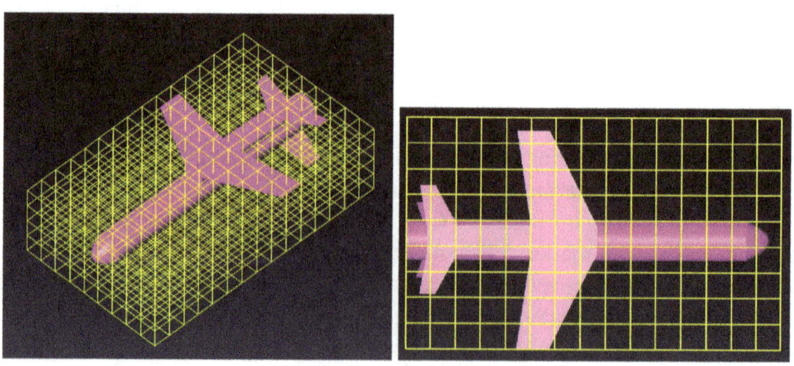

Figure 6.20 Example of a structure in the 3-D Cartesian grid. Left: 3-D view and right: 2-D view.

6.3.5 Computational complexity

The computational complexity of generating each level-l mesh is observed to be $O(M^l)$, where M^l represents the number of macro-cells at level-l. It is worth noting that the average number of connected macro-cells remains constant at each level and is not dependent on the total number of cells, M, in the initial mesh. Consequently, the number of levels $(L+1)$ is determined to be logarithmic with respect to M. Hence, the overall complexity of the proposed grouping algorithm is $O(M \log M)$.

However, to achieve the aforementioned complexity, it is crucial to handle the updating of the ranking with care. An implementation without proper consideration may require $O(M)$ operations at each step, potentially resulting in a total count approaching $O(M^2)$. However, this issue can be mitigated by recognizing that, at each application of the RA(p) algorithm, it is possible to determine which cells need updating. This leads to a more efficient update algorithm, outlined as follows.

Each cell, i, that necessitates an update, is located at a metric distance $d_{ij} = 1$ or $d_{ij} = 2$ from all the cells, j, forming the generated level-l cell (excluding the central cell, which does not impact the weights of other cells). The algorithm adjusts the weights, F_i^1 and F_i^2, of each considered cell, i, as follows:

- if $d_{ij} = 1$, F_i^1 is decremented by 1 (indicating that the cell, j, connected to cell i, is no longer free);
- if $d_{ij} = 2$, F_i^2 is incremented by 1 (representing the usage of the non-central cell, j, at a metric distance of 2 from cell i).

After each weight adjustment, the algorithm maintains a record of the best position, which corresponds to the maximum value of F^2. In cases where multiple cells have the same F^2 value, the algorithm selects the maximum F^1 among them.

6.4 Application to MoM

In the following sections, we describe how the generated change-of-basis matrix $[T]$ (6.52) can be applied in the solution of the MoM linear system. The memory allocation of the matrix $[T]$ is described in Section 6.4.1; then, the direct solution (inversion of the MoM matrix) is considered in Section 6.4.2, while its insertion in an iterative solver is investigated in Section 6.4.3. Section 6.4.4 describes how the MR preconditioner can be efficiently applied to electrically large multi-scale structures, and, finally, Section 6.4.5 details on the evaluation of the MoM system matrix at very low frequencies.

6.4.1 Change-of-basis matrix memory allocation

As derived in Section 6.2, the change-of-basis matrix $[T]$ (6.52) is a $N \times N$ highly sparse matrix, where N is the total number of unknowns of the considered problem. An efficient memory allocation of the $[T]$ matrix is to store only the non-zeros elements into a compress linear array, and, then, provide some number of auxiliary arrays to describe the locations of the non-zeros in the original matrix. The compression of the non-zeros of the sparse matrix $[T]$ into a linear array is done by walking down each column (column major format) and writing the non-zero elements to a linear array in the order that they appear in the walk. The proposed storage format for the $[T]$ matrix consists of three arrays, which are indicated with $[T_{\mathrm{value}}]$, $[T_{\mathrm{row}}]$, and $[T_{\mathrm{col}}]$.

$[T_{\mathrm{value}}]$ is a complex array containing the non-zero entries of $[T]$ that are mapped into $[T_{\mathrm{value}}]$ using the column major storage mapping described earlier. $[T_{\mathrm{row}}]$ is an integer array where each element contains the row index in $[T]$ of the corresponding element in $[T_{\mathrm{value}}]$. Finally, $[T_{\mathrm{col}}]$ is an integer array that gives the index in the $[T_{\mathrm{value}}]$ and $[T_{\mathrm{row}}]$ arrays that contain the first non-zero element in each column of $[T]$. The length of the $[T_{\mathrm{value}}]$ and $[T_{\mathrm{row}}]$ arrays is equal to the number of non-zeros in $[T]$. Since the $[T_{\mathrm{col}}]$ array gives in $[T_{\mathrm{value}}]$ and $[T_{\mathrm{row}}]$, the location of the first non-zero for each column of $[T]$ and the non-zeros are stored consecutively, then we would like to be able to compute the number of non-zeros in the ith column as the difference of $[T_{\mathrm{col}}]_i$ and $[T_{\mathrm{col}}]_{i+1}$. In order to have this relationship hold for the last column of T, we need to add an entry (dummy entry) to the end of the $[T_{\mathrm{col}}]$ array whose value is equal to the number of non-zeros in $[T]$ plus one. This makes the total length of the $[T_{\mathrm{col}}]$ array equal to $N + 1$.

To clarify the proposed storage scheme, a simple example is reported the following. Consider the matrix

$$[T] = \begin{bmatrix} 1 & -1 & 0 & -3.2 & 0 \\ -2 & 5.1 & 0 & 0 & 0 \\ 0 & 0 & 4 & 6.7 & 0.4 \\ -4 & 0 & 2.1 & 0.07 & 0 \\ 0 & 8.5 & 0 & 0 & -0.5 \end{bmatrix}; \tag{6.56}$$

the corresponding arrays are equal to

$$\begin{aligned} [T_{\mathrm{value}}] &= [1, -2, -4, -1, 5.1, 8.5, 4, 2.1, -3.2, 6.7, 0.07, 0.4, -0.5] \\ [T_{\mathrm{row}}] &= [1, 2, 4, 1, 2, 5, 3, 4, 1, 3, 4, 3, 5] \\ [T_{\mathrm{col}}] &= [1, 4, 7, 9, 12, 14] \end{aligned} \tag{6.57}$$

6.4.2 Direct solution

We recall the initial MoM system $[Z] \cdot [I] = [V]$ (6.9), derived in Section 6.1.3. In the case of direct solution, the MoM matrix $[Z]$ is a full matrix, and the unknown current $[I]$ is obtained through a direct inversion of the MoM matrix $[Z]$, that is $[I] = [Z]^{-1} \cdot [V]$

The derived change-of-basis matrix $[T]$ (6.52) is applied to (6.9) as

$$[T]^T \cdot [Z] \cdot [T] \cdot [I_{\mathrm{MR}}] = [T]^T \cdot [V], \tag{6.58}$$

where

$$[I] = [T] \cdot [I_{\mathrm{MR}}] \tag{6.59}$$

Indicating

$$[Z_{\mathrm{MR}}] = [T]^T \cdot [Z] \cdot [T] \tag{6.60}$$

and

$$[V_{\mathrm{MR}}] = [T]^T \cdot [V], \tag{6.61}$$

(6.58) can be written as

$$[Z_{\mathrm{MR}}] \cdot [I_{\mathrm{MR}}] = [V_{\mathrm{MR}}] \tag{6.62}$$

Then, a diagonal preconditioning (DP) is applied to (6.62):

$$[D] \cdot [Z_{\mathrm{MR}}] \cdot [D] \cdot \left[I_{\mathrm{MR,DP}}\right] = [D] \cdot [V_{\mathrm{MR}}] \tag{6.63}$$

where

$$[I] = [T] \cdot [D] \cdot \left[I_{\mathrm{MR,DP}}\right] \tag{6.64}$$

and $[D]$ is a diagonal matrix with each element in the main diagonal equal to

$$D_{i,i} = \frac{1}{\sqrt{Z_{\mathrm{MR},i,i}}} \qquad \text{with } i = 1, \ldots, N. \tag{6.65}$$

Indicating

$$\left[Z_{\mathrm{MR,DP}}\right] = [D] \cdot [Z_{\mathrm{MR}}] \cdot [D] \tag{6.66}$$

and

$$\left[V_{\mathrm{MR,DP}}\right] = [D] \cdot [V_{\mathrm{MR}}], \tag{6.67}$$

(6.63) can be written as

$$\left[Z_{\mathrm{MR,DP}}\right] \cdot \left[I_{\mathrm{MR,DP}}\right] = \left[V_{\mathrm{MR,DP}}\right] \tag{6.68}$$

The matrix $\left[Z_{\mathrm{MR,DP}}\right]$ in (6.68) has a low and stable conditioning decreasing the working frequency or increasing the mesh density. Finally, the unknown current $\left[I_{\mathrm{MR,DP}}\right]$ in (6.68) can be evaluated through a direct inversion of the MoM matrix $\left[Z_{\mathrm{MR,DP}}\right]$:

$$\left[I_{\mathrm{MR,DP}}\right] = \left[Z_{\mathrm{MR,DP}}\right]^{-1} \cdot \left[V_{\mathrm{MR,DP}}\right] \tag{6.69}$$

and through (6.64) the original unknown current $[I]$ is obtained.

6.4.3 Application to iterative solvers

The generated change-of-basis matrix $[T]$ can be easily applied to a MoM-based fast iterative solver, such as the fast multipole method (FMM) [23], the multi-level fast multipole algorithm (MLFMA) [24], the conjugate gradient fast Fourier transform (CG-FFT), the adaptive integral method (AIM) [25,26], Green's function interpolation with fast Fourier transform (GIFFT) [27], and the multi-level fast multipole algorithm fast Fourier transform (MLFMA-FFT) [28]. In all these methods, the MoM matrix $[Z]$ is separated into two matrices

$$[Z] = \left[Z^S\right] + \left[Z^W\right] \tag{6.70}$$

where $[Z^S]$ is called the "strong" matrix, and it is a highly sparse matrix containing the interactions between close basis functions; instead $[Z^W]$ is the "weak" matrix and collects all the other interactions. The $[Z^W]$ is expressed as the product of several sparse matrices, related to the chosen MoM-based fast method. Substituting (6.70) into (6.9), the initial system can be written as:

$$([Z^S] + [Z^W]) \cdot [I] = [V] \tag{6.71}$$

where the complexity of the product between $[Z^W]$ and $[I]$ is related to the chosen MoM-based fast method, with the lowest bound $O(N \log N)$ applying, e.g., the MLFMA [24].

The first step in the application of the proposed MR preconditioner to a MoM-based fast method is the generation of the diagonal preconditioner $[D]$. As shown in (6.65), the construction of the diagonal matrix $[D]$ should require the knowledge of the entire MoM matrix $[Z]$, which is not possible with MoM-based fast methods. Then, an approximation of the required diagonal matrix, $[\widetilde{D}]$, is conveniently generated via the strong matrix $[Z^S]$ as

$$[\widetilde{D}] = diag\left([T]^T \cdot [Z^S] \cdot [T]\right). \tag{6.72}$$

In the following, to simplify the notation, the tilde is removed, and the approximate diagonal matrix is indicated simply with $[D]$.

Once generated the diagonal matrix $[D]$, the proposed preconditioner is applied to (6.71) as follows:

$$[D] \cdot [T]^T \cdot ([Z^S] + [Z^W]) \cdot [T] \cdot [D] \cdot [I_{\text{MR,DP}}] = [V_{\text{MR,DP}}] \tag{6.73}$$

where $[I_{\text{MR,DP}}]$, and $[V_{\text{MR,DP}}]$ are related to $[I]$ and $[V]$ as reported in (6.64), (6.67), and (6.61). The matrix–vector products, in the left part of (6.73), are performed on-the-fly at each iteration of the used iterative solver. Figure 6.21 graphically shows the sequence of the matrix–vector products involving $[Z^W]$ at a generic iteration n of the iterative solution, pointing out the computational complexity of each operation. It is evident that the products are performed from right to left, and the application of the proposed preconditioner does not change the whole computational complexity, that in the case of an MLFMA is $O(N \log N)$. For the strong part, the matrix–vector products are evaluated analogously, and the computational complexity is related to the sparsity pattern of the $[Z^S]$.

6.4.4 *Application to electrically large multi-scale structures*

In the case of multiscale problems where the overall dimension of the analyzed structure is larger than the working wavelength, the linear system is solved iteratively applying the basis change matrix (6.54) described in Section 6.2.4.

The initial phase of the solution procedure involves evaluating the MoM strong matrix, $[\widetilde{Z}^S]$ in the multi-level basis. In this new basis, the generalized functions of level-$(L+1)$ are defined on the coarsest mesh, while the MR functions are defined

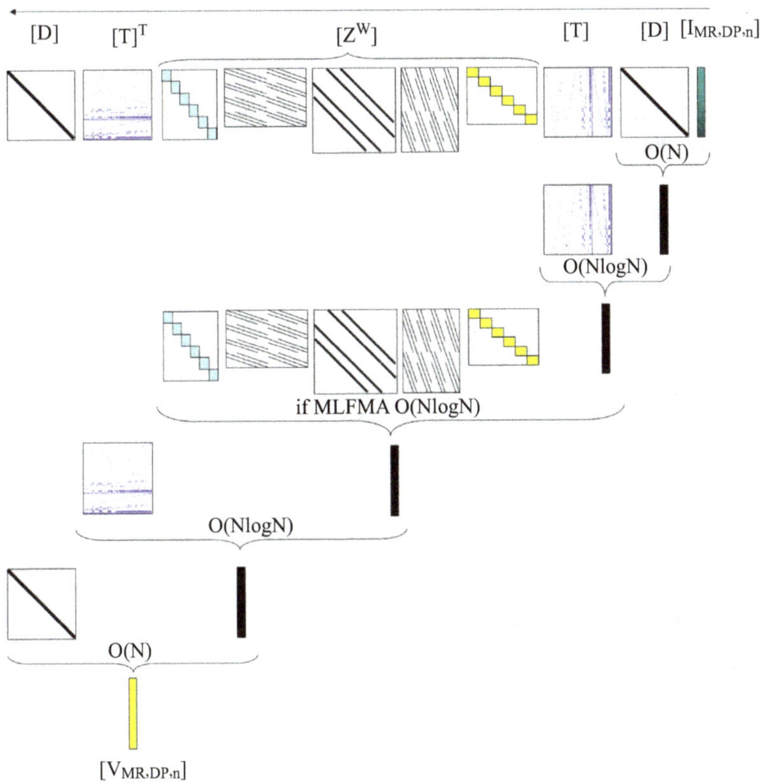

Figure 6.21 Description of the on-the-flight matrix–vector product involving $[Z^W]$ at the iteration n

on the remaining detailed levels (refer to Section 6.2.4). To accomplish this, the basis-change matrix $[T]$ in (6.54) is applied to the strong matrix associated with the standard underlying basis, $[Z^S]$:

$$[\widetilde{Z}^S] = \begin{bmatrix} [\widetilde{Z}^S_{\mathrm{MR}}] & [\widetilde{Z}^S_{\mathrm{MR,g}}] \\ [\widetilde{Z}^S_{\mathrm{g,MR}}] & [\widetilde{Z}^S_{\mathrm{g}}] \end{bmatrix} = [T]^T \cdot [Z^S] \cdot [T], \tag{6.74}$$

where the subscript "MR" indicates the interactions between only MR functions, the subscript "g" the interactions between only generalized functions, and the double subscript the interactions between an MR function and a generalized function.

Afterwards, the process continues by addressing each set of basis functions and geometrical details separately. The diagonal preconditioning $[D]$ (6.72) is then applied to the entire basis. This step is sufficient to obtain the preconditioning for the portion represented by MR functions, while it merely serves to balance the generalized functions. Consequently, we utilize the Pivoting Incomplete LU factorization with dual threshold strategy [29, pp. 312–314], referred to as “ILUTP” hereafter, on the specific segment of the MoM matrix that encompasses interactions solely between generalized functions.

With transparent notation, we call $\left[D_{\mathrm{g}}\right]$ the portion of $[D]$ that pertains to the generalized functions of the coarse level. Then, the block undergoes the approximate LU factorization through the ILUTP algorithm:

$$\left[D_{\mathrm{g}}\right]\cdot\left[\widetilde{Z}_{\mathrm{g}}^{S}\right]\cdot\left[D_{\mathrm{g}}\right]\approx[L]\cdot[U] \tag{6.75}$$

To enhance the sparsity pattern of the matrix $\left[D_{\mathrm{g}}\right]\left[\widetilde{Z}_{\mathrm{g}}^{s}\right]\left[D_{\mathrm{g}}\right]$, any elements below a selected threshold τ_d are discarded (recommended range $\tau_d\in\left[10^{-4},10^{-3}\right]$). This operation will be implied and not explicitly denoted by a different notation. It is important to note that the ILUTP decomposition is only applied to a specific segment of the MoM system, specifically the part related to the generalized function component. Compared to the standard ILUTP scheme, this approach necessitates less memory usage and facilitates a quicker generation of the $[L]$ and $[U]$ factors.

Inserting the proposed hybrid preconditioner into the MoM system (6.71), the resulting system can be algebraically written as

$$\begin{bmatrix}[1] & [0]\\ [0]^{T} & ([L]\,[U])^{-1}\end{bmatrix}[\widetilde{Z_D}][\widetilde{I}]=[\widetilde{V}] \tag{6.76}$$

with

$$[\widetilde{Z_D}]=[D]\,[T]^{T}\left(\left[Z^{S}\right]+\left[Z^{W}\right]\right)[T]\,[D] \tag{6.77}$$

$$[\widetilde{V}]=\begin{bmatrix}[1] & [0]\\ [0]^{T} & ([L]\,[U])^{-1}\end{bmatrix}[D]\,[T]^{T}\,[V] \tag{6.78}$$

$$[I]=[T]\,[D]\,[\widetilde{I}] \tag{6.79}$$

where $[1]$ is an identity matrix with dimension $N_{\mathrm{MR}}\times N_{\mathrm{MR}}$, and $[0]$ is a zero matrix with dimension $N_{\mathrm{MR}}\times N^{L+1}$.

The resulting system, as expressed in (6.76), is solved using an iterative solver. In each iteration, all the specified products are performed in a right-to-left fashion. It is worth noting that the multiplication involving the term $([L]\,[U])^{-1}$ is implemented differently. Instead, a forward substitution followed by a backward substitution is employed. Furthermore, these substitutions are only applied to the generalized function component of the vector obtained from the product $[\widetilde{Z_D}][\widetilde{I}]$ (6.76), resulting in reduced per-iteration cost compared to the standard ILUTP scheme.

6.4.5 Low-frequency matrix entries evaluation

Finally, we report some suggestions for the evaluation, at low frequencies, of the MoM entries via the EFIE, to avoid possible accuracy degradations, and apply properly the proposed preconditioner.

As previously stated, the MR basis functions are separated into a solenoidal part and the non-solenoidal reminder. Indicating with $[T^S]$ the part of the change-of-basis matrix $[T]$ that contains the coefficients of the MR solenoidal functions, and with $[T^{NS}]$ the non-solenoidal reminder,

$$[T] = [[T^S], [T^{NS}]], \tag{6.80}$$

a generic solenoidal function $\boldsymbol{w}_i^S$ can be written as

$$\boldsymbol{w}_i^S(\boldsymbol{r}) = \sum_{n=1}^{N} T_{n,i}^S \boldsymbol{f}_n^0(\boldsymbol{r}). \tag{6.81}$$

The solenoidal function has zero divergence, that is

$$\nabla \cdot \boldsymbol{w}_i^S(\boldsymbol{r}) = 0, \tag{6.82}$$

so applying $\nabla\cdot$ to (6.81), we obtain:

$$\sum_{n=1}^{N} T_{n,i}^S \ \nabla \cdot \boldsymbol{f}_n^0(\boldsymbol{r}) = 0. \tag{6.83}$$

Observing that the charge density of a standard basis function, $\nabla \cdot \boldsymbol{f}_n^0(\boldsymbol{r})$, is different from 0 in each cell of its definition domain, it is evident from (6.83) that the zero divergence of the solenoidal function $\boldsymbol{w}_i^S$ is obtained through a proper weighted sum of the involved standard function charge densities. Hence, the charge density of different basis functions $\boldsymbol{f}_n^0$ defined on the same mesh cell must be evaluated with the same accuracy.

We recall that, in the evaluation of the EFIE MoM matrix elements related to the *scalar* potential, the divergence of the basis functions is evaluated in both the source and test integrals. Usually, only the upper part of the MoM matrix is evaluated (the test function index is always lower or equal to the source function index) because the matrix is symmetrical, and the accuracy (i.e., the number of used quadrature integration points) of the source and test integral can be different. Hence, in the interaction between two cells, the divergence on the same cell can be evaluated differently, and the consequence is a not correct compensation of the charge density of the standard functions forming the solenoidal function.

An implementation trick, to simply enforce the same accuracy for the divergence evaluated in the same pair of cells, is to put as test cell (definition domain of the test integral) always a cell with the index lower than the considered source cell (definition domain of the source integral). In this way, the charge density is properly compensated, and the divergence of the considered solenoidal function, described as a linear combination of standard basis functions (6.81), is "numerically" zero.

To better describe the proposed implementation, we refer to the example in Figure 6.22. We consider the following integral I between the RWG function $\boldsymbol{f}_2^0$ and a solenoidal function $\boldsymbol{w}_i^S$

$$I = \int_{S_2} dS \, \nabla \cdot \boldsymbol{f}_2^0(\boldsymbol{r}) \int_{S_i} dS' g(\boldsymbol{r}, \boldsymbol{r}') \, \nabla \cdot \boldsymbol{w}_i^S(\boldsymbol{r}') \tag{6.84}$$

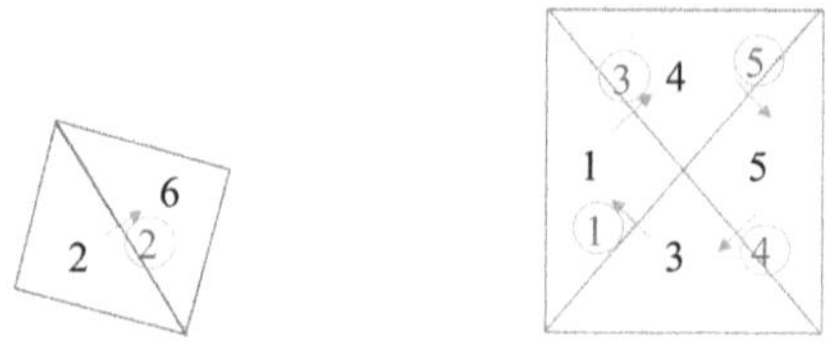

Figure 6.22 Example of mesh to describe the charge density compensation; in black the cell indices, in blue with circles the level-0 function indices

where g is the free space Green's function, $S_2 = C_2^0 \bigcup C_6^0$ is the definition domain of $\boldsymbol{f}_2^0$, and $S_i = \bigcup_{m=1,3,4,5} C_m^0$ is the definition domain of the solenoidal function $\boldsymbol{w}_i^S$, defined as:

$$\boldsymbol{w}_i^S = \sum_{n=1,3,4,5} T_{n,i}^S \boldsymbol{f}_n^0. \tag{6.85}$$

Considering that $\nabla \cdot \boldsymbol{w}_i^S = 0$, the integral (6.84) must be equal to 0. We indicate with $I_{1,2}$ the integral between the triangle cell C_2^0, in the definition domain of $\boldsymbol{f}_2^0$, and C_1^0, in the domain of $\boldsymbol{w}_i^S$, and we check if the charge density of the functions forming $\boldsymbol{w}_i^S$ is properly compensated. If any test/source cell switch is applied, the integral $I_{1,2}$ would be evaluated as:

$$I_{1,2} = T_{1,i}^S \int_{C_1^0} dS \, \nabla \cdot \boldsymbol{f}_1^0 \int_{C_2^0} dS' g \, \nabla \cdot \boldsymbol{f}_2^0 + \\ T_{3,i}^S \int_{C_2^0} dS \, \nabla \cdot \boldsymbol{f}_2^0 \int_{C_1^0} dS' g \, \nabla \cdot \boldsymbol{f}_3^0 \tag{6.86}$$

so in the first integral the cell C_2^0 is the source integral domain, instead in the second integral, the same cell C_2^0 is the test integral domain. If the number of quadrature integration points is different between test and source integrals, the density charge on the cell C_2^0 is evaluated with different accuracy, and so it does not compensate. Instead, applying the proposed implementation, in the second integral, the test integral is switched with the source one, because we force a test cell index always lower than the source cell index, and the charge density on C_2^0 is evaluated with the same accuracy as required.

6.5 Numerical results

To show the performance of the described MR preconditioner, two multi-scale realistic structures are analyzed in the following: a Ferrari Testarossa car, considering both an impinging plane wave and an excited shark-type antenna located in the car roof, and a vessel model with four patch antennas placed on the main mast [28,30]. In both test cases, the MR preconditioner is combined with the MLFMA-FFT solver [31,32]. In the Ferrari Testarossa case, the mesh is conformal, while, for the vessel case, a non-conformal mesh is used.

6.5.1 Ferrari Testarossa test case

All the simulations for the Ferrari Testarossa test case are performed using an Intel Xeon E7-8867 v3 computing server with 32 cores and 3 TB of RAM memory. Two formulations are considered: the EFIE applied to the whole structure and the CFIE applied to the closed bodies, leaving to EFIE to the open ones, e.g., the antenna. The non-metallic parts, i.e., the car windows and windshields, are modeled using the thin dielectric sheet (TDS) approximation [33,34]. The working frequency is 900 MHz, and the average mesh size is equal to $\lambda/20$, where λ is the corresponding wavelength, with some detailed parts meshed up to $\lambda/500$ due to the multi-scale nature of the test case, yielding a total of 1,051,408 unknowns. The maximum car size is 4.46 m, corresponding to around 13 λ.

Figure 6.23 reports the solution convergence in the case of a plane wave excitation impinging into the car front with incident angles $\theta = 90°$ and $\phi = 0°$. The MR preconditioner, indicated in red, is compared to the simple diagonal preconditioner (DP, green lines) and the incomplete lower-upper factorization (ILU, blue lines) applied to the whole mesh [29]. Both the MR and the ILU preconditioners are able to reduce to number of iterations needed to reach a residual of 10^{-6}, as shown in Figure 6.23(a), and, looking at the wall-clock time, the MR approach leads to the best performance, as shown in Figure 6.23(b). The obtained equivalent electric surface current density is reported in Figures 6.24 and 6.25.

Then, a radiation problem is considered exciting a shark-type antenna placed on the car roof. Figure 6.26 reports the solution convergence again comparing the DP, ILU, and MR preconditioners for both formulations (EFIE and CFIE). The obtained performance are similar to the previous scattering problem, with a significant reduction of needed iterations if the ILU and MR preconditioners are used, and the lowest wall-clock time in the case of the MR preconditioner. The obtained equivalent electric surface current density is shown in Figure 6.27.

Finally, Table 6.2 reports the simulations peak memory for the different considered preconditioners and formulations. It is evident that the MR preconditioner memory requirements are around three times lower than the ILU one.

Table 6.2 Ferrari Testarossa: peak memory (GB) for the different considered preconditioners and formulations

	Preconditioner		
Formulation	**DP**	**ILU**	**MR**
EFIE	65.64	279.41	91.34
CFIE-EFIE	65.67	280.65	92.43

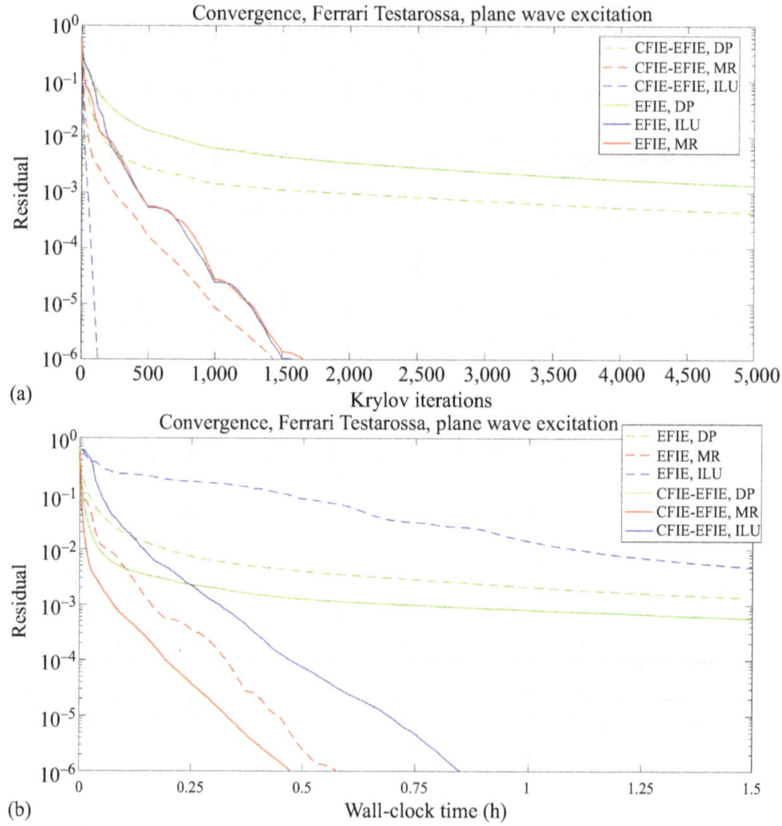

Figure 6.23 Ferrari Testarossa, plane wave excitation at 900 MHz: (a) number of iterations and (b) wall-clock time (h)

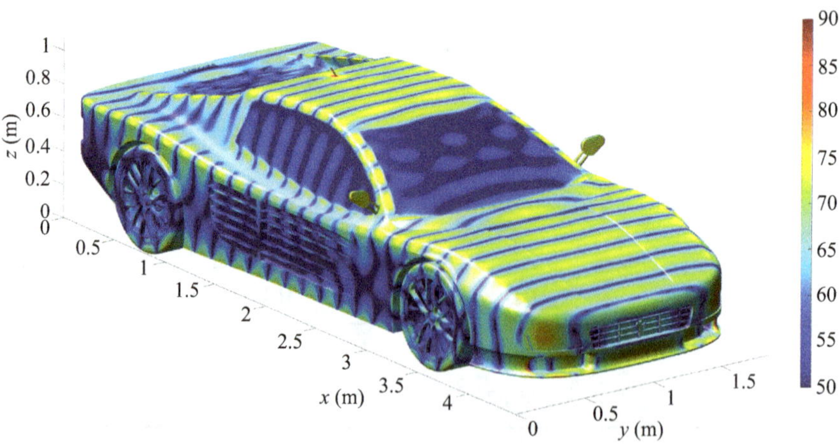

Figure 6.24 Ferrari Testarossa: real part of the equivalent electric surface current density (dBμA/m) induced by an incident plane wave at 900 MHz

Figure 6.25 *Ferrari Testarossa: details of the real part of the equivalent electric surface current density (dBμA/m) induced by an incident plane wave at 900 MHz*

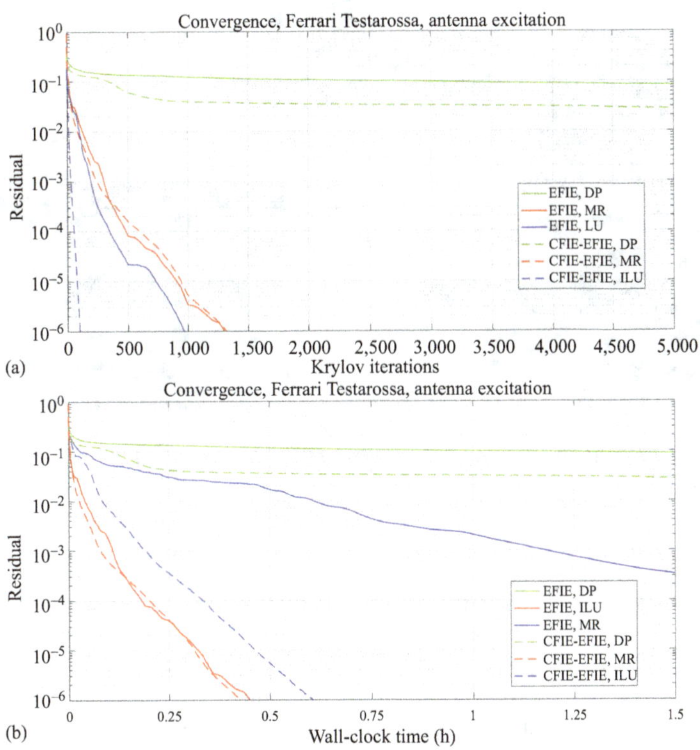

Figure 6.26 *Ferrari Testarossa, shark-type antenna excitation at 900 MHz: (a) number of iterations and (b) wall-clock time (h)*

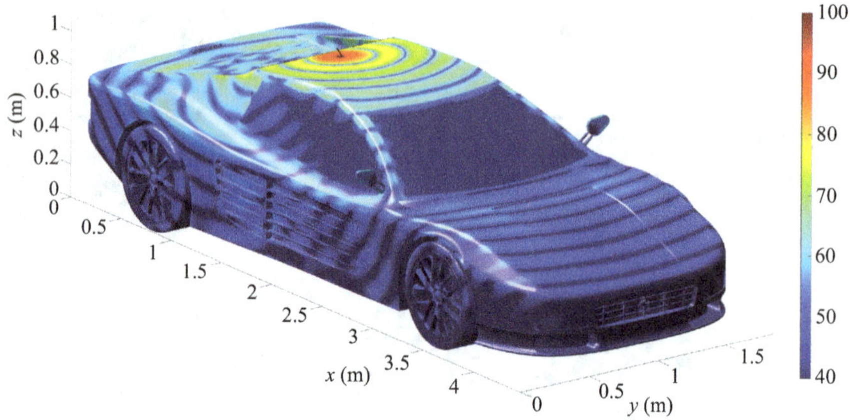

Figure 6.27 Ferrari Testarossa: real part of the equivalent electric surface current density (dBμA/m) induced by the shark-type antenna located in its roof, excited at 900 MHz

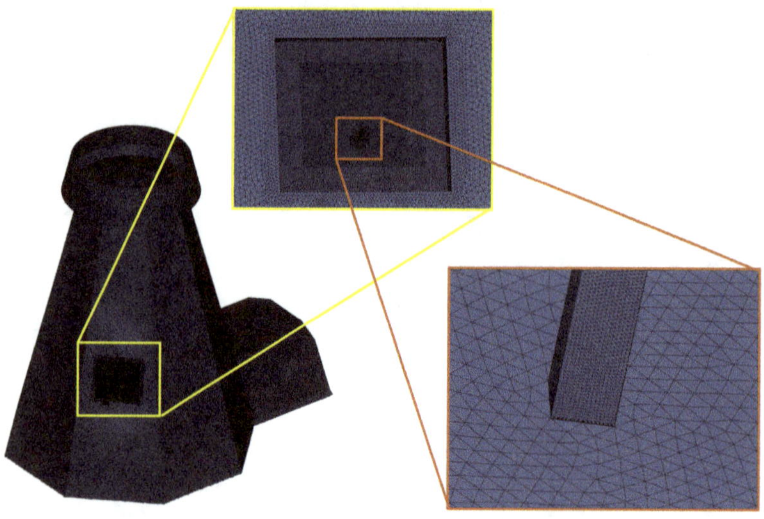

Figure 6.28 Realistic vessel: meshed mast and antenna with detailed non-conformal parts

6.5.2 Realistic vessel test case

All the simulations for the vessel test case were performed using a 2 × AMD EPYC 7H12 2.6 GHz computing server with 128 cores and 2 TB of RAM memory. The CFIE applied to the closed bodies, leaving to EFIE to the open ones.

The antennas are meshed separately, with a fine mesh related to their details, and, then, placed at the mid-level of the main mast of the vessel via a non-conformal mesh, as shown in Figure 6.28. Moreover, also the antenna feeding line is meshed

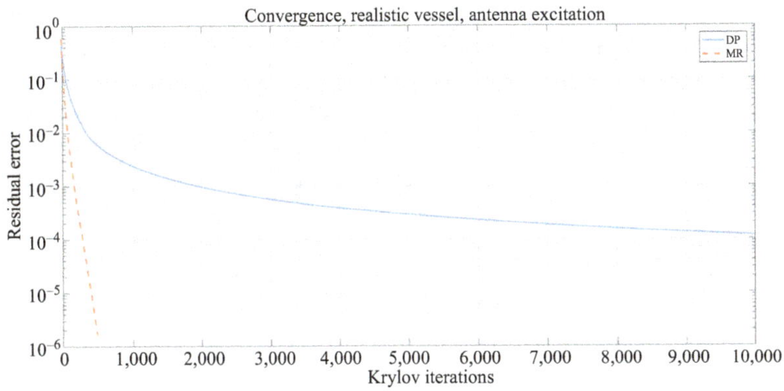

Figure 6.29 Realistic vessel: antenna excitation at 550 MHz, number of iterations

Figure 6.30 Realistic vessel: real part of the equivalent electric surface current density (dBμA/m) induced by the patch antennas on its mast at 550 MHz

non-conformal to the rest of the antenna. The final non-conformal mesh is constituted of 13,782,364 RWG basis functions and 3,468 MB-RWG basis functions. The antenna working frequency is 550 MHz that corresponds to a vessel maximum size of around 250 λ, where λ is the wavelength. The average mesh size is equal to $\lambda/15$ with small details meshed up to $\lambda/1{,}650$.

Figure 6.29 shows the solution convergence comparing the DP preconditioner to the MR one that allows a strong iteration reduction. Moreover, Figure 6.28 reports the obtained equivalent electric surface current density.

6.6 Conclusion and perspectives

In this chapter, we described how the MR basis functions can be generated on PEC surfaces and wires, discretized with conformal and non-conformal meshes, as a linear combination of the standard basis functions used in the MoM solution of the EFIE and CFIE. The chapter started from the generation of generalized basis functions that are organized in different mesh levels and are the generalization of the standard basis function on polygonal mesh cells. Then, the generalized basis functions were employed to generate the MR (hierarchical) basis functions, separated into the solenoidal and non-solenoidal parts. The obtained MR basis was applied to the solution of the MoM linear system via a change-of-basis matrix in the case of direct and iterative solutions, and combined with fast MoM solvers. Finally, the MR performance were compared to the diagonal preconditioner and the incomplete lower-upper factorization for some multi-scale realistic test cases.

Perspectives of the described MR basis functions are their application to the domain decomposition method to accelerate the solution of the sub-domain problems [35–38], and their extension to finite dielectrics.

Acknowledgments

The contribution of V. F. Martin and J. M. Taboada to this chapter has been supported in part by the European Regional Development Fund (ERDF) and the Spanish Ministerio de Ciencia, Innovacion y Universidades under Projects PID2020-116627RB-C21, and PID2020-116627RB-C22, supported by MCIN/AEI/10.13039/501100011033. The contribution of V. F. Martin was also supported by the Spanish Ministerio de Ciencia, Innovación y Universidades (FPU00550/17, EST21/00590) and a Margarita Salas grant (MS-26 RD 289/2021).

References

[1] Andriulli FP, Vipiana F, and Vecchi G. Hierarchical bases for non-hierarchic 3D triangular meshes. *IEEE Trans Antennas Propag*. 2008;56:2288–2297.

[2] Eibert TF. Iterative-solver convergence for loop-star and loop-tree decomposition in method-of-moments solutions of the electric-field integral equation. *IEEE Antennas Propag Mag*. 2004;46(3):80–85.

[3] Lee JF, Lee R, and Burkholder RJ. Loop star basis functions and a robust preconditioner for EFIE scattering problems. *IEEE Trans Antennas Propag*. 2003;51(8):1855–1863.

[4] Zhao JS and Chew WC. Integral equation solution of Maxwell's equations from zero frequency to microwave frequency. *IEEE Trans Antennas Propag*. 2000;48(10):1635–1645.

[5] Vecchi G. Loop-star decomposition of basis functions in the discretization of EFIE. *IEEE Trans Antennas Propag*. 1999;47:339–346.

[6] Burton M and Kashyap S. A study of a recent moment-method algorithm that is accurate to very low frequencies. *Appl Computat Electromagn Soc J*. 1995;10(3):58–68.

[7] Wu W, Glisson AW, and Kajfez D. A study of two numerical solution procedures for the electric field integral equation at low frequency. *Appl Computat Electromagn Soc J*. 1995;10(3):69–80.

[8] Wilton DR. Topological consideration in surface patch and volume cell modeling of electromagnetic scatterers. In: *Proceeding in URSI International Symposium on Electromagnetics Theory*. Santiago de Compostela (Spain); 1983. p. 65–68.

[9] Vipiana F, Pirinoli P, and Vecchi G. Spectral properties of the EFIE-MoM matrix for dense meshes with different types of bases. *IEEE Trans Antennas Propag*. 2007;55:3229–3238.

[10] Butler CM and Wilton DR. Analysis of various numerical techniques applied to thin wire scatterers. *IEEE Trans Antennas Propag*. 1975;23:534–540.

[11] Rao SM, Wilton DR, and Glisson AW. Electromagnetic scattering by surfaces of arbitrary shape. *IEEE Trans Antennas Propag*. 1982;30(3):409–418.

[12] Hwu SU, Wilton DR, and Rao SM. Electromagnetic scattering and radiation by arbitrary conducting wire/surface configurations. In: *Proceeding IEEE International Symposium on Antennas and Propagation*, vol. 26; 1988. p. 890–893.

[13] Champagne NJ, Johnson WA, and Wilton DR. On attaching a wire to a triangulated surface. In: *Proceeding of the IEEE International Symposium on Antennas and Propagation*, 2002. p. 54–57.

[14] Glisson AW, Rao SM, and Wilton DR. Physically-based approximation of electromagnetic field quantities. In: *Proceeding of the IEEE International Symposium on Antennas and Propagation*, 2002. p. 78–81.

[15] Huang S, Xiao G, Hu Y, *et al*. Multibranch Rao–Wilton–Glisson basis functions for electromagnetic scattering problems. *IEEE Trans Antennas Propag*. 2021 Oct;69(10):6624–6634.

[16] Vipiana F and Vecchi G. A novel, symmetrical solenoidal basis for the MoM analysis of closed surfaces. *IEEE Trans Antennas Propag*. 2009;57: 1294–1299.

[17] Vipiana F, Andriulli FP, and Vecchi G. Two-tier non-simplex grid hierarchic basis for general 3D meshes. *Waves Random Complex Media*. 2009;19(1): 126–146.

[18] Vipiana F, Vecchi G, and Wilton DR. Automatic loop-tree scheme for arbitrary conducting wire-surface structures. *IEEE Trans Antennas Propag*. 2009;57(11):3564–3574.

[19] Vipiana F, Vecchi G, and Pirinoli P. A multi-resolution system of Rao-Wilton-Glisson functions. *IEEE Trans Antennas Propag*. 2007;55:924–930.

[20] Vipiana F, Pirinoli P, and Vecchi G. A multiresolution method of moments for triangular meshes. *IEEE Trans Antennas Propag*. 2005;53(7):2247–2258.

[21] Vipiana F, Francavilla MA, and Vecchi G. EFIE modeling of high-definition multi-scale structures. *IEEE Trans Antennas Propag*. 2010;58:2362–2374.

[22] Francavilla MA, Vipiana F, Vecchi G, *et al*. Hierarchical fast MoM solver for the modeling of large multi-scale wire-surface structures. *IEEE Antennas Wireless Propag Lett*. 2012;11:1378–1381.

[23] De Vita P, Freni A, Pirinoli P, *et al*. A combined MR-FMM approach. In: *Proceeding of the URSI National Radio Science Meeting*, Washington, DC, 2005.

[24] Chew WC, Jin JM, Michielssen E, *et al. Fast and Efficient Algorithms in Computational Electromagnetics*. Boston, MA: Artech House; 2001.

[25] De Vita P, Freni A, Vipiana F, *et al.* Fast analysis of large finite arrays with a combined multiresolution SM/AIM approach. *IEEE Trans Antennas Propag.* 2006;54(12):3827–3832.

[26] Bleszynski E, Bleszynski M, and Jaroszewicz T. AIM: adaptive integral method for solving large-scale electromagnetic scattering and radiation problems. *Radio Sci.* 1996;5:1225–1251.

[27] Fasenfest BJ, Capolino F, Wilton DR, *et al.* A fast MoM solution for large arrays: Green's function interpolation with FFT. *IEEE Antennas Wireless Propag Lett.* 2004;3:161–164.

[28] Solis DM, Martin VF, Taboada JM, *et al.* Multiresolution preconditioners for solving realistic multi-scale complex problems. *IEEE Access.* 2022;10: 22038–22048.

[29] Saad Y. *Iterative Methods for Sparse Linear System*. Philadelphia, PA: Society for Industrial and Applied Mathematics (SIAM), 2003.

[30] Martin VF, Taboada JM, and Vipiana F. A multi-resolution preconditioner for non-conformal meshes in the MoM solution of large multi-scale structures. *IEEE Trans Antennas Propag.* 2023;99:1–1.

[31] Taboada JM, Gomez-Araujo M, Bertolo JM, *et al.* MLFMA-FFT parallel algorithm for the solution of large-scale problems in electromagnetics (invited paper). *Prog Electromagnet Res.* 2010;105:15–30.

[32] Taboada JM, Araujo MG, Basteiro FO, *et al.* MLFMA-FFT parallel algorithm for the solution of extremely large problems in electromagnetics. *Proc IEEE.* 2013;101(2):350–363.

[33] Harrington RF and Mautz JR. An impedance sheet approximation for thin dielectric shells. *IEEE Trans Antennas Propag.* 1975;23(4):531–534.

[34] Chiangand IT and Chew WC. Thin dielectric sheets simulation by surface integral equation using modified RWG and pulse base. *IEEE Trans Antennas Propag.* 2006;54(7):1927–1934.

[35] Peng Z, Lim KH, and Lee JF. Nonconformal domain decomposition methods for solving large multiscale electromagnetic scattering problems. *Proc IEEE.* 2013;101(2):298–319.

[36] Bautista MAE, Vipiana F, Francavilla MA, *et al.* A non-conformal domain decomposition scheme for the analysis of multi-scale structures. *IEEE Trans Antennas Propag.* 2015;63(8):3548–3560.

[37] Solis DM, Martin VF, Araujo MG, *et al.* Accurate EMC engineering on realistic platforms using an integral equation domain decomposition approach. *IEEE Trans Antennas Propag.* 2020;68(4):3002–3015.

[38] Martin VF, Larios D, Solis DM, *et al.* Tear-and-interconnect domain decomposition scheme for solving multiscale composite penetrable objects. *IEEE Access.* 2020;8:107345–107352.

Chapter 7

Calderón preconditioners for electromagnetic integral equations

Adrien Merlini[1], Simon B. Adrian[2], Alexandre Dély[3] and Francesco P. Andriulli[3]

7.1 Introduction

Integral equations (IEs), numerically solved via the boundary element method (BEM) [1–4], are particularly suited to modeling electromagnetic scattering problems because they inherently enforce the Silver–Müller radiation conditions without requiring additional boundary conditions to be imposed [5,6], unlike other popular methods such as the finite element method (FEM) or the finite-difference time-domain (FDTD) method, thus limiting the computational overhead incurred [7]. When dealing with problems involving piecewise homogeneous scatterers, IEs can push their efficiency advantage further by only requiring unknowns to be placed on the boundaries between different media, which usually significantly reduces the number of degrees of freedom to be solved for, and thus yields sensibly smaller system matrices. Finally, they are resilient to numerical dispersion [8].

These significant advantages are, however, offset by the fact that the matrices resulting from the BEM discretization of IEs are dense and often ill-conditioned which, despite the reduced matrix size, compromises the numerical efficiency and computational complexity of such schemes. Overcoming these limitations requires two key components: fast matrix–vector product (MVP) algorithms [9–14] and preconditioners. When fast MVP algorithms are combined with iterative solvers, the solution to the problem can be obtained in $\mathcal{O}(N_{\text{iter}}N \log(N))$ complexity, where N is the number of degrees of freedom and N_{iter} is the number of iterations required for the iterative solver to reach the desired precision. For this complexity to be competitive with other popular methods, N_{iter} must not depend on N. This is not true in general since N_{iter} is dramatically influenced by the conditioning of the matrix. Consequently, the BEM is usually combined with preconditioning techniques to achieve optimal or near-optimal performances [15–18].

[1]Microwave Department, IMT Atlantique, Brest, France
[2]Fakultät für Informatik und Elektrotechnik, Universität Rostock, Rostock, Germany
[3]Department of Electronics and Telecommunications, Politecnico di Torino, Turin, Italy

Very different sources of ill-conditioning can be encountered in the solution of IE: the choice of the IE itself (and of the operators implied) has the most significant impact, but factors such as mesh quality, basis function type and order, or level of detail of the geometry must be considered [18], and specific cures must be sought for each of these. For instance, working with a well-conditioned second-kind IE is preferable to working with a first-kind equation that becomes increasingly ill-conditioned as the discretization density of the geometry is increased (to capture smaller details or increase accuracy) [19]. However, such a choice might not always be possible. For instance, the magnetic field IE (MFIE), the underlying integral operator of which is of the second kind, is very appealing when modeling perfectly electrically conducting (PEC) objects. The MFIE is, however, only applicable to closed scatterers and is not uniquely solvable for all frequencies [5]. For scenario that fall outside of those that this equation can handle, other IEs such as the electric field IE (EFIE) or the combined field IE (CFIE) must be used; both alternatives, however, involve first-kind operators. In this situation, preconditioning strategies must be employed. Common preconditioning techniques include algebraic preconditioning [20–31], hierarchical bases [32–40], or domain decomposition [41–43].

Among the numerous approaches, Calderón preconditioning [15,16,19,44–47] and closely related schemes [48–53] provide optimal preconditioners. Calderón strategies can be used, among other approaches, to form second-kind formulations out of first-kind IEs by leveraging properly chosen continuous identities linking different operators together [19]. They have also been used to cure the conditioning issues related to the frequency of the scenario under study. For instance, the low-frequency issues plaguing the EFIE and CFIE can be partially addressed using the well-known Calderón identities [54], quasi-Helmholtz decompositions [55–58] or a combination of these two preconditioner families [59]. Strategies to overcome the ill-conditioning that occurs at high frequencies—that differs from the dense-discretization ill-conditioning—have also been explored [46,48,50,60].

Beyond the standard theoretical benchmark of PEC scenario, applications of Calderón strategies have also been explored for scattering by penetrable bodies. Because of its popularity, numerous contributions have focused on applying Calderón techniques to the Poggio–Miller–Chang–Harrington–Wu–Tsai (PMCHWT) formulation [61–63]: in [64], a dual Calderón preconditioner is introduced to address the dense-discretization ill-conditioning of the equation, which has later been extended for low-frequency stabilization with quasi-Helmholtz projectors [65]; another preconditioner based on the appropriate use of a mixed discretization to better exploit iterative schemes has been presented in [66] that also tackles the conditioning issues arising when the contrast between the dielectric properties of the scatterer and of the background medium increases. Beyond the PMCHWT, a Calderón-preconditioned single-source formulation that is immune to the low-frequency, high-contrast, and dense discretization breakdown has been introduced in [67]. Preconditioners in the presence of junctions between dielectric and metals are studied in [68].

This chapter intends to provide a high-level understanding of some of the mechanisms of Calderón preconditioning as well as an overview of different ways in which it has successfully been applied to widely used electromagnetic formulations. As such, some of the more intricate mathematical developments will only be alluded to and pointers to detailed analyses will be provided. First, some background material will be presented and the notations set in Section 7.2 after which the Calderón identities will be derived in Section 7.3. Standard discretization schemes for the EFIE and MFIE will be introduced in Section 7.4. We will then present the different issues plaguing the EFIE and CFIE in Sections 7.5 and 7.6, respectively, along with Calderón strategies to cure them. Finally, we will introduce the PMCHWT formulation for dielectric scattering along with its limitations and Calderón stabilization strategies in Section 7.7 before concluding in Section 7.8.

7.2 Background and notations

Let Ω^- be a PEC scatterer with a smooth boundary Γ residing in a background medium with permeability μ_0, permittivity ε_0, and impedance $\eta_0 := \sqrt{\mu_0/\varepsilon_0}$, on which is a time-harmonic electromagnetic wave $(\boldsymbol{e}^{\mathrm{i}}, \boldsymbol{h}^{\mathrm{i}})$ with angular frequency ω and wavenumber $k := \omega\sqrt{\varepsilon_0\mu_0}$ impinges. The complementary space is denoted as $\Omega^+ := \mathbb{R}^3 \setminus \Omega^-$. The total electromagnetic field $(\boldsymbol{e}, \boldsymbol{h})$ in the presence of the scatterer is the superposition of the incident field and the scattered field $(\boldsymbol{e}^{\mathrm{s}}, \boldsymbol{h}^{\mathrm{s}})$ that can be computed by radiating the electric surface current density $\boldsymbol{j}_\Gamma$ on Γ which is obtained by solving the EFIE

$$\eta_0 \mathcal{T}_k \boldsymbol{j}_\Gamma = -\hat{\boldsymbol{n}} \times \boldsymbol{e}^{\mathrm{i}} \tag{7.1}$$

or the MFIE

$$\hat{\boldsymbol{n}} \times \boldsymbol{h}^{\mathrm{i}} = \mathcal{M}_k^+ \boldsymbol{j}_\Gamma := +(\mathcal{I}/2 + \mathcal{K}_k)\boldsymbol{j}_\Gamma, \tag{7.2}$$

with

$$\mathcal{T}_k := \mathrm{i}k\mathcal{T}_{\mathrm{A},k} + 1/(\mathrm{i}k)\mathcal{T}_{\Phi,k}, \tag{7.3}$$

$$(\mathcal{T}_{\mathrm{A},k}\boldsymbol{j}_\Gamma)(\boldsymbol{r}) := \hat{\boldsymbol{n}} \times \int_\Gamma G_k(\boldsymbol{r}, \boldsymbol{r}')\boldsymbol{j}_\Gamma(\boldsymbol{r}')\mathrm{d}S(\boldsymbol{r}'), \tag{7.4}$$

$$(\mathcal{T}_{\Phi,k}\boldsymbol{j}_\Gamma)(\boldsymbol{r}) := -\hat{\boldsymbol{n}} \times \nabla_\Gamma \int_\Gamma G_k(\boldsymbol{r}, \boldsymbol{r}')\nabla'_\Gamma \cdot \boldsymbol{j}_\Gamma(\boldsymbol{r}')\mathrm{d}S(\boldsymbol{r}'), \tag{7.5}$$

$$(\mathcal{K}_k\boldsymbol{j}_\Gamma)(\boldsymbol{r}) := -\hat{\boldsymbol{n}} \times \int_\Gamma \nabla G_k(\boldsymbol{r}, \boldsymbol{r}') \times \boldsymbol{j}_\Gamma \mathrm{d}S(\boldsymbol{r}'), \tag{7.6}$$

$$G_k(\boldsymbol{r}, \boldsymbol{r}') := 4\pi \frac{\mathrm{e}^{\mathrm{i}k|\boldsymbol{r}-\boldsymbol{r}'|}}{|\boldsymbol{r} - \boldsymbol{r}'|}, \tag{7.7}$$

where $\hat{\boldsymbol{n}}$ is the unit vector field normal to Γ pointing away from Ω^-, and where $\mathcal{I}$ is the identity operator. Later on, the inner MFIE operator $\mathcal{M}_k^- := -\mathcal{I}/2 + \mathcal{K}_k$ will also be required. The scattered fields obtained following this procedure naturally satisfy the Silver–Müller radiation conditions

$$\lim_{r\to\infty} |\eta_0 \boldsymbol{h}^{\mathrm{s}} \times \boldsymbol{r} - r\boldsymbol{e}^{\mathrm{s}}| = 0. \tag{7.8}$$

7.3 Calderón identities

The Calderón identities that we will derive in this section are a powerful tool to build well-conditioned IE formulations. Consider a source-free scenario in which $(\boldsymbol{e}^{\mathrm{i}}, \boldsymbol{h}^{\mathrm{i}}) = (\mathbf{0}, \mathbf{0})$. According to the equivalence principle, the fields in $\Omega^{\pm}$ are uniquely determined by their rotated tangential traces $\hat{\boldsymbol{n}} \times \boldsymbol{e}^{\pm}$ and $\hat{\boldsymbol{n}} \times \boldsymbol{h}^{\pm}$ on Γ through the standard expressions of the radiation integrals

$$\begin{aligned}
\pm\boldsymbol{e}^{\pm}(\boldsymbol{r}) = {} & \mathrm{i}k\eta_0 \int_{\Gamma} G_k(\boldsymbol{r},\boldsymbol{r}') \left(\hat{\boldsymbol{n}} \times \boldsymbol{h}^{\pm}(\boldsymbol{r}')\right) \mathrm{d}S(\boldsymbol{r}') \\
& -\frac{\eta_0}{\mathrm{i}k} \nabla\nabla \cdot \int_{\Gamma} G_k(\boldsymbol{r},\boldsymbol{r}') \left(\hat{\boldsymbol{n}} \times \boldsymbol{h}^{\pm}(\boldsymbol{r}')\right) \mathrm{d}S(\boldsymbol{r}') \\
& -\nabla \times \int_{\Gamma} G_k(\boldsymbol{r},\boldsymbol{r}') \left(-\hat{\boldsymbol{n}} \times \boldsymbol{e}^{\pm}(\boldsymbol{r}')\right) \mathrm{d}S(\boldsymbol{r}'), \quad \boldsymbol{r} \in \Omega^{\pm},
\end{aligned} \tag{7.9}$$

$$\begin{aligned}
\pm\boldsymbol{h}^{\pm}(\boldsymbol{r}) = {} & \frac{\mathrm{i}k}{\eta_0} \int_{\Gamma} G_k(\boldsymbol{r},\boldsymbol{r}') \left(-\hat{\boldsymbol{n}} \times \boldsymbol{e}^{\pm}(\boldsymbol{r}')\right) \mathrm{d}S(\boldsymbol{r}') \\
& -\frac{1}{\mathrm{i}k\eta_0} \nabla\nabla \cdot \int_{\Gamma} G_k(\boldsymbol{r},\boldsymbol{r}') \left(-\hat{\boldsymbol{n}} \times \boldsymbol{e}^{\pm}(\boldsymbol{r}')\right) \mathrm{d}S(\boldsymbol{r}') \\
& +\nabla \times \int_{\Gamma} G_k(\boldsymbol{r},\boldsymbol{r}') \left(\hat{\boldsymbol{n}} \times \boldsymbol{h}^{\pm}(\boldsymbol{r}')\right) \mathrm{d}S(\boldsymbol{r}'), \quad \boldsymbol{r} \in \Omega^{\pm},
\end{aligned} \tag{7.10}$$

which can be rewritten into [54]

$$\begin{aligned}
\pm\boldsymbol{e}^{\pm}(\boldsymbol{r}) = {} & \mathrm{i}k\eta_0 \int_{\Gamma} G_k(\boldsymbol{r},\boldsymbol{r}') \left(\hat{\boldsymbol{n}} \times \boldsymbol{h}^{\pm}(\boldsymbol{r}')\right) \mathrm{d}S(\boldsymbol{r}') \\
& -\frac{\eta_0}{\mathrm{i}k} \nabla \int_{\Gamma} G_k(\boldsymbol{r},\boldsymbol{r}') \nabla'_{\Gamma} \cdot \left(\hat{\boldsymbol{n}} \times \boldsymbol{h}^{\pm}(\boldsymbol{r}')\right) \mathrm{d}S(\boldsymbol{r}') \\
& -\int_{\Gamma} \nabla G_k(\boldsymbol{r},\boldsymbol{r}') \times \left(-\hat{\boldsymbol{n}} \times \boldsymbol{e}^{\pm}(\boldsymbol{r}')\right) \mathrm{d}S(\boldsymbol{r}'), \quad \boldsymbol{r} \in \Omega^{\pm},
\end{aligned} \tag{7.11}$$

$$\begin{aligned}
\pm\boldsymbol{h}^{\pm}(\boldsymbol{r}) = {} & \frac{\mathrm{i}k}{\eta_0} \int_{\Gamma} G_k(\boldsymbol{r},\boldsymbol{r}') \left(-\hat{\boldsymbol{n}} \times \boldsymbol{e}^{\pm}(\boldsymbol{r}')\right) \mathrm{d}S(\boldsymbol{r}') \\
& -\frac{1}{\eta_0 \mathrm{i}k} \nabla \int_{\Gamma} G_k(\boldsymbol{r},\boldsymbol{r}') \nabla'_{\Gamma} \cdot \left(-\hat{\boldsymbol{n}} \times \boldsymbol{e}^{\pm}(\boldsymbol{r}')\right) \mathrm{d}S(\boldsymbol{r}') \\
& +\int_{\Gamma} \nabla G_k(\boldsymbol{r},\boldsymbol{r}') \times \left(\hat{\boldsymbol{n}} \times \boldsymbol{h}^{\pm}(\boldsymbol{r}')\right) \mathrm{d}S(\boldsymbol{r}'), \quad \boldsymbol{r} \in \Omega^{\pm},
\end{aligned} \tag{7.12}$$

written more compactly as

$$\begin{bmatrix} \pm\boldsymbol{e}^{\pm} \\ \pm\boldsymbol{h}^{\pm} \end{bmatrix} = \begin{bmatrix} -\mathscr{W}_k & \eta_0 \mathscr{L}_k \\ -\eta_0^{-1} \mathscr{L}_k & -\mathscr{W}_k \end{bmatrix} \begin{bmatrix} \hat{\boldsymbol{n}} \times \boldsymbol{e}^{\pm} \\ \hat{\boldsymbol{n}} \times \boldsymbol{h}^{\pm} \end{bmatrix}, \quad \boldsymbol{r} \in \Omega^{\pm}, \tag{7.13}$$

where

$$(\mathscr{L}_k \boldsymbol{x}_\Gamma)(\boldsymbol{r}) := \mathrm{i}k \int_\Gamma G_k(\boldsymbol{r},\boldsymbol{r}')\boldsymbol{x}_\Gamma(\boldsymbol{r}')\mathrm{d}S(\boldsymbol{r}') \\ -\frac{1}{\mathrm{i}k}\nabla \int_\Gamma G_k(\boldsymbol{r},\boldsymbol{r}')\nabla'_\Gamma \cdot \boldsymbol{x}_\Gamma(\boldsymbol{r}')\mathrm{d}S(\boldsymbol{r}'), \quad \boldsymbol{r} \in \Omega^\pm, \tag{7.14}$$

$$(\mathscr{W}_k \boldsymbol{x}_\Gamma)(\boldsymbol{r}) := -\int_\Gamma \nabla G_k(\boldsymbol{r},\boldsymbol{r}') \times \boldsymbol{x}_\Gamma(\boldsymbol{r}')\mathrm{d}S(\boldsymbol{r}'), \quad \boldsymbol{r} \in \Omega^\pm. \tag{7.15}$$

The Calderón projectors can be obtained from (7.13) through a limiting process in which the tangential traces of the fields are evaluated on the equivalent surface. When the limit is taken from outside, we have

$$\begin{bmatrix} \hat{\boldsymbol{n}} \times \boldsymbol{e}^+ \\ \hat{\boldsymbol{n}} \times \boldsymbol{h}^+ \end{bmatrix} = \begin{bmatrix} \mathscr{I}/2 - \mathscr{K}_k & \eta_0 \mathscr{T}_k \\ -\eta_0^{-1}\mathscr{T}_k & \mathscr{I}/2 - \mathscr{K}_k \end{bmatrix} \begin{bmatrix} \hat{\boldsymbol{n}} \times \boldsymbol{e}^+ \\ \hat{\boldsymbol{n}} \times \boldsymbol{h}^+ \end{bmatrix}, \quad \text{on } \Gamma, \tag{7.16}$$

and inside

$$\begin{bmatrix} \hat{\boldsymbol{n}} \times \boldsymbol{e}^- \\ \hat{\boldsymbol{n}} \times \boldsymbol{h}^- \end{bmatrix} = \begin{bmatrix} \mathscr{I}/2 + \mathscr{K}_k & -\eta_0 \mathscr{T}_k \\ \eta_0^{-1}\mathscr{T}_k & \mathscr{I}/2 + \mathscr{K}_k \end{bmatrix} \begin{bmatrix} \hat{\boldsymbol{n}} \times \boldsymbol{e}^- \\ \hat{\boldsymbol{n}} \times \boldsymbol{h}^- \end{bmatrix}, \quad \text{on } \Gamma. \tag{7.17}$$

This allows us to define the Calderón projectors

$$\mathscr{P}^\pm = \begin{bmatrix} \mathscr{I}/2 \mp \mathscr{K}_k & \pm\eta_0 \mathscr{T}_k \\ \mp\eta_0^{-1}\mathscr{T}_k & \mathscr{I}/2 \mp \mathscr{K}_k \end{bmatrix}, \tag{7.18}$$

for which the projector property can easily be shown: since

$$\mathscr{P}^\pm \begin{bmatrix} \hat{\boldsymbol{n}} \times \boldsymbol{e}^\pm & \hat{\boldsymbol{n}} \times \boldsymbol{h}^\pm \end{bmatrix}^{\mathrm{T}} = \begin{bmatrix} \hat{\boldsymbol{n}} \times \boldsymbol{e}^\pm & \hat{\boldsymbol{n}} \times \boldsymbol{h}^\pm \end{bmatrix}^{\mathrm{T}}, \tag{7.19}$$

we have

$$\mathscr{P}^\pm \mathscr{P}^\pm \begin{bmatrix} \hat{\boldsymbol{n}} \times \boldsymbol{e}^\pm & \hat{\boldsymbol{n}} \times \boldsymbol{h}^\pm \end{bmatrix}^{\mathrm{T}} = \mathscr{P}^\pm \begin{bmatrix} \hat{\boldsymbol{n}} \times \boldsymbol{e}^\pm & \hat{\boldsymbol{n}} \times \boldsymbol{h}^\pm \end{bmatrix}^{\mathrm{T}} \tag{7.20}$$

and since $\mathscr{P}^+ + \mathscr{P}^- = \mathscr{I}$, we have

$$\mathscr{P}^\pm \begin{bmatrix} \hat{\boldsymbol{n}} \times \boldsymbol{e}^\pm & \hat{\boldsymbol{n}} \times \boldsymbol{h}^\pm \end{bmatrix}^{\mathrm{T}} + \mathscr{P}^\mp \begin{bmatrix} \hat{\boldsymbol{n}} \times \boldsymbol{e}^\pm & \hat{\boldsymbol{n}} \times \boldsymbol{h}^\pm \end{bmatrix}^{\mathrm{T}} = \begin{bmatrix} \hat{\boldsymbol{n}} \times \boldsymbol{e}^\pm & \hat{\boldsymbol{n}} \times \boldsymbol{h}^\pm \end{bmatrix}^{\mathrm{T}}, \tag{7.21}$$

thus,

$$\mathscr{P}^\pm \begin{bmatrix} \hat{\boldsymbol{n}} \times \boldsymbol{e}^\mp & \hat{\boldsymbol{n}} \times \boldsymbol{h}^\mp \end{bmatrix}^{\mathrm{T}} = \begin{bmatrix} \boldsymbol{0} & \boldsymbol{0} \end{bmatrix}^{\mathrm{T}}, \tag{7.22}$$

and, finally,

$$\mathscr{P}^\pm \mathscr{P}^\pm \begin{bmatrix} \hat{\boldsymbol{n}} \times \boldsymbol{e}^\mp & \hat{\boldsymbol{n}} \times \boldsymbol{h}^\mp \end{bmatrix}^{\mathrm{T}} = \mathscr{P}^\pm \begin{bmatrix} \hat{\boldsymbol{n}} \times \boldsymbol{e}^\mp & \hat{\boldsymbol{n}} \times \boldsymbol{h}^\mp \end{bmatrix}^{\mathrm{T}}; \tag{7.23}$$

overall, we can conclude that $\left(\mathscr{P}^\pm\right)^2 = \mathscr{P}^\pm$. This result, combined with the identity $\mathscr{P}^+ + \mathscr{P}^- = \mathscr{I}$, demonstrates that $\mathscr{P}^\pm(\mathscr{P}^+ + \mathscr{P}^-) = \mathscr{P}^\pm$ and, consequently, that

$\mathcal{P}^{\pm}\mathcal{P}^{\mp} = \mathbf{0}$, which means that the projectors are orthogonal to each other. Using this orthogonality relationship, we can deduce that

$$\begin{bmatrix} \mathcal{I}/2 - \mathcal{K}_k & \eta_0 \mathcal{T}_k \\ -\eta_0^{-1}\mathcal{T}_k & \mathcal{I}/2 - \mathcal{K}_k \end{bmatrix} \begin{bmatrix} \mathcal{I}/2 + \mathcal{K}_k & -\eta_0 \mathcal{T}_k \\ \eta_0^{-1}\mathcal{T}_k & \mathcal{I}/2 + \mathcal{K}_k \end{bmatrix}$$
$$= \begin{bmatrix} \mathcal{I}/4 - \mathcal{K}_k^2 + \mathcal{T}_k^2 & \eta_0 \left(\mathcal{T}_k\mathcal{K}_k + \mathcal{K}_k\mathcal{T}_k\right) \\ -\eta_0^{-1}\left(\mathcal{T}_k\mathcal{K}_k + \mathcal{K}_k\mathcal{T}_k\right) & \mathcal{I}/4 - \mathcal{K}_k^2 + \mathcal{T}_k^2 \end{bmatrix} = \begin{bmatrix} \mathbf{0} & \mathbf{0} \\ \mathbf{0} & \mathbf{0} \end{bmatrix}, \tag{7.24}$$

which establishes the well-known Calderón identities

$$\mathcal{T}_k^2 = -\mathcal{I}/4 + \mathcal{K}_k^2 = \mathcal{M}_k^+\mathcal{M}_k^-, \tag{7.25}$$
$$\mathcal{T}_k\mathcal{K}_k = -\mathcal{K}_k\mathcal{T}_k \text{ or } \mathcal{T}_k\mathcal{M}_k^+ = -\mathcal{M}_k^-\mathcal{T}_k. \tag{7.26}$$

7.4 Discretization

The numerical solution of the EFIE and MFIE requires their discretization, often based on a triangulation of Γ, to form linear systems. Following a standard scheme, the unknown surface current density is expanded into N Rao–Wilton–Glisson (RWG) [1] basis functions $\{\boldsymbol{f}_n\}$

$$\boldsymbol{j}_\Gamma(\boldsymbol{r}) \approx \sum_{n=1}^{N} [j]_n \boldsymbol{f}_n(\boldsymbol{r}), \tag{7.27}$$

where N is the number of edges (of average size h) of the mesh and where for each inner edge, shared by the triangular cells c_n^+ and c_n^- and connecting the vertices r_n^+ and r_n^-, the function $\boldsymbol{f}_n$ is defined as

$$\boldsymbol{f}_n(\boldsymbol{r}) := \begin{cases} \dfrac{\boldsymbol{r} - \boldsymbol{r}_n^+}{2A_{c_n^+}} & \text{for } \boldsymbol{r} \in c_n^+, \\[2ex] \dfrac{\boldsymbol{r}_n^- - \boldsymbol{r}}{2A_{c_n^-}} & \text{for } \boldsymbol{r} \in c_n^-, \end{cases} \tag{7.28}$$

with $A_{c_n^\pm}$ being the area of the cell $c_n^\pm$. The linear systems are then obtained by Petrov–Galerkin testing. In the case of the EFIE, rotated RWG functions $\{\hat{\boldsymbol{n}} \times \boldsymbol{f}_n\}$ are used, which results in the discrete system

$$\eta_0[T_k][j] = \eta_0 \left(\mathrm{i}k\,[T_{\mathrm{A},k}] + 1/(\mathrm{i}k)\,[T_{\Phi,k}]\right)[j] = -[e^{\mathrm{i}}] \tag{7.29}$$

where

$$[T_{\mathrm{A},k}]_{nm} := \langle \hat{\boldsymbol{n}} \times \boldsymbol{f}_n, \mathcal{T}_{\mathrm{A},k}\boldsymbol{f}_m \rangle_\Gamma, \tag{7.30}$$
$$[T_{\Phi,k}]_{nm} := \langle \hat{\boldsymbol{n}} \times \boldsymbol{f}_n, \mathcal{T}_{\Phi,k}\boldsymbol{f}_m \rangle_\Gamma, \tag{7.31}$$
$$[e^{\mathrm{i}}]_n := \langle \hat{\boldsymbol{n}} \times \boldsymbol{f}_n, \hat{\boldsymbol{n}} \times \boldsymbol{e}^{\mathrm{i}} \rangle_\Gamma, \tag{7.32}$$

with $\langle \boldsymbol{f}, \boldsymbol{g} \rangle_\Gamma := \int_\Gamma \boldsymbol{f}(\boldsymbol{r}) \cdot \boldsymbol{g}(\boldsymbol{r}) \mathrm{d}S(\boldsymbol{r})$.

In the case of the MFIE, to avoid forming a singular Gram matrix when discretizing the identity operator, rotated Buffa–Christiansen (BC) [69] functions $\{\hat{\boldsymbol{n}} \times \tilde{\boldsymbol{f}}_n\}$ must be used instead of rotated RWG functions [70], which results in the linear system

$$\left[M_k^+\right][j] := \left(1/2\left[G_{\hat{\boldsymbol{n}}\times\tilde{\boldsymbol{f}},\boldsymbol{f}}\right] + [K_k]\right)[j] = [\tilde{h}^{\mathrm{i}}], \tag{7.33}$$

with

$$[K_k]_{nm} := \langle \hat{\boldsymbol{n}} \times \tilde{\boldsymbol{f}}_n, \mathscr{K}_k \boldsymbol{f}_m \rangle_\Gamma, \tag{7.34}$$

$$[\tilde{h}^{\mathrm{i}}]_n := \langle \tilde{\boldsymbol{n}} \times \tilde{\boldsymbol{f}}_n, \hat{\boldsymbol{n}} \times \boldsymbol{h}^{\mathrm{i}} \rangle_\Gamma, \tag{7.35}$$

and where we define the Gram matrix between the bases $\{\boldsymbol{f}\}$ and $\{\tilde{\boldsymbol{f}}\}$ as

$$\left[G_{\boldsymbol{f},\tilde{\boldsymbol{f}}}\right]_{mn} := \langle \boldsymbol{f}_m, \tilde{\boldsymbol{f}}_n \rangle_\Gamma. \tag{7.36}$$

A formal definition of the BC functions has been omitted due to space constraints but can be found in [69].

In addition to the definitions required for the numerical solution of the standard equations, we introduce others that will become useful later on when introducing preconditioning schemes. First, the discretization of the EFIE operator with BC functions is

$$\left[\tilde{T}_k\right]_{nm} := \langle \hat{\boldsymbol{n}} \times \tilde{\boldsymbol{f}}_n, \mathscr{T}_k \tilde{\boldsymbol{f}}_m \rangle_\Gamma. \tag{7.37}$$

As we will become clear in the following, some of the instabilities of the standard schemes are caused by a behavior of the electromagnetic operators that differ depending on whether or not the current density $\boldsymbol{j}_\Gamma$ to which they are applied is divergence free. To analyze and address these limitations, it is necessary to introduce tools capable of performing a discrete (quasi-)Helmholtz decomposition. We opt for the loop-star decomposition that decomposes the RWG space into loop, star, and global loop bases that are composed of solenoidal, non-solenoidal, and quasi-harmonic functions. Given our choice of normalization of the RWG functions, the loop-to-RWG and star-to-RWG matrices are, respectively,

$$[\Lambda]_{ij} = \begin{cases} 1 & \text{for } \boldsymbol{v}_j = \boldsymbol{v}_i^- \\ -1 & \text{for } \boldsymbol{v}_j = \boldsymbol{v}_i^+ \\ 0 & \text{otherwise,} \end{cases} \text{ and } [\Sigma]_{ij} = \begin{cases} 1 & \text{for } c_j = c_i^+ \\ -1 & \text{for } c_j = c_i^- \\ 0 & \text{otherwise,} \end{cases} \tag{7.38}$$

where $\boldsymbol{v}_j$ is the jth vertex of the mesh, c_j is the jth cell of the mesh, $\boldsymbol{v}_i^\pm$ are the vertices of the oriented segment on which the RWG function $\boldsymbol{f}_i$ is defined, and $c_i^\pm$ are the ordered triangular cells forming its support in (7.28). No such simple definition is available for the global-loop-to-RWG matrix $[H]$ because it requires cycle finding algorithms to be used.

7.5 Electric field IE

7.5.1 The original equation

The standard EFIE equation suffers from an ill-conditioning related to the discretization density of the mesh of the scatterer. This can be evidenced by studying the condition number of the EFIE discretized on a given structure and progressively increasing the discretization density (Figure 7.1). The condition number of the system matrix will grow proportionally to h^{-2} (assuming a uniform mesh). This is in contrast with the MFIE matrices whose condition number does not depend on h, since it is a second-kind IE. A particularly simple analysis of the ill-conditioning of $[T_k]$ can be obtained for a spherical scatterer. In this case, in fact, we can consider a discretization of $\mathcal{T}_k$ with vector spherical harmonics (VSHs) as basis and test functions, which leads to a diagonal stiffness matrix [54,71]. Let

$$Y_{lm}(\vartheta,\varphi) = \sqrt{\frac{2l+1}{4\pi}}\sqrt{\frac{(l-m)!}{(l+m)!}}P_l^m(\cos\vartheta)\mathrm{e}^{\mathrm{i}m\varphi} \tag{7.39}$$

denote the scalar spherical harmonics, where $l \geq 0$ and $|m| < l$ and P_l^m are the associated Legendre functions. The spherical harmonics form a complete basis of scalar functions on $\mathbb{S}^2$ and give rise to a complete basis of tangential vector fields

$$\boldsymbol{X}_{lm}(\vartheta,\varphi) = \frac{a}{\mathrm{i}m\sqrt{l(l+1)}}\hat{\boldsymbol{n}} \times \nabla Y_{lm}(\vartheta,\varphi), \tag{7.40}$$

$$\boldsymbol{U}_{lm}(\vartheta,\varphi) = -\frac{a}{\mathrm{i}\sqrt{l(l+1)}}\nabla Y_{lm}(\vartheta,\varphi), \tag{7.41}$$

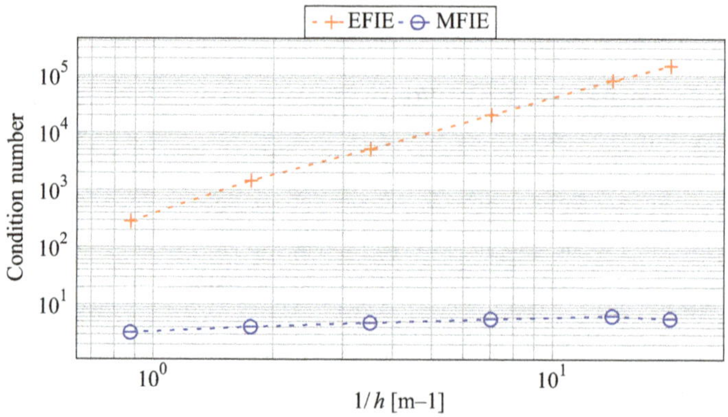

Figure 7.1 Condition number of the standard EFIE (7.29) *and MFIE* (7.33) *system matrices for an increasingly discretized cube of 1 m sides*

where, by construction, $\boldsymbol{X}_{lm}$ are solenoidal and $\boldsymbol{U}_{lm}$ are irrotational. The vector spherical harmonics are orthogonal with respect to the weighted inner product

$$\langle \boldsymbol{u}, \boldsymbol{v} \rangle = \int_{\mathbb{S}^2} \overline{\boldsymbol{u}} \cdot \boldsymbol{v} a^{-2} \mathrm{d}S(\boldsymbol{r}), \tag{7.42}$$

that is,

$$\langle \boldsymbol{X}_{lm}, \boldsymbol{X}_{l'm'} \rangle = \delta_l^{l'} \delta_m^{m'}, \tag{7.43}$$

$$\langle \boldsymbol{U}_{lm}, \boldsymbol{U}_{l'm'} \rangle = \delta_l^{l'} \delta_m^{m'}, \tag{7.44}$$

$$\langle \boldsymbol{X}_{lm}, \boldsymbol{U}_{l'm'} \rangle = 0, \tag{7.45}$$

where $\delta_m^{m'}$ denotes the Kronecker delta. Furthermore, we need to introduce the Riccati–Bessel function $\mathbb{J}_l(z) = \sqrt{\pi z/2}\, \mathrm{J}_{l+1/2}(z)$ and Riccati–Hankel function $\mathbb{H}_l(z) = \sqrt{\pi z/2} \mathrm{H}_{l+1/2}(z)$, where J_l is the Bessel function of the first kind and H_l is the Hankel function of the first kind. When the VSHs are applied to the EFIE and MFIE operator, we find

$$\mathscr{T}_k(\boldsymbol{X}_{lm}) = -\mathbb{J}_l(ka)\mathbb{H}_l(ka)\boldsymbol{U}_{lm}, \tag{7.46}$$

$$\mathscr{T}_k(\boldsymbol{U}_{lm}) = +\mathbb{J}_l'(ka)\mathbb{H}_l'(ka)\boldsymbol{X}_{lm}, \tag{7.47}$$

and

$$\mathscr{K}_k(\boldsymbol{X}_{lm}) = +\frac{\mathrm{i}}{2} \left(\mathbb{J}_l'(ka)\mathbb{H}_l(ka) + \mathbb{J}_l(ka)\mathbb{H}_l'(ka) \right) \boldsymbol{X}_{lm}, \tag{7.48}$$

$$\mathscr{K}_k(\boldsymbol{U}_{lm}) = -\frac{\mathrm{i}}{2} \left(\mathbb{J}_l(ka)\mathbb{H}_l'(ka) + \mathbb{J}_l'(ka)\mathbb{H}_l(ka) \right) \boldsymbol{U}_{lm}. \tag{7.49}$$

While obtaining an analytic expression for the condition number of the discretization of the EFIE operator on a sphere with RWGs as basis and testing functions is not straightforward, we are able to obtain such an expression if VSHs are used as basis and testing functions. To assess the condition number as a function of h, we use the correspondence of the spectral index l with the discrete spectral index which is proportional to a/h. We note that similar to the number of RWGs, which grows quadratically in $1/h$, the number of spherical harmonics grows quadratically in l as $|m| \leq l$.

Due to (7.43)–(7.45) and to the mapping properties of the operator $\mathscr{T}_k$ (7.46)–(7.47), the stiffness matrix is diagonal and the condition number can be computed from the ratio of the absolute value of the largest and smallest diagonal entries. Since we are interested in the dense-discretization breakdown, that is, when $h \to 0$, we consider the asymptotic behavior of the Riccati–Bessel and –Hankel functions for $l \to \infty$; we have [72, Section 10.19]

$$\mathbb{J}_l(ka) \to \frac{1}{\sqrt{2\mathrm{e}}}\left(\frac{\mathrm{e}ka}{2l+1}\right)^{l+1}, \tag{7.50}$$

$$\mathbb{J}_l'(ka) \to \sqrt{\frac{\mathrm{e}}{2}}\frac{l+1}{2l+1}\left(\frac{\mathrm{e}ka}{2l+1}\right)^{l}, \tag{7.51}$$

$$\mathbb{H}_l(ka) \to -\mathrm{i}\sqrt{\frac{2}{\mathrm{e}}}\left(\frac{2l+1}{\mathrm{e}ka}\right)^{l}, \tag{7.52}$$

$$\mathbb{H}_l'(ka) \to \frac{\mathrm{i}l\sqrt{2\mathrm{e}}}{2l+1}\left(\frac{2l+1}{\mathrm{e}ka}\right)^{l}. \tag{7.53}$$

From the asymptotic behavior, it is clear that

$$|\mathbb{J}_l(ka)\mathbb{H}_l(ka)| = \mathcal{O}(l^{-1}), \tag{7.54}$$

$$|\mathbb{J}_l'(ka)\mathbb{H}_l'(ka)| = \mathcal{O}(l). \tag{7.55}$$

Thus, given our previous consideration, the condition number scales as $\mathcal{O}(l^2)$ corresponding to the well-known $\mathcal{O}(h^{-2})$ ill-conditioning of the RWG-based system matrix.

The spherical harmonics analysis of the EFIE operator has also shown that its spectrum is composed of two branches that correspond to the non-solenoidal and solenoidal parts of the spectrum: the first branch contains singular values that will diverge towards infinity as $1/h$ when $h \to 0$, while the second branch clusters at 0 as h when $h \to 0$. This structure of the spectrum is also inherited by standard RWG-discretized systems (Figure 7.2) and will play a crucial role in identifying cures for this dense-discretization ill-conditioning. One can show that the two branches, respectively, correspond to the spectral contributions of the scalar and vector potentials.

An additional difficulty that arises when handling the EFIE system matrices is that they become increasingly ill-conditioned as the frequency decreases with a fixed discretization. This second source of ill-conditioning can be evidenced by studying the solenoidal and non-solenoidal parts of the spectrum. The solenoidal nullspace of the scalar potential, which is the space of divergence-free functions (or loops), causes the spectrum of the overall matrix to behave differently on the solenoidal and non-solenoidal subspaces. This phenomenon translates into a spectrum composed of two branches diverging away from each other as the frequency decreases: one branch grows as k^{-1} and the other decreases as k when $k \to 0$ (Figure 7.3).

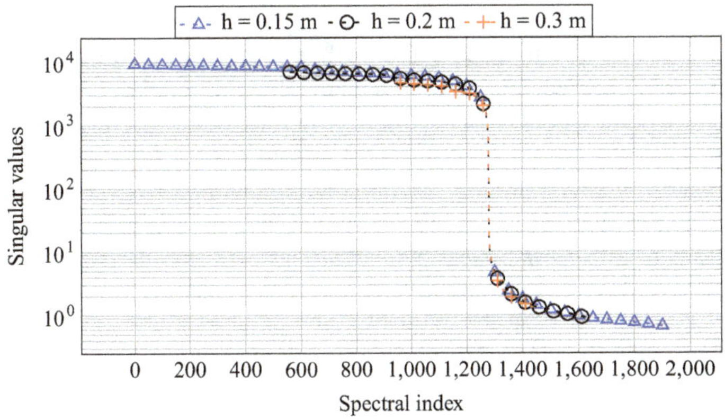

Figure 7.2 Spectra of the EFIE for three different discretization densities showing the two branches diverging away from each other. The lower discretization spectral indices have been offset for clarity. The spectra have been obtained with a sphere of 1 m radius and a frequency of 1×10^7 Hz.

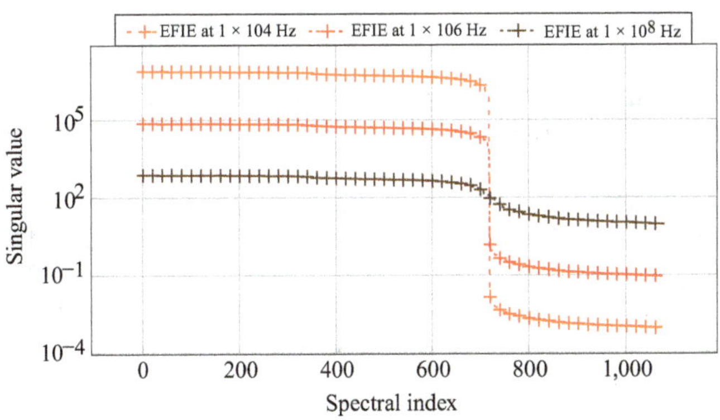

Figure 7.3 Singular value spectra of the EFIE matrix for three different frequencies showing the two branches of the spectrum diverging away from each other as the frequency decreases, in the case of a sphere of 1 m radius and edge length 0.2 m

To better illustrate this effect, we leverage the loop-star decomposition introduced earlier and apply it to the EFIE system matrix—for a simply-connected geometry—to obtain its decomposed form

$$\left[T_k^{\mathrm{LS}}\right] := [[\Lambda] \quad [\Sigma]]^{\mathrm{T}} [T_k] [[\Lambda] \quad [\Sigma]] \tag{7.56}$$

$$= \begin{bmatrix} \mathrm{i}k[\Lambda]^{\mathrm{T}}[T_{\mathrm{A},k}][\Lambda] & \mathrm{i}k[\Lambda]^{\mathrm{T}}[T_{\mathrm{A},k}][\Sigma] \\ \mathrm{i}k[\Sigma]^{\mathrm{T}}[T_{\mathrm{A},k}][\Lambda] & [\Sigma]^{\mathrm{T}}\left(\mathrm{i}k[T_{\mathrm{A},k}] + (\mathrm{i}k)^{-1}[T_{\Phi,k}]\right)[\Sigma] \end{bmatrix} \tag{7.57}$$

where we have used the property $[\Lambda]^{\mathrm{T}}[T_{\Phi,k}] = [0]$ and $[T_{\Phi,k}][\Lambda] = [0]$. Clearly, a part of the matrix elements will diverge as k^{-1} while others will converge to 0 as k. The Gershgorin circle theorem can be used to show that the condition number of $\left[T_k^{\mathrm{LS}}\right]$ behaves as $\mathcal{O}(k^{-2})$ as k goes to 0 [18]. Finally, because $[[\Lambda] \quad [\Sigma]]$ is frequency independent and invertible, we conclude that $[T_k]$ suffers from the same ill-conditioning in frequency as $\left[T_k^{\mathrm{LS}}\right]$; note, however, that the condition number of $\left[T_k^{\mathrm{LS}}\right]$ is significantly worse than that of the standard EFIE matrix because the $[\Lambda]$ and $[\Sigma]$ matrices are discretizations of differential operators that are intrinsically ill-conditioned [73,74]. This results trivially extends to multiply-connected geometries [59].

7.5.2 The preconditioned equation

Several schemes have been presented to cure the EFIE from the sources of its ill-conditioning in both frequency and refinement. For instance, the loop-star decomposition introduced to demonstrate the EFIE low-frequency breakdown can be used to regularize the original equation's behavior when it is coupled with a diagonal preconditioning. However, in this contribution, we will focus on the Calderón approaches.

The Calderón identity (7.25) indicates that applying $\mathcal{T}_k$ to itself results in a well-conditioned operator, since this yields the sum of an identity and a compact operator. As such, $\mathcal{T}_k$ can be used as its own—almost ideal—preconditioner: applying on both sides of (7.1) would yield a well-conditioned system at the price of an additional matrix–vector product. While this appears simple in practice, complications arise when discretizing the equation. Indeed, $\mathcal{T}_k$ maps $\mathcal{H}_{\mathrm{div}}^{-1/2}$ into itself and as such one could think of preconditioning (7.1) by multiplying it by an RWG discretized $\mathcal{T}_k$ operator and the inverse of the corresponding discretized identity $[G_{\hat{n}\times f,f}]^{-1}$. However, the $[G_{\hat{n}\times f,f}]^{-1}$ Gram matrix is singular and cannot be inverted stably [19]. Instead, one can use the dual-discretized counterpart $[\tilde{T}_k][G_{\hat{n}\times f,\tilde{f}}]^{-1}$ as a preconditioner, because $[G_{\hat{n}\times f,\tilde{f}}]^{-1}$ is non-singular. The preconditioned system then reads

$$\eta_0[\tilde{T}_k][G_{\hat{n}\times f,\tilde{f}}]^{-1}[T][k][j] = -[\tilde{T}_k][G_{\hat{n}\times f,\tilde{f}}]^{-1}[e^i]. \tag{7.58}$$

Another way to understand the effects of Calderón preconditioning is to develop the product

$$\mathcal{T}_k^2 = \left(\mathrm{i}k\,\mathcal{T}_{\mathrm{A},k} + (\mathrm{i}k)^{-1}\,\mathcal{T}_{\Phi,k}\right)\left(\mathrm{i}k\,\mathcal{T}_{\mathrm{A},k} + (\mathrm{i}k)^{-1}\,\mathcal{T}_{\Phi,k}\right) \tag{7.59}$$

$$= -k^2\,\mathcal{T}_{\mathrm{A},k}^2 + \mathcal{T}_{\mathrm{A},k}\mathcal{T}_{\Phi,k} + \mathcal{T}_{\Phi,k}\mathcal{T}_{\mathrm{A},k} - k^{-2}\,\mathcal{T}_{\Phi,k}^2. \tag{7.60}$$

Pseudo-differential operator theory indicates that the first term is compact, that the sum of the second and third terms is spectrally equivalent to an identity, while the last term would have derivative strength (see [18] and references therein). This last term would compromise the overall conditioning of the preconditioned operator, however, $\mathcal{T}^2_{\Phi,k}$ vanishes since

$$\left(\mathcal{T}_{\Phi,k}\mathcal{T}_{\Phi,k}\boldsymbol{x}\right)(\boldsymbol{r}) = \hat{\boldsymbol{n}} \times \nabla_\Gamma \int_\Gamma G_k\left(\boldsymbol{r},\boldsymbol{r}'\right) \nabla'_\Gamma \cdot \left(\hat{\boldsymbol{n}}' \times \nabla'_\Gamma \int_\Gamma G_k\left(\boldsymbol{r}',\boldsymbol{r}''\right) \nabla''_\Gamma \cdot \boldsymbol{x}\left(\boldsymbol{r}''\right) \mathrm{d}S\left(r''\right)\right) \mathrm{d}S\left(r'\right) \quad (7.61)$$

and simple passages show that $\nabla'_\Gamma \cdot \left(\hat{\boldsymbol{n}}' \times \nabla'_\Gamma(\mathcal{O}\boldsymbol{x})(\boldsymbol{r}')\right) = 0$—because it is the surface divergence of a surface curl—with

$$(\mathcal{O}\boldsymbol{x})\left(\boldsymbol{r}'\right) := \int_\Gamma G_k\left(\boldsymbol{r}',\boldsymbol{r}''\right) \nabla''_\Gamma \cdot \boldsymbol{x}\left(\boldsymbol{r}''\right) \mathrm{d}S\left(r''\right). \quad (7.62)$$

In addition to the h-refinement preconditioning effect of the Calderón identities, a partial low-frequency regularization also occurs. Indeed,

$$\lim_{k\to 0} \mathcal{T}_k^2(\boldsymbol{r}) = \left(\mathcal{T}_{\mathrm{A},0}\mathcal{T}_{\Phi,0} + \mathcal{T}_{\Phi,0}\mathcal{T}_{\mathrm{A},0}\right)(\boldsymbol{r}), \quad \forall \boldsymbol{r} \in \Gamma, \quad (7.63)$$

thus ensuring that the low-frequency limit is well-defined. However, despite this limit being well defined, the resulting operator admits the same magnetostatic nullspace as $\mathcal{M}_k^+\mathcal{M}_k^-$ [75]—by virtue of (7.25). In practice, this means that $\mathcal{T}_k^2$ will be well conditioned until arbitrarily low frequencies if the underlying geometry is simply connected, but, if the geometry is multiply connected, the operator will become increasingly ill-conditioned as the frequency decreases until it becomes singular in the static limit (Figure 7.4) [75]. The range of applicability of the formulation is, however, greatly increased with regard to the original EFIE. Finally, an additional effect occurs that reduces the frequency range to which this method can be applied: the computation of some of the integrals appearing in the discretization procedure becomes increasingly unstable as the frequency decreases. Solving this instability requires the use of a quasi-Helmholtz decomposition to allow the vanishing of certain integrals to be enforced [59]. For instance, consider the case of a plane-wave excitation $\boldsymbol{e}^{\mathrm{i}}_{\mathrm{PW}}$ in the EFIE, for which the components are of the form

$$\left[e^{\mathrm{i}}_{\mathrm{PW}}\right]_n = \langle \hat{\boldsymbol{n}} \times \boldsymbol{f}_n, \hat{\boldsymbol{n}} \times \boldsymbol{e}^{\mathrm{i}}_{\mathrm{PW}}\rangle_\Gamma = \boldsymbol{E}_0 \cdot \int_\Gamma \boldsymbol{f}_n(\boldsymbol{r}) \exp\left(\mathrm{i}k\hat{\boldsymbol{k}} \cdot \boldsymbol{r}\right) \mathrm{d}S(\boldsymbol{r}). \quad (7.64)$$

One can easily show that when tested with a solenoidal function, the static term—corresponding to the first term in a Taylor series expansion of the exponential in k—of this equation vanishes [59]. However, because of limitations in standard numerical integration schemes, the static contribution in this integral will not vanish and will dominate the frequency-dependent remainder. As a result, the frequency behavior of the discrete right-hand-side will be lost. To obtain the correct EFIE solution, instead, this term should be explicitly enforced to cancel out. This can be done by performing an Helmholtz decomposition of the right-hand-side. This could be achieved through loop-star decompositions, however, these techniques yield ill-conditioned Gram matrices and require the computationally burdensome detection of global loops

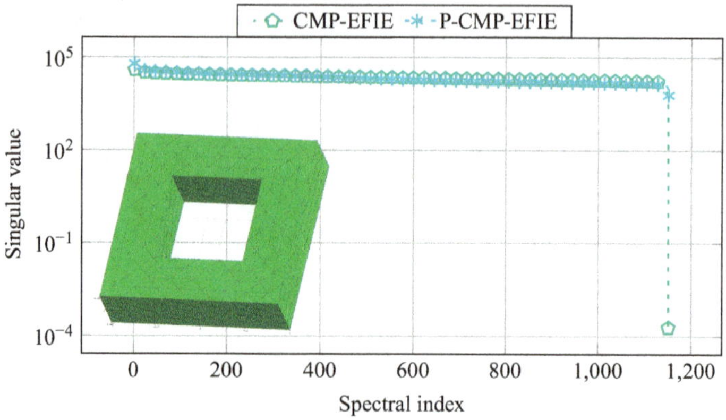

Figure 7.4 Singular values of the CMP-EFIE (7.58) *and P-CMP-EFIE* (7.72) *matrices indicating the presence of the magnetostatic nullspace in the CMP-EFIE when computed on a multiply-connected geometry—here a torus with square cross-section simulated at* 1×10^4 *Hz*

or handles. An alternative approach is to combine the Calderón preconditioning with quasi-Helmholtz projectors that do not degrade the Gram matrices conditioning and do not require the detection of the global loops.

The quasi-Helmholtz projectors are defined as [18]

$$[P_{\Sigma}] := [\Sigma]([\Sigma]^{\mathrm{T}}[\Sigma])^{+}[\Sigma]^{\mathrm{T}}, \tag{7.65}$$

$$[P_{\Lambda\mathrm{H}}] := [\mathrm{I}] - [P_{\Sigma}], \tag{7.66}$$

where $^+$ denotes the Moore–Penrose pseudoinverse. The projector $[P_{\Sigma}]$ projects any RWG coefficient vector to its non-solenoidal part, while $[P_{\Lambda\mathrm{H}}]$ projects it to its solenoidal part. The counterparts of these projectors for dual discretizations are

$$[P_{\Lambda}] := [\Lambda]([\Lambda]^{\mathrm{T}}[\Lambda])^{+}[\Lambda]^{\mathrm{T}}, \tag{7.67}$$

$$[P_{\Sigma\mathrm{H}}] := [I] - [P_{\Lambda}], \tag{7.68}$$

where $[P_{\Lambda}]$ and $[P_{\Sigma\mathrm{H}}]$ are the dual non-solenoidal and solenoidal projectors. These projectors can be used to form the preconditioners

$$[P_k] = (k/d)^{-1/2}[P_{\Lambda\mathrm{H}}] + \mathrm{i}(k/d)^{1/2}[P_{\Sigma}], \tag{7.69}$$

$$[\tilde{P}_k] = (k/d)^{-1/2}[P_{\Sigma\mathrm{H}}] + \mathrm{i}(k/d)^{1/2}[P_{\Lambda}], \tag{7.70}$$

for the primal and dual EFIE system matrices, where d is a constant introduced to ensure that the scheme is scale-invariant and that the dimensions are consistent; a typical choice for d is the diameter of the bounding sphere of Ω^-. In the following developments, we will use d =1 m. After preconditioning, the system matrices are

$$\begin{aligned}[P_k][T_k][P_k] = \; & \mathrm{i}[P_{\Lambda\mathrm{H}}][T_{\mathrm{A},k}][P_{\Lambda\mathrm{H}}] + \mathrm{i}[T_{\Phi,k}] \\ & -k\left([P_{\Lambda\mathrm{H}}][T_{\mathrm{A},k}][P_\Sigma] + [P_\Sigma][T_{\mathrm{A},k}][P_{\Lambda\mathrm{H}}]\right) \\ & -\mathrm{i}k^2[P_\Sigma][T_{A,k}][P_\Sigma] \end{aligned} \tag{7.71}$$

and its low-frequency limit $\lim_{k\to 0}[P_k][T_k][P_k] = \mathrm{i}[P_{\Lambda\mathrm{H}}][T_{\mathrm{A},0}][P_{\Lambda\mathrm{H}}] + \mathrm{i}[T_{\Phi,0}]$ is well defined and has no nullspace; for a proof of this last statement, we refer the reader to [59, Section IV]. Because the right-hand side is also decomposed, one can enforce that the relevant integrals vanish thus alleviating the issue preventing the accurate computation of (7.64). Note that the factors $k^{\pm 1/2}$ used in (7.69) and (7.70) are sufficient to cure the low-frequency breakdown but are not always optimal. A different choice of coefficients, based on operator norms, that often yields lower condition numbers for the preconditioned EFIE matrices has been presented in [18].

With these decompositions in place, a low-frequency and dense-discretization stable EFIE can be obtained by using the low-frequency-stabilized EFIE matrices in the Calderón scheme which results in the equation

$$\begin{aligned} & \eta_0[\tilde{P}_k][\tilde{T}_k][\tilde{P}_k][G_{\hat{\boldsymbol{n}}\times f,\tilde{f}}]^{-1}[P_k][T_k][P_k][y] = \\ & -[\tilde{P}_k][\tilde{T}_k][\tilde{P}_k][G_{\hat{\boldsymbol{n}}\times f,\tilde{f}}]^{-1}[P_k][e^{\mathrm{i}}] \end{aligned} \tag{7.72}$$

where $[j] = [P_k][y]$. The conditioning of this last formulation is compared against other EFIE-preconditioning strategies in Figure 7.5.

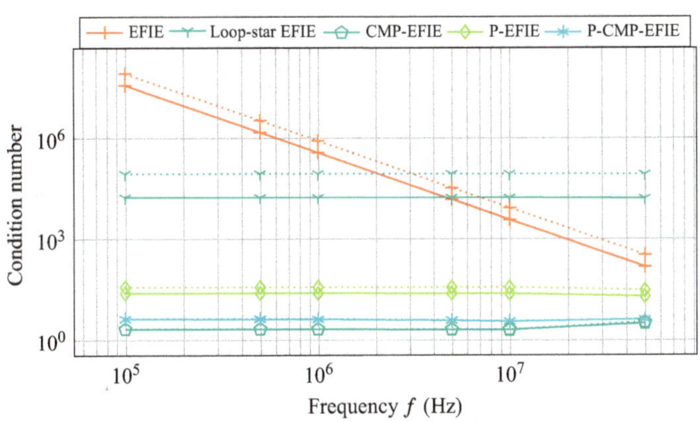

Figure 7.5 Condition number of the different EFIE formulations computed on a sphere of 1 m radius discretized with an average edge length of 0.2 m (dotted lines) and 0.3 m (solid lines), showing the curative effects of the Calderón approaches both in the low-frequency and high-refinement regimes for a simply-connected geometry

7.6 Combined field IE

7.6.1 The original equation

The combined field IE (CFIE) has been introduced as a way to obtain a uniquely solvable equation at high frequencies since it does not suffer from spurious resonances that both the EFIE and the MFIE are subject to [6]. The CFIE is formed as a linear combination of the EFIE (7.1) and the MFIE that can be expressed as

$$-\alpha\eta_0\mathcal{T}_k\boldsymbol{j}_\Gamma + (1-\alpha)\eta_0\hat{\boldsymbol{n}}\times\mathcal{M}_k^+\boldsymbol{j}_\Gamma = \alpha\hat{\boldsymbol{n}}\times\boldsymbol{e}^{\mathrm{i}} + (1-\alpha)\eta_0\hat{\boldsymbol{n}}\times\hat{\boldsymbol{n}}\times\boldsymbol{h}^{\mathrm{i}}, \quad (7.73)$$

and is discretized as [18]

$$\begin{aligned}&\left(-\alpha\eta_0[T_k] + (1-\alpha)\eta_0[G_{f,f}][G_{\hat{n}\times\tilde{f},f}]^{-1}[M_k^+]\right)[j]\\&= \alpha[e^{\mathrm{i}}] + (1-\alpha)\eta_0[G_{f,f}][G_{\hat{n}\times\tilde{f},f}]^{-1}[h^{\mathrm{i}}],\end{aligned} \quad (7.74)$$

where the linear combination factor is typically chosen to be $0 < \alpha < 1$.

While it is immune to spurious resonances, this formulation inherits the critical flaws of the operators it is built out of: it suffers from both a dense-discretization and a low-frequency breakdown.

7.6.2 The preconditioned equation

One could think of solving the ill-conditioning of the standard CFIE by replacing the EFIE with its Calderón-preconditioned counterpart, which would result in

$$-\alpha\eta_0\mathcal{T}_k^2\boldsymbol{j}_\Gamma + (1-\alpha)\eta_0\hat{\boldsymbol{n}}\times\mathcal{M}_k^+\boldsymbol{j}_\Gamma = \alpha\mathcal{T}_k\left(\hat{\boldsymbol{n}}\times\boldsymbol{e}^{\mathrm{i}}\right) + (1-\alpha)\eta_0\hat{\boldsymbol{n}}\times\hat{\boldsymbol{n}}\times\boldsymbol{h}^{\mathrm{i}}. \quad (7.75)$$

However, a closer examination of the Calderón identities reveals that this equation would not be resonance-free. In fact, consider (7.75) in the light of (7.25):

$$\mathcal{T}_k^2 = -\left(\mathcal{I}/2 - \mathcal{K}_k\right)\left(\mathcal{I}/2 + \mathcal{K}_k\right) = \mathcal{M}_k^-\mathcal{M}_k^+, \quad (7.76)$$

which indicates that any element of the nullspace of $\mathcal{M}_k^+$ will also be in the nullspace of $\mathcal{T}_k^2$, and as such the equation will be resonating.

The traditional way to prevent this from happening is to use modified wavenumbers in the preconditioning operators. For instance, in the case of the EFIE, we consider $\mathcal{T}_{\mathrm{i}k}\mathcal{T}_k$ instead of $\mathcal{T}_k^2$, and we can find that

$$\mathcal{T}_{\mathrm{i}k}\mathcal{T}_k = -\mathrm{i}k^2\mathcal{T}_{\mathrm{A},\mathrm{i}k}\mathcal{T}_{\mathrm{A},k} + \mathrm{i}\mathcal{T}_{\mathrm{A},\mathrm{i}k}\mathcal{T}_{\Phi,k} - \mathrm{i}\mathcal{T}_{\Phi,\mathrm{i}k}\mathcal{T}_{\mathrm{A},k} + \mathrm{i}k^{-2}\mathcal{T}_{\Phi,\mathrm{i}k}\mathcal{T}_{\Phi,k}. \quad (7.77)$$

The presence of the different wavenumber in the preconditioning operators does not compromise the overall conditioning because it only introduces a compact perturbation when compared to non-modified approaches [45]. For instance, we have

$$\begin{aligned}\mathcal{T}_{\Phi,\mathrm{i}k}\mathcal{T}_{\mathrm{A},k} &= \left(\mathcal{T}_{\Phi,\mathrm{i}k} - \mathcal{T}_{\Phi,k} + \mathcal{T}_{\Phi,k}\right)\mathcal{T}_{\mathrm{A},k}\\&= \left(\mathcal{T}_{\Phi,\mathrm{i}k} - \mathcal{T}_{\Phi,k}\right)\mathcal{T}_{\mathrm{A},k} + \mathcal{T}_{\Phi,k}\mathcal{T}_{\mathrm{A},k},\end{aligned} \quad (7.78)$$

and because the operator $\mathcal{T}_{\Phi,\mathrm{i}k} - \mathcal{T}_{\Phi,k}$ has a smooth, non-singular kernel, the term $\left(\mathcal{T}_{\Phi,\mathrm{i}k} - \mathcal{T}_{\Phi,k}\right)\mathcal{T}_{\mathrm{A},k}$ only introduces a compact perturbation of $\mathcal{T}_{\Phi,k}\mathcal{T}_{\mathrm{A},k}$. A similar reasoning applies for the three first terms in (7.78) and the last term vanishes, which can be demonstrated using the same arguments as in (7.61). Thus, the Calderón-preconditioned EFIE with modified wavenumber is still well-conditioned in h, but does not share the resonances of $\mathcal{M}_k^+$ nor of $\mathcal{M}_k^- \mathcal{M}_k^+$. Resonance-free and second-kind combined field IEs can then be obtained as

$$\alpha\eta_0 \mathcal{T}_{\mathrm{i}k} \mathcal{T}_k j_\Gamma + (1-\alpha)\eta_0 \hat{\boldsymbol{n}} \times \mathcal{M}_k^+ j_\Gamma = -\alpha \mathcal{T}_{\mathrm{i}k} \left(\hat{\boldsymbol{n}} \times \boldsymbol{e}^{\mathrm{i}}\right) + (1-\alpha)\eta_0 \hat{\boldsymbol{n}} \times \hat{\boldsymbol{n}} \times \boldsymbol{h}^{\mathrm{i}}, \tag{7.79}$$

or, alternatively, as its more recent variant leveraging a symmetrized MFIE operator $\mathcal{M}_k^- \mathcal{M}_k^+$

$$\alpha\eta_0 \mathcal{T}_{\mathrm{i}k} \mathcal{T}_k j_\Gamma + (1-\alpha)\eta_0 \mathcal{M}_{\mathrm{i}k}^- \mathcal{M}_k^+ j_\Gamma = -\alpha \mathcal{T}_{\mathrm{i}k} \left(\hat{\boldsymbol{n}} \times \boldsymbol{e}^{\mathrm{i}}\right) + (1-\alpha)\eta_0 \mathcal{M}_{\mathrm{i}k}^- \left(\hat{\boldsymbol{n}} \times \boldsymbol{h}^{\mathrm{i}}\right). \tag{7.80}$$

These equations can be discretized as [18]

$$\left(\alpha\eta_0[\tilde{T}_{\mathrm{i}k}][G_{\hat{\boldsymbol{n}}\times f,\tilde{f}}]^{-1}[T_k] + (1-\alpha)\eta_0[G_{f,f}][G_{\hat{\boldsymbol{n}}\times\tilde{f},f}]^{-1}[M_k^+]\right)[j] = -\alpha[\tilde{T}_{\mathrm{i}k}][G_{\hat{\boldsymbol{n}}\times f,\tilde{f}}]^{-1}[e^{\mathrm{i}}] + (1-\alpha)\eta_0[G_{f,f}][G_{\hat{\boldsymbol{n}}\times\tilde{f},f}]^{-1}[h^{\mathrm{i}}], \tag{7.81}$$

and

$$\left(\alpha\eta_0[\tilde{T}_{\mathrm{i}k}][G_{\hat{\boldsymbol{n}}\times f,\tilde{f}}]^{-1}[T_k] + (1-\alpha)\eta_0[M_{\mathrm{i}k}^-][G_{\hat{\boldsymbol{n}}\times\tilde{f},f}][M_k^+]\right)[j] = -\alpha[\tilde{T}_{\mathrm{i}k}][G_{\hat{\boldsymbol{n}}\times f,\tilde{f}}]^{-1}[e^{\mathrm{i}}] + (1-\alpha)\eta_0[M_{\mathrm{i}k}^-][G_{\hat{\boldsymbol{n}}\times f,\tilde{f}}]^{-1}[h^{\mathrm{i}}], \tag{7.82}$$

The resonant or non-resonant behavior of the different formulations we have discussed are illustrated and summarized in Figure 7.6.

Finally, to achieve a broadband framework for electromagnetic scattering by PEC objects, the low-frequency behavior of both (7.81) and (7.82) must be fixed, since neither can be stably used in this regime—for the same reasons that the Calderón preconditioned EFIE could not. A full-wave formulation can be achieved by leveraging the Calderón-preconditioned EFIE and MFIE operators stabilized by the quasi-Helmholtz projectors. For instance, (7.82) becomes

$$\eta_0[\tilde{P}_k]\left([\tilde{T}_{\mathrm{i}k}][\tilde{P}_k][G_{\hat{\boldsymbol{n}}\times f,\tilde{f}}]^{-1}[P_k][T_k][P_k] + \xi\left([M_{\mathrm{i}k}^-][G_{\hat{\boldsymbol{n}}\times\tilde{f},f}]^{-1}[M_k^+] - [P_{\Sigma\mathrm{H}}][M_0^-][G_{\hat{\boldsymbol{n}}\times\tilde{f},f}]^{-1}[M_0^+][P_{\Lambda\mathrm{H}}]\right)\right)[P_k][y] = -[\tilde{P}_k]\left([\tilde{T}_{\mathrm{i}k}][\tilde{P}_k][G_{\hat{\boldsymbol{n}}\times f,\tilde{f}}]^{-1}[P_k][e^{\mathrm{i}}] - \xi\eta_0[M_{\mathrm{i}k}^-][G_{\hat{\boldsymbol{n}}\times\tilde{f},f}][h^{\mathrm{i}}]\right), \tag{7.83}$$

where the term $[P_{\Sigma\mathrm{H}}][M_0^-][G_{\hat{\boldsymbol{n}}\times\tilde{f},f}]^{-1}[M_0^+][P_{\Lambda\mathrm{H}}]$ is explicitly removed for numerical stability (see [18, Section IV.B.1] for more details) and where we have set $\alpha = 1/2$ but other choices are valid. Note that in high-frequency scenarios, the imaginary constant must be removed from the definition of the projectors or, alternatively, a suitable complex value must be chosen for ξ [18].

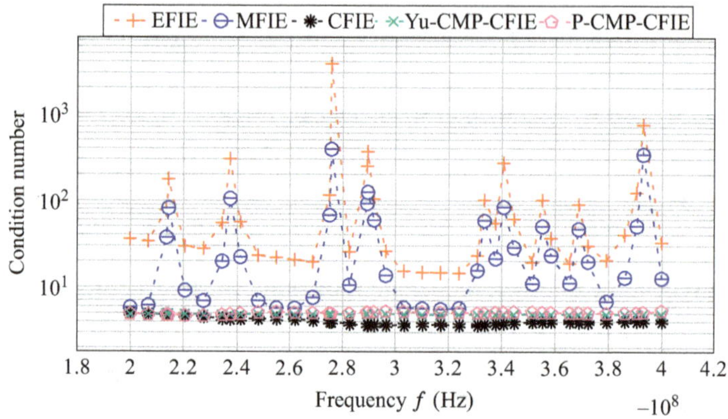

Figure 7.6 Conditioning of different formulations for increasing frequencies with a fixed discretization illustrating their resonating or non-resonating behavior. The matrices were obtained by discretizing a sphere of radius 1 m with an average edge length of 0.15 m. The label "CFIE" refers to (7.74), *"Yu-CMP-CFIE" to (7.82), and "P-CMP-CFIE" to* (7.83).

7.7 PMCHWT

7.7.1 The original equation

The Poggio-Miller-Chang-Harrington-Wu-Tsai (PMCHWT) formulation is one of the standard techniques for modeling scattering by penetrable bodies [61–63]. Consider a setting identical to that of Section 7.3, except that the space enclosed by the surface Γ now has permittivity $\varepsilon_1 := \varepsilon_0 \varepsilon_r' + \mathrm{i}\sigma/\omega$ and permeability μ_1 that differ from the permittivity and the permeability of the background ε_0 and μ_0, where $\varepsilon_r' \in \mathbb{R}$ is the relative permittivity of the medium and σ_1 is its conductivity. The wavenumbers in the media are $k_0 := \omega\sqrt{\mu_0\varepsilon_0}$ outside and $k_1 := \omega\sqrt{\mu_1\varepsilon_1}$ inside. If an incident field $\boldsymbol{e}^{\mathrm{i}}$ in the external domain Ω^+ impinges on the scatterer, the equivalent electric and magnetic current densities $\boldsymbol{j}_\Gamma$ and $\boldsymbol{m}_\Gamma$ that will radiate the scattered field $\boldsymbol{e}^{\mathrm{s}+}$ outside and $\boldsymbol{e}^{\mathrm{s}-}$ inside can be computed by solving the standard PMCHWT equation

$$\begin{bmatrix} \mathcal{K}_{k_0} + \mathcal{K}_{k_1} & \eta_0 \mathcal{T}_{k_0} + \eta_1 \mathcal{T}_{k_1} \\ \eta_0^{-1} \mathcal{T}_{k_0} + \eta_1^{-1} \mathcal{T}_{k_1} & -\left(\mathcal{K}_{k_0} + \mathcal{K}_{k_1}\right) \end{bmatrix} \begin{bmatrix} \boldsymbol{m}_\Gamma \\ \boldsymbol{j}_\Gamma \end{bmatrix} = - \begin{bmatrix} \hat{\boldsymbol{n}} \times \boldsymbol{e}^{\mathrm{i}} \\ \hat{\boldsymbol{n}} \times \boldsymbol{h}^{\mathrm{i}} \end{bmatrix}. \tag{7.84}$$

This form of the PMCHWT has, however, an intrinsically high condition number that can typically be brought down significantly by using a scaled electric current $\eta_0 \boldsymbol{j}_\Gamma$ as unknown [65]

$$\mathcal{Q}_{k_0,k_1} \begin{bmatrix} \boldsymbol{m}_\Gamma \\ \eta_0 \boldsymbol{j}_\Gamma \end{bmatrix} = - \begin{bmatrix} \hat{\boldsymbol{n}} \times \boldsymbol{e}^{\mathrm{i}} \\ \eta_0 \hat{\boldsymbol{n}} \times \boldsymbol{h}^{\mathrm{i}} \end{bmatrix}, \tag{7.85}$$

where

$$\mathscr{Q}_{k_0,k_1} := \begin{bmatrix} \mathscr{K}_{k_0} + \mathscr{K}_{k_1} & \mathscr{T}_{k_0} + \eta_1\eta_0^{-1}\mathscr{T}_{k_1} \\ \mathscr{T}_{k_0} + \eta_0\eta_1^{-1}\mathscr{T}_{k_1} & -\left(\mathscr{K}_{k_0} + \mathscr{K}_{k_1}\right) \end{bmatrix}. \tag{7.86}$$

The standard discretization of the PMCHWT equation results in the matrix system

$$[Q_{k_0,k_1}]\begin{bmatrix} [m] \\ \eta_0[j] \end{bmatrix} = -\begin{bmatrix} [e^{\mathrm{i}}] \\ \eta_0[h^{\mathrm{i}}] \end{bmatrix}, \tag{7.87}$$

with

$$[Q_{k_0,k_1}] := \begin{bmatrix} [\tilde{K}_{k0}] + [\tilde{K}_{k1}] & [T_{k_0}] + \eta_1\eta_0^{-1}[T_{k_1}] \\ [T_{k_0}] + \eta_0\eta_1^{-1}[T_{k_1}] & -\left([\tilde{K}_{k0}] + [\tilde{K}_{k1}]\right) \end{bmatrix}, \tag{7.88}$$

where $[m]$ is the unknown vector of the expansion coefficients of $\boldsymbol{m}_\Gamma$ in the basis of RWG functions, the $\mathscr{T}$ operators and $[e^{\mathrm{i}}]$ are discretized in the same way as in the EFIE (7.29), and

$$[\tilde{K}_k]_{nm} := \langle \hat{\boldsymbol{n}} \times \boldsymbol{f}_n, \mathscr{K}_k \boldsymbol{f}_m \rangle_\Gamma, \tag{7.89}$$

$$[h^{\mathrm{i}}]_n := \langle \hat{\boldsymbol{n}} \times \boldsymbol{f}_n, \hat{\boldsymbol{n}} \times \boldsymbol{h}^{\mathrm{i}} \rangle_\Gamma. \tag{7.90}$$

Similar to the EFIE, upon discretization, this equation will yield ill-conditioned system matrices for highly discretized objects or low frequencies. It also suffers from an additional source of ill-conditioning when the contrast between the materials composing the scatterer and the background increases [67]. The source of the dense-discretization breakdown can be identified by splitting the PMCHWT system matrix as

$$[Q_{k_0,k_1}] = \begin{bmatrix} [\tilde{K}_{k0}] + [\tilde{K}_{k1}] & [0] \\ [0] & -[\tilde{K}_{k0}] - [\tilde{K}_{k1}] \end{bmatrix} + \begin{bmatrix} [0] & [T_{k_0}] + \eta_1\eta_0^{-1}[T_{k_1}] \\ [T_{k_0}] + \eta_0\eta_1^{-1}[T_{k_1}] & [0] \end{bmatrix} \tag{7.91}$$

in which the first matrix has eigenvalues clustering at 0, since its spectrum is the union of that of system matrices stemming from compact operators, and the second, matrix behaves as the union of the spectra of sums of EFIEs, which, as we have seen in Section 7.5, are ill-conditioned in the dense-discretization regime; as a result, we can conclude that the overall operator will lead to ill-conditioned system matrices when discretized with an $\boldsymbol{L}^2$-stable basis [76].

The low-frequency source of ill-conditioning can be highlighted using a loop-star decomposed system matrix $[A]^{\mathrm{T}}[Q_{k_0,k_1}][A]$, where $[A] = \left[[\Lambda]\ [H]\ [\Sigma]\right]$, which exhibits the following low-frequency behaviors for lossless dielectrics ($\sigma = 0$) [65]

$$\begin{bmatrix} [A]^{\mathrm{T}} & [0] \\ [0] & [A]^{\mathrm{T}} \end{bmatrix}[Q_{k_0,k_1}]\begin{bmatrix} [A] & [0] \\ [0] & [A] \end{bmatrix} = \begin{bmatrix} \omega^2 & \omega^2 & 1 & \omega & \omega & \omega \\ \omega^2 & 1 & 1 & \omega & \omega & \omega \\ 1 & 1 & 1 & \omega & \omega & \omega^{-1} \\ \omega & \omega & \omega & \omega^2 & \omega^2 & 1 \\ \omega & \omega & \omega & \omega^2 & 1 & 1 \\ \omega & \omega & \omega^{-1} & 1 & 1 & 1 \end{bmatrix}, \tag{7.92}$$

when $\omega \to 0$ and for $\sigma \neq 0$ (and independent from ω)—which is a typical setting of eddy current modeling—[77]

$$\begin{bmatrix}[A]^{\mathrm{T}} & [0]\\ [0] & [A]^{\mathrm{T}}\end{bmatrix}[Q_{k_0,k_1}]\begin{bmatrix}[A] & [0]\\ [0] & [A]\end{bmatrix} = \begin{bmatrix}\omega & \omega & 1 & \omega & \omega & \omega\\ \omega & 1 & 1 & \omega & \omega & \omega\\ 1 & 1 & 1 & \omega & \omega & \omega^{-1}\\ \sigma & \sigma & \sigma & \omega & \omega & 1\\ \sigma & \sigma & \sigma & \omega & 1 & 1\\ \sigma & \sigma & \omega^{-1} & 1 & 1 & 1\end{bmatrix}, \tag{7.93}$$

when $\omega \to 0$. In both regimes, the absolute value of parts of the matrix terms diverges to infinity while others decay to 0, which immediately indicates that the system matrices do not admit a low-frequency limit, and rapidly become unstable as the frequency decreases. Helmholtz decomposition techniques such as loop-star decomposition and quasi-Helmholtz projector can be leveraged to address these instabilities.

7.7.2 The preconditioned equation

Once again the Calderón identities can be leveraged to form an effective preconditioner for the dense-discretization breakdown of the equation. Following the now familiar strategy, we will apply to the main operator its modified version (obtained after flipping the signs on the diagonal), yielding the left-preconditioned equation

$$\mathcal{Q}'_{k_0,k_1}\mathcal{Q}_{k_0,k_1}\begin{bmatrix}\boldsymbol{m}_\Gamma\\ \eta_0\boldsymbol{j}_\Gamma\end{bmatrix} = -\mathcal{Q}'_{k_0,k_1}\begin{bmatrix}\hat{\boldsymbol{n}}\times\boldsymbol{e}^{\mathrm{i}}\\ \eta_0\hat{\boldsymbol{n}}\times\boldsymbol{h}^{\mathrm{i}}\end{bmatrix}, \tag{7.94}$$

where

$$\mathcal{Q}'_{k_0,k_1} := \begin{bmatrix}-\left(\mathcal{K}_{k_0}+\mathcal{K}_{k_1}\right) & \mathcal{T}_{k_0}+\eta_1\eta_0^{-1}\mathcal{T}_{k_1}\\ \mathcal{T}_{k_0}+\eta_0\eta_1^{-1}\mathcal{T}_{k_1} & \mathcal{K}_{k_0}+\mathcal{K}_{k_1}\end{bmatrix}, \tag{7.95}$$

and verify the properties of the preconditioned operator [64] by first introducing the following block notation:

$$\mathcal{Q}'_{k_0,k_1}\mathcal{Q}_{k_0,k_1} =: \begin{bmatrix}[\mathcal{Q}'_{k_0,k_1}\mathcal{Q}_{k_0,k_1}]_{1,1} & [\mathcal{Q}'_{k_0,k_1}\mathcal{Q}_{k_0,k_1}]_{1,2}\\ [\mathcal{Q}'_{k_0,k_1}\mathcal{Q}_{k_0,k_1}]_{2,1} & [\mathcal{Q}'_{k_0,k_1}\mathcal{Q}_{k_0,k_1}]_{2,2}\end{bmatrix}. \tag{7.96}$$

For the first term, we have

$$[\mathcal{Q}'_{k_0,k_1}\mathcal{Q}_{k_0,k_1}]_{1,1} = \left(\mathcal{T}_{k_0}+\eta_1\eta_0^{-1}\mathcal{T}_{k_1}\right)\left(\mathcal{T}_{k_0}+\eta_0\eta_1^{-1}\mathcal{T}_{k_1}\right)-\left(\mathcal{K}_{k_0}+\mathcal{K}_{k_1}\right)^2 \tag{7.97}$$

which is composed of the following types of products: $\mathcal{T}_{k_0}^2$, which are second kind due to (7.25), $\mathcal{T}_{k_0}\mathcal{T}_{k_1}$, which are also the second kind following the same argument as in (7.77), and, finally, products of $\mathcal{K}_{k_0}$ and $\mathcal{K}_{k_1}$, which are compact. Because none of the principal parts (of the form $\mathcal{I}/4$) of the second-kind contributions delete, this first block is of the second kind. The analysis of the second block

$$\begin{aligned}\eta_0[\mathcal{Q}'_{k_0,k_1}\mathcal{Q}_{k_0,k_1}]_{1,2} &= -\left(\eta_0\mathcal{T}_{k_0}+\eta_1\mathcal{T}_{k_1}\right)\left(\mathcal{K}_{k_0}+\mathcal{K}_{k_1}\right)\\ &\quad -\left(\mathcal{K}_{k_0}+\mathcal{K}_{k_1}\right)\left(\eta_0\mathcal{T}_{k_0}+\eta_1\mathcal{T}_{k_1}\right) \qquad (7.98)\\ &= -\left(\eta_0\mathcal{T}_{k_0}\mathcal{K}_{k_0}+\eta_1\mathcal{T}_{k_1}\mathcal{K}_{k_0}\right)-\left(\eta_0\mathcal{T}_{k_0}\mathcal{K}_{k_1}+\eta_1\mathcal{T}_{k_1}\mathcal{K}_{k_1}\right)\\ &\quad -\left(\eta_0\mathcal{K}_{k_0}\mathcal{T}_{k_0}+\eta_1\mathcal{K}_{k_0}\mathcal{T}_{k_1}\right)-\left(\eta_0\mathcal{K}_{k_1}\mathcal{T}_{k_0}+\eta_1\mathcal{K}_{k_1}\mathcal{T}_{k_1}\right) \qquad (7.99)\end{aligned}$$

requires the derivation of propaedeutic results. First, we consider the composition $\mathcal{T}_{k_0}\mathcal{K}_{k_1}$ and $\mathcal{K}_{k_1}\mathcal{T}_{k_0}$, for which the second Calderón identity (7.26) does not hold because of the different wavenumbers; we have

$$\mathcal{K}_{k_1}\mathcal{T}_{k_0} = \mathcal{K}_{k_1}\left(\mathcal{T}_{k_0}-k_1/k_0\mathcal{T}_{k_1}\right)+k_1/k_0\mathcal{K}_{k_1}\mathcal{T}_{k_1}, \qquad (7.100)$$

$$\mathcal{T}_{k_0}\mathcal{K}_{k_1} = \left(\mathcal{T}_{k_0}-k_1/k_0\mathcal{T}_{k_1}\right)\mathcal{K}_{k_1}+k_1/k_0\mathcal{T}_{k_1}\mathcal{K}_{k_1}, \qquad (7.101)$$

where the choice of coefficient k_1/k_0 will become clear later on; by summing the two identities and using (7.26), we have

$$\begin{aligned}\mathcal{K}_{k_1}\mathcal{T}_{k_0}+\mathcal{T}_{k_0}\mathcal{K}_{k_1} &= \mathcal{K}_{k_1}\left(\mathcal{T}_{k_0}-k_1/k_0\mathcal{T}_{k_1}\right)\\ &\quad +\left(\mathcal{T}_{k_0}-k_1/k_0\mathcal{T}_{k_1}\right)\mathcal{K}_{k_1}. \qquad (7.102)\end{aligned}$$

Inserting (7.26) and (7.102) into (7.99), we have that

$$\begin{aligned}\left[\mathcal{Q}'_{k_0,k_1}\mathcal{Q}_{k_0,k_1}\right]_{1,2} &= -\left(\mathcal{K}_{k_1}\mathcal{T}_{k_0}+\mathcal{T}_{k_0}\mathcal{K}_{k_1}\right)-\eta_1\eta_0^{-1}\left(\mathcal{K}_{k_0}\mathcal{T}_{k_1}+\mathcal{T}_{k_1}\mathcal{K}_{k_0}\right) \qquad (7.103)\\ &= -\left(\mathcal{K}_{k_1}\left(\mathcal{T}_{k_0}-k_1/k_0\mathcal{T}_{k_1}\right)+\left(\mathcal{T}_{k_0}-k_1/k_0\mathcal{T}_{k_1}\right)\mathcal{K}_{k_1}\right) \qquad (7.104)\\ &\quad -\eta_1\eta_0^{-1}\left(\mathcal{K}_{k_0}\left(\mathcal{T}_{k_1}-k_0/k_1\mathcal{T}_{k_0}\right)+\left(\mathcal{T}_{k_1}-k_0/k_1\mathcal{T}_{k_0}\right)\mathcal{K}_{k_0}\right).\end{aligned}$$

Demonstrating that this summation is in fact a compact operator requires examining the kernel of the difference operator

$$\mathcal{T}_{k_0}-k_1/k_0\mathcal{T}_{k_1} = \mathrm{i}k_0\left(\mathcal{T}_{\mathrm{A},k_0}-(k_1/k_0)^2\,\mathcal{T}_{\mathrm{A},k_1}\right)-\mathrm{i}/k_0\left(\mathcal{T}_{\Phi,k_0}-\mathcal{T}_{\Phi,k_1}\right), \qquad (7.105)$$

where, in integral form, we have

$$\left(\left(\mathcal{T}_{\mathrm{A},k_0}-(k_1/k_0)^2\,\mathcal{T}_{\mathrm{A},k_1}\right)\boldsymbol{f}\right)(\boldsymbol{r}) = \int_\Gamma \frac{\mathrm{e}^{\mathrm{i}k_0R}-(k_1/k_0)^2\,\mathrm{e}^{\mathrm{i}k_1R}}{4\pi R}\boldsymbol{f}\left(\boldsymbol{r}'\right)\mathrm{d}S\left(\boldsymbol{r}'\right), \qquad (7.106)$$

$$\left(\left(\mathcal{T}_{\Phi,k_0}-\mathcal{T}_{\Phi,k_1}\right)\boldsymbol{f}\right)(\boldsymbol{r}) = \int_\Gamma \nabla_\Gamma\frac{\mathrm{e}^{\mathrm{i}k_0R}-\mathrm{e}^{\mathrm{i}k_1R}}{4\pi R}\nabla'_\Gamma\cdot\boldsymbol{f}\left(\boldsymbol{r}'\right)\mathrm{d}S\left(\boldsymbol{r}'\right), \qquad (7.107)$$

where $R=\|\boldsymbol{r}-\boldsymbol{r}'\|$. The kernels in (7.106) and (7.107) are at most weakly singular since

$$\frac{\mathrm{e}^{\mathrm{i}k_0R}-(k_1/k_0)^2\,\mathrm{e}^{\mathrm{i}k_1R}}{4\pi R} = \frac{1-(k_1/k_0)^2}{4\pi R}+\mathcal{O}(1), \qquad \text{when } R\to 0, \qquad (7.108)$$

$$\nabla_\Gamma\frac{\mathrm{e}^{\mathrm{i}k_0R}-\mathrm{e}^{\mathrm{i}k_1R}}{4\pi R} = \mathcal{O}(1), \qquad \text{when } R\to 0. \qquad (7.109)$$

Because of the weak singularity of the kernel of this operator, we can conclude that the off-diagonal block of the preconditioned PMCHWT is compact [6,78]. The overall PMCHWT operator is then the sum of a second-kind block-diagonal operator and of a block off-diagonal compact operator and as such is the second kind [76]. A further analysis of (7.97) (and of the corresponding lower diagonal block), after developing the product and using (7.25) and (7.77)—see, for instance, [66]—shows that the eigenvalues of the block operators will cluster at different points that diverge away from each other as the material contrast between the scatterer and the background increases, which will compromise the conditioning and the iterative convergence of the formulation.

7.7.3 Different solution strategies

The basic principle behind Calderón preconditioning for the PMCHWT operator has been successfully exploited by several research groups, and some of these schemes will be detailed below.

7.7.3.1 Direct discretization

The first approach, presented in [64], is to discretize the left preconditioned equation (7.94) following the approach used for the EFIE: the main operator is discretized as in (7.87), while the left preconditioner is discretized using dual functions, yielding

$$[\tilde{Q}_{k_0,k_1}]\begin{bmatrix}[G_{\hat{n}\times f,\tilde{f}}] & [0]\\ [0] & [G_{\hat{n}\times f,\tilde{f}}]\end{bmatrix}^{-1}[Q_{k_0,k_1}]\begin{bmatrix}[m]\\ \eta_0[j]\end{bmatrix}$$

$$= -[\tilde{Q}_{k_0,k_1}]\begin{bmatrix}[G_{\hat{n}\times f,\tilde{f}}] & [0]\\ [0] & [G_{\hat{n}\times f,\tilde{f}}]\end{bmatrix}^{-1}\begin{bmatrix}[e^{\mathrm{i}}]\\ \eta_0[h^{\mathrm{i}}]\end{bmatrix}, \quad (7.110)$$

where

$$[\tilde{Q}_{k_0,k_1}] := \begin{bmatrix}-\left([\tilde{\mathbb{K}}_{k_0}]+[\tilde{\mathbb{K}}_{k_1}]\right) & [\tilde{T}_{k_0}]+\eta_1\eta_0^{-1}[\tilde{T}_{k_1}]\\ [\tilde{T}_{k_0}]+\eta_0\eta_1^{-1}[\tilde{T}_{k_1}] & [\tilde{\mathbb{K}}_{k_0}]+[\tilde{\mathbb{K}}_{k_1}]\end{bmatrix}, \quad (7.111)$$

$$[\tilde{\mathbb{K}}_k]_{nm} := \langle \hat{\boldsymbol{n}} \times \tilde{\boldsymbol{f}}_n, \mathcal{K}_k \tilde{\boldsymbol{f}}_m \rangle_\Gamma. \quad (7.112)$$

While dense-discretization is stable, this standard Calderón approach requires further treatments at very low frequencies [65]. Alternative formulations, based on quasi-Helmholtz projectors, have been presented to address both forms of instabilities simultaneously, for lossless [65] and lossy [77] scatterers. In the lossless case, the standard PMCHWT is first stabilized in frequency using (7.69) and (7.70) yielding the discrete system

$$\begin{bmatrix} [\tilde{P}_{k_0}][G_{\hat{n}\times f,\tilde{f}}]^{-1} & [0] \\ [0] & [\tilde{P}_{k_0}][G_{\hat{n}\times f,\tilde{f}}]^{-1} \end{bmatrix} [Q_{k_0,k_1}] \begin{bmatrix} [P_{k_0}] & [0] \\ [0] & [P_{k_0}] \end{bmatrix} \begin{bmatrix} [x] \\ [y] \end{bmatrix} = - \begin{bmatrix} [\tilde{P}_{k_0}][G_{\hat{n}\times f,\tilde{f}}]^{-1}[e^{\mathrm{i}}] \\ \eta_0[\tilde{P}_{k_0}][G_{\hat{n}\times f,\tilde{f}}]^{-1}[h^{\mathrm{i}}] \end{bmatrix}, \tag{7.113}$$

where $[P_{k_0}][x] = [m]$ and $[P_{k_0}]\vec{y} = \eta_0[j]$. To simplify further developments, we introduce the notation

$$[R_{k_0,k_1}] := \begin{bmatrix} [\tilde{P}_{k_0}][G_{\hat{n}\times f,\tilde{f}}]^{-1} & [0] \\ [0] & [\tilde{P}_{k_0}][G_{\hat{n}\times f,\tilde{f}}]^{-1} \end{bmatrix} [Q_{k_0,k_1}] \begin{bmatrix} [P_{k_0}] & [0] \\ [0] & [P_{k_0}] \end{bmatrix}. \tag{7.114}$$

The second step of the preconditioning is to form the preconditioner $[\tilde{R}_{k_0,k_1}]$, dual of $[R_{k_0,k_1}]$, obtained by exchanging the roles of the RWG and BC functions, of the primal and dual projectors, and the signs of the magnetic operators on the diagonal. The preconditioned equation is then

$$[\tilde{R}_{k_0,k_1}][R_{k_0,k_1}] \begin{bmatrix} [x] \\ [y] \end{bmatrix} = -[\tilde{R}_{k_0,k_1}] \begin{bmatrix} [\tilde{P}_{k_0}][G_{\hat{n}\times f,\tilde{f}}]^{-1}[e^{\mathrm{i}}] \\ \eta_0[\tilde{P}_{k_0}][G_{\hat{n}\times f,\tilde{f}}]^{-1}[h^{\mathrm{i}}] \end{bmatrix}. \tag{7.115}$$

The low-frequency conditioning stabilization effect of the projectors is illustrated in Figure 7.7.

7.7.3.2 Mixed discretization of the PMCHWT

The work presented in [66] suggests a clever alternative mixed discretization for the PMCHWT for which the same matrix can be used as the main operator and preconditioner, which paves the way for an efficient iterative solution. The form, in which we present the equation, is slightly different from that in [66] since we solve for $\eta_0[\tilde{j}]$. This modification does not alter the core properties of the scheme but allows for a fair comparison to the other schemes. The proposed discretization of the main PMCHWT operator is

$$[A_{k_0,k_1}] := \begin{bmatrix} [K_{k_0}] + [K_{k_1}] & [\tilde{T}_{k_0}] + \eta_1\eta_0^{-1}[\tilde{T}_{k_1}] \\ [T_{k_0}] + \eta_1^{-1}\eta_0[T_{k_1}] & -\left([\mathbb{K}_{k_0}] + [\mathbb{K}_{k_1}]\right) \end{bmatrix} \tag{7.116}$$

and the right-preconditioned equation is

$$[A_{k_0,k_1}][G]^{-1}[A_{k_0,k_1}][G]^{-1}[x] = -\begin{bmatrix} [\tilde{e}^{\mathrm{i}}] \\ \eta_0[h^{\mathrm{i}}] \end{bmatrix} \tag{7.117}$$

where

$$[\tilde{e}^{\mathrm{i}}]_n := \langle \hat{n} \times \tilde{\boldsymbol{f}}_n, \hat{n} \times \boldsymbol{e}^{\mathrm{i}} \rangle_\Gamma, \tag{7.118}$$

$$[\mathbb{K}_k]_{nm} := \langle \hat{n} \times \boldsymbol{f}_n, \mathcal{K}_k \tilde{\boldsymbol{f}}_m \rangle_\Gamma, \tag{7.119}$$

$$[G] := \begin{bmatrix} [G_{\hat{n}\times\tilde{f} f}] & [0] \\ [0] & [G_{\hat{n}\times f\tilde{f}}] \end{bmatrix}, \tag{7.120}$$

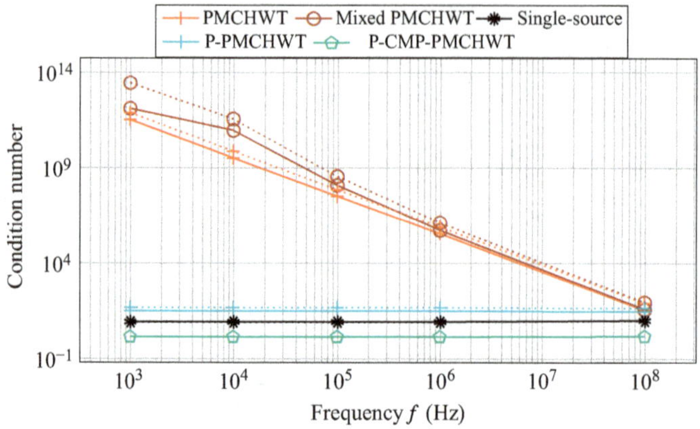

Figure 7.7 Low-frequency conditioning of different formulations for dielectric scatterers, obtained for two discretizations of a sphere of radius 1 m with relative permittivity 1.5 and average edge lengths 0.3 m (solid lines) and 0.2 m (dotted lines). The labels "P-PMCHWT" correspond to (7.113)*, "P-CMP-PMCHWT" to* (7.115)*, and "Single source" to* (7.126).

and $\left[[m]^{\mathrm{T}}\ \eta_0[j]^{\mathrm{T}}\right]^{\mathrm{T}} = [G]^{-1}[A_{k_0,k_1}][G]^{-1}[x]$, which has the advantage of only requiring the computation and compression of a single dense matrix and exhibits interesting properties when solved via iterative schemes. It might, however, be negatively impacted by the poorer interpolation properties of the BC functions used to expand the electric current [69].

The authors of the formulation have also proposed an alternative preconditioning method that cures the high-contrast ill-conditioning, in addition to solving the dense-discretization breakdown [66]. The new right preconditioner is

$$[A^{\mathrm{hc}}_{k_0,k_1}] := \begin{bmatrix} [K_{k_0}] + [K_{k_1}] & a[\tilde{T}_{k_0}] + \eta_1\eta_0^{-1}[\tilde{T}_{k_1}] \\ [T_{k_0}] + a\eta_1^{-1}\eta_0[T_{k_1}] & -\left([\mathbb{K}_{k_0}] + [\mathbb{K}_{k_1}]\right) \end{bmatrix} \tag{7.121}$$

where $a = \eta_1^2\eta_0^{-2}$ and the new equation is

$$[A_{k_0,k_1}][G]^{-1}[A^{\mathrm{hc}}_{k_0,k_1}][G]^{-1}[x] = -\begin{bmatrix} [\tilde{e}^{\mathrm{i}}] \\ \eta_0[h^{\mathrm{i}}] \end{bmatrix}, \tag{7.122}$$

with $\left[[m]^{\mathrm{T}}\ \eta_0[j]^{\mathrm{T}}\right]^{\mathrm{T}} = [G]^{-1}[A^{\mathrm{hc}}_{k_0,k_1}][G]^{-1}[x]$.

7.7.3.3 Single-source formulation

Another formulation for penetrable bodies—based on a single-source approach—has been presented in [67] that, thanks to a clever preconditioning based on Calderón identities, becomes immune to the high contrast and the dense discretization ill-conditioning away from the resonating frequencies of the EFIE and MFIE. The original equation presented in that work is

$$\begin{bmatrix} \frac{\mathcal{I}}{2} - \mathcal{K}_{k_0} & -\mathcal{I} \\ \eta_0^{-1}\mathcal{T}_{k_0} & -\eta_1^{-1}\mathcal{T}_{k_1}^{-1}\left(\frac{\mathcal{I}}{2} - \mathcal{K}_{k_1}\right) \end{bmatrix} \begin{bmatrix} \boldsymbol{m}'_\Gamma \\ -\hat{\boldsymbol{n}} \times \boldsymbol{e} \end{bmatrix} = \begin{bmatrix} \hat{\boldsymbol{n}} \times \boldsymbol{e}^{\mathrm{i}} \\ -\hat{\boldsymbol{n}} \times \boldsymbol{h}^{\mathrm{i}} \end{bmatrix}, \tag{7.123}$$

where $\boldsymbol{m}'_\Gamma$ is an equivalent surface current density that radiates the scattered fields $\boldsymbol{e}^{\mathrm{s}}$ and $\boldsymbol{h}^{\mathrm{s}}$ in Ω^+, to which the preconditioner

$$\begin{bmatrix} \mathcal{I} & 0 \\ 0 & -\eta_1\mathcal{T}_{k_1} \end{bmatrix} \tag{7.124}$$

is then applied, which yields

$$\begin{bmatrix} \frac{\mathcal{I}}{2} - \mathcal{K}_{k_0} & -\mathcal{I} \\ -\eta_1\eta_0^{-1}\mathcal{T}_{k_1}\mathcal{T}_{k_0} & \frac{\mathcal{I}}{2} - \mathcal{K}_{k_1} \end{bmatrix} \begin{bmatrix} \boldsymbol{m}'_\Gamma \\ -\hat{\boldsymbol{n}} \times \boldsymbol{e} \end{bmatrix} = \begin{bmatrix} \hat{\boldsymbol{n}} \times \boldsymbol{e}^{\mathrm{i}} \\ \eta_1\mathcal{T}_{k_1}\left(\hat{\boldsymbol{n}} \times \boldsymbol{h}^{\mathrm{i}}\right) \end{bmatrix}. \tag{7.125}$$

Finally, the equation is discretized as

$$\begin{bmatrix} \frac{\left[G_{\hat{n}\times\tilde{f},f}\right]}{2} - [K_{k_0}] & -\left[G_{\hat{n}\times\tilde{f},f}\right] \\ -\eta_1\eta_0^{-1}[\tilde{T}_{k_1}][G_{\hat{n}\times f,\tilde{f}}]^{-1}[T_{k_0}] & \frac{\left[G_{\hat{n}\times\tilde{f},f}\right]}{2} - [K_{k_1}] \end{bmatrix} \begin{bmatrix} [m'] \\ [x] \end{bmatrix} = \begin{bmatrix} [\tilde{e}^{\mathrm{i}}] \\ \eta_1[\tilde{T}_{k_1}][G_{\hat{n}\times f,\tilde{f}}]^{-1}[h^{\mathrm{i}}] \end{bmatrix}. \tag{7.126}$$

7.8 Conclusions

This chapter discussed optimal h-refinement preconditioning strategies, leveraging Calderón identities, for some of the most widespread IEs: the EFIE, CFIE, and PMCHWT. The treatment has been kept accessible by providing the main insights into the curing mechanisms and effectiveness of the schemes. Because the Calderón identities also provide a partial regularization of the low-frequency ill-conditioning and numerical issues affecting these equations, the coupling of the Calderón strategies with quasi-Helmholtz projectors—that make this partial stabilization complete—have also been presented. To ensure that the reader can make the most out of the techniques presented, we have consistently provided discretization strategies for the different schemes, with a particular focus on the PMCHWT for which several alternatives were introduced, along with numerical examples illustrating the effectiveness of the schemes.

References

[1] Rao S, Wilton D, and Glisson A. Electromagnetic scattering by surfaces of arbitrary shape. *IEEE Transactions on Antennas and Propagation*. 1982;30(3):409–418.

[2] Sauter SA and Schwab C. *Boundary Element Methods*. No. 39 in Springer Series in Computational Mathematics. Berlin: Springer; 2011.

[3] Steinbach O. *Numerical Approximation Methods for Elliptic Boundary Value Problems*. New York, NY: Springer New York; 2008.

[4] Buffa A and Hiptmair R. Galerkin boundary element methods for electromagnetic scattering. In: Barth TJ, Griebel M, Keyes DE, *et al.*, editors. *Topics in Computational Wave Propagation*. vol. 31. Berlin, Heidelberg: Springer Berlin Heidelberg; 2003. p. 83–124.

[5] Peterson AF, Ray SL, and Mittra R. *Computational Methods for Electromagnetics*. New York, NY; Oxford: IEEE Press; Oxford University Press; 1998.

[6] Nédélec JC. *Acoustic and Electromagnetic Equations: Integral Representations for Harmonic Problems*. No. 144 in Applied Mathematical Sciences. New York, NY: Springer; 2001.

[7] Jin JM. *Theory and Computation of Electromagnetic Fields*. 2nd ed. Hoboken, NJ: John Wiley & Sons, Inc; 2015.

[8] Galkowski J and Spence EA. The Helmholtz boundary element method does not suffer from the pollution effect. arXiv preprint arXiv:220109721. 2022.

[9] Chew WC, Michielssen E, Song J, *et al. Fast and Efficient Algorithms in Computational Electromagnetics*. Artech House, Inc.; 2001.

[10] Michielssen E and Boag A. A multilevel matrix decomposition algorithm for analyzing scattering from large structures. *IEEE Transactions on Antennas and Propagation*. 1996;44(8):1086–1093.

[11] Hackbusch W and Nowak ZP. On the fast matrix multiplication in the boundary element method by panel clustering. *Numerische Mathematik*. 1989;54(4):463–491.

[12] Bebendorf M and Rjasanow S. Adaptive low-rank approximation of collocation matrices. *Computing*. 2003;70(1):1–24.

[13] Rokhlin V. Diagonal forms of translation operators for the Helmholtz equation in three dimensions. *Applied and Computational Harmonic Analysis*. 1993;1(1):82–93.

[14] Coifman R, Rokhlin V, and Wandzura S. The fast multipole method for the wave equation: a pedestrian prescription. *IEEE Antennas and Propagation Magazine*. 1993;35(3):7–12.

[15] Adams RJ. Combined field integral equation formulations for electromagnetic scattering from convex geometries. *IEEE Transactions on Antennas and Propagation*. 2004;52(5):1294–1303.

[16] Adams RJ. Physical and analytical properties of a stabilized electric field integral equation. *IEEE Transactions on Antennas and Propagation*. 2004;52(2):362–372.

[17] Darbas M and Lohrengel S. Review on mathematical modelling of electroencephalography (EEG). *Jahresbericht der Deutschen Mathematiker-Vereinigung*. 2019;121(1):3–39.

[18] Adrian SB, Dély A, Consoli D, *et al.* Electromagnetic integral equations: insights in conditioning and preconditioning. *IEEE Open Journal of Antennas and Propagation*. 2021;2:1143–1174.

[19] Christiansen SH and Nédélec JC. A preconditioner for the electric field integral equation based on Calderon formulas. *SIAM Journal on Numerical Analysis*. 2002;40(3):1100–1135.

[20] Carpentieri B, Duff IS, Giraud L, *et al.* Sparse symmetric preconditioners for dense linear systems in electromagnetism. *Numerical Linear Algebra with Applications*. 2004;11(8–9):753–771.

[21] Carpentieri B, Duff IS, Giraud L, *et al.* Combining fast multipole techniques and an approximate inverse preconditioner for large electromagnetism calculations. *SIAM Journal on Scientific Computing*. 2005;27(3):774–792.

[22] Carpentieri B. A matrix-free two-grid preconditioner for solving boundary integral equations in electromagnetism. *Computing*. 2006;77(3):275–296.

[23] Carpentieri B. Algebraic preconditioners for the fast multipole method in electromagnetic scattering analysis from large structures: trends and problems. *Electronic Journal of Boundary Elements*. 2009;7(1).

[24] Carpentieri B and Bollhöfer M. Symmetric inverse-based multilevel ILU preconditioning for solving dense complex non-hermitian systems in electromagnetics. *Progress in Electromagnetics Research*. 2012;128:55–74.

[25] Carpentieri B. Preconditioning for large-scale boundary integral equations in electromagnetics. *IEEE Antennas and Propagation Magazine*. 2014;56(6):338–345.

[26] Carpentieri B. New trends in algebraic preconditioning. In: Ergul O, editor. *New Trends in Computational Electromagnetics*. Institution of Engineering and Technology; 2019. p. 535–566.

[27] Sertel K and Volakis JL. Incomplete LU preconditioner for FMM implementation. *Microwave and Optical Technology Letters*. 2000;26(4):265–267.

[28] Lee J, Zhang J, and Lu CC. Incomplete LU preconditioning for large scale dense complex linear systems from electromagnetic wave scattering problems. *Journal of Computational Physics*. 2003;185(1):158–175.

[29] Malas T and Gürel L. Incomplete LU preconditioning with the multilevel fast multipole algorithm for electromagnetic scattering. *SIAM Journal on Scientific Computing*. 2007;29(4):1476–1494.

[30] Malas T and Gürel L. Accelerating the multilevel fast multipole algorithm with the sparse-approximate-inverse (SAI) preconditioning. *SIAM Journal on Scientific Computing*. 2009;31(3):1968–1984.

[31] Eibert TF. Iterative near-zone preconditioning of iterative method of moments electric field integral equation solutions. *IEEE Antennas and Wireless Propagation Letters*. 2003;2(1):101–102.

[32] Vipiana F, Pirinoli P, and Vecchi G. A multiresolution method of moments for triangular meshes. *IEEE Transactions on Antennas and Propagation*. 2005;53(7):2247–2258.

[33] Vipiana F, Vecchi G, and Pirinoli P. A multiresolution system of Rao-Wilton-Glisson functions. *IEEE Transactions on Antennas and Propagation*. 2007;55(3):924–930.

[34] Andriulli FP, Bagci H, Vipiana F, *et al.* A marching-on-in-time hierarchical scheme for the solution of the time domain electric field integral equation. *IEEE Transactions on Antennas and Propagation*. 2007;55(12):3734–3738.

[35] Andriulli FP, Vipiana F, and Vecchi G. Hierarchical bases for nonhierarchic 3-D triangular meshes. *IEEE Transactions on Antennas and Propagation*. 2008;56(8):2288–2297.
[36] Vipiana F, Andriulli FP, and Vecchi G. Two-tier non-simplex grid hierarchic basis for general 3D meshes. *Waves in Random and Complex Media*. 2009;19(1):126–146.
[37] Echeverri Bautista MA, Francavilla MA, Vipiana F, *et al.* A hierarchical fast solver for EFIE-MoM analysis of multiscale structures at very low frequencies. *IEEE Transactions on Antennas and Propagation*. 2014;62(3):1523–1528.
[38] Adrian SB, Andriulli FP, and Eibert TF. On the hierarchical preconditioning of the combined field integral equation. *IEEE Antennas and Wireless Propagation Letters*. 2016;15:1897–1900.
[39] Adrian SB, Andriulli FP, and Eibert TF. A hierarchical preconditioner for the electric field integral equation on unstructured meshes based on primal and dual Haar bases. *Journal of Computational Physics*. 2017;330:365–379.
[40] Guzman JEO, Adrian SB, Mitharwal R, *et al.* On the hierarchical preconditioning of the PMCHWT integral equation on simply and multiply connected geometries. *IEEE Antennas and Wireless Propagation Letters*. 2017;16:1044–1047.
[41] Peng Z and Lee JF. Non-conformal domain decomposition method with second-order transmission conditions for time-harmonic electromagnetics. *Journal of Computational Physics*. 2010;229(16):5615–5629.
[42] Bautista MAE, Vipiana F, Francavilla MA, *et al.* A nonconformal domain decomposition scheme for the analysis of multiscale structures. *IEEE Transactions on Antennas and Propagation*. 2015;63(8):3548–3560.
[43] Peng Z, Hiptmair R, Shao Y, *et al.* Domain decomposition preconditioning for surface integral equations in solving challenging electromagnetic scattering problems. *IEEE Transactions on Antennas and Propagation*. 2016;64(1):210–223.
[44] Adams RJ and Brown GS. Stabilisation procedure for electric field integral equation. *Electronics Letters*. 1999;35(23):2015.
[45] Contopanagos H, Dembart B, Epton M, *et al.* Well-conditioned boundary integral equations for three-dimensional electromagnetic scattering. *IEEE Transactions on Antennas and Propagation*. 2002;50(12):1824–1830.
[46] Boubendir Y and Turc C. Well-conditioned boundary integral equation formulations for the solution of high-frequency electromagnetic scattering problems. *Computers & Mathematics with Applications*. 2014;67(10):1772–1805.
[47] Kleanthous A, Betcke T, Hewett DP, *et al.* Calderón preconditioning of PMCHWT boundary integral equations for scattering by multiple absorbing dielectric particles. *Journal of Quantitative Spectroscopy and Radiative Transfer*. 2019;224:383–395.
[48] Darbas M. Préconditioneeurs analytiques de type Calderon pour les formulations intégrales des problèmes de diffraction d'ondes [PhD Thesis]. Institut National des Sciences Appliquées de Toulouse; 2004.

[49] Borel S, Levadoux DP, and Alouges F. A new well-conditioned integral formulation for Maxwell equations in three dimensions. *IEEE Transactions on Antennas and Propagation*. 2005;53(9):2995–3004.

[50] Darbas M. Generalized combined field integral equations for the iterative solution of the three-dimensional Maxwell equations. *Applied Mathematics Letters*. 2006;19(8):834–839.

[51] Alouges F, Borel S, and Levadoux DP. A stable well-conditioned integral equation for electromagnetism scattering. *Journal of Computational and Applied Mathematics*. 2007;204(2):440–451.

[52] Levadoux DP. Some preconditioners for the CFIE equation of electromagnetism. *Mathematical Methods in the Applied Sciences*. 2008;31(17): 2015–2028.

[53] Antoine X and Darbas M. An introduction to operator preconditioning for the fast iterative integral equation solution of time-harmonic scattering problems. *Multiscale Science and Engineering*. 2021;3:1–35.

[54] Hsiao GC and Kleinman RE. Mathematical foundations for error estimation in numerical solutions of integral equations in electromagnetics. *IEEE Transactions on Antennas and Propagation*. 1997;45(3):316–328.

[55] Mautz J and Harrington R. An E-field solution for a conducting surface small or comparable to the wavelength. *IEEE Transactions on Antennas and Propagation*. 1984;32(4):330–339.

[56] Wu WL, Glisson AW, and Kajfez D. A study of two numerical solution procedures for the electric field integral equation. *Applied Computational Electromagnetics Society Journal*. 1995;10(3):69–80.

[57] Burton M and Kashyap S. A study of a recent, moment-method algorithm that is accurate to very low frequencies. *Applied Computational Electromagnetics Society Journal*. 1995;10(3):58–68.

[58] Zhao JS and Chew WC. Integral equation solution of Maxwell's equations from zero frequency to microwave frequencies. *IEEE Transactions on Antennas and Propagation*. 2000;48(10):1635–1645.

[59] Andriulli FP, Cools K, Bogaert I, *et al.* On a well-conditioned electric field integral operator for multiply connected geometries. *IEEE Transactions on Antennas and Propagation*. 2013;61(4):2077–2087.

[60] Dely A, Merlini A, Adrian SB, *et al.* On preconditioning electromagnetic integral equations in the high frequency regime via Helmholtz operators and quasi-Helmholtz projectors. In: *International Conference on Electromagnetics in Advanced Applications (ICEAA)*. Granada, Spain: IEEE; 2019. p. 1338–1341.

[61] Poggio AJ and Miller EK. Integral equation solutions of three-dimensional scattering problems. In: *Computer Techniques for Electromagnetics*. Elsevier; 1973. p. 159–264.

[62] Chang Y and Harrington R. A surface formulation for characteristic modes of material bodies. *IEEE Transactions on Antennas and Propagation*. 1977;25(6):789–795.

[63] Wu TK and Tsai LL. Scattering from arbitrarily-shaped lossy dielectric bodies of revolution. *Radio Science*. 1977;12(5):709–718.

[64] Cools K, Andriulli FP, and Michielssen E. A Calderón multiplicative preconditioner for the PMCHWT integral equation. *IEEE Transactions on Antennas and Propagation*. 2011;59(12):4579–4587.

[65] Beghein Y, Mitharwal R, Cools K, *et al*. On a low-frequency and refinement stable PMCHWT integral equation leveraging the quasi-Helmholtz projectors. *IEEE Transactions on Antennas and Propagation*. 2017;65(10):5365–5375.

[66] Niino K and Nishimura N. Calderón preconditioning approaches for PMCHWT formulations for Maxwell's equations. *International Journal of Numerical Modelling: Electronic Networks, Devices and Fields*. 2012; 25(5–6):558–572.

[67] Gossye M, Huynen M, Vande Ginste D, *et al*. A Calderón preconditioner for high dielectric contrast media. *IEEE Transactions on Antennas and Propagation*. 2018;66(2):808–818.

[68] Yla-Oijala P, Kiminki SP, and Jarvenpaa S. Calderon preconditioned surface integral equations for composite objects with junctions. *IEEE Transactions on Antennas and Propagation*. 2010;59(2):546–554.

[69] Buffa A and Christiansen SH. A dual finite element complex on the barycentric refinement. *Mathematics of Computation*. 2007;76(260):1743–1770.

[70] Cools K, Andriulli FP, De Zutter D, *et al*. Accurate and conforming mixed discretization of the MFIE. *IEEE Antennas and Wireless Propagation Letters*. 2011;10:528–531.

[71] Vico F, Greengard L, and Gimbutas Z. Boundary integral equation analysis on the sphere. *Numerische Mathematik*. 2014;128(3):463–487.

[72] Olver FWJ, Lozier DW, Boisvert RF, *et al*., editors. *Nist Handbook of Mathematical Functions*. Cambridge; New York, NY: Cambridge University Press; NIST; 2010.

[73] Eibert TF. Iterative-solver convergence for loop-star and loop-tree decompositions in method-of-moments solutions of the electric-field integral equation. *IEEE Antennas and Propagation Magazine*. 2004;46(3):80–85.

[74] Andriulli FP. Loop-star and loop-tree decompositions: analysis and efficient algorithms. *IEEE Transactions on Antennas and Propagation*. 2012;60(5):2347–2356.

[75] Cools K, Andriulli FP, Olyslager F, *et al*. Nullspaces of MFIE and Calderón preconditioned EFIE operators applied to toroidal surfaces. *IEEE Transactions on Antennas and Propagation*. 2009;57(10):3205–3215.

[76] Pillain A. Line, surface, and volume integral equations for the electromagnetic modelling of the electroencephalography forward problem [PhD Thesis]. *Ecole Nationale Supérieure des Télécommunications de Bretagne-ENSTB*. Brest, France; 2016.

[77] Chhim TL, Merlini A, Rahmouni L, *et al*. Eddy current modeling in multiply connected regions via a full-wave solver based on the quasi-Helmholtz projectors. *IEEE Open Journal of Antennas and Propagation*. 2020;1:534–548.

[78] Colton D and Kress R. *Integral Equation Methods in Scattering Theory*, vol. 72. SIAM; 2013.

Chapter 8
Decoupled potential integral equation

Felipe Vico[1] *and Miguel Ferrando-Bataller*[1]

In this chapter, we study an experimental formulation called decoupled potential integral equation (DPIE) [1]. The aim of this formulation is to obtain a method that is robust at all frequencies, in particular at low frequencies for multiply connected geometries. We also discuss experimental discretization methods that are high-order, adaptive and fast.

8.1 Scattering problem and boundary conditions

The electromagnetic scattering problem time-harmonic regime consists of finding the scattered electromagnetic field produced by an object (called scatterer) in the presence of an incoming field. In this chapter, we will assume that the scatterer is the perfect electric conductor. We will always consider that the scatterer is a closed surface and has interior. We denote the interior by D and the surface ∂D (the boundary of the interior). The case of opened surfaces (thin metallic scatterers) is beyond the scope of this chapter.

The problem can be formulated in the following mathematical form:

Given an incoming electromagnetic field $\boldsymbol{E}^{\text{in}}(\boldsymbol{r}), \boldsymbol{H}^{\text{in}}(\boldsymbol{r})$, find the scattered electric and magnetic fields $\boldsymbol{E}^{\text{scat}}(\boldsymbol{r}),\boldsymbol{H}^{\text{scat}}(\boldsymbol{r})$ for $\boldsymbol{r} \in \mathbb{R}^3\backslash\overline{D}$ such that:

$$\begin{aligned} \nabla \times \boldsymbol{E}^{\text{scat}}(\boldsymbol{r}) &= -j\omega\mu\boldsymbol{H}^{\text{scat}}(\boldsymbol{r}) \\ \nabla \times \boldsymbol{H}^{\text{scat}}(\boldsymbol{r}) &= j\omega\varepsilon\boldsymbol{E}^{\text{scat}}(\boldsymbol{r}) \end{aligned} \tag{8.1}$$

with $\omega > 0$ and the boundary conditions:

$$\begin{aligned} &\hat{\boldsymbol{n}} \times \boldsymbol{E}|_{\partial D} = 0, \ \hat{\boldsymbol{n}} \times \boldsymbol{H}|_{\partial D} = \boldsymbol{J} \\ &\hat{\boldsymbol{n}} \cdot \boldsymbol{H}|_{\partial D} = 0, \ \ \hat{\boldsymbol{n}} \cdot \boldsymbol{E}|_{\partial D} = \rho \end{aligned} \tag{8.2}$$

[1]Departamento de Comunicaciones, Universitat Politècnica de València, Valencia, Spain

and verifies the Silver–Müller radiation condition:

$$\boldsymbol{H}^{\text{scat}}(\boldsymbol{r}) \times \hat{\boldsymbol{r}} - \sqrt{\frac{\mu}{\varepsilon}}\boldsymbol{E}^{\text{scat}}(\boldsymbol{r}) = o\left(\frac{1}{r}\right), r \to \infty \tag{8.3}$$

The total electric and magnetic fields are:

$$\begin{aligned} \boldsymbol{E}(\boldsymbol{r}) &= \boldsymbol{E}^{\text{in}} + \boldsymbol{E}^{\text{scat}} \\ \boldsymbol{H}(\boldsymbol{r}) &= \boldsymbol{H}^{\text{in}} + \boldsymbol{H}^{\text{scat}} \end{aligned}$$

And the conducting scatterer is D with boundary ∂D and normal vector $\hat{\boldsymbol{n}}$.

The notion of boundary condition is different when reading a physics book than when reading a math book. In mathematics, the notion of boundary condition happens in the context of boundary value problems. A boundary value problem contains two ingredients, a partial differential equation and a boundary condition on the solution. In this context, it is interesting and desirable that the boundary condition implies existence and uniqueness of the solution, that is, among the infinite set of functions that verify the partial differential equation, there is one and only one that also verifies the boundary condition. There are other desirable properties related to the stability of the problem. The existence and uniqueness should happen for any boundary data in a suitable function space. The dependence of the solution with the boundary data should be continuous for a suitable norm in the function space, and finally, in the presence of parameters in the problem (like ω in time-harmonic electromagnetism), the continuity should be uniform within the set of values of the parameters of interest. All these requirements make a boundary value problem well posed. In physics, boundary conditions are conditions that happen at the boundary and that are true, regardless of whether they identify uniquely the solution, or if the boundary conditions are redundant, or if the boundary conditions are redundant for some values of some parameters but not redundant for others. When dealing with numerical methods, it is our task to select carefully the boundary condition (or set of boundary conditions) among the physically true possibilities to obtain a well-posed boundary value problem. The conditions in 8.2 are boundary conditions in a physics sense.

The scalar condition $\hat{\boldsymbol{n}} \cdot \boldsymbol{H}|_{\partial D} = 0$ can be obtained from the vector condition $\hat{\boldsymbol{n}} \times \boldsymbol{E}|_{\partial D} = 0$ for $\omega > 0$ by taking the surface divergence:

$$\begin{aligned} \nabla_s \cdot \hat{\boldsymbol{n}} \times \boldsymbol{E} &= 0 \\ -\hat{\boldsymbol{n}} \cdot \nabla \times \boldsymbol{E} &= 0 \\ j\omega\mu\hat{\boldsymbol{n}} \cdot \boldsymbol{H} &= 0 \\ \hat{\boldsymbol{n}} \cdot \boldsymbol{H} &= 0 \end{aligned} \tag{8.4}$$

Notice that we have used the identity $\nabla_s \cdot \hat{\boldsymbol{n}} \times \boldsymbol{F} = -\hat{\boldsymbol{n}} \cdot \nabla \times \boldsymbol{F}$. Notice also that the last identity in 8.4 is only implied if $\omega \neq 0$. Similarly, one can proof that $\hat{\boldsymbol{n}} \cdot \boldsymbol{E}|_{\partial D} = \rho$ from $\hat{\boldsymbol{n}} \times \boldsymbol{H}|_{\partial D} = 0$ for $\omega \neq 0$.

It is well known that the tangent electric field on the boundary of the scatterer specifies uniquely the scattered field:

Definition 1: *Given a vector field* $\boldsymbol{f} \in C_t^{0,\alpha}(\nabla_s \cdot, \partial D)$ *defined on* ∂D *and tangent to that surface, the* ***perfect electric conducting scattering problem*** *consists of finding the fields* $\boldsymbol{E}^{\text{scat}}(\boldsymbol{r}), \boldsymbol{H}^{\text{scat}}(\boldsymbol{r}) \in C^1(\mathbb{R}^3 \backslash \overline{D}) \cap C^0(\mathbb{R}^3 \backslash D)$ *such that they verify Maxwell's equations 8.1 for* $\omega > 0$ *and the radiation condition 8.3 and the boundary condition*

$$\hat{\boldsymbol{n}} \times \boldsymbol{E}^{\text{scat}}|_{\partial D} = \boldsymbol{f}$$

Proposition 1 (Existence and uniqueness): *The perfect electric conducting scattering problem has one and only one solution for any boundary data* $\boldsymbol{f} \in C_t^{0,\alpha}(\nabla_s \cdot, \partial D)$.

See Appendix E for definitions of Hölder function spaces.

There are many different integral formulations to solve this problem, like the MFIE, EFIE, or CFIE. Most of the formulations have problems of inaccuracy or instability in low frequency. Those problems are often called low-frequency breakdown. To understand better the source of these problems, it is interesting to study the low-frequency limit of the boundary value.

8.2 Low-frequency limit boundary value problems

In this section, we study the zero frequency limit ($\omega \to 0$) of the perfect electric conducting scattering problem. This limit is interesting for the understanding of the issues that most numerical methods have when dealing with very low frequency $\omega << 1$. When we take zero frequency limit of the time-harmonic electromagnetic scattering problem $\omega \to 0$, we observe that the electric and magnetic problems are decoupled.

Definition 2: *The* ***electrostatic boundary value problem*** *consists of finding a static scattered vector field* $\boldsymbol{E}_0^{\text{scat}}(\boldsymbol{r})$ *that verifies the equations*

$$\begin{aligned} \nabla \times \boldsymbol{E}_0^{\text{scat}} &= 0 \\ \nabla \cdot \boldsymbol{E}_0^{\text{scat}} &= 0 \end{aligned}$$

and the static radiation condition

$$\boldsymbol{E}_0^{\text{scat}}(\boldsymbol{r}) = o(1), \quad r \to \infty$$

and the boundary condition

$$\hat{\boldsymbol{n}} \times \boldsymbol{E}_0^{\text{scat}}|_{\partial D} = \boldsymbol{f}$$

where $\boldsymbol{f} \in C_t^{0,\alpha}(\nabla_s \cdot, \partial D)$ *is a tangent vector field on the surface* ∂D.

Definition 3: *The* ***magnetostatic boundary value problem*** *consists of finding a static scattered vector field* $\boldsymbol{H}_0^{\text{scat}}(\boldsymbol{r})$ *that verifies the equations*

$$\begin{aligned} \nabla \times \boldsymbol{H}_0^{\text{scat}} &= 0 \\ \nabla \cdot \boldsymbol{H}_0^{\text{scat}} &= 0 \end{aligned}$$

and the static radiation condition

$$\boldsymbol{H}_0^{\text{scat}}(\boldsymbol{r}) = o(1), \quad r \to \infty$$

and the boundary condition

$$\hat{\boldsymbol{n}} \cdot \boldsymbol{H}_0^{\text{scat}}|_{\partial D} = h$$

where $h \in C^{0,\alpha}(\partial D)$ *is a scalar function on the surface* ∂D.

It is important to notice that those boundary value problems are the natural zero frequency limit of the time-harmonic electromagnetic scattering problem for perfect electric conductors, but are not well-posed boundary value problems from a mathematical viewpoint. Next we show why.

The electrostatic boundary value problem has no unique solution. To show why we consider a spherical surface $D = \{\boldsymbol{r} \in \mathbb{R}^3, r < 1\}$. The following vector field:

$$\boldsymbol{E}_0^{\text{scat}}(\boldsymbol{r}) = \frac{\boldsymbol{r}}{r^3}$$

By construction, $\boldsymbol{E}_0^{\text{scat}}(\boldsymbol{r})$ verifies $\nabla \times \boldsymbol{E}_0^{\text{scat}}(\boldsymbol{r}) = 0$ and $\nabla \cdot \boldsymbol{E}_0^{\text{scat}}(\boldsymbol{r}) = 0$ in the exterior region $r > 1$ and boundary condition $\hat{\boldsymbol{n}} \times \boldsymbol{E}_0^{\text{scat}}|_{\partial D} = 0$ (see Figure 8.1). This counterexample can be built for any arbitrary geometry and can be added to any solution with arbitrary boundary data $\boldsymbol{f}$ to obtain a different solution.

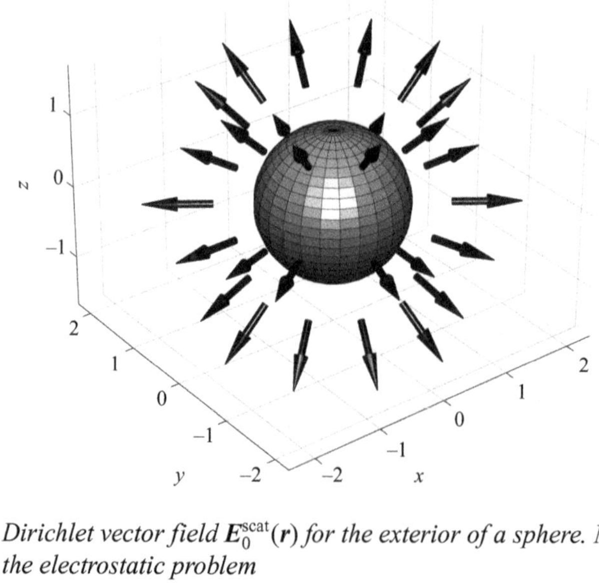

Figure 8.1 Dirichlet vector field $\boldsymbol{E}_0^{\text{scat}}(\boldsymbol{r})$ *for the exterior of a sphere. Null vector of the electrostatic problem*

The problem of the non-uniqueness is not the only one. We can prove that there is non-existence in general. To show why we consider the same geometry $D = \{\boldsymbol{r} \in \mathbb{R}^3, r < 1\}$ and the following boundary data:

$$\boldsymbol{f}(\boldsymbol{r}) = -\hat{\boldsymbol{n}} \times (-y\hat{\boldsymbol{x}} + x\hat{\boldsymbol{y}})$$

Assuming that we can find a solution $\boldsymbol{E}_0^{\text{scat}}(\boldsymbol{r})$ for $r > 1$ with $\nabla \times \boldsymbol{E}_0^{\text{scat}}(\boldsymbol{r}) = 0$ and $\nabla \cdot \boldsymbol{E}_0^{\text{scat}}(\boldsymbol{r}) = 0$ and $\hat{\boldsymbol{n}} \times \boldsymbol{E}_0^{\text{scat}}|_{\partial D} = \boldsymbol{f}$, therefore, $\boldsymbol{E}_0^{\text{scat}}|_{\partial D}^{tan} = -y\hat{\boldsymbol{x}} + x\hat{\boldsymbol{y}}$. This is obviously impossible. To see this, let us take a closed loop on the surface $l \subset \{\boldsymbol{r}, r = 1\}$ such as $l = \{\cos(t)\sin(\theta_0)\hat{\boldsymbol{x}} + \sin(t)\sin(\theta_0)\hat{\boldsymbol{y}} + \cos(\theta_0)\hat{\boldsymbol{z}},\ t \in [0, 2\pi]\}$. This closed loop is the boundary of the surface $S = \{\boldsymbol{r} \in \mathbb{R}^3, |\boldsymbol{r} - \hat{\boldsymbol{z}}| = r_0 \wedge \boldsymbol{r} \cdot \hat{\boldsymbol{z}} > \cos(\theta_0)\}$; that is, $\partial S = l$. See Figure 8.2; in blue the loop l and the surface S transparent. In the picture, we have $r_0 = 0.7$ and $\theta_0 = 0.72$ for plotting reasons, but the particular choice of these constants within a certain range is not important.

And applying the Stokes theorem on $\boldsymbol{E}_0^{\text{scat}}(\boldsymbol{r})$, we get:

$$\int_S \hat{\boldsymbol{n}} \cdot \nabla \times \boldsymbol{E}_0^{\text{scat}} dS = \int_l \boldsymbol{E}_0^{\text{scat}} \cdot d\mathbf{l} > 0$$

$$0 > 0$$

which is a contradiction. In the last line, we made use of the fact that $\nabla \times \boldsymbol{E}_0^{\text{scat}} = 0$ and that $\boldsymbol{E}_0^{\text{scat}} \cdot \frac{d\mathbf{l}}{dt} = 1$.

In the magnetostatic problem, similar problems are observed in terms of non-uniqueness. For the detailed description of the counterexample we refer to [2]. Here

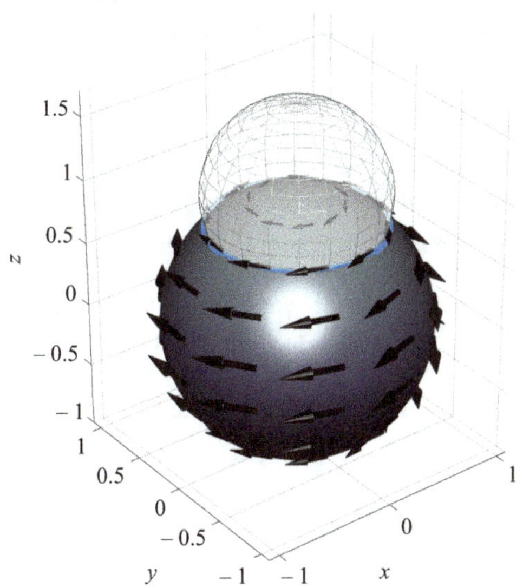

Figure 8.2 Vector boundary data with no solution for the electrostatic boundary value problem

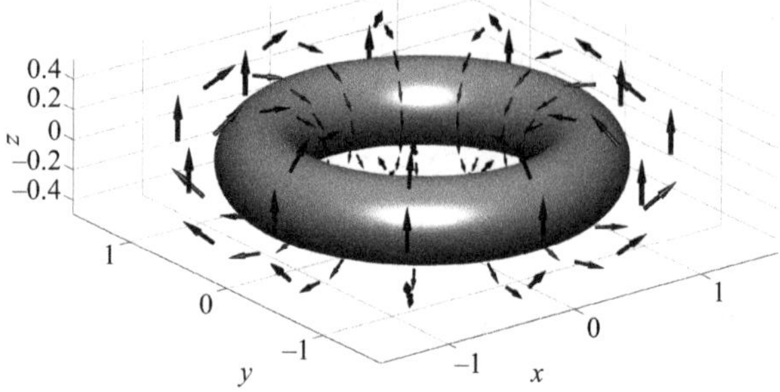

Figure 8.3 Neumann vector field for the exterior of a toroidal surface. Null vector of the magnetostatic problem

we give a qualitative description of a counterexample. Given a standard toroidal surface, there is a vector field $Z(\boldsymbol{r})$ called Neumann vector field such that $\nabla \times Z = 0$, $\nabla \cdot Z = 0$ and $\hat{\boldsymbol{n}} \cdot Z = 0$ (see Figure 8.3).

Physically, it corresponds to the magnetostatic field produced by a steady-state current circulating around the perfect electric conducting torus. In theory, the amplitude of the current is undetermined (without external force) as the conductivity of the object is considered infinite and there is no self-inductance effect at zero frequency that prevents the current from flowing creating an impedance.

In general, the dimension of the null-space of the electrostatic problem equals to the number of connected components of the surface (number of pieces, $D = \cup_{j=1}^{N} D_j$), and they are called *Dirichlet fields*, while the dimension of the null-space of the magnetostatic problem equals to the genus g of the surface (number of "holes"), and they are called *Neumann fields* (see Appendix A).

Notice that the non-existence and non-uniqueness of the electrostatic problem is not a major issue in physics but a consequence of applying the wrong way of thinking. In statics $\omega = 0$, one should change his mind and think in terms of electrostatic potentials, postulating that $\boldsymbol{E}_0^{\text{scat}} = -\nabla\phi_0^{\text{scat}}$ and also for the incoming field $\boldsymbol{E}_0^{\text{in}} = -\nabla\phi_0^{\text{in}}$. Scattered and incoming are not the best denomination for the fields, as they do not propagate, nevertheless we will keep these denominations to simplify and build a unified theory along the chapter. By postulating that, the equation $\nabla \times \boldsymbol{E}_0^{\text{scat}} = 0$ is automatically verified by the construction, and $\nabla \cdot \boldsymbol{E}_0^{\text{scat}} = 0$ implies $\Delta\phi_0^{\text{scat}} = 0$

The boundary condition on the tangent fields implies a boundary condition in the potentials:

$$\hat{\boldsymbol{n}} \times \nabla\phi_0^{\text{scat}} = -\hat{\boldsymbol{n}} \times \nabla\phi_0^{\text{in}}$$

implies that:

$$\phi_0^{\text{scat}} + \phi_0^{\text{in}}|_{\partial D_j} = V_j$$

where V_j are arbitrary constant values (one per connected component of the scatterer). To determine those constants, additional information is needed. Here there are two main options. The first one is to consider that all the conductors D_j are grounded, and, therefore, $V_j = 0$. This will lead to a purely Dirichlet boundary value problem:

Definition 4: *The* ***Laplace Dirichlet problem*** *consists of finding a scalar potential* $\phi_0^{\text{scat}}(\boldsymbol{r}) \in C^2(\mathbb{R}^3 \backslash \overline{D}) \cap C(\mathbb{R}^3 \backslash D)$ *that verifies*

$$\Delta\phi_0^{\text{scat}} = 0$$

and the static radiation condition

$$\phi_0^{\text{scat}}(\boldsymbol{r}) = o(1), \quad r \to \infty$$

and the boundary condition

$$\phi_0^{\text{scat}}|_{\partial D} = f$$

where $f \in C(\partial D)$ *is a scalar function on the surface* ∂D.

In contrast, when the conductors are isolated, the values of V_j are unknown while the total charge on each conductor remains constant (equal to the initial charge). The extra N unknowns are balanced by N extra conditions on the flux of $\boldsymbol{E}_0^{\text{scat}}$; that is $\int_{\partial D_j} \hat{\boldsymbol{n}} \cdot \boldsymbol{E}_0^{\text{scat}} dS = Q_j$. This leads to the modified Laplace Dirichlet boundary value problem:

Definition 5: *The* ***Modified Laplace Dirichlet problem*** *consists of finding a scalar potential* $\phi_0^{\text{scat}}(\boldsymbol{r}) \in C^2(\mathbb{R}^3 \backslash \overline{D}) \cap C(\mathbb{R}^3 \backslash D)$ *and constants* $V_j, j = 1, \ldots, N$ *that verifies*

$$\Delta\phi_0^{\text{scat}} = 0$$

and the static radiation condition

$$\phi_0^{\text{scat}}(\boldsymbol{r}) = o(1), \quad r \to \infty$$

and the boundary conditions

$$\phi_0^{\text{scat}}|_{\partial D_j} = f + V_j$$

and flux conditions on each connected component ∂D_j *of* ∂D

$$\int_{\partial D_j} \hat{\boldsymbol{n}} \cdot \nabla\phi_0^{\text{scat}} dS = Q_j$$

where $f \in C(\partial D)$ *is a scalar function on the surface* ∂D *and* Q_j *are known constants.*

Both problems are well posed in the following sense:

Proposition 2 (Existence and uniqueness): *The Laplace Dirichlet problem has one and only one solution for any boundary data* $f \in C(\partial D)$.

Proposition 3 (Existence and uniqueness): *The Modified Laplace Dirichlet problem has one and only one solution for any boundary data* $f \in C(\partial D)$ *and constants* $Q_j, j = 1, \ldots, N$.

The approach to take at zero frequency depends on the physical setup of the problem.

Next section we explore a particular static regime that is the zero frequency limit of time-harmonic regime.

8.3 Stabilizing conditions

To fix the low-frequency breakdown of Maxwell's equations, we define the solution at zero frequency as the zero frequency limit $\omega \to 0$ of time-harmonic regime. This will lead us to extra conditions to fix the non-uniqueness in a way that is meaningful and useful when solving time-harmonic problems using numerical methods at very low frequency.

Let us consider the perfect electric conducting scattering problem, at $\omega > 0$, with solution $\boldsymbol{E}^{\mathrm{scat}}, \boldsymbol{H}^{\mathrm{scat}}$ for certain arbitrary boundary condition $\boldsymbol{f}$. Let us consider one connected component ∂D_j of the scatterer ∂D. Taking the flux of the electric field, we get:

$$\int_{\partial D_j} \hat{\boldsymbol{n}} \cdot \boldsymbol{E}^{\mathrm{scat}} dS = Q_j$$

Using Maxwell's equations, we get:

$$\int_{\partial D_j} \hat{\boldsymbol{n}} \cdot \boldsymbol{E}^{\mathrm{scat}} dS = \frac{1}{j\omega\varepsilon} \int_{\partial D_j} \hat{\boldsymbol{n}} \cdot \nabla \times \boldsymbol{H}^{\mathrm{scat}} dS = \frac{1}{-j\omega\varepsilon} \int_{\partial D_j} \nabla_s \cdot \hat{\boldsymbol{n}} \times \boldsymbol{H}^{\mathrm{scat}} dS = 0$$

The last equality is obtained using the surface divergence theorem (see Appendix A) and the fact that the surface ∂D_j has no boundary, that is $\partial\partial D_j = \emptyset$. For any $\omega > 0$, we finally obtain $\int_{\partial D_j} \hat{\boldsymbol{n}} \cdot \boldsymbol{E}^{\mathrm{scat}} dS = 0$, and, therefore, $Q_j = 0$. At $\omega = 0$, we cannot infer that yet this result shows that the correct zero frequency limit for the perfect electric conducting scattering problem is to consider that $Q_j = 0$ and, therefore, consider the Modified Laplace Dirichlet problem with the following set of extra conditions:

$$\int_{\partial D_j} \hat{\boldsymbol{n}} \cdot \boldsymbol{E}^{\mathrm{scat}} dS = 0 \tag{8.5}$$

for each connected component ∂D_j of the scatterer. Notice that this extra conditions at zero frequency prevents the non-uniqueness as it kills the Dirichlet fields, yet it does not address the non-existence issue. The non-existence issue will be fixed with the explicit use of boundary conditions on the potentials at $\omega \geq 0$.

Next we find additional boundary conditions for the magnetostatic boundary value problem (see [3]).

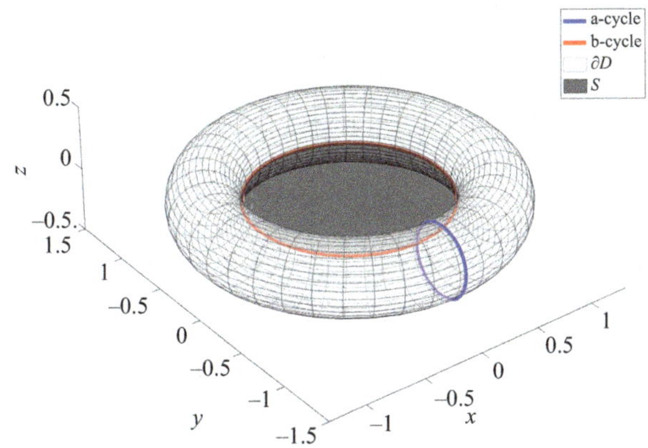

Figure 8.4 *a-cycle and b-cycle on a toroidal surface* ∂D *(genus 1). The surface S where the stokes theorem is applied is the span surface of the b-cycle*

Let us consider again the perfect electric conducting scattering problem, at $\omega > 0$, with solution $\boldsymbol{E}^{\text{scat}}, \boldsymbol{H}^{\text{scat}}$. Here we introduce the notion of a-cycles and b-cycles of a multiply connected surface ∂D. We introduce these notions in an intuitive way, a more rigorous approach would require to introduce more complicated homotropy concepts. Given a toroidal surface (see Figure 8.4) with genus 1, we define an a-cycle as a loop contained in ∂D that cannot be contracted to a point and whose spanning surface is contained in the interior of the domain D. Similarly, a b-cycle is a loop contained in the surface ∂D whose spanning surface lies in the exterior of the surface. The spanning surface S of the b-cycle is plotted in the picture in gray. A loop and its spanning surface are the two geometrical ingredients needed to apply the Stokes theorem.

Remark 1 (a-cycle versus b-cycle): *If we consider the torus as a geometrical surface (not formed by any p.e.c. material), then a steady current (at* $\omega = 0$*) looping in the direction of the a-cycle would produce a magnetic field in the interior of the torus and zero field in the exterior. Conversely, a steady current looping in the direction of a b-cycle would produce a magnetic field in the exterior of the torus and zero in the interior.*

Applying the Stokes theorem, we get:

$$\int_S \hat{\boldsymbol{n}} \cdot \nabla \times (\boldsymbol{E}^{\text{scat}} + \boldsymbol{E}^{\text{in}}) dS = \int_l (\boldsymbol{E}^{\text{scat}} + \boldsymbol{E}^{\text{in}}) \cdot d\mathbf{l}$$

$$-j\omega\mu \int_S \hat{\boldsymbol{n}} \cdot (\boldsymbol{H}^{\text{scat}} + \boldsymbol{H}^{\text{in}}) dS = \int_l (\boldsymbol{E}^{\text{scat}} + \boldsymbol{E}^{\text{in}}) \cdot d\mathbf{l} = 0$$

we conclude that for any $\omega > 0$

$$\int_S \hat{\boldsymbol{n}} \cdot \boldsymbol{H}^{\mathrm{scat}} dS = -\int_S \hat{\boldsymbol{n}} \cdot \boldsymbol{H}^{\mathrm{in}} dS \tag{8.6}$$

Here l is the b-cycle and S is its spanning surface. The convenient zero frequency limit from time-harmonic regime ($\omega \to 0$). In case of having a surface ∂D with genus greater than one, we can obtain one such condition for each b-cycle l_j and its spanning surface S_j. We can, therefore, assume this set of extra conditions as a way to uniquely determine the zero frequency limit of the magnetostatic boundary value problem.

$$\int_{S_j} \hat{\boldsymbol{n}} \cdot \boldsymbol{H}^{\mathrm{scat}} dS = \Phi_j \tag{8.7}$$

where Φ_j are known constants with value $\Phi_j = -\int_{S_j} \hat{\boldsymbol{n}} \cdot \boldsymbol{H}^{\mathrm{in}} dS$

With the extra set of conditions 8.5 and 8.7, we obtain enough data to have uniqueness in both the electrostatic boundary value problem and the magnetostatic boundary value problem. It is important to notice that the existence of the electrostatic boundary value problem is not guaranteed unless the boundary data is produced by an electrostatic incoming field, that is, it is the gradient of a scalar function. Without assuming the formulation in terms of scalar potentials, existence cannot be guaranteed. Notice that when the formulation in terms of scalar potentials for the electrostatic fields is assumed, the vector boundary condition $\hat{\boldsymbol{n}} \times \boldsymbol{E}|_{\partial D} = 0$ is transformed into a scalar boundary condition $\phi|_{\partial D_j} = V_j$.

To obtain a formulation that is stable at any frequency including the static limit ($\omega \geq 0$), we need to reformulate the scattering perfect electric conducting boundary value problem in a way that make sense for $\omega > 0$ and also for $\omega = 0$ without changing from electric field to electrostatic potential. This is the idea behind the decoupled potential integral equation.

8.4 Decoupled potentials and different Lorenz gauge fixings

Let us consider that we write the incoming electric field in terms of the standard vector and scalar potentials in the Lorenz gauge:

$$\boldsymbol{E}^{\mathrm{in}} = -j\omega \boldsymbol{A}^{\mathrm{in}} - \nabla \phi^{\mathrm{in}}$$

If we write also the scattered field in terms of the vector and scalar potentials $\boldsymbol{E}^{\mathrm{scat}} = i\omega \boldsymbol{A}^{\mathrm{scat}} - \nabla \phi^{\mathrm{scat}}$, we can rewrite the boundary condition on the tangent electric field as:

$$-j\omega \hat{\boldsymbol{n}} \times \boldsymbol{A}^{\mathrm{scat}} - \hat{\boldsymbol{n}} \times \nabla \phi^{\mathrm{scat}} = j\omega \hat{\boldsymbol{n}} \times \boldsymbol{A}^{\mathrm{in}} + \hat{\boldsymbol{n}} \times \nabla \phi^{\mathrm{in}}$$

It is very tempting to consider boundary conditions separately for the vector and scalar potentials as

$$\hat{\boldsymbol{n}} \times \boldsymbol{A}^{\text{scat}} = -\hat{\boldsymbol{n}} \times \boldsymbol{A}^{\text{in}}$$
$$\hat{\boldsymbol{n}} \times \nabla\phi^{\text{scat}} = -\hat{\boldsymbol{n}} \times \nabla\phi^{\text{in}}$$

Nevertheless, this step is extremely delicate, next we show why. The Lorenz gauge is not a complete gauge fixing, therefore, one can write the same electromagnetic field using two different sets of potentials $\boldsymbol{A}_1, \phi_1$ and $\boldsymbol{A}_2, \phi_2$. We can show that in the following example: let us consider a plane wave traveling in the $\hat{z}$ direction with the electric field polarized in $\hat{\boldsymbol{x}}$:

$$\boldsymbol{E}^{\text{in}} = E_p e^{-jkz}\hat{\boldsymbol{x}}$$
$$\boldsymbol{H}^{\text{in}} = \frac{E_p}{Z_0} e^{-jkz}\hat{\boldsymbol{y}}$$

where $k = \omega\sqrt{\mu\varepsilon}$ and $Z_0 = \sqrt{\frac{\mu}{\varepsilon}}$, and E_0 is a scalar. This plane wave can be written in terms of the following potentials:

$$\boldsymbol{A}_1^{\text{in}} = \frac{E_p}{-j\omega} e^{-jkz}\hat{\boldsymbol{x}}$$
$$\phi_1^{\text{in}} = 0$$

as $\boldsymbol{E}^{\text{in}} = -j\omega\boldsymbol{A}_1^{\text{in}} - \nabla\phi_1^{\text{in}}, \boldsymbol{H}^{\text{in}} = \frac{1}{\mu}\nabla \times \boldsymbol{A}_1^{\text{in}}$. Obviously $\boldsymbol{A}_1^{\text{in}}$ and ϕ_1^{in} are in the Lorenz gauge, that is $\nabla \cdot \boldsymbol{A}_1^{\text{in}} = i\omega\varepsilon\mu\phi_1^{\text{in}}$ (one can also check that $\Delta\boldsymbol{A}_1^{\text{in}} + k^2\boldsymbol{A}_1^{\text{in}} = 0$ and $\Delta\phi_1^{\text{in}} + k^2\phi_1^{\text{in}} = 0$). The same plane wave can be written in terms of the alternative set of potentials:

$$\boldsymbol{A}_2^{\text{in}} = -E_p x\sqrt{\varepsilon\mu} e^{-jkz}\hat{\boldsymbol{z}}$$
$$\phi_2^{\text{in}} = -E_p x e^{-jkz}$$

It is easy to verify that $\boldsymbol{E}^{\text{in}} = -j\omega\boldsymbol{A}_2^{\text{in}} - \nabla\phi_2^{\text{in}}, \boldsymbol{H}^{\text{in}} = \frac{1}{\mu}\nabla \times \boldsymbol{A}_2^{\text{in}}$ and also that $\boldsymbol{A}_2^{\text{in}}$ and ϕ_2^{in} is in the Lorenz gauge. Notice that once we have two different potentials that generate the same electromagnetic fields, any linear combination will also generate the same fields:

$$\boldsymbol{A}^{\text{in}} = q\boldsymbol{A}_1^{\text{in}} + (1-q)\boldsymbol{A}_2^{\text{in}}$$
$$\phi^{\text{in}} = q\phi_1^{\text{in}} + (1-q)\phi_2^{\text{in}}$$

for any complex scalar $q \in \mathbb{C}$. Therefore, in general, there are infinitely many potentials in the Lorenz gauge that generates the same electromagnetic fields. Of course, the same thing happens for the scattered field. Next we study the simple example of the scattered field produced by an infinite conducting plane produced by a plane wave. We use different choices of potentials (all in the Lorenz gauge) for the total field, and we see that the choice implies different boundary conditions on the potentials.

Considering an infinite perfect electric conducting plate located at $z = 0$, the scattered field will be:

$$
\begin{aligned}
\boldsymbol{E}^{\text{scat}} &= -E_p e^{jkz}\hat{\boldsymbol{x}} \\
\boldsymbol{H}^{\text{scat}} &= \frac{E_p}{Z_0} e^{jkz}\hat{\boldsymbol{y}}
\end{aligned}
$$

The total field will be:

$$
\begin{aligned}
\boldsymbol{E} = \boldsymbol{E}^{\text{in}} + \boldsymbol{E}^{\text{scat}} &= -E_p 2j \sin{(kz)}\hat{\boldsymbol{x}} \\
\boldsymbol{H} = \boldsymbol{H}^{\text{in}} + \boldsymbol{H}^{\text{scat}} &= \frac{2E_p}{Z_0} \cos{(kz)}\hat{\boldsymbol{y}}
\end{aligned}
\tag{8.8}
$$

This total field can be written using the potentials $\boldsymbol{A}_1, \phi_1$ in the Lorenz gauge:

$$
\begin{aligned}
\boldsymbol{A}_1 &= \frac{2E_p}{\omega} \sin{(kz)}\hat{\boldsymbol{x}} \\
\phi_1 &= 0
\end{aligned}
$$

but can be also written using the alternative potentials $\boldsymbol{A}_2, \phi_2$ also in the Lorenz gauge:

$$
\begin{aligned}
\boldsymbol{A}_2 &= -E_p 2x\sqrt{\varepsilon\mu} \cos{(kz)}\hat{\boldsymbol{z}} \\
\phi_2 &= E_p 2jx \sin{(kz)}
\end{aligned}
$$

Notice that the choice $\boldsymbol{A}_2, \phi_2$ is more suitable as it does not blow-up for $\omega \to 0$, while $\boldsymbol{A}_1, \phi_1$ does. This unpleasant blow-up implies that the electrostatic field $\boldsymbol{E}_0$ (zero frequency limit of the electric field) is not described by the electrostatic potential $\boldsymbol{E}_0 = -\nabla\phi_0$, but by the limit $\boldsymbol{E}_0 = \lim_{\omega\to 0} i\omega\boldsymbol{A}_1(\omega)$.

Notice also that in both cases $\boldsymbol{A}_1, \phi_1$ and $\boldsymbol{A}_2, \phi_2$, we obtain nice boundary conditions for the potentials:

$$
\begin{aligned}
\hat{\boldsymbol{n}} \times \boldsymbol{A}_{1,2}|_{\partial D} &= 0 \\
\phi_{1,2}|_{\partial D} &= 0
\end{aligned}
\tag{8.9}
$$

Unfortunately, the absence of blow-up is not a sufficient condition to obtain nice boundary conditions on the potentials separately, as we show in the next example:

$$
\begin{aligned}
\boldsymbol{A}_3 &= E_p\sqrt{\varepsilon\mu}(-2x\cos{(kz)}\hat{\boldsymbol{z}} - 2j\sin{(kx)}\hat{\boldsymbol{x}}) \\
\phi_3 &= E_p(2jx\sin{(kz)} + 2\cos{(kx)})
\end{aligned}
$$

It is easy to check that the potentials $\boldsymbol{A}_3, \phi_3$ are in the Lorenz gauge and that they generate the total field in 8.8. The reader can also notice that the boundary conditions 8.9 are not verified in this case $\hat{\boldsymbol{n}} \times \boldsymbol{A}_3|_{\partial D} \neq 0, \phi_3|_{\partial D} = 2\cos{(kx)} \neq 0$.

In summary, the Lorenz gauge is not a complete gauge fixing and we can obtain different representations of the same field using different potentials that verify different boundary conditions and different radiation conditions. If we want to write down a scattering theory of decoupled potentials, we need to use a complete gauge fixing. This means that, for each electromagnetic field, we need to specify the particular choice of potentials used to describe it among the class of all potentials in the Lorenz gauge that produces the same fields.

8.5 Incoming potentials in a low-frequency stable Lorenz gauge

The first ingredient we need to develop a scattering theory of decoupled potentials is to write down the incoming potentials in the Lorenz gauge in a way that is low-frequency stable. If the incoming field is produced by some known impressed sources $\boldsymbol{J}^{\text{imp}}$, ρ^{imp} located in the region V, the most convenient choice for the potentials is the most common one:

$$\begin{aligned} \boldsymbol{A}^{\text{in}}(\boldsymbol{r}) &= \mu \int_V \frac{e^{-jk|\boldsymbol{r}-\boldsymbol{r}'|}}{4\pi|\boldsymbol{r}-\boldsymbol{r}'|} \boldsymbol{J}^{\text{imp}}(\boldsymbol{r}')dV' \\ \phi^{\text{in}}(\boldsymbol{r}) &= \int_V \frac{e^{-jk|\boldsymbol{r}-\boldsymbol{r}'|}}{4\pi|\boldsymbol{r}-\boldsymbol{r}'|} \rho^{\text{imp}}(\boldsymbol{r}')dV' \end{aligned} \tag{8.10}$$

This choice is satisfactory as there is no blow-up in the vector potential when $\omega \to 0$. The expressions of the potentials at zero frequency are:

$$\begin{aligned} \boldsymbol{A}_0^{\text{in}}(\boldsymbol{r}) &= \mu \int_V \frac{1}{4\pi|\boldsymbol{r}-\boldsymbol{r}'|} \boldsymbol{J}^{\text{imp}}(\boldsymbol{r}')dV' \\ \phi_0^{\text{in}}(\boldsymbol{r}) &= \int_V \frac{1}{4\pi|\boldsymbol{r}-\boldsymbol{r}'|} \rho^{\text{imp}}(\boldsymbol{r}')dV' \end{aligned}$$

And the corresponding expressions for the static fields are:

$$\begin{aligned} \boldsymbol{H}_0^{\text{in}}(\boldsymbol{r}) &= \frac{1}{\mu}\nabla \times \boldsymbol{A}_0^{\text{scat}} = \nabla \times \int_V \frac{1}{4\pi|\boldsymbol{r}-\boldsymbol{r}'|} \boldsymbol{J}_0^{\text{imp}}(\boldsymbol{r}')dV' \\ \boldsymbol{E}_0^{\text{in}}(\boldsymbol{r}) &= -\nabla\phi_0^{\text{scat}} = -\nabla \int_V \frac{1}{4\pi|\boldsymbol{r}-\boldsymbol{r}'|} \rho_0^{\text{imp}}(\boldsymbol{r}')dV' \end{aligned}$$

These are ordinary expressions for the electric, magnetic fields and the vector and scalar potentials so, there is nothing new about that, we just observe that if the incoming field is produced by some impressed sources $\boldsymbol{J}^{\text{imp}}$, ρ^{imp}, then the ordinary expressions 8.10 are suitable to feed the right-hand side of the decoupled potential integral equation that we are going to develop in this chapter.

If the incoming field is a plane wave with arbitrary polarization vector $\boldsymbol{E}_p$ and direction $\hat{\boldsymbol{u}}$:

$$\boldsymbol{E}^{\text{in}} = \boldsymbol{E}_p e^{-jk\hat{\boldsymbol{u}}\cdot\boldsymbol{r}}$$

$$\boldsymbol{H}^{\text{in}} = \frac{\hat{\boldsymbol{u}} \times \boldsymbol{E}_p}{Z_0} e^{-jk\hat{\boldsymbol{u}}\cdot\boldsymbol{r}}$$

A suitable choice for the incoming potentials is

$$\boldsymbol{A}^{\text{in}} = -\hat{\boldsymbol{u}}(\boldsymbol{r}\cdot\boldsymbol{E}_p)\sqrt{\varepsilon\mu}e^{-jk\hat{\boldsymbol{u}}\cdot\boldsymbol{r}}$$

$$\phi^{\text{in}} = -(\boldsymbol{r}\cdot\boldsymbol{E}_p)e^{-jk\hat{\boldsymbol{u}}\cdot\boldsymbol{r}}$$

Notice that in this case, the zero frequency limit for the fields and the potentials is

$$\boldsymbol{E}_0^{\text{in}} = \boldsymbol{E}_p$$

$$\boldsymbol{H}_0^{\text{in}} = \frac{\hat{\boldsymbol{u}} \times \boldsymbol{E}_p}{Z_0}$$

(constant electrostatic and magnetostatic vectors) and

$$\boldsymbol{A}_0^{\text{in}} = -\hat{\boldsymbol{u}}(\boldsymbol{r}\cdot\boldsymbol{E}_p)\sqrt{\varepsilon\mu}$$

$$\phi_0^{\text{in}} = -(\boldsymbol{r}\cdot\boldsymbol{E}_p)$$

And, therefore, we obtain the standard representation for the static fields in terms of the static potentials as:

$$\boldsymbol{H}_0^{\text{in}}(\boldsymbol{r}) = \frac{1}{\mu}\nabla\times\boldsymbol{A}_0^{\text{scat}} = \frac{\hat{\boldsymbol{u}} \times \boldsymbol{E}_p}{Z_0}$$

$$\boldsymbol{E}_0^{\text{in}}(\boldsymbol{r}) = -\nabla\phi_0^{\text{scat}} = \boldsymbol{E}_p$$

Next we consider the situation where the incoming electromagnetic wave is described as a partial wave expansion

$$\boldsymbol{E}^{\text{in}} = \sum_{m,n}[a_{mn}\nabla\times\nabla\times(\boldsymbol{r}f_n(k,r)Y_n^m) - j\omega\mu b_{mn}\nabla\times(\boldsymbol{r}f_n(k,r)Y_n^m)]$$

$$\boldsymbol{H}^{\text{in}} = \sum_{m,n}[b_{mn}\nabla\times\nabla\times(\boldsymbol{r}f_n(k,r)Y_n^m) + j\omega\varepsilon a_{mn}\nabla\times(\boldsymbol{r}f_n(k,r)Y_n^m)]$$

where the functions $f_n(k,r)$ are defined as modified spherical Bessel/Hankel functions:

$$f_n(k,r) := \begin{cases} \widetilde{h}_n(kr) := h_n^{(2)}(kr)\frac{k^{n+1}}{j(2n-1)(2n-3)\ldots5\cdot3\cdot1} \\ \widetilde{j}_n(kr) := j_n(kr)\frac{(2n+1)(2n-1)\ldots5\cdot3\cdot1}{k^{n+1}} \end{cases}$$

The reason for this re-scale is that we get the right static limit without blow-up

$$\lim_{\omega\to 0} f_n(k,r) = \begin{cases} \lim_{k\to 0} h_n(k,r) = \frac{1}{r^{n+1}} \\ \lim_{k\to 0} j_n(k,r) = r^n \end{cases}$$

The fields corresponding to a magnetic multipole of degree n and order m are defined to be

$$\dot{\boldsymbol{E}}_{nm}^{\text{in}} = -j\omega\mu\nabla \times (\boldsymbol{r}f_n(k,r)Y_n^m)$$

$$\dot{\boldsymbol{H}}_{nm}^{\text{in}} = \nabla \times \nabla \times (\boldsymbol{r}f_n(k,r)Y_n^m)$$

The corresponding vector and scalar potentials can be defined by

$$\dot{\boldsymbol{A}}_{nm}^{\text{in}} = \mu\nabla \times (\boldsymbol{r}f_n(k,r)Y_n^m)$$

$$\dot{\phi}_{nm}^{\text{in}} = 0$$

This is a perfectly fine representation of the magnetic multipoles in terms of potentials in the Lorenz gauge, as in the static limit, the potentials do not blow-up and can recover the static fields in a standard way.

The analog result for the electric multipoles is not so straightforward. The fields of an electric multipole of degree n and order m are defined to be

$$\ddot{\boldsymbol{E}}_{nm}^{\text{in}} = \nabla \times \nabla \times (\boldsymbol{r}f_n(k,r)Y_n^m),$$

$$\ddot{\boldsymbol{H}}_{nm}^{\text{in}} = j\omega\varepsilon\nabla \times (\boldsymbol{r}f_n(k,r)Y_n^m).$$

A suitable choice for the potentials in the Lorenz gauge is the following:

$$\ddot{\boldsymbol{A}}_{nm}^{\text{in}} = -i\omega\mu\varepsilon\boldsymbol{r}\widetilde{h}_n(k,r)Y_n^m + \frac{j\omega\mu\varepsilon}{2n-1}\nabla(r\widetilde{h}_{n-1}(k,r)Y_n^m),$$

$$\ddot{\phi}_{nm}^{\text{in}} = n\widetilde{h}_n(k,r)Y_n^m$$

for the outgoing waves, and

$$\ddot{\boldsymbol{A}}_{nm}^{\text{in}} = j\omega\mu\varepsilon\boldsymbol{r}\widetilde{j}_n(k,r)Y_n^m + \frac{-j\omega\mu\varepsilon}{2n+3}\nabla(r\widetilde{j}_{n-1}(k,r)Y_n^m)$$

$$\ddot{\phi}_{nm}^{\text{in}} = -(n+1)\widetilde{j}_n(k,r)Y_n^m$$

for the incoming waves.

A proof and a detailed discussion of that are available in [1].

With this, we conclude the discussion about the incoming potentials in the Lorenz gauge needed to write down the right-hand side of the decoupled potential integral equation.

8.6 Decoupled potential boundary value problems

In this section, we describe the boundary conditions and radiation conditions for the scattered vector and scalar potentials in the Lorenz gauge. The idea is that the complete gauge fixing is obtained by postulating boundary value problems for the scattered vector and scalar potentials that have a unique solution. The idea is that the uniqueness for the potentials happens down to $\omega = 0$, and that, by construction, at $\omega = 0$, the stabilizing conditions described in Section 1.3 are verified implicitly by construction.

The appropriateness of the definitions of the following boundary value problems has to be understood through the theorems that those definitions imply and that will be described in this and subsequent sections.

Definition 6: *By the* scalar modified Dirichlet problem, *we mean the calculation of a scalar Helmholtz or Laplace potential* $\phi^{\text{scat}} \in C^2(\mathbb{R}^3\backslash\overline{D}) \cap C(\mathbb{R}^3\backslash D)$ *which satisfies standard radiation conditions at infinity. Letting N denote the number of connected components of the boundary* ∂D*, we introduce extra unknown degrees of freedom* $\{V_j\}_{j=1}^N$ *and the boundary data* $f \in C(\partial D)$ *is supplemented with additional (known) constants* $\{Q_j\}_{j=1}^N$*. For the scalar Helmholtz potential,*

$$\Delta\phi^{\text{scat}} + k^2\phi^{\text{scat}} = 0, \quad \phi^{\text{scat}}|_{\partial D_j} = f + V_j, \tag{8.11}$$

$$\tfrac{\boldsymbol{r}}{|\boldsymbol{r}|} \cdot \nabla\phi^{\text{scat}}(\boldsymbol{r}) - ik\phi^{\text{scat}}(\boldsymbol{r}) = o\left(\tfrac{1}{|\boldsymbol{r}|}\right), \ |\boldsymbol{r}| \to \infty, \tag{8.12}$$

(uniformly for all directions $\frac{\boldsymbol{r}}{|\boldsymbol{r}|}$*), with the flux condition:*

$$\int_{\partial D_j} \frac{\partial\phi^{\text{scat}}}{\partial n} dS = Q_j \tag{8.13}$$

For the scalar Laplace potential,

$$\Delta\phi_0^{\text{scat}} = 0, \quad \phi_0^{\text{scat}}|_{\partial D_j} = f + V_j, \tag{8.14}$$

$$\phi_0^{\text{scat}}(\boldsymbol{r}) = o(1), \ |\boldsymbol{r}| \to \infty, \tag{8.15}$$

8.15 (uniformly for all directions $\frac{\boldsymbol{r}}{|\boldsymbol{r}|}$*), with the same flux condition (8.13):*

$$\int_{\partial D_j} \frac{\partial\phi_0^{\text{scat}}}{\partial n} dS = Q_j$$

Definition 7: *By the* vector modified Dirichlet problem, *we mean the calculation of a vector Helmholtz or Laplace potential* $\boldsymbol{A}^{\text{scat}} \in C^2(\mathbb{R}^3\backslash\overline{D})$ *with* $\nabla \cdot \boldsymbol{A}^{\text{scat}} \in C(\mathbb{R}^3\backslash D)$ *which satisfies standard radiation conditions at infinity and tangent boundary data* $\boldsymbol{f} \in C_t(\partial D)$*. Letting N denote the number of connected components of the boundary* ∂D*, we introduce extra unknown degrees of freedom* $\{v_j\}_{j=1}^N$ *and the boundary data* $h \in C(\partial D)$ *is supplemented with additional (known) constants* $\{q_j\}_{j=1}^N$*. For the vector Helmholtz potential,*

$$\Delta\boldsymbol{A}^{\text{scat}} + k^2\boldsymbol{A}^{\text{scat}} = 0, \quad \hat{\boldsymbol{n}} \times \boldsymbol{A}^{\text{scat}}|_{\partial D} = \boldsymbol{f}, \quad \nabla \cdot \boldsymbol{A}^{\text{scat}}|_{\partial D_j} = h + v_j, \tag{8.16}$$

$$\nabla \times \boldsymbol{A}^{\text{scat}}(\boldsymbol{r}) \times \tfrac{\boldsymbol{r}}{|\boldsymbol{r}|} + \tfrac{\boldsymbol{r}}{|\boldsymbol{r}|}\nabla \cdot \boldsymbol{A}^{\text{scat}}(\boldsymbol{r}) - ik\boldsymbol{A}^{\text{scat}}(\boldsymbol{r}) = o\left(\tfrac{1}{|\boldsymbol{r}|}\right),\ |\boldsymbol{r}| \to \infty, \tag{8.17}$$

with

$$\int_{\partial D_j} \hat{\boldsymbol{n}} \cdot \boldsymbol{A}^{\text{scat}} dS = q_j. \tag{8.18}$$

For the vector Laplace potential,

$$\Delta \boldsymbol{A}_0^{\text{scat}} = 0, \quad \hat{\boldsymbol{n}} \times \boldsymbol{A}_0^{\text{scat}}|_{\partial D^+} = \boldsymbol{f}, \quad \nabla \cdot \boldsymbol{A}_0^{\text{scat}}|_{\partial D_j} = h + v_j, \tag{8.19}$$

$$\boldsymbol{A}_0^{\text{scat}}(\boldsymbol{r}) = o(1),\ |\boldsymbol{r}| \to \infty, \tag{8.20}$$

8.20 (uniformly for all directions $\frac{\boldsymbol{r}}{|\boldsymbol{r}|}$), with the same flux condition 8.18.

We take these two definitions (scalar and vector modified Dirichlet problems) as the starting point. The next step is to study the properties they have about uniqueness, existence, continuity of the solution with respect to the boundary data and uniform continuity of the solution for $\omega \to 0$. In particular, by the uniqueness, we will get a complete gauge fixing of the scattered scalar and vector potentials. The third step is to show the connection between these two boundary value problems and the original Maxwell electromagnetic scattering problem. By doing this, we will show that the fields obtained by the potential solution of these boundary value problems are the scattered electromagnetic fields of the original problem.

Next we give without poof the following results about the previous boundary value problems (for a proof, see [1]).

Theorem 1: *(Uniqueness) The scalar modified Dirichlet problem has at most one solution for any $\omega \geq 0$.*

Theorem 2: *(Uniqueness) The vector-modified Dirichlet problem has at most one solution for any $\omega \geq 0$.*

Theorem 3: *(Existence) The scalar modified Dirichlet problem has one solution for any boundary data $f \in C(\partial D)$ and scalars $\{Q_j\}_{j=1}^{N}$ for any $\omega \geq 0$.*

Theorem 4: *(Existence) The vector-modified Dirichlet problem has one solution for any boundary data $\boldsymbol{f} \in C_t(\partial D)$, $h \in C(\partial D)$ and scalars $\{q_j\}_{j=1}^{N}$ for any $\omega \geq 0$.*

Next we show two technical theorems about the stability of the solution in the low-frequency regime:

Theorem 5: *The scalar-modified Dirichlet problem has a unique solution for $\omega \geq 0$.*

For continuous boundary data $f \in C(\partial D)$, the solution can be extended continuously up to the boundary. The following stability condition holds uniformly on the interval $[0, \omega_{\max}]$*:*

$$\|\phi^{\text{scat}}\|_{\infty,\mathbb{R}^3\backslash D} \leq K_{(\partial D,k_{\max})}\left(\|f\|_{\infty,\partial D} + \sum_{j=1}^{N} |Q_j|^2\right)$$

For uniformly Hölder continuous boundary data $f \in C^{0,\alpha}(\partial D)$, the solution can be extended in a uniformly Hölder continuous way up to the boundary $\phi^{\text{scat}} \in C^{0,\alpha}(\mathbb{R}^3\backslash D)$. The following stability condition holds uniformly on the interval $[0, \omega_{\max}]$*:*

$$\|\phi^{\text{scat}}\|_{\alpha,\mathbb{R}^3\backslash D} \leq K_{(\alpha,\partial D,\omega_{\max})}\left(\|f\|_{\alpha,\partial D} + \sum_{j=1}^{N} |Q_j|^2\right)$$

For uniformly Hölder continuous differentiable boundary data $f \in C^{1,\alpha}(\partial D)$, the solution can be extended in a uniformly Hölder continuous differentiable way up to the boundary $\phi^{\text{scat}} \in C^{1,\alpha}(\mathbb{R}^3\backslash D)$. The following stability condition holds uniformly on the interval $[0, \omega_{\max}]$*:*

$$\|\phi^{\text{scat}}\|_{\alpha,\mathbb{R}^3\backslash D} \leq K_{(\alpha,\partial D,\omega_{\max})}\left(\|f\|_{\alpha,\partial D} + \sum_{j=1}^{N} |Q_j|^2\right)$$

$$\|\nabla\phi^{\text{scat}}\|_{\alpha,\mathbb{R}^3\backslash D} \leq K_{(\alpha,\partial D,\omega_{\max})}\left(\|f\|_{\alpha,\partial D} + \|\nabla_s f\|_{\alpha,\partial D} + \sum_{j=1}^{N} |Q_j|^2\right)$$

The constants K only depend on the subindexes shown.

Theorem 6: *The vector-modified Dirichlet problem has a unique solution for* $\omega \geq 0$.

For continuous boundary data $\boldsymbol{f} \in C_t(\partial D), h \in C(\partial D)$, the divergence and tangent components of the solution can be extended continuously up to the boundary. The following stability condition holds uniformly on the interval $[0, \omega_{\max}]$*:*

$$\|\boldsymbol{A}^{\text{scat}}\|_{\infty,\mathfrak{M}} \leq K_{(\partial D,\omega_{\max},\mathfrak{M})}\left(\|\boldsymbol{f}\|_{\infty,\partial D} + \|h\|_{\infty,\partial D} + \sum_{j=1}^{N} |q_j|^2\right)$$

for any closed $\mathfrak{M} \subset \mathbb{R}^3\backslash\overline{D}$

$$\|\nabla\cdot\boldsymbol{A}^{\text{scat}}\|_{\infty,\mathbb{R}^3\backslash D} \leq K_{(\partial D,\omega_{\max})}\left(\|\boldsymbol{f}\|_{\infty,\partial D} + \|h\|_{\infty,\partial D} + \sum_{j=1}^{N} |q_j|^2\right)$$

For uniformly Hölder continuous boundary data $\boldsymbol{f} \in C_t^{0,\alpha}(\partial D), h \in C^{0,\alpha}(\partial D)$, *the solution and its divergence can be extended in a uniformly Hölder continuous way up to the boundary* $\boldsymbol{A} \in C^{0,\alpha}(\mathbb{R}^3 \backslash D), \nabla \cdot \boldsymbol{A} \in C^{0,\alpha}(\mathbb{R}^3 \backslash D)$. *The following stability condition holds uniformly on the interval* $[0, \omega_{\max}]$:

$$\|\boldsymbol{A}^{\text{scat}}\|_{\alpha,\mathbb{R}^3\backslash D} \le K_{(\alpha,\partial D,\omega_{\max})}\left(\|\boldsymbol{f}\|_{\alpha,\partial D} + \|h\|_{\alpha,\partial D} + \sum_{j=1}^{N} |q_j|^2\right)$$

$$\|\nabla \cdot \boldsymbol{A}^{\text{scat}}\|_{\alpha,\mathbb{R}^3\backslash D} \le K_{(\alpha,\partial D,\omega_{\max})}\left(\|\boldsymbol{f}\|_{\alpha,\partial D} + \|h\|_{\alpha,\partial D} + \sum_{j=1}^{N} |q_j|^2\right)$$

For boundary data $\boldsymbol{f} \in C_t^{0,\alpha}(Div, \partial D), h \in C^{0,\alpha}(\partial D)$, *the solution and its divergence and curl can be extended in a uniformly Hölder continuous way up to the boundary* $\boldsymbol{A} \in C^{0,\alpha}(\mathbb{R}^3 \backslash D)$, $\nabla \times \boldsymbol{A} \in C^{0,\alpha}(\mathbb{R}^3 \backslash D), \nabla \cdot \boldsymbol{A} \in C^{0,\alpha}(\mathbb{R}^3 \backslash D)$. *The following stability condition holds uniformly on the interval* $[0, \omega_{\max}]$:

$$\|\boldsymbol{A}^{\text{scat}}\|_{\alpha,\mathbb{R}^3\backslash D} \le K_{(\alpha,\partial D,\omega_{\max})}\left(\|\boldsymbol{f}\|_{\alpha,\partial D} + \|h\|_{\alpha,\partial D} + \sum_{j=1}^{N} |q_j|^2\right)$$

$$\|\nabla \cdot \boldsymbol{A}^{\text{scat}}\|_{\alpha,\mathbb{R}^3\backslash D} \le K_{(\alpha,\partial D,\omega_{\max})}\left(\|\boldsymbol{f}\|_{\alpha,\partial D} + \|h\|_{\alpha,\partial D} + \sum_{j=1}^{N} |q_j|^2\right)$$

$$\|\nabla \times \boldsymbol{A}^{\text{scat}}\|_{\alpha,\mathbb{R}^3\backslash D} \le K_{(\alpha,\partial D,\omega_{\max})}\left(\|\boldsymbol{f}\|_{\alpha,\partial D} + \|\nabla_s \cdot \boldsymbol{f}\|_{\alpha,\partial D} + \|h\|_{\alpha,\partial D} + \sum_{j=1}^{N} |q_j|^2\right)$$

Next definition formalizes the relation between the incoming potentials and the scattered potentials.

Definition 8: *Let* ϕ^{in}, $\boldsymbol{A}^{\text{in}}$ *denote incoming scalar and vector potentials and assume that D is a perfect conductor. The* scattered scalar potential ϕ^{scat} *is the solution to the scalar-modified Dirichlet problem with boundary data:*

$$f := -\phi^{\text{in}}|_{\partial D_j},\ Q_j := -\int_{\partial D_j} \frac{\partial \phi^{\text{in}}}{\partial n} ds.$$

Likewise, the scattered vector potential $\boldsymbol{A}^{\text{scat}}$ *is the solution to the vector-modified Dirichlet problem with boundary data:*

$$\boldsymbol{f} := -\hat{\boldsymbol{n}} \times \boldsymbol{A}^{\text{in}}|_{\partial D},\ h := -\nabla \cdot \boldsymbol{A}^{\text{in}}|_{\partial D},\ q_j := -\int_{\partial D_j} \hat{\boldsymbol{n}} \cdot \boldsymbol{A}^{\text{in}} ds.$$

Once we have seen that the two boundary value problems are well posed, we show the relationship of them with the electromagnetic scattering problem of perfect electric conductors.

Theorem 7: *For any $\omega \geq 0$, let $\boldsymbol{E}^{\text{in}}, \boldsymbol{H}^{\text{in}}$ be an incoming electromagnetic field described by the potentials $\boldsymbol{A}^{\text{in}}, \phi^{\text{in}}$ in the Lorenz gauge:*

$$\begin{aligned}\boldsymbol{H}^{\text{in}} &= \frac{1}{\mu}\nabla \times \boldsymbol{A}^{\text{in}} \\ \boldsymbol{E}^{\text{in}} &= -j\omega \boldsymbol{A}^{\text{in}} - \nabla\phi^{\text{in}} \\ \nabla \cdot \boldsymbol{A}^{\text{in}} &= -j\omega\mu\varepsilon\phi^{\text{in}}\end{aligned}$$

and let $\boldsymbol{A}^{\text{scat}}, \phi^{\text{scat}}$ denote the corresponding scattered vector and scalar potentials (Definition 8). Then the electromagnetic fields $\boldsymbol{E}^{\text{scat}}, \boldsymbol{H}^{\text{scat}}$ scattered from a perfect conductor are given by

$$\begin{aligned}\boldsymbol{H}^{\text{scat}} &= \frac{1}{\mu}\nabla \times \boldsymbol{A}^{\text{scat}} \\ \boldsymbol{E}^{\text{scat}} &= -j\omega \boldsymbol{A}^{\text{scat}} - \nabla\phi^{\text{scat}}\end{aligned}$$

with

$$\begin{aligned}\nabla \cdot \boldsymbol{A}^{\text{scat}} &= -j\omega\mu\varepsilon\phi^{\text{scat}} \\ \hat{\boldsymbol{n}} \times \boldsymbol{E}^{\text{scat}} &= -\hat{\boldsymbol{n}} \times \boldsymbol{E}^{\text{in}}|_{\partial D} \\ \hat{\boldsymbol{n}} \cdot \boldsymbol{H}^{\text{scat}} &= -\hat{\boldsymbol{n}} \cdot \boldsymbol{H}^{\text{in}}|_{\partial D}\end{aligned}$$

At zero frequency, the scattered electromagnetic fields are the low-frequency limit of time-harmonic regime, that is:

$$\begin{aligned}&\lim_{\omega\to 0} \|\boldsymbol{E}^{\text{scat}}_{\omega} - \boldsymbol{E}^{\text{scat}}_{0}\|_{\alpha,\mathbb{R}^3/D} \to 0 \\ &\lim_{\omega\to 0} \|\boldsymbol{H}^{\text{scat}}_{\omega} - \boldsymbol{H}^{\text{scat}}_{0}\|_{\alpha,\mathbb{R}^3/D} \to 0\end{aligned}$$

This last theorem is fundamental in this theory, as it shows that the scattered electromagnetic field can be obtained (at arbitrarily low frequencies, in a stable way) by solving the scalar- and vector-modified Dirichlet problems associated to the potentials. This theorem is what justifies the definitions of the boundary value problems associated with the scalar and vector potentials.

8.7 Second-kind integral equation

In this section, we propose a second kind of integral equation to solve the boundary values associated to the scalar and vector potentials. The goal is to obtain an integral equation that is stable under refinement.

Next we define classical operators in potential theory (see [4]):

$$
\begin{aligned}
S_k\sigma &= \int_{\partial D} g_k(\boldsymbol{r}-\boldsymbol{r}')\sigma(\boldsymbol{r}')dA_{\boldsymbol{r}'}\\
D_k\sigma &= \int_{\partial D} \frac{\partial g_k}{\partial n_y}(\boldsymbol{r}-\boldsymbol{r}')\sigma(\boldsymbol{r}')dS_{\boldsymbol{r}'}\\
S_k'\sigma &= \int_{\partial D} \frac{\partial g_k}{\partial n_x}(\boldsymbol{r}-\boldsymbol{r}')\sigma(\boldsymbol{r}')dS_{\boldsymbol{r}'}\\
D_k'\sigma &= \frac{\partial}{\partial n_r}\int_{\partial D} \frac{\partial g_k}{\partial n_{r'}}(\boldsymbol{r}-\boldsymbol{r}')\sigma(\boldsymbol{r}')dS_{\boldsymbol{r}'}
\end{aligned}
$$

where $\boldsymbol{r} \in \partial D$ and Green's function on the free space is:

$$
g_k(\boldsymbol{r}) = \frac{e^{ik|\boldsymbol{r}|}}{4\pi|\boldsymbol{r}|}
$$

For off-surface evaluations $\boldsymbol{r} \in \mathbb{R}^3\backslash\partial D$, we have:

$$
\begin{aligned}
S_k[\sigma](\boldsymbol{r}) &= \int_{\partial D} g_k(\boldsymbol{r}-\boldsymbol{r}')\sigma(\boldsymbol{r}')dS_{\boldsymbol{r}'}\\
D_k[\sigma](\boldsymbol{r}) &= \int_{\partial D} \frac{\partial g_k}{\partial n_{r'}}(\boldsymbol{r}-\boldsymbol{r}')\sigma(\boldsymbol{r}')dS_{\boldsymbol{r}'}
\end{aligned}
$$

Next theorem describes the appropriate integral equation and representation for the scattered scalar potential to solve the modified Dirichlet problem.

Theorem 8: *The scalar field*

$$
\phi^{\text{scat}}(\boldsymbol{r}) = D_k[\sigma](\boldsymbol{r}) + j\alpha S_k[\sigma](\boldsymbol{r})
$$

with scalar density σ of class $C(\partial D)$ and $\alpha > 0$ and the constants $\{V_j\}_{j=1}^N$ solve the modified Dirichlet problem ($\omega \geq 0$) provided σ and $\{V_j\}_{j=1}^N$ solve the equation:

$$
\begin{aligned}
&\frac{\sigma}{2} + D_k\sigma + j\alpha S_k\sigma - \sum_{j=1}^{N} V_j\chi_j = f\\
&\int_{\partial D_j}\left((D_k' - D_0')\sigma - j\alpha\frac{\sigma}{2} + j\alpha S_k'\sigma\right)ds = Q_j
\end{aligned}
\tag{8.21}
$$

where χ_j denotes the characteristic function for boundary ∂D_j. The resulting integral equation is of the second kind on $C(\partial D) \times \mathbb{C}^N$ and invertible.

To study the vector-modified Dirichlet problem, we define the following dyadic operators:

$$\overline{\overline{L}}\begin{pmatrix} \boldsymbol{a} \\ \rho \end{pmatrix} = \begin{pmatrix} L_{11}\boldsymbol{a} + L_{12}\rho \\ L_{21}\boldsymbol{a} + L_{22}\rho \end{pmatrix}$$

where

$$\begin{aligned} L_{11}\boldsymbol{a} &= M_k\boldsymbol{a} = \int_{\partial D} \hat{\boldsymbol{n}}_r \times \nabla_r \times (g_k(\boldsymbol{r}-\boldsymbol{r}')\boldsymbol{a}(\boldsymbol{r}'))dS_{r'} \\ L_{12}\rho &= -\hat{\boldsymbol{n}}_r \times S_k(\hat{\boldsymbol{n}}_{r'}\rho) \\ L_{21}\boldsymbol{a} &= 0 \\ L_{22}\rho &= D_k\rho \end{aligned}$$

and

$$\overline{\overline{R}}\begin{pmatrix} \boldsymbol{a} \\ \rho \end{pmatrix} = \begin{pmatrix} R_{11}\boldsymbol{a} + R_{12}\rho \\ R_{21}\boldsymbol{a} + R_{22}\rho \end{pmatrix}$$

where

$$\begin{aligned} R_{11}\boldsymbol{a} &= \hat{\boldsymbol{n}}_r \times S_k(\hat{\boldsymbol{n}}_{r'} \times \boldsymbol{a}) \\ R_{12}\rho &= \hat{\boldsymbol{n}}_r \times \nabla S_k(\rho) \\ R_{21}\boldsymbol{a} &= \nabla \cdot S_k(\hat{\boldsymbol{n}}_{r'} \times \boldsymbol{a}) \\ R_{22}\rho &= -k^2 S_k\rho \end{aligned}$$

Theorem 9: *The vector field*

$$\boldsymbol{A}^{\text{scat}} = \nabla \times S_k[\boldsymbol{a}](\boldsymbol{r}) - S_k[\hat{\boldsymbol{n}}_{r'}\rho](\boldsymbol{r}) - j\alpha\left(S_k[\hat{\boldsymbol{n}}_{r'} \times \boldsymbol{a}](\boldsymbol{r}) + \nabla S_k[\rho](\boldsymbol{r})\right)$$

with tangential density $\boldsymbol{a}$ *of the class* $C_t(\partial D)$ *and scalar function* ρ *of class* $C(\partial D)$ *and* $\alpha > 0$ *and the constants* $\{v_j\}_{j=1}^N$ *solve the vector-modified Dirichlet problem* ($\omega \geq 0$) *provided* $\boldsymbol{a}$, ρ *and* $\{v_j\}_{j=1}^N$ *solve the equation:*

$$\begin{aligned} &\frac{1}{2}\begin{pmatrix} \boldsymbol{a} \\ \rho \end{pmatrix} + \overline{\overline{L}}\begin{pmatrix} \boldsymbol{a} \\ \rho \end{pmatrix} - j\alpha\overline{\overline{R}}\begin{pmatrix} \boldsymbol{a} \\ \rho \end{pmatrix} + \begin{pmatrix} 0 \\ \sum_{j=1}^N v_j\chi_j \end{pmatrix} = \begin{pmatrix} \boldsymbol{f} \\ h \end{pmatrix} \\ &\int_{\partial D_j} \left(-\hat{\boldsymbol{n}}_r \cdot S_k(\hat{\boldsymbol{n}}_{r'}\rho) - j\alpha(\hat{\boldsymbol{n}}_r \cdot S_k(\hat{\boldsymbol{n}}_{r'} \times \boldsymbol{a}) - \frac{\rho}{2} + S_k'(\rho))\right) ds = q_j \end{aligned} \tag{8.22}$$

where χ_j *denotes the characteristic function for boundary* ∂D_j*. The resulting integral equation is of the second kind on* $C_t(\partial D) \times C(\partial D) \times \mathbb{C}^N$

Definition 9: *We will refer to (8.21) and (8.22) as the scalar- and vector-decoupled potential integral equations. The former will be abbreviated by* DPIEs *and the latter by* DPIEv. *Together, they form the* DPIE.

The DPIEs has one scalar function unknown $\sigma(\boldsymbol{r}), \boldsymbol{r} \in \partial D$ and N scalar unknowns V_j where N is the number of connected components of the boundary ∂D. It also contains one scalar functional equality with right-hand side $f(\boldsymbol{r}), \boldsymbol{r} \in \partial D$ and N scalar equalities with right-hand side Q_j.

On the other hand, the DPIEv has one vector function unknown $\boldsymbol{a}(\boldsymbol{r}), \boldsymbol{r} \in \partial D$ where $\boldsymbol{a}$ is tangent to the surface ∂D, one scalar function unknown function $\rho(\boldsymbol{r}), \boldsymbol{r} \in \partial D$ and also has N scalar unknowns v_j. Similarly the DPIEv has one functional vector identity with right-hand side $\boldsymbol{f}(\boldsymbol{r}), \boldsymbol{r} \in \partial D$, one functional scalar identity with right hand side $h(\boldsymbol{r}), \boldsymbol{r} \in \partial D$ and N scalar identities with right-hand side q_j.

The relation between the right-hand side and the incoming potentials in the DPIEs and DPIEv is given in definition 8.

8.8 Discretization of an integral equation of the second kind

In this section, we describe the way to obtain a high order and adaptive discretization of an integral equation of the second kind. This can be applied to the DPIE as well as other integral equations of the same type in electromagnetism like decoupled field integral equation (see [1]), charge current integral equations [5–10], and regularized combined source integral equations (see [11]). These techniques can also be used in other physics like acoustics [12] or fluid dynamics [13,14].

Remark 2: *It is important to notice that the discretization method described in this section and in this chapter can be applied to any second kind of integral equation, but it does* **not** *include hypersingular integral equations like EFIE or PMCHWT.*

The discretization technique described here is based on the locally corrected Nyström method (see [15]). You need essentially two ingredients to discretize properly an integral equation: the ability of **numerical integration** and **interpolation**. The numerical integration is obviously important when computing the integral operator and the interpolation is needed in the local correction for the self and near terms.

It is important to notice that in this section, we do not address the geometric problem, therefore, here we will assume that the surface (perfect electric conducting scatterer) is well defined and the user has access to a high fidelity description of the surface in terms of its charts $\boldsymbol{r}_m(u, v)$:

$$\begin{aligned}
\partial D &= \cup T_m \\
T_m &= \boldsymbol{r}_m(T_0) \\
\boldsymbol{r}_m(u, v) &: T_0 \to T_m \subset \mathbb{R}^3
\end{aligned}$$

where $T_0 = \{(u, v), u \geq 0, v \geq 0, u + v \leq 1\}$ is the basic canonical triangle. We assume that the user of this discretization method has the ability to evaluate the functions $\boldsymbol{r}_m(u, v)$ and its derivatives $\mathbf{u}_m(u, v), \mathbf{v}_m(u, v)$ (and therefore $dS_m(u, v), \hat{\boldsymbol{n}}_m(u, v)$).

The DPIE contains two surface density unknowns $\boldsymbol{a}, \sigma$, where $\boldsymbol{a}$ is a vector function defined on the surface ∂D and tangent to it, and σ is a scalar function defined on the surface. Once we consider a triangulation of the surface, we can consider that $\boldsymbol{a}$ and σ is a set of functions on the basic triangle T_0:

$$\boldsymbol{a}(\boldsymbol{r}) = \boldsymbol{a}(\boldsymbol{r}_m(u,v)) = \boldsymbol{a}_m(u,v)$$
$$\sigma(\boldsymbol{r}) = \sigma(\boldsymbol{r}_m(u,v)) = \sigma_m(u,v)$$

If the reader is not familiarized with the Vioreanu–Rokhlin nodes and the Koornwinder polynomials, we suggest to read first Appendix C for a brief tutorial on numerical integration and interpolation in 2D. For a general introduction about interpolation and integration in 1D, we refer to the Appendix B.

First we will describe the scalar case (scalar integral equation) like the DPIEs, where the unknown is a scalar function defined on the surface and the right-hand side is a prescribed scalar function on the surface.

To discretize the scalar function on a basic triangle, we sample it on the Vioreanu–Rokhlin nodes and consider the values of the function at those nodes as degrees of freedom to describe σ, and, therefore, unknowns of the resulting discrete linear system:

$$\sigma_m(u_i, v_i) = \sigma_m^i$$

Here (u_i, v_i) are the Vioreanu–Rokhlin nodes on the basic unit triangle T_0. The number of nodes depends on the order of discretization used. σ_m^i are the values of the function on those sampling points ($i = 1, \ldots, n_p$, where $n_p = (p+1)(p+2)/2$ is the number of sampling points of order p). We can obtain the value of σ at any other point inside the triangle by interpolation using the Koornwinder polynomials ($K_{n_1 n_2}(u,v)$):

$$\sigma_m(u,v) \approx \sum_{\substack{n_1=0 \\ n_2 \le n_1}}^{n_1=p} c_{n_1 n_2} K_{n_1 n_2}(u,v)$$

$c_{n_1 n_2}$ are the coefficients of the Koornwinder polynomials. For certain order p, we have n_p orthogonal polynomials of order less or equal to p. We can write $c^m_{n_1 n_2}$ to denote explicitly that for each triangle m on the surface ∂D we have a different set of coefficients for the orthogonal polynomials on that triangle. We will remove the explicit dependency on the triangle m when there is no risk of confusion.

For a reader familiar with the method of moments where the unknowns are the coefficients of the basis functions (typically RWG basis functions), in the Nyström method, the unknowns of the system are the samples σ_m^i of the function (not the coefficients $c^m_{n_1,n_2}$ of the orthogonal polynomials). Considering σ_m^i or $c^m_{n_1,n_2}$ as unknowns of the discrete linear system is a minor difference as one can jump from one to the other with a matrix C.5 of very low condition number.

Next thing to do is to define how to **test** the integral equation. For this, Nyström method that uses "delta" test functions on the same Vioreanu–Rokhlin nodes on each triangular patch of the surface ∂D. In the case of testing a scalar quantity, we just sample the value at the Vioreanu–Rokhlin nodes on each triangle:

$$\int_{\partial D} \mathfrak{K}(\boldsymbol{r}_{m_t}(u_j, v_j), \boldsymbol{r}')\sigma(\boldsymbol{r}')dS' =$$

$$= \sum_m \int_{T_0} \mathfrak{K}(\boldsymbol{r}_{m_t}(u_j, v_j), \boldsymbol{r}_m(u', v'))\sigma_m(\boldsymbol{r}_m(u', v'))dS'_m(u', v') \approx$$

$$\approx \sum_m \int_{T_0} \mathfrak{K}(\boldsymbol{r}_{m_t}(u_j, v_j), \boldsymbol{r}_m(u', v')) \sum_{\substack{n_1=0 \\ n_2 \le n_1}}^{n_1=p} c^m_{n_1 n_2} K_{n_1 n_2}(u', v')ds'_m(u', v')$$

where $\mathfrak{K}(\boldsymbol{r}, \boldsymbol{r}')$ is the scalar kernel:

$$\mathfrak{K}(\boldsymbol{r}, \boldsymbol{r}') = \left(\hat{\boldsymbol{n}}(\boldsymbol{r}') \cdot (\boldsymbol{r}' - \boldsymbol{r})\frac{-jkr^2 - 1}{r^2} + j\alpha\right)\frac{e^{-jkr}}{4\pi r}$$

$\boldsymbol{r}_m(u, v)$ is the triangular patch on the source triangle and $\boldsymbol{r}_{m_t}(u, v)$ is the triangular patch on the target triangle. The nodes (u_j, v_j) for $j = 1, \ldots, n_p$ are the Vioreanu–Rokhlin nodes on the standard canonical triangle T_0.

When the source triangle and target point are far, one can consider that the integrand is a smooth function and uses the Vioreanu–Rokhlin quadrature rule with the exact same nodes, obtaining the following expression:

$$\begin{aligned} &\int_{T_0} \mathfrak{K}(\boldsymbol{r}_{m_t}(u_j, v_j), \boldsymbol{r}_m(u', v')) \sum_{\substack{n_1=0 \\ n_2 \le n_1}}^{n_1=p} c^m_{n_1 n_2} K_{n_1 n_2}(u', v')ds'_m(u', v') \approx \\ &\approx \sum_{i=1}^{n_p} \mathfrak{K}(\boldsymbol{r}_{m_t}(u_j, v_j), \boldsymbol{r}_m(u_i, v_i))\sigma^i_m |\mathbf{u}_m(u_i, v_i) \times \mathbf{v}_m(u_i, v_i)| w_i \end{aligned} \tag{8.23}$$

As we can see in this last expression in 8.23, the matrix of the resulting discrete linear system that corresponds to the far-field interactions is just a sampling of Green's function (together with the corresponding metric elements dS of the surface at the source point). In particular, the matrix entry that corresponds to the interaction of the source point number i in the m triangle with the target point j in the triangle m_t (where i and j are Vioreanu–Rokhlin nodes) is given by:

$$\mathfrak{K}(\boldsymbol{r}_{m_t}(u_j, v_j), \boldsymbol{r}_m(u_i, v_i))|\mathbf{u}_m(u_i, v_i) \times \mathbf{v}_m(u_i, v_i)|w_i$$

and the source unknown that corresponds to this matrix entry is σ^i_m (the ith Vioreanu–Rokhlin node of the mth triangle). This fact makes the Nyström method particularly attractive, as no coupling integral has to be done in the far field. This fact also makes the fast multipole method (FMM in the mathematics literature, MLFMA in the electrical engineering literature) simpler to apply (see [16–23]).

Due to the singularity of the kernel $\mathfrak{K}(\boldsymbol{r}, \boldsymbol{r}')$, different numerical integration techniques have to be applied for the near and self-interaction terms. This will depend on the distance of the source triangle $\boldsymbol{r}_m(u, v)$ and the target point $\boldsymbol{r}_{m_t}(u_j, v_j)$ as well as the total accuracy required by the user. In general, the self-interaction term and the adjacent triangles have to be integrated using a different method. Beyond that, on a given neighborhood of each target point, a special integration method may also be needed if high accuracy is required. The self-interaction and near interactions will be described.

The interactions that are near or self (target point contained in the source triangle) require a special treatment that will be explained later. The near and self-interaction will be described in a further section.

Next we treat the vector case, that is, integral equations where the source term or the right-hand side (or both) are tangent vector fields.

Let us consider, for example, the operator $\overline{\overline{L}}_{11} - j\alpha\overline{\overline{R}}_{11}$ that maps the tangent vector field $\boldsymbol{a}$ (unknown in the DPIEv) onto the tangent vector field $\hat{\boldsymbol{n}} \times \boldsymbol{A}$. To decompose a tangent vector field on two scalar components, we need to define an orthonormal basis on the tangent space of the surface ∂D. Unfortunately, the vectors $\mathbf{u}, \mathbf{v}$ are not always orthogonal (see Appendix A on differential geometry), therefore, an orthonormalization process is needed. Next we define two orthonormal vectors $\hat{\boldsymbol{u}}_m(u, v), \hat{\boldsymbol{v}}_m(u, v)$ on each triangle of the surface:

$$\hat{\boldsymbol{u}}_m(u, v) := \frac{\boldsymbol{u}_m(u, v)}{|\boldsymbol{u}_m(u, v)|}$$

$$\hat{\boldsymbol{v}}_m(u, v) := \hat{\boldsymbol{n}}_m(u, v) \times \hat{\boldsymbol{u}}_m(u, v)$$

Figure 8.5 shows the $\hat{\boldsymbol{u}}_m(u, v)$ and $\hat{\boldsymbol{v}}_m(u, v)$ vectors on a curved triangle sampled at the Vioreanu–Rokhlin nodes.

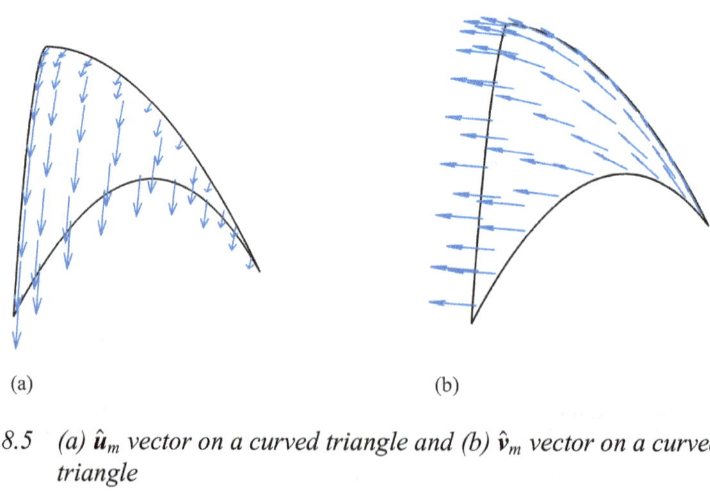

(a) (b)

Figure 8.5 (a) $\hat{\boldsymbol{u}}_m$ vector on a curved triangle and (b) $\hat{\boldsymbol{v}}_m$ vector on a curved triangle

Notice that there is no relation whatsoever between the $\hat{\boldsymbol{u}}_m$ vector defined on one triangle of the surface and the $\hat{\boldsymbol{u}}_{m'}$ vector defined on the contiguous triangle, no continuity of any type is satisfied. The tangent vector fields $\hat{\boldsymbol{u}}(\boldsymbol{r})$ and $\hat{\boldsymbol{v}}(\boldsymbol{r})$ defined on the entire surface $\boldsymbol{r} \in \partial D$ patch by patch as $\hat{\boldsymbol{u}}_m(u,v), \hat{\boldsymbol{v}}_m(u,v)$ are not (in general) continuous vector functions on ∂D (only piecewise continuous).

Using this basis, one can write the tangent vector field $\boldsymbol{a}$ on each chart m as:

$$\boldsymbol{a}_m(u,v) = a_{Um}(u,v)\hat{\boldsymbol{u}}_m(u,v) + a_{Vm}(u,v)\hat{\boldsymbol{v}}_m(u,v)$$

Now $a_m^U(u,v)$ and $a_m^V(u,v)$ are two scalar functions on each triangle m of the surface. They can be discretized in the same way as the scalar function σ_m by sampling them on the Vioreanu–Rokhlin nodes:

$$a_{Um}(u_i, v_i) = a^i_{Um}$$
$$a_{Vm}(u_i, v_i) = a^i_{Vm}$$

The functions $a_{Um}(u,v)$ and $a_{Vm}(u,v)$ are scalar functions defined on the mth patch. We can obtain the values of those two scalar functions at any other point by interpolation using the Koornwinder polynomials:

$$a_{Um}(u,v) \approx \sum_{\substack{n_1=0\\ n_2\le n_1}}^{n_1=p} c^{U_m}_{n_1n_2} K_{n_1n_2}(u,v)$$

$$a_{Vm}(u,v) \approx \sum_{\substack{n_1=0\\ n_2\le n_1}}^{n_1=p} c^{V_m}_{n_1n_2} K_{n_1n_2}(u,v)$$

Figures 8.6 and 8.7 show the resulting vector basis on the unit triangle T_0.

If we write down the operator of the DPIEv in a symbolic form as $\overline{\overline{M}} := \overline{\overline{L}} + i\alpha\overline{\overline{R}}$, the first block is:

$$M_{11} = L_{11}\boldsymbol{a} - j\alpha R_{11}\boldsymbol{a} =$$

$$= \int_{\partial D} \hat{\boldsymbol{n}}(\boldsymbol{r}) \times \nabla_r \times \left(g_k(\boldsymbol{r}-\boldsymbol{r}')\boldsymbol{a}(\boldsymbol{r}')\right) - j\alpha\hat{\boldsymbol{n}}(\boldsymbol{r}) \times \left(g_k(\boldsymbol{r}'-\boldsymbol{r})\hat{\boldsymbol{n}}(\boldsymbol{r}') \times \boldsymbol{a}(\boldsymbol{r}')\right) dS =$$

$$= \int_{\partial D} \hat{\boldsymbol{n}}(\boldsymbol{r}) \times \left(\nabla_r g_k(\boldsymbol{r}-\boldsymbol{r}') \times \boldsymbol{a}(\boldsymbol{r}')\right) - j\alpha\hat{\boldsymbol{n}}(\boldsymbol{r}) \times \left(g_k(\boldsymbol{r}'-\boldsymbol{r})\hat{\boldsymbol{n}}(\boldsymbol{r}') \times \boldsymbol{a}(\boldsymbol{r}')\right) dS$$

Now we write the tangent vector field $\boldsymbol{a}$ in the $\hat{\boldsymbol{u}}, \hat{\boldsymbol{v}}$ basis and test the resulting tangent vector field in that same basis:

$$L_{11}\boldsymbol{a} - j\alpha R_{11}\boldsymbol{a} =$$

$$= \sum_m \int_{T_m} \hat{\boldsymbol{n}}(\boldsymbol{r}) \times \left(\nabla_r g_k(\boldsymbol{r}-\boldsymbol{r}') \times (a_{Um}(u,v)\hat{\boldsymbol{u}}_m(u,v) + a_{Vm}(u,v)\hat{\boldsymbol{v}}_m(u,v))\right) dS -$$

$$j\alpha\hat{\boldsymbol{n}}(\boldsymbol{r}) \times \left(g_k(\boldsymbol{r}'-\boldsymbol{r})\hat{\boldsymbol{n}}(\boldsymbol{r}') \times \boldsymbol{a}(\boldsymbol{r}')\right) dS =$$

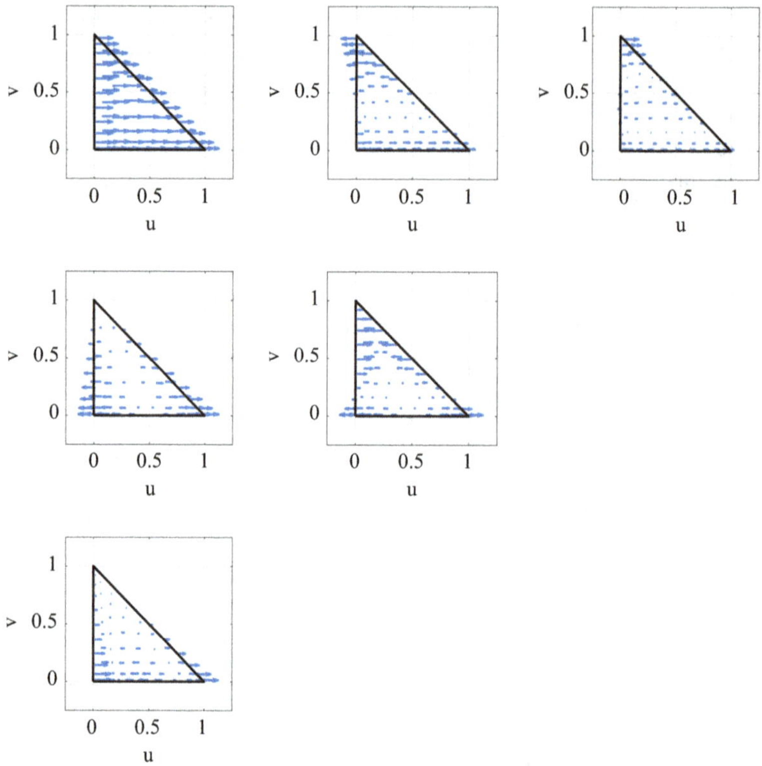

Figure 8.6 Orthogonal vector basis on the standard unit triangle T_0 up to order 2. $K_{nm}(u,v)\hat{\boldsymbol{u}}(u,v)$ vector basis

$$M^{UU}(\boldsymbol{r},\boldsymbol{r}') = \hat{\boldsymbol{v}}(\boldsymbol{r}) \cdot \left(\left(\frac{jkr^2+1}{r^2}(\boldsymbol{r}'-\boldsymbol{r}) + j\alpha\hat{\boldsymbol{n}}(\boldsymbol{r}') \right) \times \hat{\boldsymbol{u}}(\boldsymbol{r}') \right) \frac{e^{-jkr}}{4\pi r}$$

$$M^{UV}(\boldsymbol{r},\boldsymbol{r}') = \hat{\boldsymbol{v}}(\boldsymbol{r}) \cdot \left(\left(\frac{jkr^2+1}{r^2}(\boldsymbol{r}'-\boldsymbol{r}) + j\alpha\hat{\boldsymbol{n}}(\boldsymbol{r}') \right) \times \hat{\boldsymbol{v}}(\boldsymbol{r}') \right) \frac{e^{-jkr}}{4\pi r}$$

$$M^{VU}(\boldsymbol{r},\boldsymbol{r}') = -\hat{\boldsymbol{u}}(\boldsymbol{r}) \cdot \left(\left(\frac{jkr^2+1}{r^2}(\boldsymbol{r}'-\boldsymbol{r}) + j\alpha\hat{\boldsymbol{n}}(\boldsymbol{r}') \right) \times \hat{\boldsymbol{u}}(\boldsymbol{r}') \right) \frac{e^{-jkr}}{4\pi r}$$

$$M^{VV}(\boldsymbol{r},\boldsymbol{r}') = -\hat{\boldsymbol{u}}(\boldsymbol{r}) \cdot \left(\left(\frac{jkr^2+1}{r^2}(\boldsymbol{r}'-\boldsymbol{r}) + j\alpha\hat{\boldsymbol{n}}(\boldsymbol{r}') \right) \times \hat{\boldsymbol{v}}(\boldsymbol{r}') \right) \frac{e^{-jkr}}{4\pi r}$$

$$M^{UU}[a_U] = \sum_m \int_{T_0} M^{UU}(\boldsymbol{r}_{m_t}(u,v), \boldsymbol{r}_m(u',v')) a_{Um}(u',v') dS'_m(u',v')$$

$$M^{UV}[a_V] = \sum_m \int_{T_0} M^{UV}(\boldsymbol{r}_{m_t}(u,v), \boldsymbol{r}_m(u',v')) a_{Vm}(u',v') dS'_m(u',v')$$

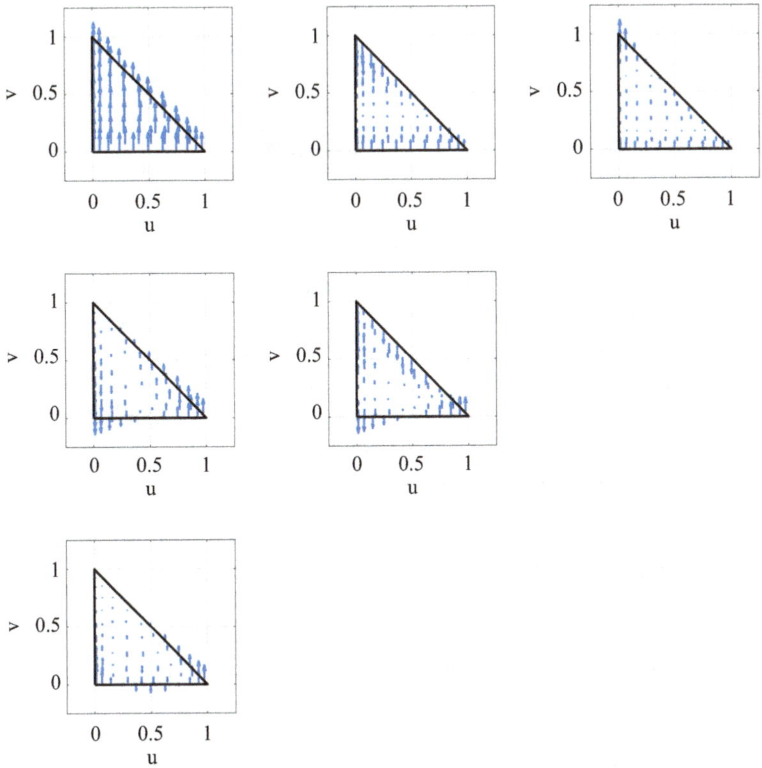

Figure 8.7 Orthogonal vector basis on the standard unit triangle T_0 up to order 2. $K_{nm}(u,v)\hat{\boldsymbol{v}}(u,v)$ vector basis

$$M^{VU}[a_U] = \sum_m \int_{T_0} M^{VU}(\boldsymbol{r}_{m_t}(u,v), \boldsymbol{r}_m(u',v'))a_{Um}(u',v')dS'_m(u',v')$$

$$M^{VV}[a_V] = \sum_m \int_{T_0} M^{VV}(\boldsymbol{r}_{m_t}(u,v), \boldsymbol{r}_m(u',v'))a_{Vm}(u',v')dS'_m(u',v')$$

After doing that we can discretize the vector operator using the same strategy as used in the scalar case. We expand the functions a_{Um}, a_{Vm} using Koornwinder polynomials that interpolate the function at Vioreanu–Rokhlin nodes and test it using delta functions at the Vioreanu–Rokhlin nodes for the tangent $\hat{\boldsymbol{u}}$ and $\hat{\boldsymbol{v}}$ directions.

$$M^{UU}[a_U] \approx \sum_m \int_{T_0} M^{UU}(\boldsymbol{r}_{m_t}(u_j,v_j), \boldsymbol{r}_m(u',v')) \sum_{\substack{n_1=0 \\ n_2\leq n_1}}^{n_1=p} a^m_{Un_1n_2} K_{n_1n_2}(u',v')dS'_m(u',v')$$

$$M^{UV}[a_V] \approx \sum_m \int_{T_0} M^{UV}(\boldsymbol{r}_{m_t}(u_j,v_j), \boldsymbol{r}_m(u',v')) \sum_{\substack{n_1=0 \\ n_2\leq n_1}}^{n_1=p} a^m_{Vn_1n_2} K_{n_1n_2}(u',v')dS'_m(u',v')$$

$$M^{VU}[a_U] \approx \sum_m \int_{T_0} M^{VU}(\boldsymbol{r}_{m_t}(u_j, v_j), \boldsymbol{r}_m(u', v')) \sum_{\substack{n_1=0 \\ n_2 \le n_1}}^{n_1=p} a^m_{Un_1n_2} K_{n_1n_2}(u', v') dS'_m(u', v')$$

$$M^{VV}[a_V] \approx \sum_m \int_{T_0} M^{VV}(\boldsymbol{r}_{m_t}(u_j, v_j), \boldsymbol{r}_m(u', v')) \sum_{\substack{n_1=0 \\ n_2 \le n_1}}^{n_1=p} a^m_{Vn_1n_2} K_{n_1n_2}(u', v') dS'_m(u', v')$$

where $a^m_{Un_1n_2}$ and $a^m_{Vn_1n_2}$ are the Koornwinder polynomials associated with the functions $a_{Um}(u', v')$ and $a_{Vm}(u', v')$. Again, the relation between the Koornwinder coefficients and the sampled values of the function $a^i_{Um} = a_{Um}(u_i, v_i)$ and $a^i_{Vm} = a_{Vm}(u_i, v_i)$ is given by the interpolant matrix that has a very low condition number.

Similarly as in the scalar case, if the source triangle and the target point are far, then the full integrand is smooth (including Green's function) and we can use the Vioreanu–Rokhlin quadrature rule [24] to obtain an accurate approximation of the contribution of that triangle $\boldsymbol{r}_m$ to that target point $\boldsymbol{r}_{m_t}(u_j, v_j)$:

$$\sum_{i=1}^{n_p} M^{UU}(\boldsymbol{r}_{m_t}(u_j, v_j), \boldsymbol{r}_m(u_i, v_i)) a^i_{Um} |\mathbf{u}_m(u_i, v_i) \times \mathbf{v}_m(u_i, v_i)| w_i$$

$$\sum_{i=1}^{n_p} M^{UV}(\boldsymbol{r}_{m_t}(u_j, v_j), \boldsymbol{r}_m(u_i, v_i)) a^i_{Vm} |\mathbf{u}_m(u_i, v_i) \times \mathbf{v}_m(u_i, v_i)| w_i$$

$$\sum_{i=1}^{n_p} M^{VU}(\boldsymbol{r}_{m_t}(u_j, v_j), \boldsymbol{r}_m(u_i, v_i)) a^i_{Um} |\mathbf{u}_m(u_i, v_i) \times \mathbf{v}_m(u_i, v_i)| w_i$$

$$\sum_{i=1}^{n_p} M^{VV}(\boldsymbol{r}_{m_t}(u_j, v_j), \boldsymbol{r}_m(u_i, v_i)) a^i_{Vm} |\mathbf{u}_m(u_i, v_i) \times \mathbf{v}_m(u_i, v_i)| w_i$$

similarly we obtain the expressions for the other blocks of the operator $\overline{\overline{M}}$ that map the scalar part to the scalar part, the scalar part to the vector part and the vector part to the scalar part. The resulting expressions of the kernels are the following:

$$L_{12} - j\alpha R_{12} =$$

$$= \int_{\partial D} -\hat{\boldsymbol{n}}(\boldsymbol{r}) \times g_k(\boldsymbol{r} - \boldsymbol{r}')\hat{\boldsymbol{n}}(\boldsymbol{r}')\rho(\boldsymbol{r}') - j\alpha\hat{\boldsymbol{n}}(\boldsymbol{r}) \times \nabla_r g_k(\boldsymbol{r} - \boldsymbol{r}')\rho(\boldsymbol{r}')dS'$$

$$L_{21} - j\alpha R_{21} =$$

$$= \int_{\partial D} -j\alpha\nabla_r \cdot (g_k(\boldsymbol{r} - \boldsymbol{r}')\boldsymbol{a}(\boldsymbol{r}'))dS' = \int_{\partial D} -j\alpha\nabla_r g_k(\boldsymbol{r} - \boldsymbol{r}') \cdot \boldsymbol{a}(\boldsymbol{r}')dS' =$$

$$L_{22} - j\alpha R_{22} =$$

$$= \int_{\partial D} \hat{\boldsymbol{n}}(\boldsymbol{r}') \cdot \nabla_{r'} g_k(\boldsymbol{r} - \boldsymbol{r}')\rho(\boldsymbol{r}') + j\alpha k^2 g_k(\boldsymbol{r} - \boldsymbol{r}')\rho(\boldsymbol{r}'))dS'$$

Therefore,

$$M^{U\rho}(\boldsymbol{r},\boldsymbol{r}') = \hat{\boldsymbol{v}}(\boldsymbol{r}) \cdot \left(\hat{\boldsymbol{n}}(\boldsymbol{r}') - j\alpha \frac{jkr^2+1}{r^2}(\boldsymbol{r}' - \boldsymbol{r}) \right) \frac{e^{-jkr}}{4\pi r}$$

$$M^{V\rho}(\boldsymbol{r},\boldsymbol{r}') = \hat{\boldsymbol{u}}(\boldsymbol{r}) \cdot \left(-\hat{\boldsymbol{n}}(\boldsymbol{r}') + j\alpha \frac{jkr^2+1}{r^2}(\boldsymbol{r}' - \boldsymbol{r}) \right) \frac{e^{-jkr}}{4\pi r}$$

$$M^{\rho U}(\boldsymbol{r},\boldsymbol{r}') = j\alpha \frac{jkr^2+1}{r^2}(\boldsymbol{r}' - \boldsymbol{r}) \cdot \hat{\boldsymbol{u}}(\boldsymbol{r}') \frac{e^{-jkr}}{4\pi r}$$

$$M^{\rho V}(\boldsymbol{r},\boldsymbol{r}') = j\alpha \frac{jkr^2+1}{r^2}(\boldsymbol{r}' - \boldsymbol{r}) \cdot \hat{\boldsymbol{v}}(\boldsymbol{r}') \frac{e^{-jkr}}{4\pi r}$$

$$M^{\rho\rho}(\boldsymbol{r},\boldsymbol{r}') = -\left(\frac{jkr^2+1}{r^2} \hat{\boldsymbol{n}}(\boldsymbol{r}') \cdot (\boldsymbol{r}' - \boldsymbol{r}) + j\alpha k^2 \right) \frac{e^{-jkr}}{4\pi r}$$

therefore, we get:

$$M^{U\rho}[\rho] = \sum_m \int_{T_0} M^{U\rho}(\boldsymbol{r}_{m_t}(u,v), \boldsymbol{r}_m(u',v'))\rho_m(u',v')dS'_m(u',v')$$

$$M^{V\rho}[\rho] = \sum_m \int_{T_0} M^{V\rho}(\boldsymbol{r}_{m_t}(u,v), \boldsymbol{r}_m(u',v'))\rho_m(u',v')dS'_m(u',v')$$

$$M^{\rho U}[a_U] = \sum_m \int_{T_0} M^{\rho U}(\boldsymbol{r}_{m_t}(u,v), \boldsymbol{r}_m(u',v'))a_{Um}(u',v')dS'_m(u',v')$$

$$M^{\rho V}[a_V] = \sum_m \int_{T_0} M^{\rho V}(\boldsymbol{r}_{m_t}(u,v), \boldsymbol{r}_m(u',v'))a_{Vm}(u',v')dS'_m(u',v')$$

$$M^{\rho\rho}[\rho] = \sum_m \int_{T_0} M^{\rho\rho}(\boldsymbol{r}_{m_t}(u,v), \boldsymbol{r}_m(u',v'))\rho_m(u',v')dS'_m(u',v')$$

expanding the sources in Koornwinder interpolant polynomials, we get:

$$M^{U\rho}[\rho] = \sum_m \int_{T_0} M^{U\rho}(\boldsymbol{r}_{m_t}(u,v), \boldsymbol{r}_m(u',v')) \sum_{\substack{n_1=0 \\ n_2 \le n_1}}^{n_1=p} \rho^m_{Un_1n_2} K_{n_1n_2}(u',v')dS'_m(u',v')$$

$$M^{V\rho}[\rho] = \sum_m \int_{T_0} M^{V\rho}(\boldsymbol{r}_{m_t}(u,v), \boldsymbol{r}_m(u',v')) \sum_{\substack{n_1=0 \\ n_2 \le n_1}}^{n_1=p} \rho^m_{Un_1n_2} K_{n_1n_2}(u',v')dS'_m(u',v')$$

$$M^{\rho U}[a_U] = \sum_m \int_{T_0} M^{\rho U}(\boldsymbol{r}_{m_t}(u,v), \boldsymbol{r}_m(u',v')) \sum_{\substack{n_1=0 \\ n_2 \le n_1}}^{n_1=p} a^m_{Un_1n_2} K_{n_1n_2}(u',v')dS'_m(u',v')$$

$$M^{\rho V}[a_V] = \sum_m \int_{T_0} M^{\rho V}(\boldsymbol{r}_{m_t}(u,v), \boldsymbol{r}_m(u',v')) \sum_{\substack{n_1=0\\ n_2\le n_1}}^{n_1=p} a^m_{Vn_1n_2} K_{n_1n_2}(u',v')dS'_m(u',v')$$

$$M^{\rho\rho}[\rho] = \sum_m \int_{T_0} M^{\rho\rho}(\boldsymbol{r}_{m_t}(u,v), \boldsymbol{r}_m(u',v')) \sum_{\substack{n_1=0\\ n_2\le n_1}}^{n_1=p} \rho^m_{Un_1n_2} K_{n_1n_2}(u',v')dS'_m(u',v')$$

Now the far-field contribution of each triangle will be given by:

$$\sum_{i=1}^{n_p} M^{U\rho}(\boldsymbol{r}_{m_t}(u_j,v_j), \boldsymbol{r}_m(u_i,v_i))\rho^i_m |\mathbf{u}_m(u_i,v_i) \times \mathbf{v}_m(u_i,v_i)| w_i$$

$$\sum_{i=1}^{n_p} M^{V\rho}(\boldsymbol{r}_{m_t}(u_j,v_j), \boldsymbol{r}_m(u_i,v_i))\rho^i_m |\mathbf{u}_m(u_i,v_i) \times \mathbf{v}_m(u_i,v_i)| w_i$$

$$\sum_{i=1}^{n_p} M^{\rho U}(\boldsymbol{r}_{m_t}(u_j,v_j), \boldsymbol{r}_m(u_i,v_i))a^i_{Um} |\mathbf{u}_m(u_i,v_i) \times \mathbf{v}_m(u_i,v_i)| w_i$$

$$\sum_{i=1}^{n_p} M^{\rho V}(\boldsymbol{r}_{m_t}(u_j,v_j), \boldsymbol{r}_m(u_i,v_i))a^i_{Vm} |\mathbf{u}_m(u_i,v_i) \times \mathbf{v}_m(u_i,v_i)| w_i$$

$$\sum_{i=1}^{n_p} M^{\rho\rho}(\boldsymbol{r}_{m_t}(u_j,v_j), \boldsymbol{r}_m(u_i,v_i))\rho^i_m |\mathbf{u}_m(u_i,v_i) \times \mathbf{v}_m(u_i,v_i)| w_i$$

We finally can write the operator $\overline{\overline{M}}$ as a 3×3 system of scalar operators:

$$\overline{\overline{M}} = \begin{pmatrix} M^{UU} & M^{UV} & M^{U\rho} \\ M^{VU} & M^{VV} & M^{V\rho} \\ M^{\rho U} & M^{\rho V} & M^{\rho\rho} \end{pmatrix}$$

The terms that correspond to the flux condition on the DPIEs and DPIEv for each connected component ∂D_j can be computed in a similar way with the appropriate kernel. The terms associated to the flux of the DPIEs are:

$$\begin{aligned}\mathfrak{K}^n(\boldsymbol{r},\boldsymbol{r}') =& \hat{\boldsymbol{n}}(\boldsymbol{r}) \cdot (\boldsymbol{r}-\boldsymbol{r}') \frac{\hat{\boldsymbol{n}}(\boldsymbol{r}') \cdot (\boldsymbol{r}'-\boldsymbol{r})}{4\pi r^5}((3+3jkr-k^2r^2)e^{-jkr}-3) \\ &+ \frac{\hat{\boldsymbol{n}}(\boldsymbol{r}) \cdot \hat{\boldsymbol{n}}(\boldsymbol{r}')}{4\pi r^3}((1+jkr)e^{-jkr}-1) - j\alpha\hat{\boldsymbol{n}}(\boldsymbol{r})(\boldsymbol{r}'-\boldsymbol{r})\frac{jkr^2+1}{r^2}\frac{e^{-jkr}}{4\pi r}\end{aligned}$$

The flux condition is:

$$\int_{\partial D_j} \left(\frac{-j\alpha\sigma(\boldsymbol{r})}{2} + \int_{\partial D} \mathfrak{K}^n(\boldsymbol{r},\boldsymbol{r}')\sigma(\boldsymbol{r}')dS' \right) dS = Q_j$$

Or equivalently:

$$\int_{\partial D_j} \frac{\partial \phi^{\text{scat}}}{\partial n} dS = Q_j$$

with:

$$\frac{\partial \phi^{\text{scat}}(\boldsymbol{r})}{\partial n} = \frac{-j\alpha\sigma(\boldsymbol{r})}{2} + \int_{\partial D} \mathfrak{K}^n(\boldsymbol{r}, \boldsymbol{r}')\sigma(\boldsymbol{r}')dS'$$

After discretizing, we obtain:

$$\frac{\partial \phi^{\text{scat}}(\boldsymbol{r}_{m_t}(u,v))}{\partial n} \approx \frac{-j\alpha}{2} \sum_{\substack{n_1=0 \\ n_2 \leq n_1}}^{n_1=p} \sigma^{m_t}_{Un_1n_2} K_{n_1n_2}(u,v) +$$

$$+ \sum_m \int_{T_0} \mathfrak{K}^n_{m_tm}(\boldsymbol{r}_{m_t}(u,v), \boldsymbol{r}_m(u',v')) \sum_{\substack{n_1=0 \\ n_2 \leq n_1}}^{n_1=p} \sigma^m_{Un_1n_2} K_{n_1n_2}(u',v')dS'_m(u',v')$$

and the following discretized flux conditions:

$$\int_{\partial D_j} \frac{\partial \phi^{\text{scat}}(\boldsymbol{r})}{\partial n} dS \approx \sum_{m \in \partial D_j} \sum_{i=1}^{n_p} \frac{\partial \phi^{\text{scat}}(\boldsymbol{r}_m(u_i,v_i))}{\partial n} |\mathbf{u}_m(u_i,v_i) \times \mathbf{v}_m(u_i,v_i)| w_i = Q_j$$

Again, for the calculation of the far-field contribution on the integral that defines $\int_{\partial D} \mathfrak{K}^n(\boldsymbol{r}, \boldsymbol{r}')\sigma(\boldsymbol{r}')dS'$, we use the same Vioreanu–Rokhlin quadrature rule for smooth functions on triangles:

$$\sum_{i=1}^{n_p} \mathfrak{K}^n_{m_tm}(\boldsymbol{r}_{m_t}(u_j,v_j), \boldsymbol{r}_m(u_i,v_i))\sigma^i_m |\mathbf{u}_m(u_i,v_i) \times \mathbf{v}_m(u_i,v_i)| w_i \tag{8.24}$$

The self and near interaction of that integral are computed in a different way that will be described in upcoming sections.

The flux condition on the DPIEv can be computed in a similar way:

$$\int_{\partial D_j} \left(-\hat{\boldsymbol{n}}_r \cdot S_k(\hat{\boldsymbol{n}}_{r'}\rho) - j\alpha(\hat{\boldsymbol{n}}_r \cdot S_k(\hat{\boldsymbol{n}}_{r'} \times \boldsymbol{a}) - \frac{\rho}{2} + S'_k(\rho)) \right) ds = q_j$$

The kernel associated to the $\boldsymbol{a}$ term is:

$$-j\alpha\hat{\boldsymbol{n}}(\boldsymbol{r}) \cdot (\hat{\boldsymbol{n}}(\boldsymbol{r}') \times \boldsymbol{a}(\boldsymbol{r}'))\frac{e^{ikr}}{4\pi r}$$

Taking the $\hat{\boldsymbol{u}}$ and $\hat{\boldsymbol{v}}$ components, we get:

$$M^{nU}(\boldsymbol{r}, \boldsymbol{r}') = i\alpha\frac{e^{ikr}}{4\pi r}\hat{\boldsymbol{n}}(\boldsymbol{r}) \cdot (\hat{\boldsymbol{n}}(\boldsymbol{r}') \times \hat{\boldsymbol{u}}(\boldsymbol{r}'))$$

$$M^{nV}(\boldsymbol{r}, \boldsymbol{r}') = i\alpha\frac{e^{ikr}}{4\pi r}\hat{\boldsymbol{n}}(\boldsymbol{r}) \cdot (\hat{\boldsymbol{n}}(\boldsymbol{r}') \times \hat{\boldsymbol{v}}(\boldsymbol{r}'))$$

The kernel associated to the ρ variable is:

$$M^{n\rho}(\boldsymbol{r},\boldsymbol{r}') = \left(j\alpha \frac{jkr^2+1}{r} \hat{\boldsymbol{n}}(\boldsymbol{r}) \cdot (\boldsymbol{r}' - \boldsymbol{r}) - \hat{\boldsymbol{n}}(\boldsymbol{r}) \cdot \hat{\boldsymbol{n}}(\boldsymbol{r}') \right) \frac{e^{ikr}}{4\pi r}$$

The outer integral on each connected component ∂D_j of ∂D and the rest of the details can be computed in the similar way as in the DPIEs.

8.8.1 High-order accurate self-interaction integral

In this section, we consider the problem of self-interaction in the Nyström method. Let us consider the following singular integral on a triangular patch of a surface:

$$\int_{T_m} M(\boldsymbol{r}_m(u,v), \boldsymbol{r}_m(u',v'))\sigma_m(u',v')dS'_m(u',v') \tag{8.25}$$

where the kernel $M_{mm}(\boldsymbol{r},\boldsymbol{r}')$ can be any singular kernel described in the previous section like: $M^{UU}, M^{UV}, M^{VU}, M^{VV}, M^{U\rho}, \ldots$, where the target point is located inside the source triangle, that is $m_t = m$. For each case, the source σ_m is defined on the triangular patch, T_m is considered to be a scalar function expanded in Koornwinder polynomials and can be $\rho_m(\boldsymbol{r}'), a_{Um}(\boldsymbol{r}'), a_{Vm}(\boldsymbol{r}')$. In contrast to what happened for far interactions, where the full integrand, source and kernel (and ds') can be considered as smooth functions and integrated accurately via Vioreanu–Rokhlin quadrature rule, now the kernel is singular and something has to be done to accurately compute the integral. There are many possibilities here. Many techniques are related to a change of variable (polar, Duffin, ...) [25–29] to change the singular integrand into a smooth integrand, and then apply a standard general purpose quadrature rule for smooth functions on the resulting integrand (typically Gauss–Legendre). Depending on the particular change of variables, the resulting integrand has different *spectral content* and a variable amount of nodes are needed to reach certain accuracy. What is described in this section is based on the concept of generalized Gaussian quadrature, that, in principle, allows to obtain the optimal number of weights and nodes to integrate certain class of functions. The reader not familiarized with generalized Gaussian quadrature can go to the corresponding appendix for an introduction.

The following observation about the singularity of the standard Helmholtz Green's function is important:

$$g_k(\boldsymbol{r}) = \frac{e^{-jkr}}{4\pi r}$$

If we consider a flat surface as a toy model, then $r = \sqrt{x^2+y^2}$. The function, as a two-variable function of x and y is not smooth at $x=0, y=0$ for two reasons, one is the obvious denominator $\frac{1}{r}$ and the other is that the numerator e^{-jkr} contains both odd and even powers of r. Writing down the standard series expansion for the exponential, we get $\frac{e^{-jkr}}{4\pi r} = \frac{1}{4\pi}\sum_{n=0}^{\infty}\frac{(-jk)^n r^{n-1}}{n!}$. The even powers are smooth functions of x, y, yet the odd powers are not. For that reason, integrating the function r^n with n odd on a triangle with an ordinary Vioreanu–Rokhlin quadrature is not a good idea if high accuracy is required. (Notice that $r = \sqrt{x^2 + y^2}$ has partial derivatives which are not continuous, and $r^3 = (x^2 + y^2)^{3/2}$ has second partial derivatives which are not continuous.) This is one reason why the standard singularity subtraction method is not a good idea for high-order high-accuracy codes. The method of singularity subtraction consists of removing the singularity adding and subtracting one to the numerator

$$g_k(\boldsymbol{r}) = \frac{e^{-jkr}}{4\pi r} = \frac{e^{-jkr} - 1}{4\pi r} + \frac{1}{4\pi r}$$

After splitting the kernel in two parts, the term $\frac{e^{-jkr}-1}{4\pi r}$ is considered as smooth and integrated using a quadrature rule suitable for smooth functions, and integrating the term $\frac{1}{4\pi r}$ using a closed form expression. For low-order codes, these two assumptions are okay, as the surface is typically described by a set of flat triangles, therefore, the accuracy obtained is not going to be high anyway, and on such patches, one can obtain closed expressions to integrate the second term. In the situation covered in this chapter, where high-order high-accuracy discretization is the main goal, the non-smoothness of $\frac{e^{-jkr}-1}{4\pi r}$ is a big problem. On the other hand, the remaining term $\frac{1}{4\pi r}$ integrated on a curved triangle is as challenging as the original Helmholtz Green's function, as no analytical solution is known for arbitrary curved triangular patches.

Other methods that imply changes in variables (polar, Duffin, ...) can be used with high accuracy. In those methods, the issue is the total number of quadrature nodes needed for certain precision. This depends on the spectral content of the resulting integrand once the change of variables is made. In general, Duffin and polar transforms are fine if the singularity is located at a point u, v not very close to the boundary of the triangle T_0. If that happens, once the change of variables is made, a standard Gaussian quadrature rule can be used on the resulting smooth triangle. If that is not the case, and the singularity is near the boundary of a triangle T_0, the resulting integrand after the change of variables (Duffin, polar, ...) is highly distorted and the spectral content is high, this makes standard Gaussian quadrature inefficient. To optimize the total number of quadrature nodes, **generalized Gaussian quadrature method** for singular integrals was developed. To see an introduction for 1D, see the Appendix. Next we describe the generalized Gaussian quadrature method applied to the calculation of singular integrals of the type 8.25.

Going back to the singular integral problem, let us consider an integral of the form 8.25. We consider the singularity located at a target point (typically a Vioreanu–Rokhlin node $\boldsymbol{r}_m(u_0, v_0)$ in the context of the locally corrected Nyström method). We can write this integral as:

$$\int_{T_0} M(\boldsymbol{r}_m(u_0, v_0), \boldsymbol{r}_m(u, v))\sigma_m(\boldsymbol{r}_m(u, v))|\mathbf{u}(u, v) \times \mathbf{v}(u, v)|dudv$$

Here the kernel M is a scalar kernel (e.g., the kernel in the DPIEs or one of the components of the DPIEv). If we apply the reparameterization of the surface patch described in the Appendix, we can obtain a conformal parametrization at the point (u_0, v_0), the price to pay is that the integration domain is no longer the unit canonical triangle T_0 but an arbitrary triangle in the $\tilde{u}, \tilde{v}$ plane.

With respect to the new parametrization, we have:

$$\tilde{\boldsymbol{r}}_m(\tilde{u}, \tilde{v}) := \boldsymbol{r}_m(a\tilde{u} + b\tilde{v} + u_0, c\tilde{u} + d\tilde{v} + v_0)$$

The new parameterization is designed to be conformal at the target point $\boldsymbol{r}(u_0, v_0) = \tilde{\boldsymbol{r}}(0,0)$, that is: $\tilde{\mathbf{u}}(0,0) \cdot \tilde{\mathbf{u}}(0,0) = 1, \tilde{\mathbf{v}}(0,0) \cdot \tilde{\mathbf{v}}(0,0) = 1,\ \tilde{\mathbf{u}}(0,0) \cdot \tilde{\mathbf{v}}(0,0) = 0$):

$$\begin{aligned}
&\int_{\tilde{T}_0} M_{mm}(\boldsymbol{r}_m(u_0, v_0), \boldsymbol{r}_m(a\tilde{u} + b\tilde{v} + u_0, c\tilde{u} + d\tilde{v} + v_0)) \\
&\qquad \sigma_m(\boldsymbol{r}_m(a\tilde{u} + b\tilde{v} + u_0, c\tilde{u} + d\tilde{v} + v_0)) \\
&\qquad |\tilde{\mathbf{u}}(\tilde{u}, \tilde{v}) \times \tilde{\mathbf{v}}(\tilde{u}, \tilde{v})| d\tilde{u} d\tilde{v} = \\
= &\int_{\tilde{T}_0} M_{mm}(\tilde{\boldsymbol{r}}_m(0,0), \tilde{\boldsymbol{r}}_m(\tilde{u}, \tilde{v}))\sigma_m(\tilde{\boldsymbol{r}}_m(\tilde{u}, \tilde{v}))|\tilde{\mathbf{u}}(\tilde{u}, \tilde{v}) \times \tilde{\mathbf{v}}(\tilde{u}, \tilde{v})| d\tilde{u} d\tilde{v}
\end{aligned}$$

If change to polar coordinates on the $\tilde{u}, \tilde{v}$ plane with center in the target point (0, 0):

$$\begin{aligned}
\tilde{u} &= \varrho \cos \phi \\
\tilde{v} &= \varrho \sin \phi
\end{aligned}$$

If the map is conformal at the target point and if the surface is piecewise analytic, then the type of singularity that is found for those kernels appearing in the DPIE is of the form:

$$\lim_{\varepsilon \to 0} \int_0^{2\pi} \int_{\varepsilon}^{R_{max}(\phi)} \left(\frac{q_{-2}(\phi)}{\varrho^2} + \frac{q_{-1}(\phi)}{\varrho} + q_0(\phi) + q_1(\phi)\varrho + q_2(\phi)\varrho^2 + \ldots \right) \varrho d\varrho d\phi \tag{8.26}$$

The functions $q_j(\phi)$ are trigonometric polynomials of order bounded by $3(j + 1) + 2$ (e.g., $q_{-2}(\phi) = \alpha \cos \phi + \beta \sin \phi$). See [30,31] for more details. For compact operators, the term $q_{-2}(\phi) = 0$. This term is only different from zero on non-compact bounded operators like $\hat{\boldsymbol{n}} \times \nabla S_k(\rho)$ or $\nabla \cdot S_k[\boldsymbol{J}]$. These operators are not compact but has similar spectral properties [32]. In that case, the integral has to be interpreted in the Cauchy principal value sense. That is why the expression 8.26 is written as a limit for $\varepsilon \to 0$. The correct value of the limit is only obtained if the map is conformal at the target point (i.e., the first fundamental form is the identity 2×2 matrix at that point). It is also important to notice that the expression of the integrand in 8.26 is not needed explicitly (the decomposition of the integrand in terms of the $q_j(\phi)$ and ϱ^j functions is not needed). Instead, only needed to develop the right generalized Gaussian quadrature rule. Unfortunately, the integral 8.26 depends on too many parameters. To simplify it, the integration domain is divided in four parts (see Figure 8.8(c)).

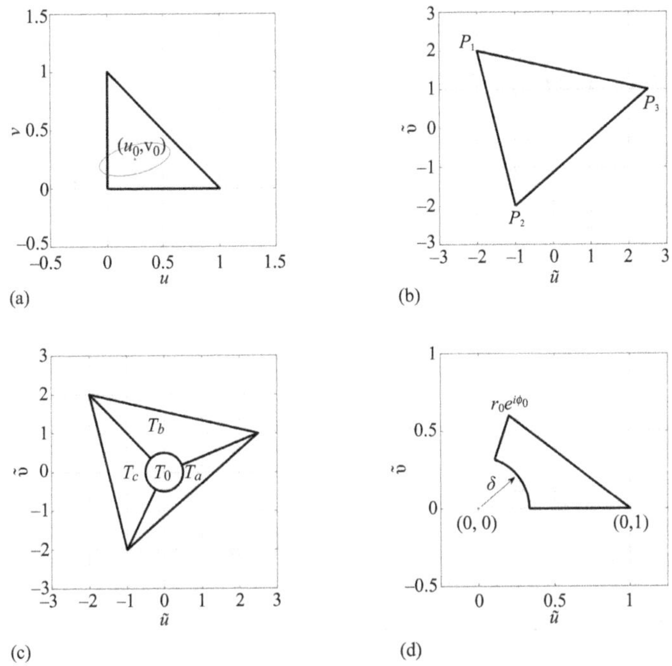

Figure 8.8 *(a) Unit triangle T_0 with arbitrarily located singularity at (u_0, v_0) with arbitrary orientation and eccentricity (notice that in the (u, v) plane, the function $\frac{1}{r}$ gets distorted with an asymptotic elliptic shape for the level sets when $r \to 0$); (b) reparametrized triangle in the variables $\tilde{u}, \tilde{v}$ with a singularity of the type $\frac{1}{\sqrt{\tilde{u}^2+\tilde{v}^2}}$; (c) division of the triangle in four regions: one circular region and three with the same shape described by three parameters; (d) shape of T_a, T_b, T_c after a minor shift and contraction. The resulting shape depends on three parameters δ, r_0 and ϕ_0*

$$\lim_{\varepsilon\to 0}\int_0^{2\pi}\int_\varepsilon^{R_{max}(\phi)}\left(\frac{q_{-2}(\phi)}{\varrho^2}+\frac{q_{-1}(\phi)}{\varrho}+q_0(\phi)+q_1(\phi)\varrho+\ldots\right)\varrho d\varrho d\phi=$$
$$=\lim_{\varepsilon\to 0}\int_0^{2\pi}\int_\varepsilon^{\gamma}\left(\frac{q_{-2}(\phi)}{\varrho^2}+\frac{q_{-1}(\phi)}{\varrho}+q_0(\phi)+q_1(\phi)\varrho+\ldots\right)\varrho d\varrho d\phi+$$
$$+\int_{\phi_1}^{\phi_2}\int_\gamma^{R_{max}^{(\phi_1,\phi_2)}(\phi)}\left(\frac{q_{-2}(\phi)}{\varrho^2}+\frac{q_{-1}(\phi)}{\varrho}+q_0(\phi)+q_1(\phi)\varrho+\ldots\right)\varrho d\varrho d\phi+$$
$$+\int_{\phi_2}^{\phi_3}\int_\gamma^{R_{max}^{(\phi_2,\phi_3)}(\phi)}\left(\frac{q_{-2}(\phi)}{\varrho^2}+\frac{q_{-1}(\phi)}{\varrho}+q_0(\phi)+q_1(\phi)\varrho+\ldots\right)\varrho d\varrho d\phi+$$
$$+\int_{\phi_3}^{\phi_1}\int_\gamma^{R_{max}^{(\phi_3,\phi_1)}(\phi)}\left(\frac{q_{-2}(\phi)}{\varrho^2}+\frac{q_{-1}(\phi)}{\varrho}+q_0(\phi)+q_1(\phi)\varrho+\ldots\right)\varrho d\varrho d\phi=$$
$$=I_{T_0}+I_{T_a}+I_{T_b}+I_{T_c}$$

where I_{T_0} is the integral on the circular region, and $I_{T_a}, I_{T_b}, I_{T_c}$ are integrals on the corresponding regions. The functions $R^{(\phi_1,\phi_2)}(\phi), \ldots$ are just the sides of the triangles in the $(\tilde{u}, \tilde{v})$ space in polar coordinates. The first integral is on a circle and can be obtained accurately with a simple tensor product of two quadrature rules, trapezoidal rule in the angular direction $\phi \in [0, 2\pi]$ and Gauss–Legendre in the radial direction $\varrho \in [0, \gamma]$. This will produce an accurate result of the limiting value for $\varepsilon \to 0$. The resulting quadrature rule is, for the angular direction:

$$\phi_j = \frac{j}{N}$$
$$w_j^\phi = \frac{2\pi}{N}$$

for $j = 0, 1, \ldots, N-1$ for the radial direction, standard Gauss–Legendre weights and nodes in the interval $[0, \gamma]$: ϱ_i, w_i^ϱ, for $i = 0, 1, \ldots, M-1$. The resulting quadrature rule is the tensor product:

$$(\varrho_i, \phi_j) = (\varrho_i, \frac{j}{N})$$
$$W_{ij} = w_i^\varrho w_j^\phi = w_i^\varrho \frac{2\pi}{N}$$

The resulting expression to evaluate that first integral is the following:

$$\lim_{\varepsilon \to 0} \iint_{\varepsilon < \sqrt{\tilde{u}^2+\tilde{v}^2} < \gamma} M_{mm}(\boldsymbol{r}_m(u_0, v_0), \boldsymbol{r}_m(a\tilde{u} + b\tilde{v} + u_0, c\tilde{u} + d\tilde{v} + v_0))$$
$$\sigma_m(\boldsymbol{r}_m(a\tilde{u} + b\tilde{v} + u_0, c\tilde{u} + d\tilde{v} + v_0))|\tilde{\mathbf{u}}(\tilde{u}, \tilde{v}) \times \tilde{\mathbf{v}}(\tilde{u}, \tilde{v})| d\tilde{u} d\tilde{v} \approx$$
$$\approx \sum_i \sum_j M_{mm}(\boldsymbol{r}_m(u_0, v_0), \boldsymbol{r}_m(a\tilde{u}_{ij} + b\tilde{v}_{ij} + u_0, c\tilde{u}_{ij} + d\tilde{v}_{ij} + v_0))$$
$$\sigma_m(\boldsymbol{r}_m(a\tilde{u}_{ij} + b\tilde{v}_{ij} + u_0, c\tilde{u}_{ij} + d\tilde{v}_{ij} + v_0))|\tilde{\mathbf{u}}(\tilde{u}_{ij}, \tilde{v}_{ij}) \times \tilde{\mathbf{v}}(\tilde{u}_{ij}, \tilde{v}_{ij})| \varrho_i W_{ij}$$

where

$$\tilde{u}_{ij} = \varrho_i \cos \phi_j$$
$$\tilde{v}_{ij} = \varrho_i \sin \phi_j$$

The number of nodes in the ϕ and ϱ direction M and N depends on the accuracy desired.

The remaining three integrals are more complicated. To develop a generalized Gaussian quadrature rule, the number of parameters must be reduced. To do that, one can always rescale the integral in such a way that one vertex of the triangle is $(\tilde{u} = 0, \tilde{v} = 0)$ (containing the singularity), the second vertex is $(\tilde{u} = 1, \tilde{v} = 0)$ and the third vertex is $(\tilde{u} = \varrho_0 \cos \phi_0, \tilde{v} = \varrho_0 \sin \phi_0)$, that depend on two parameters. The third parameter is γ. The goal now is to develop a generalized Gaussian quadrature rule for these kind of integrals in these domains. The quadrature rule will provide nodes and weights and will depend on the two parameters ϱ_0, ϕ_0. After its calculation through a complicated procedure, the values can be stored and reused. For each particular triangle, the weights and nodes to use will have to be interpolated for the particular value of the parameters ϱ_0, ϕ_0 based on the samples stored for other values of the parameters on a suitable grid. The generalized Gaussian quadrature method can provide a set of weights and nodes that integrate accurately all functions of the type:

$$\frac{q_{-2}(\phi)}{\varrho}$$
$$q_{-1}(\phi)$$
$$q_0(\phi)\varrho$$
$$,\ldots,$$
$$q_{N-1}(\phi)\varrho^N,$$

where $q_j(\phi)$ is a trigonometric polynomial of degree bounded by $3(j+1)+2$. In the region described by the parameters $\gamma, \varrho_0, \phi_0$. These weights and nodes will provide quadrature rule that is optimal in the sense of accuracy and number of nodes. With this quadrature rule $\{(\varrho_j, \phi_j), W_j\}_{j=1}^N$, the calculation of the integral can be done in a trivial way:

$$\int_0^{\phi_0}\int_\gamma^{R_{max}(\phi,\varrho_0,\phi_0)} M_{mm}(\boldsymbol{r}_m(u_0, v_0),$$
$$,\boldsymbol{r}_m(a\varrho\cos\phi + b\varrho\sin\phi + u_0, c\varrho\cos\phi + d\varrho\sin\phi + v_0))\cdot$$
$$\cdot\,\sigma_m(\boldsymbol{r}_m(a\varrho\cos\phi + b\varrho\sin\phi + u_0, c\varrho\cos\phi + d\varrho\sin\phi + v_0))\cdot$$
$$\cdot\,|\tilde{\mathbf{u}}(\varrho,\phi)\times\tilde{\mathbf{v}}(\varrho,\phi)|\varrho d\varrho d\phi \approx$$
$$\approx \sum_j M_{mm}(\boldsymbol{r}_m(u_0, v_0), \boldsymbol{r}_m(a\varrho_j\cos\phi_j + b\varrho_j\sin\phi_j + u_0, c\varrho_j\cos\phi_j + d\varrho_j\sin\phi_j + v_0))$$
$$\cdot\,\sigma_m(\boldsymbol{r}_m(a\varrho_j\cos\phi_j + b\varrho_j\sin\phi_j + u_0, c\varrho_j\cos\phi_j + d\varrho_j\sin\phi_j + v_0))\cdot$$
$$\cdot\,|\tilde{\mathbf{u}}(\varrho_j,\phi_j)\times\tilde{\mathbf{v}}(\varrho_j,\phi_j)|\varrho_j W_j$$

The number of quadrature weights and nodes N will depend on the accuracy required.

Once again, notice that the quadrature weights and nodes will depend on the parameters $\gamma, \varrho_0, \phi_0$. To be more explicit, we can write:

$$\{(\varrho_j^{(\gamma,\varrho_0,\phi_0)}, \phi_j^{(\gamma,\varrho_0,\phi_0)}), W_j^{(\gamma,\varrho_0,\phi_0)}\}_{j=1}^N$$

Those values are precomputed on a Chebyshev tensor product grid $\gamma^{n1}, \varrho_0^{n2}, \phi_0^{n3}$ and stored:

$$\{(\varrho_j^{(\gamma^{n1},\varrho_0^{n2},\phi_0^{n3})}, \phi_j^{(\gamma^{n1},\varrho_0^{n2},\phi_0^{n3})}), W_j^{(\gamma^{n1},\varrho_0^{n2},\phi_0^{n3})}\}_{j=1}^N$$

From those values, for each integral, the values of 8.8.1 are interpolated. Because of the difficulty of finding this quadrature rule and its universality, we provide a link to public code that allows to compute the desired quadrature weights and nodes and the resulting singular integral on the full triangle T_0 using a very simple interface and with all the weights and nodes stored (see https://github.com/JamesCBremerJr/Surfaces/). To see more details about the generation of the generalized Gaussian quadrature rule for these types of functions, see [30,31].

8.9 Near interaction quadrature

In this section, we describe briefly how to compute the near interaction integrals. Those correspond to the case where the target point $\boldsymbol{r}_m(u_0, v_0)$ and the source triangle $\boldsymbol{r}_{m'}(u, v)$ are located in different triangles which are adjacent or nearby. The notion of proximity entirely depends on the accuracy desired in the result. The more accuracy is needed in the final result of the integral equation (induced sources on the scatterer), the wider the region around a target point has to be considered as near interaction. One way to do it is through the use of the parameter η (see Figure 8.9). R_0 is the diameter of a given triangle T, and the near-field region defined near the triangle T is limited by ηR_0. This implies that smaller triangles will have smaller near regions of interaction. In the near interaction region, one way to do the integral accurately is via adaptive quadrature (see [12,34]). Figure 8.9 shows the source triangle T_s divided adaptively to obtain the radiated field of this triangle on the quadrature nodes of the target triangle T_t. Notice that there is no need to do any additional integration on T_t as we are describing Nyström discretization instead of Galerkin.

Appendix A: Differential geometry of surfaces

Let D be a bounded open region in $\mathbb{R}^3$ with boundary ∂D consisting of a finite number of disjoint closed bounded surfaces of class C^2. We also assume that the complement $\mathbb{R}^3\backslash\overline{D}$ is connected. A surface ∂D is of class C^2 when, for each point $\boldsymbol{r} \in \partial D$, there is a small ball B_ε of radius ε such that the intersection $\partial D \cap B_\varepsilon$ can be mapped with a bijective map onto an open domain $U \subset \mathbb{R}^2$. The map can be described as:

$$\boldsymbol{r}(u, v) = x(u, v)\hat{\boldsymbol{r}} + y(u, v)\hat{\boldsymbol{r}'} + z(u, v)\hat{\boldsymbol{z}}$$

This map is also called a chart of the surface. An atlas of the surface is a set of charts that cover the surface ∂D.

For a given surface ∂D, a given atlas, and a given chart $\boldsymbol{r}(u, v)$, the **u** and **v** vectors are defined as:

$$\mathbf{u}(u, v) := \frac{\partial \boldsymbol{r}}{\partial u}(u, v) = \frac{\partial x}{\partial u}(u, v)\hat{\boldsymbol{x}} + \frac{\partial y}{\partial u}(u, v)\hat{\boldsymbol{y}} + \frac{\partial z}{\partial u}(u, v)\hat{\boldsymbol{z}}$$
$$\mathbf{v}(u, v) := \frac{\partial \boldsymbol{r}}{\partial v}(u, v) = \frac{\partial x}{\partial v}(u, v)\hat{\boldsymbol{x}} + \frac{\partial y}{\partial v}(u, v)\hat{\boldsymbol{y}} + \frac{\partial z}{\partial v}(u, v)\hat{\boldsymbol{z}}$$

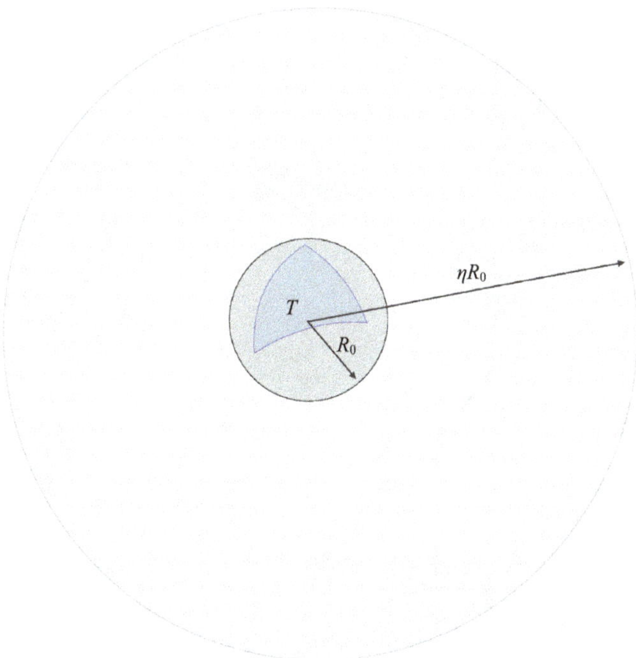

Figure 8.9 Near-field region associated with triangle T. The parameter η depends on the accuracy required, and R_0 is the "diameter" of the triangle T

The surface element at a given point can be calculated as:

$$dS = |\mathbf{u}(u, v) \times \mathbf{v}(u, v)|dudv$$

This allows the notion of surface integral of a function $f(\boldsymbol{r})$ defined on the surface $\boldsymbol{r} \in \partial D$ where the result is independent of the atlas and set of charts used to represent the surface.

$$\int_{\partial D} f(\boldsymbol{r})dS := \sum_m \int_{T_m} f(\boldsymbol{r}_m(u, v))|\mathbf{u}_m(u, v) \times \mathbf{v}_m(u, v)|dudv$$

Here the sum is extended to all the charts $\boldsymbol{r}_m(u, v)$ in the atlas used to represent the surface ∂D. Without loss of generality, we can assume that the domains of parameters $T_m \subset \mathbb{R}^3$ for each chart is the unit triangle T_0 of vertices $\{(0, 0), (0, 1), (1, 0)\}$. In that case, the image $\boldsymbol{r}_m(T_0) \subset \partial D$ is a curved triangle with three curved edges and three vertices on the surface ∂D. In that case, the atlas based on triangular charts is called a triangulation of the surface ∂D.

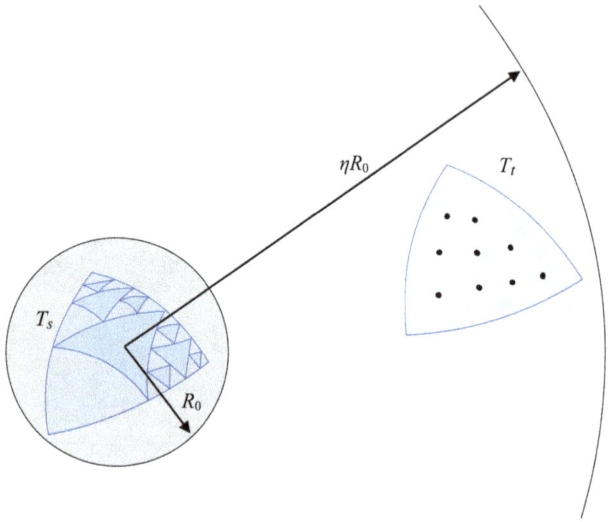

Figure 8.10 Adaptive refinement of triangle T for target points in the near region

The unit normal vector on the surface can be obtained by:

$$\hat{\boldsymbol{n}}(u, v) = \frac{\mathbf{u}(u, v) \times \mathbf{v}(u, v)}{|\mathbf{u}(u, v) \times \mathbf{v}(u, v)|}$$

Notice that, as every chart $\boldsymbol{r}(u, v)$ is a bijective map, the product $\mathbf{u}(u, v) \times \mathbf{v}(u, v)$ is never zero.

The first fundamental form of a surface at an arbitrary point is defined by the following symmetric matrix:

$$\begin{pmatrix} E & F \\ F & G \end{pmatrix}$$

where:

$$E(u, v) := \mathbf{u}(u, v) \cdot \mathbf{u}(u, v)$$
$$F(u, v) := \mathbf{u}(u, v) \cdot \mathbf{v}(u, v)$$
$$G(u, v) := \mathbf{v}(u, v) \cdot \mathbf{v}(u, v)$$

One can always find an alternative parametrization such that, at a given point $\boldsymbol{r}_0 = \boldsymbol{r}(u_0, v_0)$, the first fundamental form is the unit matrix. We call this map, conformal at point (u_0, v_0). To do that, we start with an arbitrary parametrization, we compute the eigenvalues and eigenvectos of the first fundamental form at point (u_0, v_0) and diagonalize the matrix:

$$\begin{pmatrix} E_0 & F_0 \\ F_0 & G_0 \end{pmatrix} = V^{-1}DV$$

Where D is diagonal and $V^{-1} = V^T$. Now we consider the following change of variables in the u, v domain:

$$\begin{aligned} u &= a\tilde{u} + b\tilde{v} + u_0 \\ v &= c\tilde{u} + d\tilde{v} + v_0 \end{aligned} \tag{A.1}$$

If we make the following choice for the constants a, b, c, d, then the new parametrization $\boldsymbol{r}(\tilde{u}, \tilde{v})$ at $(\tilde{u} = 0, \tilde{v} = 0)$ has the identity as the first fundamental form:

$$\begin{pmatrix} a & b \\ c & d \end{pmatrix} = V^T D^{-\frac{1}{2}}$$

Using this change of variables, we obtain the following chart:

$$\boldsymbol{r}(a\tilde{u} + b\tilde{v} + u_0, c\tilde{u} + d\tilde{v} + v_0)$$

Notice that the map is no longer from the unit triangle T_0 but from the inverse image of T_0 through the map A.1. Applying the chain rule, we get:

$$\begin{aligned} \frac{\partial \boldsymbol{r}}{\partial \tilde{u}} &= a\frac{\partial \boldsymbol{r}}{\partial u} + c\frac{\partial \boldsymbol{r}}{\partial v} \\ \frac{\partial \boldsymbol{r}}{\partial \tilde{v}} &= b\frac{\partial \boldsymbol{r}}{\partial u} + d\frac{\partial \boldsymbol{r}}{\partial v} \end{aligned}$$

and we can obtain:

$$\begin{aligned} \frac{\partial \boldsymbol{r}}{\partial \tilde{u}} \cdot \frac{\partial \boldsymbol{r}}{\partial \tilde{u}} &= \begin{pmatrix} a & c \end{pmatrix} \begin{pmatrix} E & F \\ F & G \end{pmatrix} \begin{pmatrix} a \\ c \end{pmatrix} \\ \frac{\partial \boldsymbol{r}}{\partial \tilde{v}} \cdot \frac{\partial \boldsymbol{r}}{\partial \tilde{v}} &= \begin{pmatrix} b & d \end{pmatrix} \begin{pmatrix} E & F \\ F & G \end{pmatrix} \begin{pmatrix} b \\ d \end{pmatrix} \\ \frac{\partial \boldsymbol{r}}{\partial \tilde{u}} \cdot \frac{\partial \boldsymbol{r}}{\partial \tilde{v}} &= \begin{pmatrix} a & c \end{pmatrix} \begin{pmatrix} E & F \\ F & G \end{pmatrix} \begin{pmatrix} b \\ d \end{pmatrix} \end{aligned}$$

Therefore, the first fundamental form with respect to the new variables is:

$$\begin{pmatrix} \tilde{E} & \tilde{F} \\ \tilde{F} & \tilde{G} \end{pmatrix} = \begin{pmatrix} \frac{\partial \boldsymbol{r}}{\partial \tilde{u}} \cdot \frac{\partial \boldsymbol{r}}{\partial \tilde{u}} & \frac{\partial \boldsymbol{r}}{\partial \tilde{u}} \cdot \frac{\partial \boldsymbol{r}}{\partial \tilde{v}} \\ \frac{\partial \boldsymbol{r}}{\partial \tilde{u}} \cdot \frac{\partial \boldsymbol{r}}{\partial \tilde{v}} & \frac{\partial \boldsymbol{r}}{\partial \tilde{v}} \cdot \frac{\partial \boldsymbol{r}}{\partial \tilde{v}} \end{pmatrix} = \begin{pmatrix} a & c \\ b & d \end{pmatrix} \begin{pmatrix} E & F \\ F & G \end{pmatrix} \begin{pmatrix} a & b \\ c & d \end{pmatrix}$$

we obtain:

$$\begin{pmatrix} \tilde{E} & \tilde{F} \\ \tilde{F} & \tilde{G} \end{pmatrix} = D^{-\frac{1}{2}} V \begin{pmatrix} E & F \\ F & G \end{pmatrix} V^T D^{-\frac{1}{2}}$$

and at the point $(u, v) = (u_0, v_0)$, we get:

$$\begin{pmatrix} \tilde{E}_0 & \tilde{F}_0 \\ \tilde{F}_0 & \tilde{G}_0 \end{pmatrix} = D^{-\frac{1}{2}} V \begin{pmatrix} E_0 & F_0 \\ F_0 & G_0 \end{pmatrix} V^T D^{-\frac{1}{2}} = D^{-\frac{1}{2}} V V^T D V V^T D^{-\frac{1}{2}} = I$$

where I is the identity matrix.

The second fundamental form of a surface ∂D is a quadratic form defined by the following matrix:

$$\begin{pmatrix} L & M \\ M & N \end{pmatrix}$$

where:

$$L(u, v) := \frac{\partial^2 \boldsymbol{r}}{\partial u^2}(u, v) \cdot \hat{\boldsymbol{n}}(u, v)$$

$$M(u, v) := \frac{\partial^2 \boldsymbol{r}}{\partial u \partial v}(u, v) \cdot \hat{\boldsymbol{n}}(u, v)$$

$$N(u, v) := \frac{\partial^2 \boldsymbol{r}}{\partial v^2}(u, v) \cdot \hat{\boldsymbol{n}}(u, v)$$

The Gauss curvature of a surface at a given point $\boldsymbol{r} = \boldsymbol{r}(u, v) \in \partial D \subset \mathbb{R}^3$ can be defined as:

$$K(u, v) = \frac{L(u, v)N(u, v) - M(u, v)^2}{E(u, v)G(u, v) - F(u, v)^2}$$

where the result is independent of the parametrization $\boldsymbol{r}(u, v)$ of the surface used.

Given a triangulation of a surface ∂D, we define v as the total number of vertices, e the number of edges and f the number of triangles. We define the Euler characteristic of a surface by:

$$\chi := v - e + f$$

It can be proved that the Euler characteristic of a surface is independent of the triangulation used.

For any compact surface, the following identity is known as the Gauss–Bonnet theorem:

$$\int_{\partial D} K dS = 2\pi \chi(\partial D)$$

The genus g of a surface is the number of holes on the surface, for example, in a sphere, $g = 0$ and in a torus, $g = 1$. In general, the genus can be defined in a rigorous way as:

$$g := 1 - \frac{1}{2}\chi(\partial D)$$

It can be defined in alternative equivalent ways using the homology group theory. It can be proved that the genus is always an integer number with $g \geq 0$ and that two different surfaces with the same genus are dipheomorphic, that is, equivalent under smooth deformations.

Given a smooth scalar function f defined on a surface ∂D, the surface gradient can be computed by:

$$\nabla_s f(u, v) = \frac{1}{EG - F^2}\left(\left(G\frac{\partial f}{\partial u} - F\frac{\partial f}{\partial v}\right)\hat{\mathbf{u}} + \left(-F\frac{\partial f}{\partial u} + E\frac{\partial f}{\partial v}\right)\hat{\mathbf{v}}\right)$$

The vector surface curl of a scalar function f can be computed by:

$$\hat{\boldsymbol{n}} \times \nabla_s f$$

Given a smooth tangent vector field, that we denote by $\boldsymbol{a}(u, v)$ to resemble the induced tangent vector current on a scatterer. The surface divergence can be obtained by:

$$\nabla_s \cdot \boldsymbol{a}(u, v) = \frac{1}{\sqrt{EG - F^2}}\left(\frac{\partial(\sqrt{EG - F^2}\boldsymbol{a}_u)}{\partial u} + \frac{\partial(\sqrt{EG - F^2}\boldsymbol{a}_v)}{\partial v}\right)$$

The surface curl of a smooth tangent vector field can be defined by:

$$\nabla_s \cdot (\hat{\boldsymbol{n}} \times \boldsymbol{a})(u, v)$$

The surface Laplacian of a smooth scalar function f can be defined by:

$$\Delta_s f(u, v) := \nabla_s \cdot (\nabla_s f(u, v))$$

The surface Laplacian of a smooth vector tangent field $\boldsymbol{a}$ can be defined by:

$$\Delta_s \boldsymbol{a} := -\hat{\boldsymbol{n}} \times \nabla_s(\nabla_s \cdot (\hat{\boldsymbol{n}} \times \boldsymbol{a})) + \nabla_s(\nabla_s \cdot \boldsymbol{a})$$

With these definitions, one can generalize the Stokes theorem on arbitrary curved surfaces.

Next we consider a smooth curve l on the surface ∂D parametrized by $\mathbf{t}(t)$ a function $t \in [0, 1] \rightarrow \mathbf{t}(t) \in l \subset \partial D \subset \mathbb{R}^3$. We also consider a surface $\Gamma \subset \partial D$ with boundary $\partial\Gamma$ parametrized by $\mathbf{d}(t)$ a function $d \in [0, 1] \rightarrow \mathbf{d}(t) \in \partial\Gamma \subset \partial D \subset \mathbb{R}^3$

$$\int_l \nabla_s f dl = f(\mathbf{t}(1)) - f(\mathbf{t}(0))$$

$$\int_\Gamma \nabla_s \cdot \boldsymbol{a} dS = \int_{\partial\Gamma} \boldsymbol{a} \cdot \hat{\mathbf{v}} dl$$

$$-\int_\Gamma \nabla_s \cdot (\hat{\boldsymbol{n}} \times \boldsymbol{a}) dS = \int_{\partial\Gamma} \boldsymbol{a} \cdot \hat{\tau} dl$$

Where $\hat{\mathbf{v}}$ is the unit normal vector to $\partial\Gamma$ that is perpendicular to the surface normal $\hat{\boldsymbol{n}}$ to ∂D and $\hat{\tau}$ is the unit tangent vector to $\partial\Gamma$ with counterclockwise orientation with respect to the unit normal vector $\hat{\boldsymbol{n}}$ to ∂D.

The following Green's identities can also be derived:

$$\int_{\partial D} f\nabla_s \cdot \boldsymbol{a} dS = -\int_{\partial D} \nabla_s f \cdot \boldsymbol{a} dS$$

$$\int_{\Gamma} (f\nabla_s \cdot \boldsymbol{a} + \nabla_s f \cdot \boldsymbol{a}) dS = \int_{\partial\Gamma} f \boldsymbol{a} \cdot \hat{\mathbf{v}} dl$$

The following identities are true for every tangent vector $\boldsymbol{a}$ or scalar field f defined on a surface:

$$\nabla_s \cdot (\hat{\boldsymbol{n}} \times \nabla_s f) = 0$$

$$\nabla_s \cdot \nabla_s (\hat{\boldsymbol{n}} \times \boldsymbol{a}) = 0$$

The surface Laplace equation can be defined as acting on scalar functions f and tangent vector fields $\boldsymbol{a}$ on a surface ∂D in the following way:

$$\Delta_s f = 0$$

$$\Delta_s \boldsymbol{a} = 0$$

It can be proved that the nullspace of the scalar surface Laplace equation has a dimension equal to the number of connected components of the surface. Indeed, the nullspace can be described as the set of functions piecewise constant that are constant on each connected component of the surface ∂D. In the case of the vector surface Laplace equation, the dimension of the nullspace is equal to $2g$ where g is the genus of the surface. (Notice the close relation of these two nullspaces and the nullspaces of the electrostatic and magnetostatic Maxwell problem.) These results connect the nullspace of the surface Laplace equation with certain aspects of the topology of the surface ∂D. These results are two particular cases of De Rham's theorem that relates the nullspace of the Laplace Beltrami operator acting on k forms and the kth Homology group of the manifold.

The following decomposition is known as the Hodge decomposition theorem and can be obtained for any tangent vector field on a surface ∂D:

$$\boldsymbol{a} = \nabla_s f + \hat{\boldsymbol{n}} \times \nabla_s g + \mathbf{h} \tag{A.2}$$

where f and g are two scalar functions and $\mathbf{h}$ is a harmonic vector field $\nabla_s \cdot \mathbf{h} = 0, \nabla_s \cdot (\hat{\boldsymbol{n}} \times \mathbf{h})$. Notice that by virtue of equation (1.28), this implies that $\Delta_s \mathbf{h} = 0$. This is called Hodge decomposition theorem. Notice that the components f and g can be obtained from the surface divergence and surface curl of $\boldsymbol{a}$ by solving surface Laplace equations. Taking the surface divergence and surface curl of A.2, we get:

$$\Delta_s f = \nabla_s \cdot \boldsymbol{a}$$

$$\Delta_s g = \nabla_s \cdot (\hat{\boldsymbol{n}} \times \boldsymbol{a})$$

Appendix B: Numerical integration and interpolation in 1D

There are two important tools in standard numerical analysis that are Gauss–Legendre numerical integration and Chebyshev interpolation. The Gauss–Legendre integration method in 1D consists of approximating the integral of a function by a finite sum of weights times samples of the function:

$$\int_a^b f(x)dx \approx \sum_{n=1}^{p} f(x_n)w_n$$

where the nodes x_n and weights are designed so that polynomial functions of degree smaller than $2p$ are integrated exactly. If the function $f(x)$ has $2p$ continuous derivatives on $[a, b]$, the following error estimate is obtained:

$$\left| \int_a^b f(x)dx - \sum_{n=1}^{p} f(x_n)w_n \right| \leq \frac{(b-a)^{2p+1}(p!)^4}{(2p+1)((2p)!)^3} f^{2p}(\xi) \tag{B.1}$$

The optimal nodes x_n end up being the roots of the nth Legendre polynomial on the interval $[a, b]$.

Regarding polynomial interpolation, the following result is standard. Given a function $f \in C^{p+1}$ and a polynomial $q_p(x)$ of order p that interpolates f at nodes $x_m \in [-1, 1], f(x_m) = q_p(x_m)$ with $0 \leq m \leq p$, the error $\|f - q_p\|_\infty$ is bounded by:

$$\|f - q_p\|_\infty \leq \max_{x\in[-1,1]} \left(\frac{|f^{(p+1)}(x)|}{(p+1)!} \right) \max_{x\in[-1,1]} \prod_{m=0}^{p} |x - x_m|$$

The value $\max_{x\in[-1,1]} \prod_{m=0}^{p} |x - x_m|$ is minimal if the nodes x_m are the Chebyshev nodes (roots of the $(p+1)$th degree Chebyshev polynomial). Therefore, the optimal choice of nodes for interpolation is Chebyshev nodes. Unfortunately, the optimal choice of nodes for interpolation is not the same as the optimal choice of nodes for numerical integration. To develop a high-order Nyström method, one would like to use the same set of nodes for interpolation and for numerical integration. Fortunately, using the Legendre nodes for integration is not optimal but nearly optimal. The following result can be obtained:

Given an interval $[a, b]$ and a function $f \in C^p[a, b]$, given $q_p(x)$ the polynomial of order p that interpolates f at Gauss–Legendre nodes $x_m \in [a, b]$ for $0 \leq n \leq p$, then the following error estimate can be obtained:

$$\|f - q_p\|_\infty \leq \max_{x\in[a,b]} \left(\frac{|f^{(p)}(x)|}{p!} \right) (b-a)^p$$

This result makes Legendre nodes suitable for both numerical integration and interpolation. One drawback of using Legendre rules for interpolation and integration is that you lose the ability of adaptivity as the optimal nodes x_m are given. To fix that and make high-order rules more adaptive, it is very common to use high-order piecewise rules, that is, divide the interval into subintervals of variable adaptive length and use a high-order rule on each interval. If we divide the unit interval [0, 1] in N intervals of variable length proportional to h and we use a Legendre rule with p nodes on each interval, we will obtain an error estimate for the integral $C_1 h^{2p+1}$ and an error estimate for the interpolation $C_2 h^P$ where C_1 and C_2 are constants that depend on high-order derivatives of the function f (assuming f is smooth enough). We can reduce the error of the numerical integration and of the interpolation by reducing h or increasing p. In numerical analysis, this is called *hp* refinement. In general, each interval can have a different order p (and a different length h).

Polynomials (especially to high order) should never be evaluated using monomials, catastrophic cancellation would destroy the accuracy. Instead, the orthogonal Legendre set of polynomials should be used. To evaluate them, the following well-known recurrence relation should be used:

$$(n+1)P_{n+1}(x) = (2n+q)xP_n(x) - nP_{n-1}(x) \tag{B.2}$$

where $P_0(x) = 1$ and $P_1(x) = x$. This recurrence relation B.2 is for Legendre polynomials defined on the interval $[-1, 1]$, for other intervals appropriate change of variable is needed. Using this basis, the interpolation polynomial $q_p(x)$ can be written as:

$$q_p(x) = \sum_{n=0}^{p} c_n P_n(x)$$

The coefficients c_n can be obtained as the solution of the linear system:

$$\mathbf{Mc} = \boldsymbol{f} \tag{B.3}$$

where $\mathbf{M}$ is a matrix of size $(p+1) \times (p+1)$ and $\mathbf{M}_{mn} = P_n(x_m)$, $\mathbf{c}$ is a column vector of unknown coefficients $\mathbf{c}_n = c_n$ and $\boldsymbol{f}$ is the right-hand side column vector $\boldsymbol{f}_m = f(x_m)$, and x_m is the mth root of the $p+1$th Legendre polynomial. Notice that the matrix $\mathbf{M}$ can be obtained (and should be obtained) using the recurrence relation B.2 avoiding the use of monomial basis x^n to avoid numerical instabilities. The evaluation of the interpolant polynomial q_p at an arbitrary point x_0 should also be done using that recurrence relation to evaluate all Legendre polynomials $P_n(x_0)$ and then take the linear combination with coefficients c_m to obtain the interpolated value $q_p(x_0) = \sum_{n=0}^{p} c_n P_n(x_0)$.

It is worth notice that in case of having a line integral on a curve l parametrized by $\mathbf{t}(t)$: $[0, 1] \rightarrow \mathbb{R}^3$, that is:

$$\int_l f dl = \int_0^1 f(\mathbf{t}(t))|\mathbf{v}(t)|dt$$

where $\mathbf{v}(t) = \frac{dt}{dt}(t)$, if we need to apply a numerical quadrature rule to estimate that integral, we will need to sample the function $f(\mathbf{t}(t))|\mathbf{v}(t)|$ as integrand on the integration domain $[0, 1]$:

$$\int_l f dl \approx \sum_{m=0}^{p} w_m f(\mathbf{t}(t_m))|\mathbf{v}(t_m)|$$

Here t_m, w_m is the Gauss–Legendre quadrature rule on the interval $[0, 1]$. The accuracy of the result will depend on the smoothness of both the function f and the curve $\mathbf{t}$. Indeed one should notice that, to obtain accurate results, one should have the possibility of sampling the velocity vector $\mathbf{v}$ of the curve, as it is part of the integrand. If all you have about the curve is a set of sampling points (even if they are properly located at Gauss–Legendre nodes), spectral differentiation would be needed to estimate $|\mathbf{v}|$, that means, polynomial interpolation and analytical differentiation of the polynomial expansion to estimate the derivative of the sampled function (in this case, the sampled curve). That would lead to instabilities and loss of accuracy, especially if the order of discretization p is high.

Appendix C: Numerical integration and interpolation in 2D

Next we describe high-order numerical integration and interpolation of a smooth function on a triangular region. This should be understood as the 2D analog of Gauss–Legendre integration and Chebyscheff/Gauss–Legendre interpolation.

First we describe a set of orthogonal polynomials on the basic canonical triangle T_0:

$$T_0 = \{(u, v), u \geq 0, v \geq 0, u + v \leq 1\}$$

Those polynomials are known as Koornwinder polynomials:

$$K_{nm}(u, v) = c_{nm}(1 - v)^m P_{n-m}^{(0,2m+1)}(1 - 2v) P_m\left(\frac{2u + v - 1}{1 - v}\right), \qquad m \leq n, \tag{C.1}$$

where $P_n^{(a,b)}$ is the Jacobi polynomial of degree n with parameters (a, b), P_m is the Legendre polynomial of degree m, and c_{nm} is a normalization constant such that

$$\int_{T_0} |K_{nm}(u, v)|^2 \, du \, dv = 1.$$

Figure C.2 shows the Koornwinder polynomials of order 0 to 5.

It is easy to see that there are $n_p = p(p + 1)/2$ Koornwinder polynomials of total degree less than p. To evaluate the Koornwinder polynomials, it is not adequate to use the formula C.1 via the evaluation of Jacobi and Legendre polynomials and using their recurrence formulas; this would lead to a loss of accuracy when the point (u, v)

is near some axis of the unit triangle T_0 (and of course, it should neither be used monomials). Instead, the following modified recurrence formula should be used. If we define the polynomial function $q_m(z, y)$ as:

$$q_m(z,y) := P_m\left(\frac{z}{y}\right)y^m$$

where $P_m(x)$ are the Legendre polynomials of degree m, then $q_m(z, y)$ verifies the following recurrence formula:

$$(m+1)q_{m+1}(z,y) = (2m+1)zq_m(z,y) - mq_{m-1}(z,y)y^2 \tag{C.2}$$

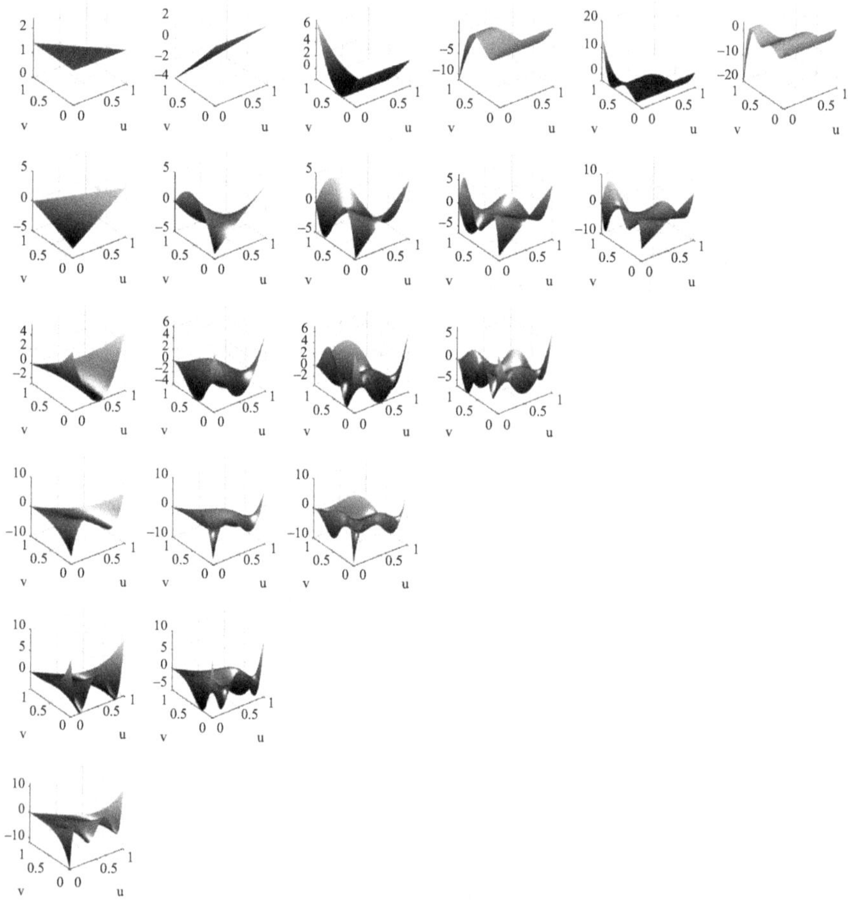

Figure C.1 Koornwinder polynomials of order 0 *to* 5. *Orthogonal polynomials on the standard unit triangle* T_0

Table C.1 Quadrature weights and nodes of order 0 *for the standard triangle* T_0

u_n	v_n	w_n
0.3333333333333333e + 00	0.3333333333333333e + 00	0.5000000000000000e + 00

Moreover, $q_0(z,y) = 1$ and $q_1(z,y) = z$. This provides a numerically stable algorithm to obtain $q_m(z,y)$. The remaining part to obtain the Koornwinder polynomials K_{mn} is $P_{n-m}^{(0,2m+1)}(1-2v)$, which can be obtained safely by using the standard recurrence relations for the Jacobi polynomials. To obtain partial derivatives of Koornwinder polynomials with respect to u and v, one can obtain appropriate recurrence relations for the partial derivatives of the polynomial function $q_m(z,y)$ by taking derivatives of the recurrence relation described earlier (equation C.2):

$$(m+1)\frac{\partial q_{m+1}}{\partial z}(z,y) = (2m+1)q_m(z,y) + (2m+1)z\frac{\partial q_m}{\partial z}(z,y) - m\frac{\partial q_{m-1}}{\partial z}(z,y)y^2$$
$$(m+1)\frac{\partial q_{m+1}}{\partial y}(z,y) = (2m+1)z\frac{\partial q_m}{\partial y}(z,y) - m\frac{\partial q_{m-1}}{\partial y}(z,y)y^2 - mq_{m-1}(z,y)2y \tag{C.3}$$

Similarly we can obtain recurrence relations for the standard Jacobi polynomials.

It is also important to notice that the set of Koornwinder polynomials is a complete basis for functions on $L^2(T_0)$ (square-integrable functions on the triangle T_0). Once we have defined a suitable set of orthogonal polynomials, we need a suitable set of interpolant nodes. Those nodes, together with appropriate weights, will also define a high-order quadrature rule for smooth functions on the triangle T_0.

Unfortunately, the theory to obtain a high-order quadrature rule for interpolation and integration on a triangle is not as clean as in 1D (see Appendix B). Yet, great effort has been made to obtain them from complicated nonlinear optimization algorithms (see [24,34]). In this chapter, we describe and recommend the use of Vioreanu–Rokhlin nodes obtained in [24] for both high-order interpolation and numerical integration.

The Vioreanu–Rokhlin quadrature rule described in [24] consist of approximating the integral of a smooth function on a triangle T_0 by using the following rule:

$$\int_{T_0} f(u,v)dS \approx \sum_{i=1}^{n_p} f(u_i,v_i)w_i \tag{C.4}$$

Table C.2 Quadrature weights and nodes of order 1 *for the standard triangle* T_0

u_n	v_n	w_n
0.6666666666666666e + 00	0.1666666666666667e + 00	0.1666666666666667e + 00
0.1666666666666666e + 00	0.6666666666666667e + 00	0.1666666666666667e + 00
0.1666666666666666e + 00	0.1666666666666667e + 00	0.1666666666666667e + 00

Table C.3 Quadrature weights and nodes of order 2 *for the standard triangle* T_0

u_n	v_n	w_n
0.9157621350977069e − 01	0.8168475729804586e + 00	0.5497587182766092e − 01
0.9157621350977072e − 01	0.9157621350977081e − 01	0.5497587182766092e − 01
0.4459484909159648e + 00	0.1081030181680703e + 00	0.1116907948390057e + 00
0.1081030181680702e + 00	0.4459484909159650e + 00	0.1116907948390057e + 00
0.8168475729804584e + 00	0.9157621350977081e − 01	0.5497587182766092e − 01
0.4459484909159649e + 00	0.4459484909159650e + 00	0.1116907948390057e + 00

Table C.4 Quadrature weights and nodes of order 3 *for the standard triangle* T_0

u_n	v_n	w_n
0.3333333333333333e + 00	0.3333333333333334e + 00	0.9948268522981829e − 01
0.8577740109331180e + 00	0.7111299453344105e − 01	0.3013949020357029e − 01
0.7111299453344089e − 01	0.8577740109331182e + 00	0.3013949020357029e − 01
0.6149261666431254e + 00	0.7109443567989363e − 01	0.5168314069324515e − 01
0.7111299453344097e − 01	0.7111299453344105e − 01	0.3013949020357029e − 01
0.3139793976769810e + 00	0.6149261666431255e + 00	0.5168314069324515e − 01
0.6149261666431254e + 00	0.3139793976769811e + 00	0.5168314069324515e − 01
0.7109443567989354e − 01	0.6149261666431255e + 00	0.5168314069324515e − 01
0.3139793976769810e + 00	0.7109443567989363e − 01	0.5168314069324515e − 01
0.7109443567989351e − 01	0.3139793976769811e + 00	0.5168314069324515e − 01

There are sets of nodes and weights for different orders up to $p = 20$ with $n_p = 231$ nodes. Next we list them up to order 4:

Higher order rules (up to 20) can be found in this repository [FALTA REPOSITORIO]

Table C.5 Quadrature weights and nodes of order 4 for the standard triangle T_0

u_n	v_n	w_n
0.2421847563444860e + 00	0.5156304873110280e + 00	0.6350446284716560e − 01
0.2421847563444861e + 00	0.2421847563444861e + 00	0.6350446284716560e − 01
0.5156304873110278e + 00	0.2421847563444861e + 00	0.6350446284716560e − 01
0.5177841189772941e − 01	0.4741107940511354e + 00	0.3243991877352587e − 01
0.4741107940511352e + 00	0.4741107940511354e + 00	0.3243991877352587e − 01
0.4741107940511353e + 00	0.5177841189772949e − 01	0.3243991877352587e − 01
0.4816980611752458e − 01	0.7184833008976065e + 00	0.2735780884051768e − 01
0.4816980611752460e − 01	0.2333468929848691e + 00	0.2735780884051768e − 01
0.2333468929848689e + 00	0.7184833008976065e + 00	0.2735780884051768e − 01
0.2333468929848690e + 00	0.4816980611752469e − 01	0.2735780884051768e − 01
0.7184833008976064e + 00	0.2333468929848691e + 00	0.2735780884051768e − 01
0.7184833008976064e + 00	0.4816980611752469e − 01	0.2735780884051768e − 01
0.5125307984145178e − 01	0.8974938403170964e + 00	0.1600666736493985e − 01
0.5125307984145182e − 01	0.5125307984145190e − 01	0.1600666736493985e − 01
0.8974938403170963e + 00	0.5125307984145190e − 01	0.1600666736493985e − 01

The interpolation problem of a smooth function on a triangle can be solved with an accuracy using the Vioreanu–Rokhyn nodes and the Koornwinder polynomials. Given a smooth function f defined on T_0, one can write the linear system:

$$\sum_{\substack{n=0\\m\leq n}}^{n=p} c_{nm}K_{nm}(u_i, v_i) = f(u_i, v_i) \tag{C.5}$$

where the unknowns are the coefficients c_{nm}, the sum contains Koornwinder polynomials up to order p, that is, n_p polynomials, with $n_p = p(p+1)/2$. The set of nodes are those of the same order that contains n_p nodes. By doing this, one can solve this linear system and obtain the unknown coefficients c_{nm}. After that, one can evaluate the interpolant function at any arbitrary point u_0, v_0 inside the triangle T_0

$$\sum_{\substack{n=0\\m\leq n}}^{n=p} c_{nm}K_{nm}(u_0, v_0) \approx f(u_0, v_0)$$

It can be shown empirically that the condition number of the interpolation matrix obtained in the linear system C.5 is small (of order 1). That, together with the possibility of evaluating the Koornwinder polynomials in a stable way using recurrence relations, makes the process of high-order interpolation on a triangle numerically stable and accurate. Figure C.2 shows the error rate in numerical integration and interpolation of the function $f(u,v) = sin(10u(1-v)^2)$ for different orders and refinements:

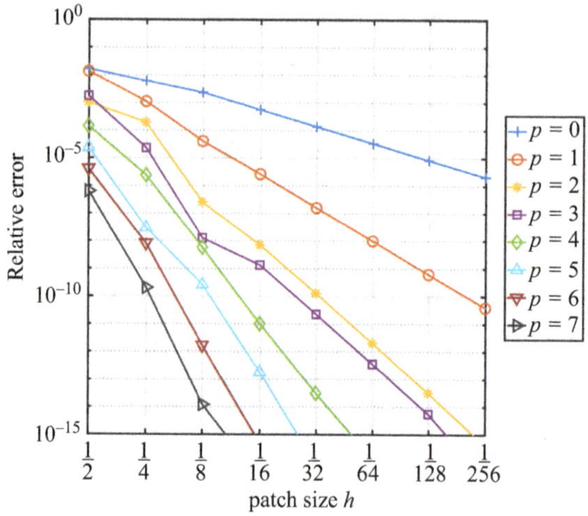

Figure C.2 Convergence of the Vioreanu–Rokhlin quadrature rule for different orders p

It is worth notice that in the case of having a surface integral of a smooth function, where the surface is parametrized by $\boldsymbol{r}(u, v) : T_0 \rightarrow \Gamma \subset \mathbb{R}^3$, that is:

$$\int_\Gamma f dS = \int_{T_0} f(\mathbf{r}(u, v))|\mathbf{u}(u, v) \times \mathbf{v}(u, v)| du dv$$

If we want to apply a numerical quadrature rule to estimate that integral, we will need to sample the function $f(\mathbf{r}(u, v))|\mathbf{u}(u, v) \times \mathbf{v}(u, v)|$ as integrand on the integration domain T_0:

$$\int_\Gamma f dS \approx \sum_{i=1}^{n_p} f(\mathbf{r}(u_i, v_i))|\mathbf{u}(u_i, v_i) \times \mathbf{v}(u_i, v_i)| w_i$$

If one needs to take the surface integral of a more complicated surface, (∂D), one has to apply the same rule for each triangular chart that contains the atlas:

$$\int_{\partial D} f dS = \sum_m \int_{T_m = \boldsymbol{r}_m(T_0)} f dS \approx \sum_m \sum_{i=1}^{n_p} f(\boldsymbol{r}_m(u_i, v_i))|\mathbf{u}_m(u_i, v_i) \times \mathbf{v}_m(u_i, v_i)| w_i$$

where $\boldsymbol{r}_m : T_0 \rightarrow T_m \subset \partial D \subset \mathbb{R}^3$ is the set of charts used to describe the surface.

Notice that even integrating the simplest function (e.g., $f = 1$, which is useful to obtain the area of the surface) can be challenging if the surface is complicated (e.g., with high curvature regions) or the parametrization chosen for the surface is deficient (e.g., with high aspect ratio triangles).

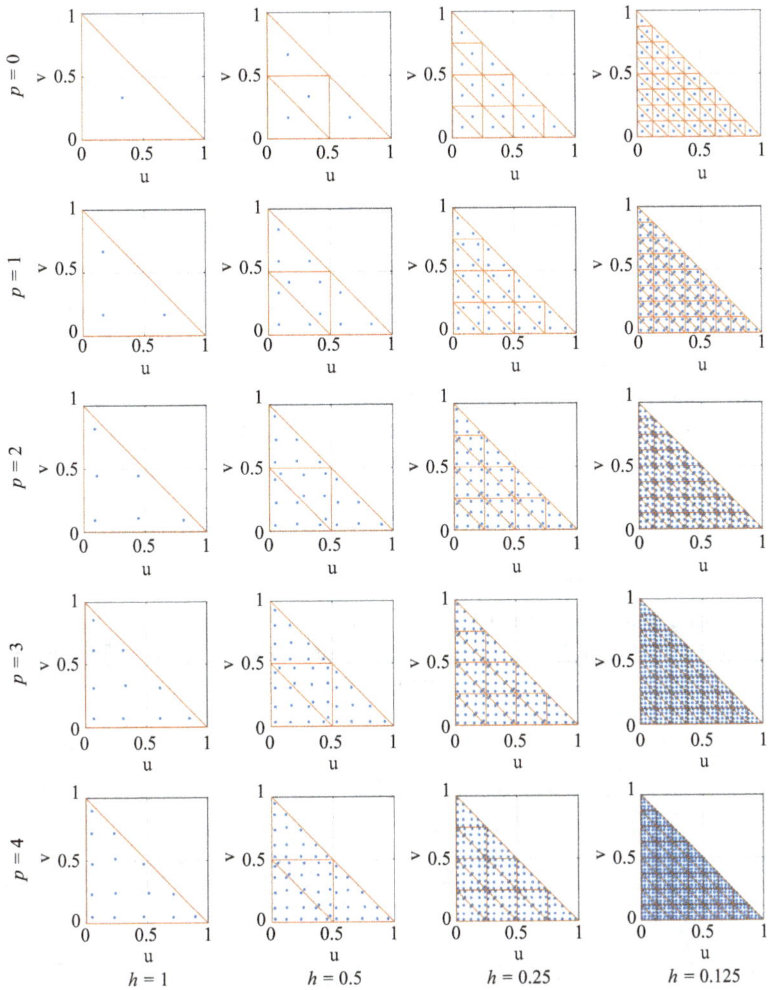

Figure C.3 Nodes of the Vioreanu–Rokhlin quadrature rule for different values of p and h

Appendix D: Generalized Gaussian quadrature for arbitrary non-smooth functions

In this section, we review the notion of generalized Gaussian quadrature rule first introduced to obtain a high-order rule on a triangle for smooth functions. As we described in the Appendix [num integration], the Vioreanu–Rokhlin quadrature rule is obtained from a nonlinear multivariable Newton root finding algorithm that obtains nodes (u_j, v_j) on the standard canonical triangle and weights w_j such that the quadrature rule:

$$\int_{T_0} f(u,v)dudv \approx \sum_{i=1}^{n_p} f(u_i, v_i)w_i$$

is exact for every polynomial in two variables u, v of order p with $n_p = (p+1)(p+2)/2$. This process can be generalized to impose exact answers on arbitrary sets of functions. For example, let us consider the following 1D problem; let $f(x)$ be a 1D function with logarithmic singularity in $x = 0$, to be precise, we know that there are two functions $g_1(x), g_2(x)$ such that $f(x) = \log(|x|)g_1(x) + g_2(x)$. Let us assume that the explicit decomposition is not known (i.e., the functions g_1 and g_2 are not known and there is only access to evaluate the function $f(x)$). Find a quadrature rule:

$$\int_{-1}^{1} f(x)dx \approx \sum_{i=1}^{n} f(x_i)w_i \tag{D.1}$$

that is accurate to a certain order for all functions f of this type.

Applying the generalized Gaussian quadrature ideas, one can define the optimal x_i, w_i nodes and weights as the ones that return an exact value of the integral for any $f(x)$ in the following set:

$$\{1, x, x^2, x^3, ..., x^m, \log(|x|), \log(|x|)x, \log(|x|)x^2, \log(|x|)x^3, ... \log(|x|)x^m\}$$

for certain order m. The nonlinear system of equations to be solved is:

$$\begin{aligned}
\int_{-1}^{1} 1dx &= \sum_{i=1}^{n} w_i \\
\int_{-1}^{1} xdx &= \sum_{i=1}^{n} x_i w_i \\
\int_{-1}^{1} x^2dx &= \sum_{i=1}^{n} x_i^2 w_i \\
&\cdots \\
\int_{-1}^{1} x^m dx &= \sum_{i=1}^{n} x_i^m w_i \\
\int_{-1}^{1} \log(|x|)dx &= \sum_{i=1}^{n} \log(|x_i|)w_i
\end{aligned} \tag{D.2}$$

$$\int_{-1}^{1} \log(|x|)x dx = \sum_{i=1}^{n} \log(|x_i|)x_i w_i$$

$$\int_{-1}^{1} \log(|x|)x^2 dx = \sum_{i=1}^{n} \log(|x_i|)x_i^2 w_i$$

$$\cdots$$

$$\int_{-1}^{1} \log(|x|)x^m dx = \sum_{i=1}^{n} \log(|x_i|)x_i^m w_i$$

The integrals in D.2 can be evaluated accurately in general by brute force, analytical techniques (if available) as the computation time is not important at the stage of finding the quadrature rule. The solution of the nonlinear system D.2 is complicated and requires high-dimensional Newton zero root finding algorithm, the entire process of finding the optimal quadrature rule x_j, w_j for each order is not easy, can be very time consuming, some parts are delicate, and there is no general recipe that guarantees success (in particular, the starting point in the Newton algorithm is delicate).

Notice that once the quadrature rule is obtained, the integral of f on the interval $[-1, 1]$ can be obtained very fast and accurately by just applying the rule D.1. Notice also that the decomposition $f(x) = \log(|x|)g_1(x) + g_2(x)$ and the explicit finding of the functions g_1, g_2 are not needed as the rule D.1 only needs the values $f(x_n)$ (there is no need to subtract the singularity). Because of the difficulty in finding the generalized Gaussian quadrature rule, it only makes sense to apply this technique when one needs to evaluate may integrals of the same type and one can amortize the time by storing the quadrature rule x_i, w_i.

Generalized Gaussian quadrature can be generalized further to any arbitrary set of functions. This technique is an important tool in numerical analysis. It can be extended even further in the case of having one parameter in the set of basis functions. Let us consider the following problem; let $f(x)$ be a function that can be expressed as $f(x) = \log^\gamma(|x|)g_1(x) + g_2(x)$ where the constant γ is known (but different every time and $\gamma \in [1, 2]$). If γ were fixed, one can apply the same strategy for the following set of functions:

$$\{1, x, x^2, x^3, ..., x^m, \log^\gamma(|x|), \log^\gamma(|x|)x, \log^\gamma(|x|)x^2, \log^\gamma(|x|)x^3, ... \log^\gamma(|x|)x^m\}$$

and obtain a quadrature rule x_j^γ, w_j^γ. The problem is that, this quadrature rule would not be amortized as every time we will have a different value of γ. The solution here is to compute different quadrature rules for different values of $\gamma_i \in [1, 2]$ using γ_i in a Chebyshev grid on the interval where the parameter will be (in this case, $[1, 2]$) and store all those quadrature rules: $x_j^{\gamma_i}, w_j^{\gamma_i}$. Now, when the quadrature rule for a particular value γ_0 is needed, one can interpolate, for each j the values of the optimal node and weight $x_j^{\gamma_0}, w_j^{\gamma_0}$ from the stored values at Chebyshev γ_i values of the parameter: $x_j^{\gamma_i}, w_j^{\gamma_i}$ (see Appendix B for Chebyshev interpolation).

Appendix E: Function spaces

In this section, we establish a suitable framework of function spaces specifically designed for addressing boundary value problems in electromagnetism.

Definition 10: *A scalar function $f(\mathbf{r})$ defined on a set $G \subset \mathbb{R}^3$ is said to be uniformly Hölder continuous with Hölder exponent $0 < \alpha \leq 1$ if there is a constant C such that:*

$$|f(\mathbf{r}) - f(\mathbf{r}')| \leq C|\mathbf{r} - \mathbf{r}'|^{\alpha}$$

for all $\mathbf{r}, \mathbf{r}' \in G$.

G can be a bounded domain $\overline{D}$, the unbounded domain $\mathbb{R}^3 \backslash D$,or the boundary ∂D.

Definition 11: *The Hölder space $C^{0,\alpha}(G)$ is defined as the linear space of all functions defined on G that are bounded and uniformly Hölder continuous with exponent α. It is a Banach space with the norm:*

$$||f(\mathbf{r})||_{\alpha,G} = \sup_{\mathbf{r}\in G}|f(\mathbf{r})| + \sup_{\mathbf{r},\mathbf{r}'\in G,\mathbf{r}\neq\mathbf{r}'} \frac{|f(\mathbf{r}) - f(\mathbf{r}')|}{|\mathbf{r} - \mathbf{r}'|^{\alpha}} \tag{E.1}$$

For vector fields, Hölder continuity and the Hölder norm are defined analogously (by replacing absolute values with Euclidian norms).

Definition 12: *The Hölder space $C^{1,\alpha}(G)$ of uniformly Hölder continuous differentiable functions is the space of differentiable functions $f(\mathbf{r})$ for which $\nabla f(\mathbf{r})$ belongs to $C^{0,\alpha}(G)$. It is equipped with the norm:*

$$||f(\mathbf{r})||_{1,\alpha,G} = \sup_{\mathbf{r}\in G}|f(\mathbf{r})| + \sup_{\mathbf{r}\in G}|\nabla f(\mathbf{r})| + \sup_{\mathbf{r},\mathbf{r}'\in G,\mathbf{r}\neq\mathbf{r}'} \frac{|\nabla f(\mathbf{r}) - \nabla f(\mathbf{r}')|}{|\mathbf{r} - \mathbf{r}'|^{\alpha}}$$

Here, the gradient ∇ is replaced by the surface gradient ∇_s in the case of $G = \partial D$

In certain cases, the requirement of additional smoothness, characterized by the parameter α, may not be necessary. In such situations, we can work with more commonly used function spaces. The space $C(G)$, also denoted as $C^0(G)$, consists of continuous functions equipped with the supremum norm $||f(\mathbf{r})||_{\infty}$. Similarly, the space $C^1(G)$ comprises functions possessing continuous derivatives and is endowed with the norm $||f(\mathbf{r})||_{C^1} = ||f(\mathbf{r})||_{\infty} + ||\nabla f(\mathbf{r})||_{\infty}$.

We denote by $C_t(\partial D)$ and $C_t^{0,\alpha}(\partial D)$ the space of all continuous and uniformly Hölder continuous tangent fields equipped with the supremum norm and the Hölder norm, respectively.

Definition 13: *We define the following vector-valued function space $C_t(\nabla_s\cdot, \partial D)$ by:*

$$C_t(\nabla_s\cdot, \partial D) = \{\mathbf{f} \in C_t(\partial D) : \nabla_s \cdot \mathbf{f} \in C(\partial D)\}$$

This is a Banach space equipped with the norm:

$$||\mathbf{f}|| = ||\mathbf{f}||_{\infty,\partial D} + ||\nabla_s \cdot \mathbf{f}||_{\infty,\partial D}$$

Definition 14: *We define* $C_t^{0,\alpha}(\nabla_s \cdot, \partial D)$ *as:*

$$C_t^{0,\alpha}(\nabla_s \cdot, \partial D) = \{\mathbf{f} \in C_t(\partial D) : \nabla_s \cdot \mathbf{f} \in C(\partial D)\}$$

This is a Banach space equipped with the norm:

$$||\mathbf{f}|| = ||\mathbf{f}||_{\alpha,\partial D} + ||\nabla_s \cdot \mathbf{f}||_{\alpha,\partial D}$$

In the above definitions, $|| \cdot ||_{\infty,\partial D}$ represents the supremum norm on ∂D. $|| \cdot ||_{\alpha,\partial D}$ denotes the Hölder norm on ∂D (Equation (E.1)).

These definitions provide the basis for understanding the spaces $C_t(\nabla_s \cdot, \partial D)$ and $C_t^{0,\alpha}(\nabla_s \cdot, \partial D)$, which are essential in the study of electromagnetism and its boundary value problems.

References

[1] Vico F, Ferrando M, Greengard L, *et al.* The decoupled potential integral equation for time-harmonic electromagnetic scattering. *Communications on Pure and Applied Mathematics*. 2016;69(4):771–812.

[2] Cools K, Andriulli FP, Olyslager F, *et al.* Nullspace of the static MFIE operator and its effect on the numerical solution of BIE's. In: *2008 IEEE Antennas and Propagation Society International Symposium*. IEEE; 2008. p. 1–4.

[3] Epstein CL, Gimbutas Z, Greengard L, *et al.* A consistency condition for the vector potential in multiply-connected domains. *IEEE Transactions on Magnetics*. 2012;49(3):1072–1076.

[4] Colton D and Kress R. *Integral Equation Methods in Scattering Theory*. SIAM; 2013.

[5] Vico F, Ferrando-Bataller M, Valero-Nogueira A, *et al.* A non-resonant current and charge integral equation. In: *2013 7th European Conference on Antennas and Propagation (EuCAP)*. IEEE; 2013. p. 407–410.

[6] Vico F, Ferrando-Bataller M, Valero-Nogueira A, *et al.* A high-order locally corrected nyström scheme for charge-current integral equations. *IEEE Transactions on Antennas and Propagation*. 2015;63(4):1678–1685.

[7] Vico F, Gimbutas Z, Greengard L, *et al.* Overcoming low-frequency breakdown of the magnetic field integral equation. *IEEE Transactions on Antennas and Propagation*. 2012;61(3):1285–1290.

[8] Taskinen M and Yla-Oijala P. Current and charge integral equation formulation. *IEEE Transactions on Antennas and Propagation*. 2006;54(1):58–67.

[9] Taskinen M and Vänskä S. Current and charge integral equation formulations and Picard's extended Maxwell system. *IEEE Transactions on Antennas and Propagation*. 2007;55(12):3495–3503.

[10] Bendali A, Collino F, Fares M, *et al.* Extension to nonconforming meshes of the combined current and charge integral equation. *IEEE Transactions on Antennas and Propagation*. 2012;60(10):4732–4744.

[11] Vico F, Greengard L, Ferrando-Bataller M, *et al.* An augmented regularized combined source integral equation for nonconforming meshes. *IEEE Transactions on Antennas and Propagation*. 2019;67(4):2513–2521.
[12] Greengard L, O'Neil M, Rachh M, *et al.* Fast multipole methods for the evaluation of layer potentials with locally-corrected quadratures. *Journal of Computational Physics: X*. 2021;10:100092.
[13] Corona E, Greengard L, Rachh M, *et al.* An integral equation formulation for rigid bodies in Stokes flow in three dimensions. *Journal of Computational Physics*. 2017;332:504–519.
[14] Ojala R and Tornberg AK. An accurate integral equation method for simulating multi-phase Stokes flow. *Journal of Computational Physics*. 2015;298: 145–160.
[15] Atkinson KE. The Numerical Solution of Integral Equations of the Second Kind. In: *Cambridge Monographs on Applied and Computational Mathematics*. Cambridge University; 1996.
[16] Greengard L and Rokhlin V. A new version of the fast multipole method for the Laplace equation in three dimensions. *Acta Numerica*. 1997;6:229–269.
[17] Greengard L and Rokhlin V. A fast algorithm for particle simulations. *Journal of Computational Physics*. 1987;73(2):325–348.
[18] Cheng H, Crutchfield WY, Gimbutas Z, *et al.* A wideband fast multipole method for the Helmholtz equation in three dimensions. *Journal of Computational Physics*. 2006;216(1):300–325.
[19] Rokhlin V. Diagonal forms of translation operators for the Helmholtz equation in three dimensions. *Applied and Computational Harmonic Analysis*. 1993;1(1):82–93.
[20] Greengard L, Huang J, Rokhlin V, *et al.* Accelerating fast multipole methods for the Helmholtz equation at low frequencies. *IEEE Computational Science and Engineering*. 1998;5(3):32–38.
[21] Chew WC, Michielssen E, Song J, *et al. Fast and Efficient Algorithms in Computational Electromagnetics*. Artech House, Inc.; 2001.
[22] Cheng H, Greengard L, and Rokhlin V. A fast adaptive multipole algorithm in three dimensions. *Journal of Computational Physics*. 1999;155(2): 468–498.
[23] Greengard LF and Huang J. A new version of the fast multipole method for screened Coulomb interactions in three dimensions. *Journal of Computational Physics*. 2002;180(2):642–658.
[24] Vioreanu B and Rokhlin V. Spectra of multiplication operators as a numerical tool. *SIAM Journal on Scientific Computing*. 2014;36(1):A267–A288.
[25] Duffy MG. Quadrature over a pyramid or cube of integrands with a singularity at a vertex. *SIAM journal on Numerical Analysis*. 1982;19(6):1260–1262.
[26] Khayat MA and Wilton DR. Numerical evaluation of singular and near-singular potential integrals. *IEEE Transactions on Antennas and Propagation*. 2005;53(10):3180–3190.
[27] Graglia RD and Lombardi G. Machine precision evaluation of singular and nearly singular potential integrals by use of Gauss quadrature formulas

for rational functions. *IEEE Transactions on Antennas and Propagation*. 2008;56(4):981–998.

[28] Helsing J. Integral equation methods for elliptic problems with boundary conditions of mixed type. *Journal of Computational Physics*. 2009;228(23): 8892–8907.

[29] Alpert BK. Hybrid Gauss-trapezoidal quadrature rules. *SIAM Journal on Scientific Computing*. 1999;20(5):1551–1584.

[30] Bremer J and Gimbutas Z. On the numerical evaluation of the singular integrals of scattering theory. *Journal of Computational Physics*. 2013;251:327–343.

[31] Schwab C and Wendland WL. Kernel properties and representations of boundary integral operators. *Mathematische Nachrichten*. 1992;156(1):187–218.

[32] Bruno O, Elling T, Paffenroth R, *et al.* Electromagnetic integral equations requiring small numbers of Krylov-subspace iterations. *Journal of Computational Physics*. 2009;228(17):6169–6183.

[33] Vico F, Greengard L, O'Neil M, *et al.* Fast adaptive high-order integral equation methods for electromagnetic scattering from smooth perfect electric conductors. arXiv preprint arXiv:230604473. 2023.

[34] Dunavant D. High degree efficient symmetrical Gaussian quadrature rules for the triangle. *International Journal for Numerical Methods in Engineering*. 1985;21(6):1129–1148.

Chapter 9
Conclusion and perspectives

Zhen Peng[1] and Francesca Vipiana[2]

The computer simulation of electromagnetic (EM) phenomena has emerged as a powerful and indispensable tool in both the design and analysis stages of many engineering endeavors. In the time-harmonic regime, integral equation (IE) methods are widely used for such simulations [1–3]. Applications range from advanced antenna design [4,5], stealth technologies [6–8], integrated circuits [9,10] to optics and photonics [11,12]. One advantage of solving an integral equation is that both the analysis and unknowns reside only on the boundary surfaces of the targets. It often requires fewer unknowns to solve compared to a differential equation, where the unknowns scale volumetrically. More significantly, since surface-based modeling is used, it is easier to prepare analysis-suitable models from computer-aided design (CAD) geometries. However, the application of IE methods to the Maxwell equations leads to a dense, complex, and indefinite matrix equation. Techniques such as the fast multipole method [13,14], fast Fourier transform [15,16], interpolative decomposition [17,18], and $\mathscr{H}$-matrix compression [19,20] have been developed to compress dense IE matrices.

Nowadays, increasing demands are placed on enhancements of the model fidelity and improvements of the simulation accuracy. Despite their apparent importance for simulation-based engineering, high-fidelity modeling and high-resolution computing present significant mathematical and computational challenges in existing IE solvers due to the growing sophistication of real-world EM applications. One of the major difficulties comes from the multi-scale nature of the geometry, in which the spatial scales differ by order of magnitude. It will be daunting and time-consuming to generate a good quality discretization for the high-definition model, let alone solve it efficiently. Even if such a discretization can be constructed, the coexistence of electrically large aircraft frames and electrically small fine features usually reflects into disparate mesh sizes. It may cause the "mixed-frequency" problem [10] and result in an extremely ill-conditioned matrix equation.

[1]Electromagnetics Lab and Center for Computational Electromagnetics, Department of Electrical and Computer Engineering University of Illinois at Urbana-Champaign, USA
[2]Wavision Research Group, Department of Electronics and Telecommunications, Politecnico di Torino, Italy

The main purpose of the book is to provide the reader with relevant and current advancements in accurate, stable, and well-conditioned IE solvers for real-life EM problems. Starting from a comprehensive review of IE formulations in Chapter 2, we introduce the kernel-based and kernel-independent fast factorization techniques for compressing dense IE matrices in Chapters 3 and 4. To solve the ill-conditioned IE matrix equations, a variety of preconditioning techniques, including the domain decomposition preconditioner (Chapter 5), multi-resolution preconditioner (Chapter 6), and Calderon preconditioner (Chapter 7), are discussed. Finally, a new and emerging IE formulation based on decoupled potential is discussed in Chapter 8. The capability of these algorithms is illustrated in different types of real-world applications ranging from high-definition jet aircraft to radar stealth objects to nano-optical and plasmonic nanoantennas.

In this book, we primarily focus on the IE solutions for time-harmonic problems involving linear, piecewise homogeneous, and isotropic materials. However, the above-discussed algorithms can be extended to tackle more complex scenarios, such as problems with arbitrarily distributed materials and shapes. For such cases, the frequency-domain volume integral equation approach [21–23] and the hybrid finite-element and boundary integral equation methods [24,25] can be employed. Additionally, when dealing with EM simulations that involve multiple right-hand-side vectors (e.g., monostatic scattering), computationally efficient direct solvers utilizing kernel-independent fast factorization methods [26–28] can be beneficial. Furthermore, when investigating transient EM phenomena and time-varying fields, the time-domain integral equation method [29–32] emerges as a natural and compelling numerical technique.

Finally, it has been assumed in this book that we have complete and exact information on the target geometry and materials. In practice, even though we are seeking the highest possible fidelity, the computer representation will not be exactly the same compared to the real world. These uncertainties may arise from imprecise knowledge of the system and small differences in manufacturing. When these uncertainties can be characterized by a few input parameters (e.g., material properties, geometric uncertainties), stochastic methods such as Monte Carlo, polynomial chaos expansion (PCE) [33–35], and Gaussian process regression (GPR) [36] can be used to capture the relationship between the input uncertainties and the output solution. The outcome provides not only statistical results but also valuable insights into the uncertainties associated with the solution, enabling a more comprehensive understanding of the system's behavior and robust design considerations [37].

Another challenging scenario is the IE-based solution of confined EM environments. Wave propagation may undergo multiple reflections/scattering from boundaries and internal structures, thus leading to randomized phase, polarization, and direction of wave fields. In the short wavelength limit, the wave scattering process may exhibit chaotic ray trajectories, albeit the underlying wave equation is linear [38]. Given the sensitive nature of wave physics to environmental details, it is crucial to develop stochastic models accounting for the large variations in local field responses. In this regard, a stochastic Green's function (SGF) approach is introduced in [39,40] as an effective statistical solution to the wave equation in wave-chaotic environments. As a statistical wave model, the SGF interprets the universal properties of wave-chaotic systems through the physics of wave chaos [41] and the mathematics of random matrix theory (RMT) [42]. The SGF statistically describes the coherent and incoherent components between a pair of source and receiving points. It can then be used to solve self-consistently for the wave propagation in wave-chaotic media.

Today, with the advancements in computing hardware, machine intelligence, quantum computing, and algorithms, there are significant opportunities to continue advancing integral equation-based solutions toward next-generation technology innovations and applications. For example, the use of hardware accelerators, such as graphics processing units (GPUs), field-programmable gate arrays (FPGAs), or application-specific integrated circuits (ASICs), can significantly speed up the numerical solution of integral equations. These hardware solutions can be tailored specifically to the requirements of the integral equation algorithms, allowing for optimized implementations that take advantage of parallelism and reduce computational bottlenecks. Moreover, machine learning models can be trained on a dataset of integral equation-based numerical solutions, allowing them to capture the underlying patterns and relationships in the integral equation problem. Once trained, these models can rapidly estimate the solution for new inputs, offering a significant speedup compared to traditional numerical solvers. Continued research and advancements in these fields are necessary to fully explore their potential in integral equation-based numerical solutions for real-life electromagnetics problems.

References

[1] Harrington RF. *Field Computation by Moment Methods*. 2nd ed., Hoboken, NJ: Wiley-IEEE Press; 1993.

[2] Wilton DR. *Computational Methods*. New York, NY: Academic Press; 2002. p. 316–365.

[3] Chew WC, Tong MS, and Hu B. *Integral Equation Methods for Electromagnetic and Elastic Waves*. Morgan & Claypool Publishers; 2008. Synthesis Lectures on Computational Electromagnetics.

[4] Wang XC, Peng Z, and Lee JF. Multi-solver domain decomposition method for modeling EMC effects of multiple antennas on a large air platform. *IEEE Transactions on Electromagnetic Compatibility*. 2012;54(2):375–388.

[5] Hesford AJ and Chew WC. On preconditioning and the eigensystems of electromagnetic radiation problems. *IEEE Transactions on Antennas and Propagation*. 2008;56(8):2413–2420.

[6] Michielssen E and Boag A. A multilevel matrix decomposition algorithm for analyzing scattering from large structures. *IEEE Transactions on Antennas Propagation*. 1996;44(8):1086–1093.

[7] Ylä-Oijala P, Taskinen M, and JäRvenpää S. Surface integral equation formulations for solving electromagnetic scattering problems with iterative methods. *Radio Science*. 2005;40(6):RS6002.

[8] Peng Z, Lim KH, and Lee JF. Non-conformal domain decomposition methods for solving large multi-scale electromagnetic scattering problems. *Proceedings of IEEE*. 2013;101(2):298–319.
[9] Eleftheriades GV, Mosig JR, and Guglielmi M. A fast integral equation technique for shielded planar circuits defined on nonuniform meshes. *IEEE Transactions on Microwave Theory and Techniques*. 1996;44(12):2293–2296.
[10] Li MK and Chew WC. Multiscale simulation of complex structures using equivalence principle algorithm with high-order field point sampling scheme. *IEEE Transactions on Antennas and Propagation*. 2008;56(8):2389–2397.
[11] Solis DM, Taboada JM, Obelleiro F, *et al*. Optimization of an optical wireless nano link using directive nanoantennas. *Optics Express*. 2013;21(2):2369–2377.
[12] Karaosmanoglu B, Yilmaz AE, and Ergül Ö. A comparative study of surface integral equations for accurate and efficient analysis of plasmonic structures. *IEEE Transactions on Antennas and Propagation*. 2017;65:3049–3057.
[13] Greengard L and Rokhlin V. A fast algorithm for particle simulations. *Journal of Computational Physics*. 1987;73(2):325–348.
[14] Song J, Lu CC, and Chew WC. Multilevel fast multipole algorithm for electromagnetic scattering by large complex objects. *IEEE Transactions on Antennas Propagation*. 1997;45(10):1488–1493.
[15] Seo SM and Lee JF. A fast IE-FFT algorithm for solving PEC scattering problems. *IEEE Transactions on Magnetics*. 2004;41(5):1476–1479.
[16] Schobert D and Eibert TF. Low-frequency surface integral equation solution by multilevel Green's function interpolation with fast Fourier transform acceleration. *IEEE Transactions on Antennas and Propagation*. 2012;60:1440–1449.
[17] Cheng H, Gimbutas Z, Martinsson PG, *et al*. On the compression of low rank matrices. *SIAM Journal of Scientific Computing*. 2005;26:1389–1404.
[18] Greengard L, Gueyffier D, Martinsson PG, *et al*. Fast direct solvers for integral equations in complex three-dimensional domains. *Acta Numerica*. 2009;18:243–275.
[19] Hackbusch W. *Hierarchische Matrizen: Algorithmen und Analysis*. Dordrecht: Springer; 2009.
[20] Chai W and Jiao D. An $\mathcal{H}^2$-matrix-based integral-equation solver of reduced complexity and controlled accuracy for solving electrodynamic problems. *IEEE Transactions on Antennas and Propagation*. 2009;57(10):3147–3159.
[21] Schaubert D, Wilton D, and Glisson A. A tetrahedral modeling method for electromagnetic scattering by arbitrarily shaped inhomogeneous dielectric bodies. *IEEE Transactions on Antennas and Propagation*. 1984;32(1):77–85.
[22] Sun LE and Chew WC. A novel formulation of the volume integral equation for electromagnetic scattering. *Waves in Random and Complex Media*. 2009;19(1):162–180.
[23] Markkanen J, Yla-Oijala P, and Sihvola A. Discretization of volume integral equation formulations for extremely anisotropic materials. *IEEE Transactions on Antennas and Propagation*. 2012;60(11):5195–5202.

[24] Sheng XQ, Jin JM, Song J, *et al.* On the formulation of hybrid finite-element and boundary-integral methods for 3-D scattering. *IEEE Transactions on Antennas and Propagation*. 1998;46(3):303–311.

[25] Gao HW, Peng Z, and Sheng XQ. A geometry-aware domain decomposition preconditioning for hybrid finite element-boundary integral method. *IEEE Transactions on Antennas and Propagation*. 2017;65(4):1875–1885.

[26] Ho KL and Greengard L. A fast direct solver for structured linear systems by recursive skeletonization. *SIAM Journal of Scientific Computing*. 2012;34(5):A2507–A2532.

[27] Wei JG, Peng Z, and Lee JF. A fast direct matrix solver for surface integral equation methods for electromagnetic wave scattering from non-penetrable targets. *Radio Science*. 2012;47:RS5003.

[28] Guo H, Liu Y, Hu J, *et al.* A butterfly-based direct integral-equation solver using hierarchical LU factorization for analyzing scattering from electrically large conducting objects. *IEEE Transactions on Antennas and Propagation*. 2016;65:4742–4750.

[29] Bluck MJ and Walker SP. Time-domain BIE analysis of large three-dimensional electromagnetic scattering problems. *IEEE Transactions on Antennas and Propagation*. 1997;45(5):894–901.

[30] Bagci H, Yilmaz AE, Lomakin V, *et al.* Fast solution of mixed-potential time-domain integral equations for half-space environments. *IEEE Transactions on Geoscience and Remote Sensing*. 2005;43(2):269–279.

[31] Shanker B, Lu M, Yuan J, *et al.* Time domain integral equation analysis of scattering from composite bodies via exact evaluation of radiation fields. *IEEE Transactions on Antennas and Propagation*. 2009;57(5):1506–1520.

[32] Wang X and Weile DS. Implicit Runge–Kutta methods for the discretization of time domain integral equations. *IEEE Transactions on Antennas and Propagation*. 2011;59(12):4651–4663.

[33] Soize C and Ghanem RG. Physical systems with random uncertainties: chaos representations with arbitrary probability measure. *SIAM Journal of Scientific Computing*. 2005;26:395–410.

[34] Zubac Z, De Zutter D, and Vande Ginste D. Scattering from two-dimensional objects of varying shape combining the method of moments with the stochastic Galerkin method. *IEEE Transactions on Antennas and Propagation*. 2014;62(9):4852–4856.

[35] Austin ACM and Sarris CD. Efficient analysis of geometrical uncertainty in the FDTD method using polynomial chaos with application to microwave circuits. *IEEE Transactions on Microwave Theory and Techniques*. 2013;61(12):4293–4301.

[36] Williams C and Rasmussen C. Gaussian processes for regression. In: Touretzky D, Mozer MC, Hasselmo M, editors. *Advances in Neural Information Processing Systems*. vol. 8. New York, NY: MIT Press; 1995.

[37] Bagci H, Yucel AC, Hesthaven JS, *et al.* A fast Stroud-based collocation method for statistically characterizing EMI/EMC phenomena on complex platforms. *IEEE Transactions on Electromagnetic Compatibility*. 2009;51(2):301–311.

[38] Galdi V, Pinto IM, and Felsen LB. Wave propagation in ray-chaotic enclosures: paradigms, oddities and examples. *Antennas and Propagation Magazine, IEEE*. 2005;47(1):62–81.

[39] Lin S, Peng Z, and Antonsen TM. A stochastic Green's function for solution of wave propagation in wave-chaotic environments. *IEEE Transactions on Antennas and Propagation*. 2020;68(5):3919–3933.

[40] Lin S, Luo S, Ma S, *et al.* Predicting statistical wave physics in complex enclosures: a stochastic dyadic Green's function approach. *IEEE Transactions on Electromagnetic Compatibility*. 2023;65(2):436–453.

[41] Berry MV. Regular and irregular semiclassical wavefunctions. *Journal of Physics A: Mathematical and General*. 1977;10(12):2083.

[42] Wigner E. Random matrices in physics. *SIAM Review*. 1967;9(1):1–23.

Index

www.ingramcontent.com/pod-product-compliance
Lightning Source LLC
LaVergne TN
LVHW011834200926
848484LV00002B/354